D0583889

ELEMENTARY
LINEAR ALGEBRA

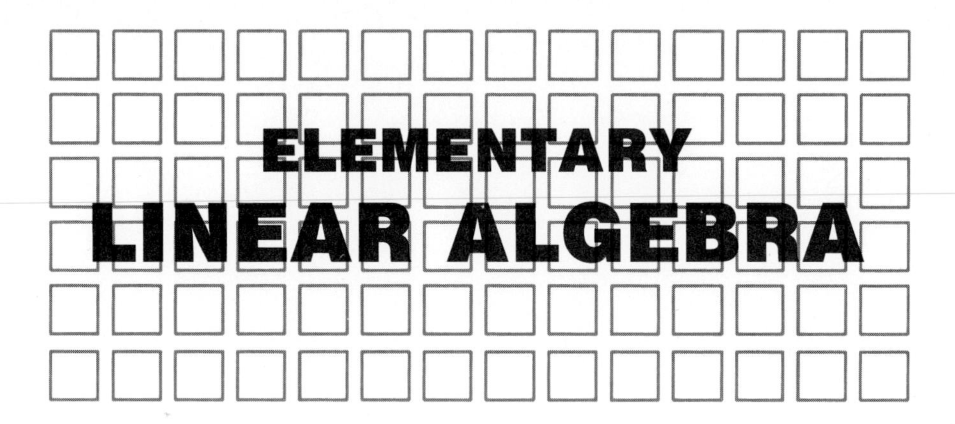

WILLIAM L. PERRY

Texas A & M University

McGraw-Hill Book Company

New York St. Louis San Francisco Auckland Bogotá Caracas Colorado Springs Hamburg Lisbon London Madrid Milan Mexico Montreal New Delhi Oklahoma City Panama Paris San Juan São Paulo Singapore Sydney Tokyo Toronto

ELEMENTARY LINEAR ALGEBRA

1 2 3 4 5 6 7 8 9 0 DOCDOC 8 9 3 2 1 0 9 8

ISBN 0-07-049431-2

This book was set in Times Roman.
The editors were Robert A. Weinstein and Jack Maisel;
the designer was Eliott Epstein;
the production supervisor was Leroy A. Young.
R. R. Donnelley & Sons Company was printer and binder.

Library of Congress Cataloging-in-Publication Data

Perry, William L. (William Leon)
Elementary linear algebra.

Includes index.
1. Algebras, Linear. I. Title.
QA184.P47 1988 512′.5 87-17098
ISBN 0-07-049431-2

ABOUT THE AUTHOR

Born and raised in Missouri, Bill Perry received a Ph.D. in mathematics from the University of Illinois, Urbana-Champaign, in 1972. He took a position with Texas A&M University, where he is now professor of mathematics. His research is in differential equations, especially those applied to physics and engineering. He has received University and College of Science awards for distinguished teaching.

When away from mathematics, the author enjoys time with his family, especially on hiking trips in Rocky Mountain National Park.

To LINDA

CONTENTS

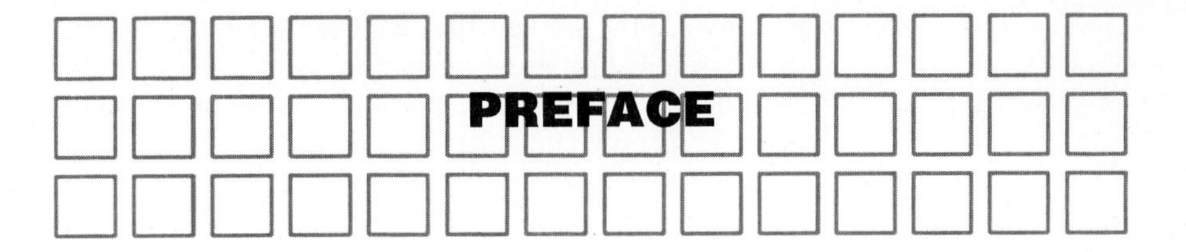

PREFACE

Linear algebra, as we know it today, furnishes a beautiful mathematical structure; the integral parts of the structure are powerful tools for engineers, scientists, and mathematicians. Students in these diverse fields come to linear algebra with a common background: They have learned mathematics by working problems and studying examples. With this idea in mind, I have drawn the reader's attention in this text to what I consider to be the five basic problems of linear algebra:

1. The problem of solving linear equations
2. The problem of constructing a basis for a vector space
3. The problem of constructing a matrix which represents a linear transformation
4. The eigenvalue-eigenvector problem
5. The diagonalization problem

These problems give the students landmarks to look for and skills to master in the course.

Of course, the solutions of the five basic problems result in the important theorems of linear algebra. These are stated accurately and then proved unless the proof lies beyond the scope of elementary linear algebra. My hope is that the reader will come to see the structure of linear algebra as supported by the five main beams of the five basic problems with the detail fleshed out by examples, applications, and theorems.

I have given a wide variety of applications of linear algebra. The lion's share are slanted toward science and engineering, however. Presenting applications can be tricky in that enough flavor of the origin of the application must be given to convince the student that the example is not bogus, yet not too much detail of the application must appear to scare away a novice. For example, the presentation of some linear equations arising in structures is the important point, not a minicourse in statics and dynamics. I believe that the applications have been given in appropriate detail and that the students will be able to see and handle the linear algebraic aspects of the applied examples once they are laid out.

Except for those sections with the word *calculus* in the section title, the reader need only be equipped with knowledge of the derivative and integral in an elementary fashion. I have put the deeper connections of calculus and linear algebra in these separate sections so that the interested reader can find them easily.

Chapter 1 is a matrix algebra, linear equations chapter. The first basic problem of linear algebra is presented here. Numerical methods appear already in Section 1.3, rather than in a separate chapter. Numerical methods come naturally here, and my students have responded to this organization very well. However, the numerical methods can be omitted with no loss of continuity. Because students of science and engineering often use complex numbers, we do not restrict our attention solely to real matrices.

Chapter 2 is concerned with concrete two- and three-space. Section 2.3 introduces linear functions and their geometric analysis. I have tried to emphasize geometric aspects of linear algebra throughout the text.

Chapter 3 treats real and complex vector spaces. The standard spaces are covered, and real vector spaces get the larger part of the attention. Students in this class will be working with complex numbers in their professional lives. Including the complex case in turn allows the natural use of complex eigenvalues and eigenvectors later. Moreover, certain linear operators in science, such as the momentum operator of quantum mechanics, are easy to introduce when linear transformations are presented. Also in Chapter 3 there is an example on binary-channel linear codes. This example involves the vector space F^n, where $F = \{0, 1\}$ with addition and multiplication defined mod 2 (but it is just done with a table in the text); this allows the instructor to introduce number fields in the context of an application, if desired. In Section 3.9, applications to calculus are found. I did this for two reasons: to make it easy for the students to find the applications and to make the discussion compact, because the necessary linear algebra groundwork has gone before. Similar sections exist in Chapter 4 and 5.

In Chapter 4, I emphasize the matrix representation problem for linear transformations and the invariants under similarity.

The diagonalization problem and the eigenvalue-eigenvector problem are discussed in Chapter 5. Because complex numbers have been used all along, the real and complex cases are not seen as fundamentally different.

I have put numerical methods for eigenvalues in Chapter 6 because I wanted to discuss the stability of the problem as well as the QR algorithm, which some instructors may wish to omit.

Inclusion of the short chapter (7) on linear programming came as a result of discussions with some chemical engineering students one semester. Rigorous proofs of results are not given; however, it is shown that the main result in linear programming rests on the concept of linear independence of sets of vectors. The chapter ends with some simple table manipulations.

At the ends of Chapters 1 through 6, I have appended some problems. These sets are not overly long and are, strictly speaking, not "review" problems. Rather, they furnish the reader with some problems that cross chapter divisions, anticipate developments, or whet the appetite for learning more advanced applications. Also at the end of each chapter I have written narrative summaries of the "where have we been and where are we going" type. They remind the student of the key ideas developed in the preceding chapter and how these ideas might come to bear on the next chapter.

This book has been used in manuscript form in a one-semester course with 45 class meetings. The material in Chapters 1 through 5 and in Sections 6.1 and 6.2 was covered with no difficulty. Answers to selected problems (usually the odd-numbered ones) are provided. With the two-color format, an effort was made to use the second color and shading to make topics such as gaussian elimination, row reduction, and matrix multiplication easier to learn.

I owe much to many people who made the successful completion of the project possible. John Corrigan, who initiated the idea, Peter Devine, who helped me through the first and second round of reviews, and Robert Weinstein, who helped me with the "endgame," all are offered my sincere thanks. I am grateful to Susan Trussell for the beautiful and professional typing of the first manuscript. I am also very grateful to the reviewers Ezra Brown, Virginia Tech; Richard A. Brualdi, University of Wisconsin-Madison; Bruce Edwards, University of Florida; John Gregory, Southern Illinois University at Carbondale; Robert Hartwig, North Carolina State University; Joseph Kitchen, Duke University; Stephen Pennel, University of Lowell; Clifford Queen, Lehigh University and Kenneth C. Washinger, Shippensburg University of Pennsylvania for their frank comments. Their efforts led to a stronger manuscript and a better learning experience for my students the second and third time I taught from the book. Special mention goes to Professor Kitchen, who performed one of the most complete, revealing, and helpful textbook reviews an author could hope for. Finally, my family deserves my deepest thanks for giving me the time to devote to the writing.

TO THE STUDENT

In a mathematics course, you have basically three resources: the lecturer, the text, and your time spent in hard work. The most important is the last. It is during those later hours when you are trying to work problems and using the text to figure out what is going on that much learning takes place. This text is designed to be used by studying the examples to learn techniques of linear algebra. A good method is to see whether you can work the examples with the book closed; this is probably the only way in which a book can be of more use closed than open. By correlating lecture notes with the text, working through the examples, studying statements and proofs of theorems, and working the problems, you will come to see linear algebra as a field that contains powerful mathematical tools to apply to science, engineering, and other areas.

I believe that linear algebra rests on a foundation of five major problems, which are listed in the Preface. I also believe that mastery of the techniques in the solutions of these problems is as essential to successful application of linear algebra as mastery of differentiation and integration is to applying calculus. Work hard to understand the five basic problems of linear algebra.

ELEMENTARY
LINEAR ALGEBRA

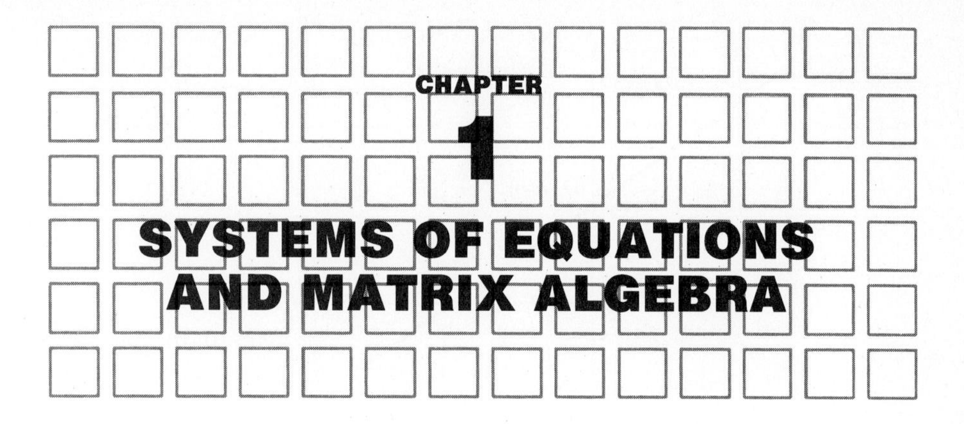

CHAPTER 1 SYSTEMS OF EQUATIONS AND MATRIX ALGEBRA

1.1 SOURCES OF LINEAR EQUATIONS

At the heart of linear algebra and much of applied mathematics is the problem of solving systems of linear equations. What are linear equations?

A simple example of a linear equation is

$$ax + by = c \tag{1.1.1}$$

where x and y are real variables and a, b, and c are given real numbers. If a and b are not both zero, then the graph of this equation is a straight line; this is one reason why the equation is called *linear* (see Fig. 1.1.1). Equation (1.1.1) is called *one linear equation in the two unknowns x and y.*

The linear equation with three real variables x, y, and z

$$ax + by + cz = d$$

where a, b, c, and d are real numbers, not all zero, represents a plane in the standard xyz coordinate system. A graph of $2x + 2y + 3z = 6$ is shown in Fig. 1.1.2.

Of course, an equation can contain more than three variables. For a non-scientific example, a person may have five employees, each paid a different hourly wage and each working a variable number of hours each week, as shown in Table 1.1.1. The weekly payroll P is given by

$$P = 4.5x_1 + 3.75x_2 + 5x_3 + 6.15x_4 + 5.75x_5$$

This is one linear equation in five variables.

The preceding are all examples of the general case of n variables, contained in Definition 1.1.1.

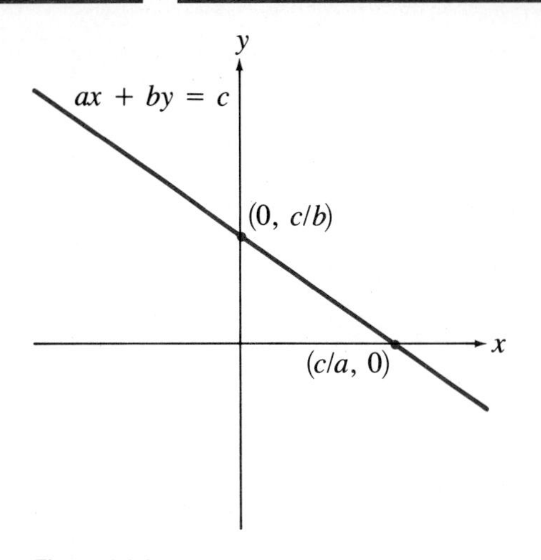

Figure 1.1.1
Graph of $ax + by = c$ (a, b, c all positive).

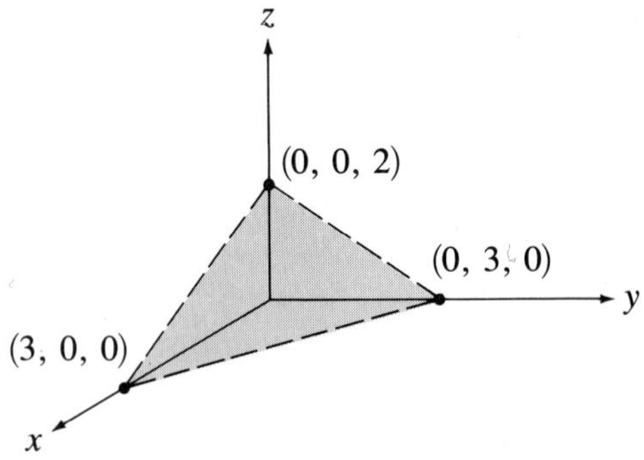

Figure 1.1.2
Graph of $2x + 2y + 3z = 6$; first octant view.

Table 1.1.1 SCHEDULE OF WAGES

EMPLOYEE	HOURLY WAGE	HOURS WORKED
1	\$4.50	x_1
2	\$3.75	x_2
3	\$5.00	x_3
4	\$6.15	x_4
5	\$5.75	x_5

DEFINITION 1.1.1 A *linear equation* in the n variables $x_1, x_2, \ldots, x_n$ is any equation of the form

$$c_1x_1 + c_2x_2 + \cdots + c_nx_n = r$$

where $c_1, c_2, \ldots, c_n$ and r are real or complex[1] numbers.

An example of linear equations involving complex numbers is

$$(h + i)x_n + x_{n-1} = 0 \qquad n = 1, 2, 3, \ldots$$

where h is real and i is the imaginary unit ($i^2 = -1$). These equations arise in

[1] See App. I for a review of $\mathbb{C}$, the set of complex numbers.

a discretization of an equation involving the momentum operator of quantum mechanics.

An equation which is not in the form given in Definition 1.1.1 is called *nonlinear*. For example,

$$x_1^2 + x_2^2 = 4$$

$$-\sqrt{x} + y = 2$$

and

$$x_1 + |x_2| - 3x_3 = 7$$

are nonlinear equations. Graphs of the first two are drawn in Fig. 1.1.3*a* and *b*, respectively. Notice that the graphs are not straight lines.

The problem of "solving" systems of linear equations is an important one, as is seen in the examples at the end of this section. First the notion of solution must be defined.

DEFINITION 1.1.2 A *solution* of the single linear equation

$$c_1x_1 + c_2x_2 + \cdots + c_nx_n = r$$

is an ordered n-tuple of complex numbers[2] $(s_1, s_2, \ldots, s_n)$ which when substituted (s_1 for x_1, s_2 for x_2, and so on) into the equation make it a true mathematical statement.

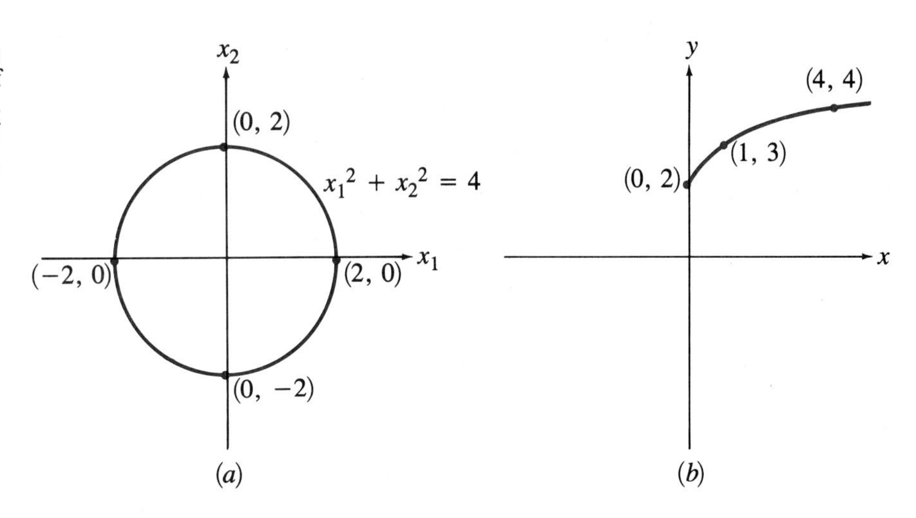

Figure 1.1.3 (*a*) Graph of $x_1^2 + x_2^2 = 4$. (*b*) Graph of $-\sqrt{x} + y = 2$.

[2] Of course, the set could consist of purely real numbers, because the real numbers are a subset of the set of complex numbers.

EXAMPLE 1 The equation $2x_1 - 4x_2 = 8$ has several sets of solutions. One is $x_1 = 4$, $x_2 = 0$; another is $x_1 = 6$, $x_2 = 1$; and yet another is $x_1 = -4$, $x_2 = -4$. Written as ordered pairs, these are $(4, 0)$, $(6, 1)$, and $(-4, -4)$, respectively.

EXAMPLE 2 Write a general solution of the equation $2x_1 - 4x_2 = 8$.

Solution Let $x_2 = s$, where s represents an arbitrary number. Then substitute into the equation. We have

$$2x_1 - 4s = 8$$

Solving for x_1 with ordinary algebra gives

$$x_1 = 4 + 2s$$

The solution can be written

$$\begin{aligned} x_1 &= 4 + 2s \\ x_2 &= s \end{aligned} \qquad \text{or} \qquad (4 + 2s, s)$$

Note that by letting s take on the specific values 0, 1, or -4, we obtain the solutions listed in Example 1. However, if we put $s = 2 + i$, then $x_1 = 8 + 2i$, $x_2 = 2 + i$, and we have a complex solution. If the unknowns were restricted beforehand to be real numbers, then complex solutions would be ruled out.

In Example 2 we could have set $x_1 = s$ and solved for x_2, to find $x_2 = x_1/2 - 2 = s/2 - 2$.

EXAMPLE 3 Show that $x_1 = 1 + i$, $x_2 = 1 - i$ is a solution of

$$\begin{aligned} (3 + i)x_1 + (1 + i)x_2 &= 4 + 4i \\ x_1 - x_2 &= 2i \end{aligned}$$

Solution Substituting the given numbers into the given equations yields

$$\begin{aligned} (3 + i)(1 + i) + (1 + i)(1 - i) &= (2 + 4i) + (2 + 0i) = 4 + 4i \\ (1 + i) - (1 - i) &= 2i \end{aligned}$$

EXAMPLE 4 Solve

$$\begin{aligned} ix_1 + (2 - i)x_2 &= 4 - 3i \\ (1 + i)x_1 - x_2 &= -3 - i \end{aligned} \tag{1.1.2}$$

Solution We can use the elementary *method of substitution.*
Solving the second equation for x_2 and substituting the resulting expression for x_2 in the first equation, we find

$$ix_1 + (2 - i)[3 + i + (1 + i)x_1] = 4 - 3i$$
$$ix_1 + 7 - i + (3 + i)x_1 = 4 - 3i$$
$$(3 + 2i)x_1 = -3 - 2i$$
$$x_1 = -1$$

Then substitution of $x_1 = -1$ into the second equation in (1.1.2) leads to $x_2 = 2$.

These are all examples of a very important general problem:

The First Fundamental Problem of Linear Algebra (Solution of Equations)

Given m equations in the n unknowns $x_1, x_2, \ldots, x_n$,

$$\begin{aligned} a_{11}x_1 + a_{12}x_2 + \cdots + a_{1n}x_n &= b_1 \\ a_{21}x_1 + a_{22}x_2 + \cdots + a_{2n}x_n &= b_2 \\ \cdots\cdots\cdots\cdots\cdots\cdots\cdots\cdots & \\ a_{m1}x_1 + a_{m2}x_2 + \cdots + a_{mn}x_n &= b_m \end{aligned} \tag{1.1.3}$$

1. Determine whether solutions exist, [A *solution* is an n-tuple of numbers which satisfies *all the equations in* (1.1.3) *simultaneously*.]
2. If solutions exist, compute all of them.

The system (1.1.2) fits the form of (1.1.3):

$$a_{11}x_1 + a_{12}x_2 = b_1$$
$$a_{21}x_1 + a_{22}x_2 = b_2$$

with $a_{11} = i$, $a_{12} = 2 - i$, $a_{21} = 1 + i$, $a_{22} = -1$, $b_1 = 4 - 3i$, and $b_2 = -3 - i$.

Note: The numbers

$$\begin{matrix} a_{11}, a_{12}, \ldots, a_{1n} \\ a_{21}, a_{22}, \ldots, a_{2n} \\ \cdots\cdots\cdots\cdots \\ a_{m1}, a_{m2}, \ldots, a_{mn} \end{matrix}$$

and $b_1, b_2, \ldots, b_m$ in (1.1.3) are given complex numbers[3]. The subscripts give information shown on the next page.

---

[3] As always, when we use the term *complex numbers*, we allow the possibility that all the numbers under consideration are purely real. In fact, in most applications this is the case.

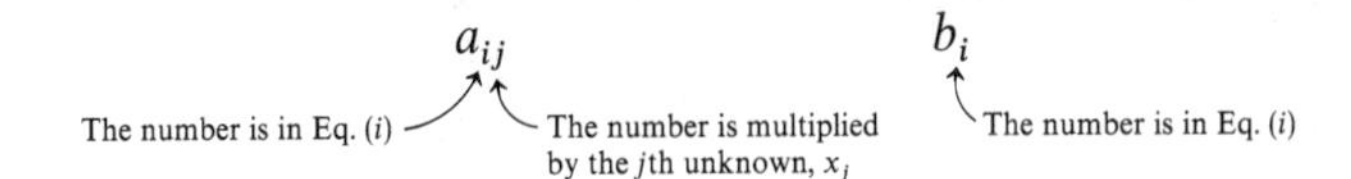

The problem of solving linear equations is important in many areas. In the following examples, the equations are not derived.

EXAMPLE 5 (Geometry) Find the point of intersection of the straight lines $3x - 2y = 6$ and $4x + 4y = 8$.

Solution Any point of intersection must satisfy both equations simultaneously. This system can be solved by substitution, as in Example 4. The solution is $x = 2$, $y = 0$. Thus the point of intersection is (2, 0). See Fig. 1.1.4.

EXAMPLE 6 (Markov chains, component reliability) In certain situations, what happens at one observation time depends only on what happens during the previous observation time. For example, in an electric circuit, intermittent failure of a component may have been observed through experience to behave as follows.

1. If the component works at time t, it works at the next observation time 99 percent of the time.
2. If the component fails at time t, it fails at the next observation time 5 percent of the time.

These facts can be represented schematically in Table 1.1.2.

The methods of Markov chains, discussed in more detail in Chap. 5, can be used to show that (on average) the chance of successful operation x_1 and the chance of failure x_2 must satisfy the equations

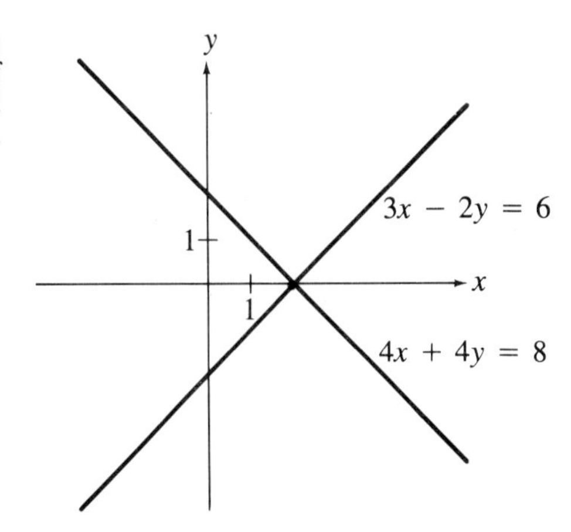

Figure 1.1.4 Intersection of $3x - 2y = 6$ and $4x + 4y = 8$.

Table 1.1.2 COMPONENT RELIABILITY TABLE

SUCCESS	FAILURE	
.99	.95	Success next time
.01	.05	Failure next time

$$\begin{aligned} 0.99x_1 + 0.95x_2 &= x_1 \\ 0.01x_1 + 0.05x_2 &= x_2 \\ x_1 + \quad x_2 &= 1 \end{aligned}$$

To solve, we collect terms and add the first two equations, to find

$$\begin{aligned} -0.01x_1 + 0.95x_2 &= 0 \\ 0 &= 0 \\ x_1 + x_2 &= 1 \end{aligned}$$

(One of the equations was redundant!)

Solving by substitution, we find

$$x_1 = \tfrac{95}{96} \qquad x_2 = \tfrac{1}{96}$$

The component fails about 1.04 percent of the time (on average).

EXAMPLE 7 (Optimal planning, industrial engineering) A soap company produces two lotion soaps, Creamy and Creamy Plus, in large batches. A batch of Creamy contains 1 ton of oil and 16 tons of base, while a batch of Creamy Plus contains 2 tons of oil and 16 tons of base. In inventory, the company has 12 tons of oil and 112 tons of base. Assuming the company wants to use all its inventory, how many batches of each type of soap should be produced?

Solution Let x_1 be the number of batches of Creamy and x_2 be the number of batches of Creamy Plus. The amount of oil used is

$$1x_1 + 2x_2 \qquad \text{(tons of oil per batch times batches)}$$

and the amount of soap base is

$$16x_1 + 16x_2 \qquad \text{(tons of base per batch times batches)}$$

To deplete the inventory completely, we must have

$$\begin{aligned} x_1 + \ 2x_2 &= 12 \\ 16x_1 + 16x_2 &= 112 \end{aligned}$$

The solution is $x_1 = 2$, $x_2 = 5$. Thus 2 batches of Creamy and 5 batches of Creamy Plus will deplete the inventory completely.

Note that if the inventory figures in Example 7 had been different, a "nice" solution might not exist. For instance, if the inventory were 12 tons of oil and 116 tons of base, we would have as the solution $x_1 = 2.5$, $x_2 = 4.75$. If fractional batches cannot be processed, other methods could be employed to determine a production schedule which would use up the most inventory. The point here is that we may not be able to implement the mathematical solution. This example shows that problems requiring integer solutions may require different solution methods.

EXAMPLE 8 (Structural analysis, civil engineering) Internal forces and node deflections of a frame structure can sometimes be determined by solving systems of linear equations (a node is a place where frame members are joined). The systems of equations can be quite complicated; for a simple example, however, consider the frame[4] in Fig. 1.1.5.

The reduced aggregate stiffness equations are (for a certain choice of Young's modulus and other parameters)

$$\begin{aligned} 45.018x_1 \qquad\qquad\quad + 0.9x_3 &= 6 \\ 45.018x_2 + 0.9x_3 &= -3 \\ 0.9x_1 + \quad 0.9x_2 + 120x_3 &= 0 \end{aligned}$$

The solution is $x_1 = 0.13329\cdots$, $x_2 = -0.06663\cdots$, $x_3 = -0.0004995\cdots$. These numbers represent deflection of the node in the x and y directions (x_1 and x_2) in inches and the change in the frame angle (x_3) at the node in radians. We note that if the solution is rounded to

$$x_1 = 0.1333 \qquad x_2 = -0.0666 \qquad x_3 = -0.0005$$

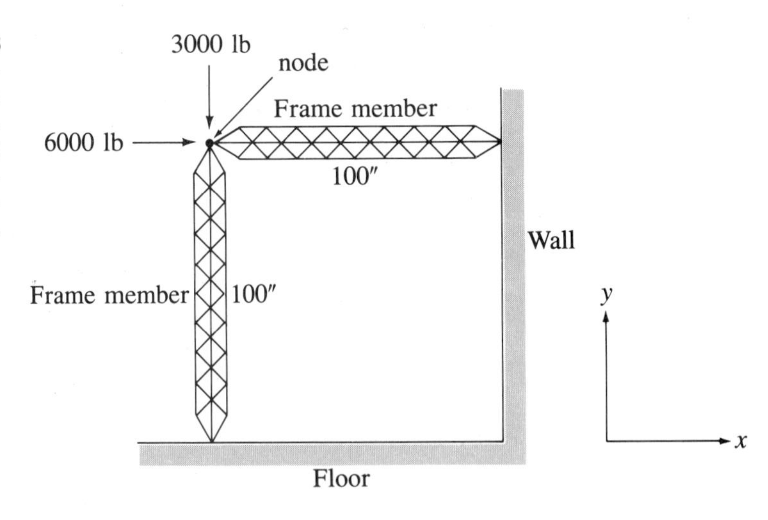

Figure 1.1.5 A frame loaded horizontally and vertically at the joint. The node (connecting point) is moved from the unloaded position by the given forces.

[4] J. J. Azar, *Matrix Structural Analysis*, Pergamon, Oxford, 1972.

then we have only an approximate solution to the problem. The reduced equations are only a part of the total aggregate stiffness equations which have nine unknowns. However, the larger system is solved efficiently when the reduced system is solved first.

The examples given do not exhaust the fields of applications for linear equations; they are merely a sample. Many more applications are given later.

PROBLEMS 1.1

1. Classify the following equations as linear or nonlinear.
   (*a*) $x^2 + y - z = 7$ (*b*) $-|x| + y = 0$
   (*c*) $-x + y = 0$ (*d*) $3x + 2y - z = 6$
   (*e*) $y + \sqrt{x^2} = 4$ (*f*) $(1 + i)x = y^2$
   (*g*) $\sin x + \cos y = 0.5$ (*h*) $ix_1 - 3x_2 = 5 - 2i$

2. The following systems of linear equations have solutions. Compute them. If an equation has more than one solution, write the general solution.
   (*a*) $\begin{aligned} 3x - 2y &= 7 \\ 2x + y &= 14 \end{aligned}$ (*b*) $\begin{aligned} x + y &= 4 \\ 3x + 3y &= 12 \end{aligned}$
   (*c*) $\begin{aligned} 2x_1 - 3x_2 &= -1 \\ 4x_1 - 6x_2 &= -2 \\ x_1 + x_2 &= 1 \end{aligned}$ (*d*) $2x_1 - 5x_2 = 10$
   (*e*) $\begin{aligned} ix_1 - 2ix_2 &= 1 \\ (1 + i)x_1 - x_2 &= -1 \end{aligned}$ (*f*) $\begin{aligned} (1 + i)x_1 + (1 - i)x_2 &= 2 \\ 2x_1 - 2ix_2 &= 2 - 2i \end{aligned}$

3. A meat packer sells lean and extra lean ground beef. A batch of lean contains $1\frac{1}{2}$ lb of fat and $8\frac{1}{2}$ lb of red meat, while a batch of extra lean contains 1 lb of fat and 9 lb of red meat. The grocer has 10 lb of fat and 80 lb of red meat. How many batches of lean and extra lean ground beef should the packer produce to use up all the meat and fat (to avoid waste)?

4. A baker sells two kinds of sweet rolls, regular and extra sweet. Each batch of dough for regular sweet rolls uses 50 lb of flour and 2 lb of sugar. Each batch of dough for extra sweet rolls uses 49 lb of flour and 4 lb of sugar. The baker has 690 lb of flour and 48 lb of sugar. How many batches of each type of roll should be made to use all the flour and sugar?

5. Use the method of substitution and a calculator to solve

$$\begin{aligned} x + 0.2y &= 1.92 \\ -1.4x + 0.6y &= -3.48 \end{aligned}$$

6. Use the method of substitution and a calculator to solve

$$\begin{aligned} 1.02x - 3.2y &= -2.18 \\ 0.3x + 10.1y &= 10.4 \end{aligned}$$

7. Find the point of intersection of $x - 3y = 7$ and $2x + 7y = 9$.

8. Find the point of intersection of $x + 4y = 3$ and $y - 2x = 0$.

9. A consulting firm bills clients for computer time C, automobile travel T, and hours of actual work H. The rates are, respectively, \$400 per hour, \$0.35 per mile, and \$120 per hour. Write a linear equation for the billing amount B in terms of C, T, and H.

10. The revenue for a self-service gas station depends on the number of gallons of premium, leaded regular, and unleaded gasoline sold. In December 1986, the average prices per gallon were, respectively, 83.9, 73.9, and 76.9 cents. Letting P, L, and U represent the number of gallons of premium, leaded, and unleaded, respectively, sold in that month, write a linear equation to estimate the December revenue.

1.2 METHOD OF ELIMINATION

We consider the problem of solving the system of linear equations

$$\begin{aligned} a_{11}x_1 + \cdots + a_{1n}x_n &= b_1 \\ a_{21}x_1 + \cdots + a_{2n}x_n &= b_2 \\ \cdots\cdots\cdots\cdots\cdots&\cdots\cdots \\ a_{m1}x_1 + \cdots + a_{mn}x_n &= b_m \end{aligned}$$

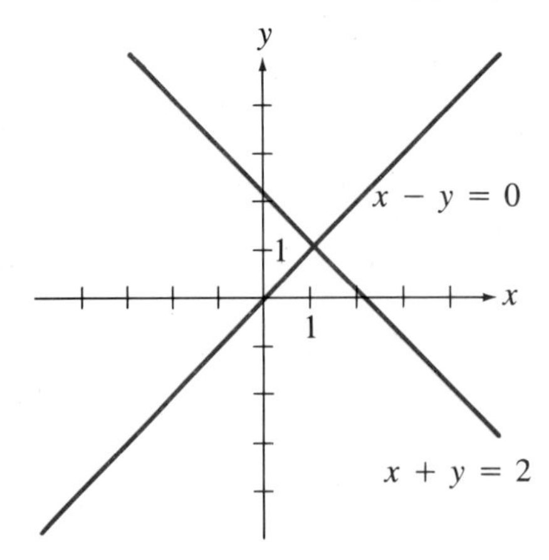

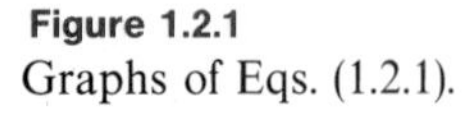

Figure 1.2.1
Graphs of Eqs. (1.2.1).

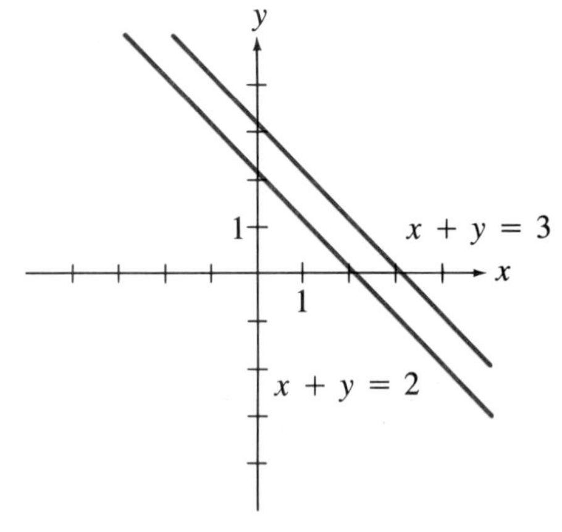

Figure 1.2.2
Graphs of Eqs. (1.2.2).

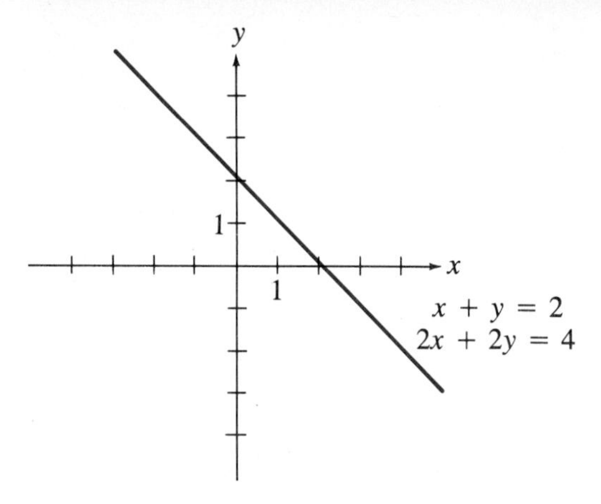

Figure 1.2.3
Graphs of Eqs. (1.2.3).

by the process of elimination of variables. When at least one of the "right-hand side" values $b_1, b_2, \ldots, b_m$ is not equal to zero, the system of equation is called *nonhomogeneous.* If $b_1 = b_2 = \cdots = b_m = 0$, the system of equations is called *homogeneous.*

Nonhomogeneous linear equations may have no solutions, exactly one solution, or infinitely many solutions. All three possibilities can occur even in the simple case of two equations in two unknowns. For example, the system

$$\begin{aligned} x - y &= 0 \\ x + y &= 2 \end{aligned} \tag{1.2.1}$$

has only the solution $x = 1$, $y = 1$ (see Fig. 1.2.1). Yet the system

$$\begin{aligned} x + y &= 2 \\ x + y &= 3 \end{aligned} \tag{1.2.2}$$

has no solution since the lines represented by the equations do not intersect (see Fig. 1.2.2). The system

$$\begin{aligned} x + \ \ y &= 2 \\ 2x + 2y &= 4 \end{aligned} \tag{1.2.3}$$

has an infinite number of solutions since each equation actually represents the same line. Each point on the common line is a solution (see Fig. (1.2.3).

Homogeneous linear equations have either exactly one solution (namely, $x_1 = x_2 = \cdots = x_n = 0$, called the *trivial solution*[5]) *or an infinite number of solutions* (including the trivial solution). For example, the system

$$\begin{aligned} x + y &= 0 \\ x - y &= 0 \end{aligned}$$

[5] A solution is *nontrivial* if any one of the unknowns is not equal to zero.

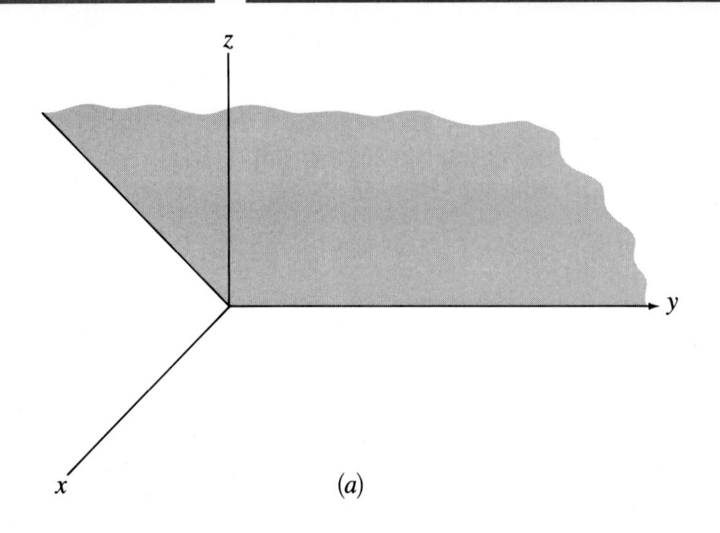

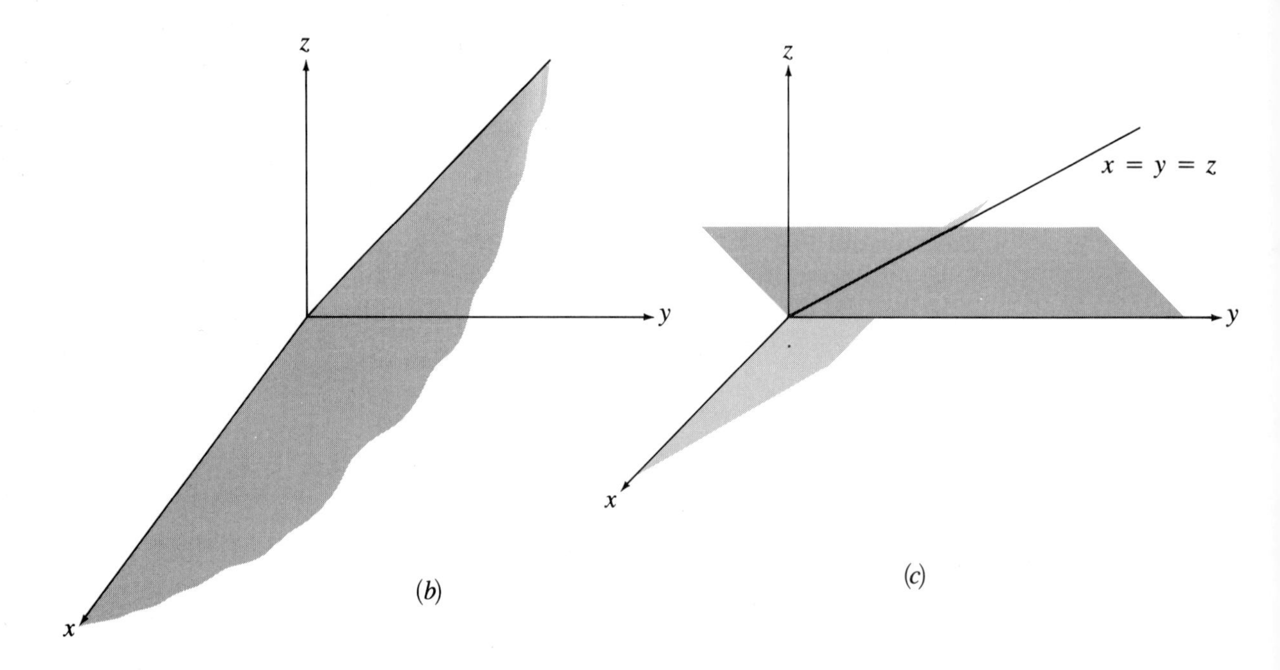

Figure 1.2.4
(*a*) The plane $x - z = 0$; first octant view. (*b*) The plane $y - z = 0$; first octant view.
(*c*) Intersection of $x - z = 0$ and $y - z = 0$; first octant view.

has only the solution $x = 0$, $y = 0$, while the system

$$\begin{aligned} x + y &= 0 \\ 2x + 2y &= 0 \end{aligned}$$

has solutions $x = k$, $y = -k$, for any number k (if $k = 0$, we get the trivial solution).

For an example involving three unknowns, consider

$$\begin{aligned} x - z &= 0 \\ y - z &= 0 \end{aligned}$$

where we require the solution to be real. Each equation is that of a plane passing through the origin. The solution is geometrically the straight line of the intersection of the planes, which is analytically $x = y = z = k$, where k is any real number. See Fig. 1.2.4.

These examples are simpler than those in four or more unknowns, when graphing may be difficult or impossible. We must have another way of determining whether solutions of linear equations exist and find the solutions when they do exist. We use a method commonly called *gaussian*[6] *elimination*. The method is illustrated in the next example.

EXAMPLE 1 Solve

$$\begin{aligned} 2x + y - z &= -1 \\ x - 2y + z &= 5 \\ 3x - y - 2z &= 0 \end{aligned}$$

Solution The plan is to eliminate the x variable first. The elementary algebraic operations we will use lead to equivalent equations. Sets of equations are *equivalent* if they have the same solution set. The operations we will use are

1. Multiplication of any equation by a nonzero constant
2. Interchange of any two equations
3. Addition to an equation the result of multiplying another equation by a constant

---

[6] Named after C. F. Gauss (1777–1855), German physicist, astronomer, and mathematician, who was a pioneer in the field of statistics and theory of magnetism. His mathematical work on curves and surfaces led to the principles of curved space time. His name is lent to a unit of magnetic flux density. In World War II so-called degaussing girdles were used to protect steel ships passing through waters containing magnetic mines.

There are many ways to eliminate variables to solve a system of linear equations. The method we use is useful for later problems in the text. First we interchange the first and second equations (operation 2), so that the arithmetic involved in eliminating x will be easier. The result is

$$\begin{aligned} x - 2y + z &= 5 \\ 2x + y - z &= -1 \\ 3x - y - 2z &= 0 \end{aligned}$$

Now add to the second equation the result of multiplying the first equation by -2 to obtain

$$\begin{array}{r} 2x + y - z = -1 \\ -2x + 4y - 2z = -10 \\ \hline \mathbf{5y - 3z = -11} \end{array}$$

and add to the third equation the result of multiplying the first equation by -3 (operation 3):

$$\begin{array}{r} 3x - y - 2z = 0 \\ -3x + 6y - 3z = -15 \\ \hline \mathbf{5y - 5z = -15} \end{array}$$

We obtain the equivalent system

$$\begin{aligned} x - 2y + z &= 5 \\ \mathbf{5y - 3z} &\mathbf{= -11} \\ \mathbf{5y - 5z} &\mathbf{= -15} \end{aligned}$$

Now we eliminate y in the third equation by adding to the third equation the result of multiplying the second equation by -1 (operation 3):

$$\begin{array}{r} \mathbf{5y - 5z = -15} \\ -5y + 3z = 11 \\ \hline \mathbf{-2z = -4} \end{array}$$

This obtains as an equivalent system

$$\begin{aligned} x - 2y + z &= 5 \\ \mathbf{5y - 3z} &\mathbf{= -11} \\ \mathbf{-2z} &\mathbf{= -4} \end{aligned}$$

The last of these equations tells us that $z = 2$. Now we use the method of *back substitution* to solve for x and y. Substituting $z = 2$ into the second equation, we have $5y - 6 = -11$ and thus $y = -1$. Now these values for z and y can be substituted into the first equation, to find $x + 2 + 2 = 5$ and $x = 1$. Therefore, the solution is

$$x = 1 \qquad y = -1 \qquad z = 2$$

as can be checked by substitution in the original equations.

The key to gaussian elimination is that the operations used yield new equations which are equivalent to the old equations.

THEOREM 1.2.1 If a set of linear equations undergoes any of the following operations, then the resulting set of equations has exactly the same solution set as the original set of equations:

(*a*) Multiplication of any equation by a nonzero constant

(*b*) Interchange of two equations

(*c*) Addition to an equation the result of multiplying another equation by a constant

Proof Parts (*a*) and (*b*) are left to the reader. We prove part (*c*). Let $(s_1, s_2, \ldots, s_n)$ be a solution to

$$\begin{aligned}
a_{11}x_1 + a_{12}x_2 + \cdots + a_{1n}x_n &= b_1 \\
\cdots\cdots&\cdots\cdots \\
a_{i1}x_1 + a_{i2}x_2 + \cdots + a_{in}x_n &= b_i \\
\cdots\cdots&\cdots\cdots \\
a_{p1}x_1 + a_{p2}x_2 + \cdots + a_{pn}x_n &= b_p \\
\cdots\cdots&\cdots\cdots \\
a_{m1}x_1 + a_{m2}x_2 + \cdots + a_{mn}x_n &= b_m
\end{aligned}$$

Suppose we add to Eq. (p) the result of multiplying Eq. (i) by r. Then the new system has a different Eq. (p):

$$(ra_{i1} + a_{p1})x_1 + (ra_{i2} + a_{p2})x_2 + \cdots + (ra_{in} + a_{pn})x_n = rb_i + b_p$$

Since the rest of the equations are unchanged, $(s_1, \ldots, s_n)$ is still a solution for those equations. Substituting $s_1 = x_1, \ldots, s_n = x_n$ in the new Eq. (p) and multiplying, we have

$$r(a_{i1}s_1 + \cdots + a_{in}s_n) + (a_{p1}s_1 + \cdots + a_{pn}s_n) = rb_i + b_p$$

and $(s_1, \ldots, s_n)$ is a solution of the new Eq. (p).

Conversely, suppose that $(t_1, t_2, \ldots, t_n)$ is any solution of the new system. That is, suppose that

$$a_{11}t_1 + \cdots + a_{1n}t_n = b_1$$

$$\cdots\cdots\cdots\cdots\cdots\cdots\cdots\cdots$$

$$(ra_{i1} + a_{p1})t_1 + \cdots (ra_{in} + a_{pn})t_n = rb_i + b_p$$

$$\cdots\cdots\cdots\cdots\cdots\cdots\cdots\cdots$$

$$a_{m1}t_1 + \cdots + a_{mn}t_n = b_m$$

Multiply Eq. (i) by $-r$ and add the result to Eq. (p) to obtain the original system with $(t_1, t_2, \ldots, t_n)$ as a solution.

To streamline our work, in the next example we use abbreviations to indicate the algebraic operations done. For example,

Abbreviation	Meaning
3E2	Multiply Eq. (2) by 3.
E2 ↔ E5	Interchange Eq. (2) and (5).
−2E4 + E7	Add the result of multiplying Eq. (4) by −2 to Eq. (7), and replace Eq. (7) by the result.

EXAMPLE 2 Solve

$$\begin{aligned} x_1 + 2x_2 + 3x_3 &= 4 \\ 4x_1 + 5x_2 + 6x_3 &= 7 \\ 7x_1 + 8x_2 + 9x_3 &= 10 \end{aligned}$$

Solution

$$\begin{aligned} x_1 + 2x_2 + 3x_3 &= 4 \\ 4x_1 + 5x_2 + 6x_3 &= 7 \\ 7x_1 + 8x_2 + 9x_3 &= 10 \end{aligned} \xrightarrow[-7E1 + E3]{-4E1 + E2} \begin{aligned} x_1 + 2x_2 + 3x_3 &= 4 \\ -3x_2 - 6x_3 &= -9 \\ -6x_2 - 12x_3 &= -18 \end{aligned}$$

$$\xrightarrow{-2E_2 + E3} \begin{aligned} x_1 + 2x_2 + 3x_3 &= 4 \\ -3x_2 - 6x_3 &= -9 \\ 0 &= 0 \end{aligned} \xrightarrow{-\frac{1}{3}E_2} \begin{aligned} x_1 + 2x_2 + 3x_3 &= 4 \\ x_2 + 2x_3 &= 3 \\ 0 &= 0 \end{aligned}$$

Now we have only two equations in three unknowns. In the second equation, we can let $x_3 = k$, where k is any complex number. Then $x_2 = 3 - 2k$. Substituting $x_3 = k$ and $x_2 = 3 - 2k$ into the first equation, we have

$$x_1 = 4 - 2x_2 - 3x_3 = 4 - 2(3 - 2k) - 3(k) = -2 + k$$

Thus the general solution is

$$(-2+k,\ 3-2k,\ k) \qquad \text{or} \qquad \begin{aligned} x_1 &= -2+k \\ x_2 &= 3-2k \\ x_3 &= k \end{aligned}$$

And we see that the system has an infinite number of solutions. Specific solutions can be generated by choosing specific values for k.

EXAMPLE 3 Solve

$$\begin{aligned} x_1 + 2x_2 + 3x_3 &= 4 \\ 4x_1 + 5x_2 + 6x_3 &= 7 \\ 7x_1 + 8x_2 + 9x_3 &= 12 \end{aligned}$$

Solution Since the coefficients of x_1, x_2, and x_3 are the same as in Example 2, the same algebraic operations are used. This time, to save more space, we do not copy the unknowns each time; care will be taken to keep the coefficients in vertical line.

$$\begin{aligned} x_1 + 2x_2 + 3x_3 &= 4 \\ 4x_1 + 5x_2 + 6x_3 &= 7 \\ 7x_1 + 8x_2 + 9x_3 &= 12 \end{aligned}$$

$$\Downarrow$$

$$\left(\begin{array}{ccc|c} 1 & 2 & 3 & 4 \\ 4 & 5 & 6 & 7 \\ 7 & 8 & 9 & 12 \end{array}\right) \xrightarrow[-7E1+E3]{-4E1+E2} \left(\begin{array}{ccc|c} 1 & 2 & 3 & 4 \\ 0 & -3 & -6 & -9 \\ 0 & -6 & -12 & -16 \end{array}\right)$$

$$\xrightarrow{-2E2+E3} \left(\begin{array}{ccc|c} 1 & 2 & 3 & 4 \\ 0 & -3 & -6 & -9 \\ 0 & 0 & 0 & 2 \end{array}\right)$$

$$\Downarrow$$

$$\begin{aligned} x_1 + 2x_2 + 3x_3 &= 4 \\ 0x_1 - 3x_2 - 6x_3 &= -9 \\ 0x_1 + 0x_2 + 0x_3 &= 2 \end{aligned}$$

The last "equation," $0 = 2$, can never hold regardless of the values assigned to x_1, x_2, and x_3. Because the last (equivalent) system has no solution, the original system of equations has no solution.

DEFINITION 1.2.1 A *matrix* is an array of numbers, arranged in horizontal rows and vertical columns.

The arrangement of the coefficients and right-hand sides of the system in Example 3 is an example of a matrix. It is called the *augmented matrix of the system.* The array on the left-hand side of the vertical line is called the *coefficient matrix* of the system.

EXAMPLE 4 Solve

$$\begin{aligned} x_1 + \ 2x_2 + \ 3x_3 &= 0 \\ 4x_1 + \ 5x_2 + \ 6x_3 &= 0 \\ 7x_1 + \ 8x_2 + \ 9x_3 &= 0 \\ 10x_1 + 11x_2 + 12x_3 &= 0 \end{aligned}$$

Solution Use gaussian elimination with the augmented matrix.

$$\left(\begin{array}{ccc|c} 1 & 2 & 3 & 0 \\ 4 & 5 & 6 & 0 \\ 7 & 8 & 9 & 0 \\ 10 & 11 & 12 & 0 \end{array}\right) \xrightarrow[-10E1 + E4]{\substack{-4E1 + E2 \\ -7E1 + E3}} \left(\begin{array}{ccc|c} 1 & 2 & 3 & 0 \\ \mathbf{0} & -3 & -6 & 0 \\ \mathbf{0} & -6 & -12 & 0 \\ \mathbf{0} & -9 & -18 & 0 \end{array}\right)$$

$$\xrightarrow{-\frac{1}{3}E2} \left(\begin{array}{ccc|c} 1 & 2 & 3 & 0 \\ \mathbf{0} & 1 & 2 & 0 \\ \mathbf{0} & -6 & -12 & 0 \\ \mathbf{0} & -9 & -18 & 0 \end{array}\right)$$

$$\xrightarrow[9E2 + E3]{6E2 + E3} \left(\begin{array}{ccc|c} 1 & 2 & 3 & 0 \\ \mathbf{0} & 1 & 2 & 0 \\ \mathbf{0} & \mathbf{0} & 0 & 0 \\ \mathbf{0} & \mathbf{0} & \mathbf{0} & 0 \end{array}\right)$$

Therefore

$$\begin{aligned} x_1 + 2x_2 + 3x_3 &= 0 \\ x_2 + 2x_3 &= 0 \end{aligned}$$

and setting $x_3 = k$ gives

$$\begin{aligned} x_2 &= -2k \\ x_1 &= -2x_2 - 3x_3 = 4k - 3k = k \end{aligned}$$

So we have

$$\begin{aligned} x_1 &= k \\ x_2 &= -2k \\ x_3 &= k \end{aligned}$$

EXAMPLE 5 Solve

$$
\begin{aligned}
x_1 - 2x_2 + x_3 &= 0 \\
3x_1 - x_2 + x_3 &= 0 \\
-x_1 + 4x_2 - x_3 &= 0
\end{aligned}
$$

Solution As before, we use gaussian elimination with the augmented matrix of the system:

$$
\left(\begin{array}{rrr|r} 1 & -2 & 1 & 0 \\ 3 & -1 & 1 & 0 \\ -1 & 4 & -1 & 0 \end{array}\right)
\xrightarrow[\text{E1 + E3}]{-3\text{E1 + E2}}
\left(\begin{array}{rrr|r} 1 & -2 & 1 & 0 \\ \mathbf{0} & 5 & -2 & 0 \\ \mathbf{0} & 2 & 0 & 0 \end{array}\right)
$$

$$
\xrightarrow{\frac{1}{5}\text{E2}}
\left(\begin{array}{rrr|r} 1 & -2 & 1 & 0 \\ \mathbf{0} & 1 & -\frac{2}{5} & 0 \\ \mathbf{0} & 2 & 0 & 0 \end{array}\right)
$$

$$
\xrightarrow{-2\text{E2 + E3}}
\left(\begin{array}{rrr|r} 1 & -2 & 1 & 0 \\ \mathbf{0} & 1 & -\frac{2}{5} & 0 \\ \mathbf{0} & \mathbf{0} & \frac{4}{5} & 0 \end{array}\right)
$$

$$
\xrightarrow{\frac{5}{4}\text{E2}}
\left(\begin{array}{rrr|r} 1 & -2 & 1 & 0 \\ \mathbf{0} & 1 & -\frac{2}{5} & 0 \\ \mathbf{0} & \mathbf{0} & 1 & 0 \end{array}\right)
$$

From the last row, we see that $x_3 = 0$. Substituting this into the equation represented by the second row gives $x_2 = \frac{2}{5}x_3 = 0$. Finally substituting $x_2 = 0$, $x_3 = 0$ into the equation represented by the first row, we find $x_1 = 0$. Therefore this system has only the trivial solution.

What these examples *indicate* is the following:

> A system of linear equations may have no solutions, have exactly one solution, or have an infinite number of solutions.

and

> To determine the solvability or nonsolvability of a system of linear equations, gaussian elimination can be used.

We have not yet proved these statements. This will be done as we develop matrix algebra.

We now summarize the procedure for gaussian elimination, carrying out an example alongside.

GAUSSIAN ELIMINATION	EXAMPLE
A system is given:	A system is given:
$$\begin{aligned} a_{11}x_1 + a_{12}x_2 + \cdots + a_{1n}x_n &= b_1 \\ a_{21}x_2 + a_{22}x_2 + \cdots + a_{2n}x_2 &= b_2 \\ \cdots\cdots\cdots\cdots\cdots\cdots\cdots\cdots \\ a_{m1}x_1 + a_{m2}x_2 + \cdots + a_{mn}x_n &= b_m \end{aligned}$$	$$\begin{aligned} x_2 - x_3 + x_4 &= 0 \\ 2x_1 - x_2 + 3x_3 - x_4 &= 6 \\ 4x_1 + 4x_2 \qquad + 4x_4 &= 12 \\ 3x_1 - 3x_2 + 6x_3 - 3x_4 &= 9 \end{aligned}$$
1. Represent the system in an augmented matrix. $$\left(\begin{array}{cccc\|c} a_{11} & a_{12} & \cdots & a_{1n} & b_1 \\ a_{21} & a_{22} & \cdots & a_{2n} & b_2 \\ \cdots & \cdots & \cdots & \cdots & \cdots \\ a_{m1} & a_{m2} & \cdots & a_{mn} & b_m \end{array}\right)$$	$$\left(\begin{array}{cccc\|c} 0 & 1 & -1 & 1 & 0 \\ 2 & -1 & 3 & -1 & 6 \\ 4 & 4 & 0 & 4 & 12 \\ 3 & -3 & 6 & -3 & 9 \end{array}\right)$$
2. Make sure that the number in the first row and first column is nonzero, by interchanging two rows if necessary. (This corresponds to interchanging two equations.) This number is called a *pivot number*. If the first column contains all zeros, move to the second column.	Interchange rows 1 and 2. Pivot number $$\left(\begin{array}{cccc\|c} 2 & -1 & 3 & -1 & 6 \\ 0 & 1 & -1 & 1 & 0 \\ 4 & 4 & 0 & 4 & 12 \\ 3 & -3 & 6 & -3 & 9 \end{array}\right)$$
3. Obtain a pivot number of 1 by multiplication.[7]	Multiply row 1 by $\frac{1}{2}$. Pivot $$\left(\begin{array}{cccc\|c} 1 & -\frac{1}{2} & \frac{3}{2} & -\frac{1}{2} & 3 \\ \mathbf{0} & 1 & -1 & 1 & 0 \\ 4 & 4 & 0 & 4 & 12 \\ 3 & -3 & 6 & -3 & 9 \end{array}\right)$$
4. Obtain a zero below the pivot in each row below the pivot's row. To do this in a row (called the *object row*), multiply the pivot row by a constant and add the result to the object row.	Multiply row 1 by -4, and add to row 3. Multiply row 1 by -3, and add to row 4. $$\left(\begin{array}{cccc\|c} 1 & -\frac{1}{2} & \frac{3}{2} & -\frac{1}{2} & 3 \\ \mathbf{0} & 1 & -1 & 1 & 0 \\ \mathbf{0} & 6 & -6 & 6 & 0 \\ \mathbf{0} & -\frac{3}{2} & \frac{3}{2} & -\frac{3}{2} & 0 \end{array}\right)$$

[7] Actually this step is not necessary. However, it makes the arithmetic simpler in many problems. More important, this step is used when we discuss echelon forms of matrices, defined later in this section.

5. Restrict attention to the subarray of the augmented matrix obtained by ignoring all previous pivot rows and columns.

$$\left(\begin{array}{cccc|c} 1 & -\frac{1}{2} & \frac{3}{2} & -\frac{1}{2} & 3 \\ \mathbf{0} & \mathbf{1} & \mathbf{-1} & \mathbf{1} & \mathbf{0} \\ \mathbf{0} & \mathbf{6} & \mathbf{-6} & \mathbf{6} & \mathbf{0} \\ \mathbf{0} & \mathbf{-\frac{3}{2}} & \mathbf{\frac{3}{2}} & \mathbf{-\frac{3}{2}} & \mathbf{0} \end{array}\right)$$

Work with this.

6. Go back to step 2 and begin working on the subarray, unless the subarray has no nonzero pivots.

Multiply row 2 by -6, and add to row 3. Multiply row 2 by 3/2, and add to row 4.

$$\left(\begin{array}{cccc|c} 1 & -\frac{1}{2} & \frac{3}{2} & -\frac{1}{2} & 3 \\ \mathbf{0} & 1 & -1 & 1 & 0 \\ \mathbf{0} & \mathbf{0} & \mathbf{0} & \mathbf{0} & \mathbf{0} \\ \mathbf{0} & \mathbf{0} & \mathbf{0} & \mathbf{0} & \mathbf{0} \end{array}\right)$$

No nonzero pivots

7. When finished, write the solution.

We have two equations in four unknowns. Thus two of the unknowns can be arbitrarily assigned. Let $x_4 = t$ and $x_3 = s$. Then

$$\begin{aligned} x_1 - \tfrac{1}{2}x_2 + \tfrac{3}{2}s - \tfrac{1}{2}t &= 3 \\ x_2 - \ s + \ t &= 0 \end{aligned}$$

$$\begin{aligned} x_2 &= s - t \\ x_1 &= \tfrac{1}{2}(s - t) - \tfrac{3}{2}s + \tfrac{1}{2}t + 3 \\ &= -s + 3 \end{aligned}$$

8. Check your solution.

The idea is, roughly speaking, to find pivots and obtain zeros below the pivots, working from the upper left-hand corner toward the right and down. When gaussian elimination is complete, the augmented matrix is said to be in *row echelon form.* That is,

1. The leftmost nonzero entry in any row is 1.

2. In row i if the first 1 is in column j, then the first 1 in row $i + 1$ is in a column to the right of column j.

3. Rows containing all zeros are at the bottom.

The matrix alongside step 6 in the explanation of gaussian elimination is in row echelon form.

To be precise, we need some definitions.

DEFINITION 1.2.2 The *elementary row operations* on a matrix are

(*a*) Multiplication of a row by a nonzero number c (that is, multiplication of each number in the row by c)

(*b*) Interchange of two rows

(*c*) Replacement of a row (say the rth row) by row r plus a constant times another row

The elementary row operations correspond to the algebraic operations which obtain equivalent equations from a given set of equations (see Theorem 1.2.1).

DEFINITION 1.2.3 Two matrices are *row-equivalent* if one can be obtained from the other by a finite number of elementary row operations.

EXAMPLE 6 Consider the matrices

$$A = \begin{pmatrix} 1 & 1 & -2 & 0 \\ 2 & 1 & 1 & 1 \\ 1 & 1 & 2 & -2 \end{pmatrix} \qquad B = \begin{pmatrix} 2 & 1 & 1 & 1 \\ 1 & 1 & -2 & 0 \\ 1 & 1 & 2 & -2 \end{pmatrix}$$

$$C = \begin{pmatrix} 1 & 0 & 3 & 1 \\ 1 & 1 & -2 & 0 \\ 1 & 1 & 2 & -2 \end{pmatrix}$$

Matrix A is row-equivalent to B because B is obtained from A by the interchange of rows 1 and 2. Matrix C is row-equivalent to B because C is obtained from B by the row operation $-\text{R}2 + \text{R}1$. Matrix A is row-equivalent to C because C is obtained from A by performing the given row operations in sequence.

DEFINITION 1.2.4 A matrix is in *row echelon form* if

(*a*) All rows with all zero entries are below all rows which have at least one nonzero entry.

(*b*) The first nonzero entry (reading from the left) in a row (with not all zeros) is 1.

(*c*) In row i, if the first 1 is in column j, then in row $i + 1$ the first 1 is in a column to the right of column j.

DEFINITION 1.2.5 A matrix is in *reduced row echelon form* if it is in row echelon form and each column which contains the leading nonzero entry of a row has all other column entries equal to zero.

EXAMPLE 7 Use row operations to reduce this matrix to a row echelon form:

$$\begin{pmatrix} 1 & -1 & 2 \\ 3 & 4 & -3 \\ 9 & 5 & 0 \end{pmatrix}$$

Solution Using the standard operations yields

$$\begin{pmatrix} 1 & -1 & 2 \\ 3 & 4 & -3 \\ 9 & 5 & 0 \end{pmatrix} \xrightarrow[-9R1 + R3]{-3R1 + R2} \begin{pmatrix} 1 & -1 & 2 \\ 0 & 7 & -9 \\ 0 & 14 & -18 \end{pmatrix}$$

$$\xrightarrow[\frac{1}{14}R3]{\frac{1}{7}R2} \begin{pmatrix} 1 & -1 & 2 \\ 0 & 1 & -\frac{9}{7} \\ 0 & 1 & -\frac{9}{7} \end{pmatrix} \xrightarrow{-1R2 + R3} \begin{pmatrix} 1 & -1 & 2 \\ \mathbf{0} & 1 & -\frac{9}{7} \\ \mathbf{0} & \mathbf{0} & \mathbf{0} \end{pmatrix}$$

The last matrix is in row echelon form. Note that an additional row operation could be performed to further reduce the matrix to reduced row echelon form:

$$\xrightarrow{R2 + R1} \begin{pmatrix} 1 & 0 & \frac{5}{7} \\ \mathbf{0} & 1 & -\frac{9}{7} \\ \mathbf{0} & \mathbf{0} & \mathbf{0} \end{pmatrix}$$

EXAMPLE 8 Solve

$$\begin{aligned} x_2 - x_3 &= -9 \\ 2x_1 - x_2 + 4x_3 &= 29 \\ x_1 + x_2 - 3x_3 &= -20 \end{aligned}$$

by reducing the augmented matrix of the system to reduced row echelon form.

Solution

$$\left(\begin{array}{ccc|c} 0 & 1 & -1 & -9 \\ 2 & -1 & 4 & 29 \\ 1 & 1 & -3 & -20 \end{array}\right) \xrightarrow{R1 \leftrightarrow R3} \left(\begin{array}{ccc|c} 1 & 1 & -3 & -20 \\ 2 & -1 & 4 & 29 \\ 0 & 1 & -1 & -9 \end{array}\right)$$

$$\xrightarrow{-2R1 + R2} \left(\begin{array}{ccc|c} 1 & 1 & -3 & -20 \\ \mathbf{0} & -3 & 10 & 69 \\ \mathbf{0} & 1 & -1 & -9 \end{array}\right)$$

$$\xrightarrow{-\frac{1}{3}R2} \left(\begin{array}{ccc|c} 1 & 1 & -3 & -20 \\ \mathbf{0} & 1 & -\frac{10}{3} & -23 \\ \mathbf{0} & 1 & -1 & -9 \end{array}\right)$$

$$\xrightarrow{-1R2 + R3} \left(\begin{array}{ccc|c} 1 & 1 & -3 & -20 \\ \mathbf{0} & 1 & -\frac{10}{3} & -23 \\ \mathbf{0} & \mathbf{0} & \frac{7}{3} & 14 \end{array}\right)$$

$$\xrightarrow{\frac{3}{7}R3} \left(\begin{array}{ccc|c} 1 & 1 & -3 & -20 \\ \mathbf{0} & 1 & -\frac{10}{3} & -23 \\ \mathbf{0} & \mathbf{0} & 1 & 6 \end{array}\right)$$

$$\xrightarrow[\substack{3R3 + R1 \\ -1R2 + R1}]{\frac{10}{3}R3 + R2} \left(\begin{array}{ccc|c} 1 & \mathbf{0} & \mathbf{0} & 1 \\ \mathbf{0} & 1 & \mathbf{0} & -3 \\ \mathbf{0} & \mathbf{0} & 1 & 6 \end{array}\right)$$

It is easy to see that $x_1 = 1$, $x_2 = -3$, $x_3 = 6$.

The process of solving a system by reducing the augmented matrix to reduced row echelon form is called *Gauss-Jordan elimination.*

EXAMPLE 9 Solve

$$\begin{aligned} x_1 - x_2 + 2x_3 &= 4 \\ 3x_1 + 4x_2 - x_3 &= 8 \\ 5x_1 + 9x_2 - 4x_3 &= 13 \end{aligned}$$

by using Gauss-Jordan elimination.

Solution

$$\left(\begin{array}{ccc|c} 1 & -1 & 2 & 4 \\ 3 & 4 & -1 & 8 \\ 5 & 9 & -4 & 13 \end{array}\right) \xrightarrow[-5R1 + R3]{-3R1 + R2} \left(\begin{array}{ccc|c} 1 & -1 & 2 & 4 \\ 0 & 7 & -7 & -4 \\ 0 & 14 & -14 & -7 \end{array}\right)$$

$$\xrightarrow{\frac{1}{7}R2} \left(\begin{array}{ccc|c} 1 & -1 & 2 & 4 \\ \mathbf{0} & 1 & -1 & -\frac{4}{7} \\ \mathbf{0} & 14 & -14 & -7 \end{array}\right)$$

$$\xrightarrow{-14R2 + R3} \left(\begin{array}{ccc|c} 1 & -1 & 2 & 4 \\ \mathbf{0} & 1 & -1 & -\frac{4}{7} \\ \mathbf{0} & \mathbf{0} & \mathbf{0} & 1 \end{array}\right)$$

$$\xrightarrow[\substack{-4R3 + R1 \\ R2 + R1}]{\frac{4}{7}R3 + R2} \left(\begin{array}{ccc|c} 1 & \mathbf{0} & 1 & 0 \\ \mathbf{0} & 1 & -1 & 0 \\ \mathbf{0} & \mathbf{0} & \mathbf{0} & 1 \end{array}\right)$$

The last row represents $0 = 1$, which means that the system has no solution.

By using the row echelon form of a matrix, solving a system of equations seems fairly easy. Can every system of linear equations be solved by using row echelon form? That is, is *every* matrix row-equivalent to a matrix in row echelon form? The answer to this important question is yes and is the content of the next theorem.

THEOREM 1.2.2 Every matrix is row-equivalent to a matrix in reduced row echelon form.

Proof Suppose the matrix is

$$\begin{pmatrix} a_{11} & a_{12} & \cdots & a_{1n} \\ a_{21} & a_{22} & \cdots & a_{2n} \\ \cdots & \cdots & \cdots & \cdots \\ a_{m1} & a_{m2} & \cdots & a_{mn} \end{pmatrix}$$

We follow steps similar to those used in the summary of gaussian elimination on page 20. We can search column 1 for a nonzero number. Once we find one (say it is a_{i1}), we bring its row to the first row by interchange (if needed). If we do not find one, we go to column 2, and so on. (Actually if the matrix has all zeros, the search fails, but in this case the matrix is already in reduced row echelon form.) Now our matrix looks like

$$\text{row } i \rightarrow \begin{pmatrix} a_{i1} & a_{i2} & \cdots & a_{in} \\ a_{21} & a_{22} & \cdots & a_{2n} \\ \cdots & \cdots & \cdots & \cdots \\ a_{11} & a_{12} & \cdots & a_{1n} \\ \cdots & \cdots & \cdots & \cdots \\ a_{m1} & a_{m2} & \cdots & a_{mn} \end{pmatrix}$$

We can multiply row 1 by $1/a_{i1}$ and then eliminate all the numbers below in column 1. For example, with the *new* first row, $-a_{21}\text{R1} + \text{R2}$ eliminates the entry in row 2, column 1. Now we have the row-equivalent matrix

$$\begin{pmatrix} 1 & a_{12} & \cdots & a_{1n} \\ 0 & \hat{a}_{22} & \cdots & \hat{a}_{2n} \\ \cdots & \cdots & \cdots & \cdots \\ 0 & \hat{a}_{m2} & \cdots & \hat{a}_{mn} \end{pmatrix}$$

(The "hats" indicate the new entries after the algebra is completed.) We begin the procedure again on the *submatrix*

$$\begin{pmatrix} \hat{a}_{22} & \cdots & \hat{a}_{2n} \\ \cdots & \cdots & \cdots \\ \hat{a}_{m2} & \cdots & \hat{a}_{mn} \end{pmatrix}$$

and continue.

After obtaining a row echelon form, we can obtain a reduced row echelon form by eliminating constants above the first 1 in each row. Thus every matrix is row-equivalent to a matrix in reduced row echelon form.

A matrix has only one reduced row echelon form. See Prob. 17.

Solving Systems with Complex Coefficients When a set of linear equations with complex coefficients is to be solved, it can be changed to a system with real coefficients.

EXAMPLE 10 By splitting the unknowns into real and imaginary parts, show that the system

$$\begin{aligned} (3+i)x_1 + (1+i)x_2 &= 4+4i \\ x_1 - \qquad\quad x_2 &= 2i \end{aligned}$$

is equivalent to a real coefficient system of four equations in four unknowns. (Recall that this is Example 3 of Sec. 1.1.)

Solution Write $x_1 = y_1 + iz_1$ and $x_2 = y_2 + iz_2$, where y_1, y_2, z_1, and z_2 are real variables and substitute to obtain

$$\begin{aligned} (3+i)(y_1 + iz_1) + (1+i)(y_2 + iz_2) &= 4+4i \\ y_1 + iz_1 - (y_2 + iz_2) &= 2i \end{aligned}$$

which yields

$$\begin{aligned} 3y_1 - z_1 + i(y_1 + 3z_1) + (y_2 - z_2) + i(y_2 + z_2) &= 4+4i \\ (y_1 - y_2) + i(z_1 - z_2) &= 0 + 2i \end{aligned}$$

Now complex numbers are equal if and only if their real and imaginary parts are equal. Therefore we have as an equivalent system

$$\begin{aligned} 3y_1 - z_1 + y_2 - z_2 &= 4 \\ y_1 + 3z_1 + y_2 + z_2 &= 4 \\ y_1 \qquad\quad - y_2 \qquad &= 0 \\ z_1 \qquad\quad - z_2 &= 2 \end{aligned}$$

By using gaussian elimination, the solution of this system is

$$y_1 = 1 \qquad y_2 = 1 \qquad z_2 = -1 \qquad z_1 = 1$$

Therefore the solution of the original system is

$$x_1 = 1 + i \qquad x_2 = 1 - i$$

EXAMPLE 11 Solve the system from Example 10 without splitting the unknowns into real and imaginary parts, using gaussian elimination on the original system.

Solution

$$\left(\begin{array}{cc|c} 3+i & 1+i & 4+4i \\ 1 & -1 & 2i \end{array}\right) \xrightarrow{\text{R1} \leftrightarrow \text{R2}} \left(\begin{array}{cc|c} 1 & -1 & 2i \\ 3+i & 1+i & 4+4i \end{array}\right)$$

$$\xrightarrow{-(3+i)\text{R1} + \text{R2}} \left(\begin{array}{cc|c} 1 & -1 & 2i \\ 0 & 4+2i & 6-2i \end{array}\right)$$

Therefore

$$x_2 = \frac{6 - 2i}{4 + 2i} = 1 - i$$

and upon back substitution we find that $x_1 = 1 + i$.

We now wish to prove our first theorem about the theory of linear equations. This theorem tells us about the solution set of homogeneous equations.

THEOREM 1.2.3 If one has a homogeneous system of linear equations in which the number of unknowns exceeds the number of equations, then the system has a nontrivial solution.

Proof Suppose the system has m equations and n unknowns, with $m < n$. After obtaining the reduced row echelon form of the augmented matrix, we have an equivalent system with p equations, $p \leq m < n$. That is, rows $1, 2, \ldots, p$ are the nonzero rows (not all entries are zero). Now suppose the leftmost 1 in row i is in column k_i; thus the unknown x_{k_i} appears with nonzero coefficient only in row i and column k_i, $i = 1, 2, \ldots, p$. This means that each x_{k_i}, $i = 1, 2, \ldots, p$, can be written in terms of the other $n - p$ unknowns, call them $z_1, z_2, \ldots, z_{n-p}$. That is, we can give $z_1, z_2, \ldots, z_{n-p}$ any values we wish, and then $x_{k_1}, x_{k_2}, \ldots, x_{k_p}$ are also determined and a nontrivial solution exists.

EXAMPLE 12 Apply Theorem 1.2.3 to the homogeneous system of linear equations

$$\begin{aligned} 2x_1 + 4x_2 + 2x_3 + 10x_4 &= 0 \\ x_1 + 2x_2 + 2x_3 + 7x_4 &= 0 \\ x_1 + 2x_2 + x_3 + 5x_4 &= 0 \end{aligned}$$

Solve the system, identifying $x_{k_1}, x_{k_2}, z_1, z_2$ from the proof of Theorem 1.2.3.

Solution Since we have only three equations in four unknowns, Theorem 1.2.3 guarantees a nontrivial solution. The augmented matrix reduces to

$$\left(\begin{array}{cccc|c} 1 & 2 & 0 & 3 & 0 \\ 0 & 0 & 1 & 2 & 0 \\ 0 & 0 & 0 & 0 & 0 \end{array}\right)$$

Therefore, $x_{k_1} = x_1$, $x_{k_2} = x_3$, $z_1 = x_2$, and $z_2 = x_4$. We have the equivalent equations

$$\begin{aligned} x_1 &= -3x_4 - 2x_2 \\ x_3 &= -2x_4 \end{aligned}$$

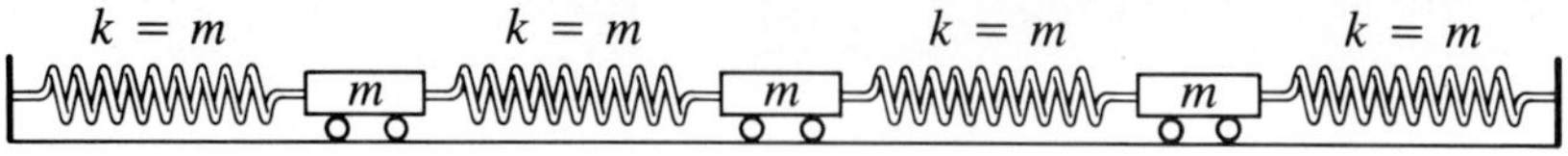

Figure 1.2.5 Trolleys of mass m connected by equivalent springs.

So, setting $x_2 = s$ and $x_4 = t$, we have solutions $x_1 = -3t - 2s$, $x_2 = s$, $x_3 = -2t$, and $x_4 = t$.

EXAMPLE 13 (Vibrating systems, mechanical engineering) An application in which homogeneous linear equations arise is that of vibrating mechanical systems. In finding the normal modes of vibration of the system in Fig. 1.2.5, the equations

$$\begin{aligned}(2 - w)x_1 - \quad x_2 \qquad\qquad &= 0\\ -x_1 + (2 - w)x_2 - \quad x_3 &= 0\\ - \quad x_2 + (2 - w)x_3 &= 0\end{aligned}$$

arise where w is the square of the characteristic frequency. We will see later how to calculate values of w for which the system has nontrivial solutions (which essentially tell us the displacements of the trolleys from equilibrium). Solve the equations for $w = 2$.

Solution Setting $w = 2$, we have the augmented matrix and reduction

$$\left(\begin{array}{ccc|c} 0 & -1 & 0 & 0\\ -1 & 0 & -1 & 0\\ 0 & -1 & 0 & 0\end{array}\right) \xrightarrow{R1 \leftrightarrow R2}$$

$$\left(\begin{array}{ccc|c} -1 & 0 & -1 & 0\\ 0 & -1 & 0 & 0\\ 0 & -1 & 0 & 0\end{array}\right) \xrightarrow[R2 + R3]{\substack{-1R1\\ -1R2}} \left(\begin{array}{ccc|c} 1 & 0 & 1 & 0\\ 0 & 1 & 0 & 0\\ 0 & 0 & 0 & 0\end{array}\right)$$

The solutions are $x_1 = -r$, $x_2 = 0$, and $x_3 = r$. We will see in a later chapter that this means that trolley 2 is stationary and trolleys 1 and 3 are always moving in opposite directions.

EXAMPLE 14 Determine conditions on a, b, and c so that

$$\begin{aligned}x_1 + 2x_2 + 3x_3 &= a\\ 4x_1 + 5x_2 + 6x_3 &= b\\ 7x_1 + 8x_2 + 9x_3 &= c\end{aligned}$$

will have no solutions or have an infinite number of solutions.

Solution Performing the same row operations as in Example 3 leads to the matrix

$$\left(\begin{array}{ccc|c} 1 & 2 & 3 & a \\ 0 & -3 & -6 & b-4a \\ 0 & 0 & 0 & c-2b+a \end{array}\right)$$

If $c - 2b + a \neq 0$, then no solution exists. If $c - 2b + a = 0$, we have two equations in three unknowns and we can set x_3 arbitrarily and then solve for x_1 and x_2.

PROBLEMS 1.2

1. Use gaussian elimination to determine whether the following systems have no solutions, exactly one solution, or an infinite number of solutions. State the solutions, if they exist.

(*a*)
$$\begin{aligned} x_1 - x_2 + x_3 &= 6 \\ x_1 + x_2 + 2x_3 &= 8 \\ 2x_1 - 3x_2 - x_3 &= 1 \end{aligned}$$

(*b*)
$$\begin{aligned} 2x_1 - 3x_2 + x_3 &= 5 \\ x_1 + x_2 - x_3 &= 3 \\ 4x_1 - x_2 - x_3 &= 11 \end{aligned}$$

(*c*)
$$\begin{aligned} x_1 - 3x_2 + x_3 &= 6 \\ 2x_1 + x_2 - 3x_3 &= -2 \\ x_1 + 4x_2 - 4x_3 &= 0 \end{aligned}$$

(*d*)
$$\begin{aligned} x_1 - x_2 &= 4 \\ 2x_1 + x_2 &= 7 \\ 5x_1 - 2x_2 &= 19 \end{aligned}$$

(*e*)
$$\begin{aligned} 3x_1 - 2x_2 + x_3 &= 12 \\ x_1 - 6x_2 + 4x_3 &= 6 \end{aligned}$$

(*f*)
$$\begin{aligned} x_1 + 2x_2 &= 6 \\ 3x_1 + 6x_2 &= 8 \\ 5x_1 + 10x_2 &= 12 \end{aligned}$$

2. Solve the systems of equations in Prob. 1 by using Gauss-Jordan elimination.

3. Which of the following matrices are in reduced row echelon form?

(*a*) $\begin{pmatrix} 0 & 1 & 2 & 0 & 3 \\ 0 & 0 & 0 & 1 & 0 \\ 0 & 0 & 0 & 0 & 0 \end{pmatrix}$ (*b*) $\begin{pmatrix} 1 & 0 & 0 & 2 & 1 \\ 0 & 1 & 0 & 0 & 0 \\ 0 & 0 & 1 & 0 & 0 \end{pmatrix}$

(*c*) $\begin{pmatrix} 1 & 1 & 1 \\ 0 & 0 & 1 \\ 0 & 0 & 0 \\ 0 & 0 & 0 \end{pmatrix}$ (*d*) $\begin{pmatrix} 0 & 0 & 0 & 1 \\ 0 & 0 & 0 & 0 \\ 0 & 0 & 0 & 0 \end{pmatrix}$

(*e*) $\begin{pmatrix} 1 & 0 & 2 & 1 \\ 0 & 1 & 0 & 1 \\ 0 & 0 & 1 & 0 \\ 0 & 0 & 0 & 0 \end{pmatrix}$ (*f*) $\begin{pmatrix} 1 & 0 & 0 & 0 \\ 0 & 1 & 0 & 0 \\ 0 & 0 & 1 & 0 \\ 0 & 0 & 0 & 2 \end{pmatrix}$

(*g*) $\begin{pmatrix} 0 & 0 & 0 \\ 0 & 0 & 0 \end{pmatrix}$ (*h*) $\begin{pmatrix} 1 \\ 0 \\ 0 \\ 0 \end{pmatrix}$

4. Put the following matrices in row echelon form and reduced row echelon form.

(*a*) $\begin{pmatrix} 1 & -1 & 2 \\ 0 & 5 & 6 \\ 4 & 7 & 5 \end{pmatrix}$ (*b*) $\begin{pmatrix} 1 & 1 & 2 & 2 \\ 2 & 2 & 1 & 1 \\ 0 & 0 & 2 & 2 \end{pmatrix}$ (*c*) $\begin{pmatrix} 1 & 3 \\ 2 & 4 \\ -1 & -6 \\ 2 & 7 \end{pmatrix}$

(*d*) $\begin{pmatrix} 0 & 0 & i \\ 0 & 1 & 0 \\ 1 & 0 & 0 \end{pmatrix}$ (*e*) $\begin{pmatrix} 1 & -1 & 2 & 2 \\ 0 & 1 & 3 & 6 \\ 2 & -1 & 7 & 10 \\ -1 & 2 & 1 & 4 \end{pmatrix}$ (*f*) $\begin{pmatrix} 1 & 0 & 0 \\ 2 & 1 & 0 \\ 3 & 4 & 1 \end{pmatrix}$

5. For each system state conditions (if any) on a, b, and c which force either no solution or an infinite number of solutions for the system.

(*a*) $\begin{aligned} 2x - 3y &= a \\ x + 4y &= b \end{aligned}$ (*b*) $\begin{aligned} x + y &= a \\ 2x + 2y &= b \end{aligned}$

(*c*) $\begin{aligned} x - y + z &= 3 \\ x + y - z &= a \\ 2x - y + 2z &= b \end{aligned}$ (*d*) $\begin{aligned} x - 2y + z &= a \\ -x + y + 3z &= b \\ -2x + y + 10z &= c \end{aligned}$

6. Find the values of k (if any) for which the system has (i) only one solution, (ii) no solutions, (iii) an infinite number of solutions.

(*a*) $\begin{aligned} x_1 - x_2 &= 3 \\ 3x_1 + kx_2 &= 6 \end{aligned}$ (*b*) $\begin{aligned} x_1 + 2kx_2 &= 7 \\ x_1 + (k^2 + 1)x_2 &= 3 \end{aligned}$

(*c*) $\begin{aligned} x + 2y + 3z &= 4 \\ 4x + 5y + kz &= 3 \\ 7x + 8y + \frac{3k}{2} z &= 0 \end{aligned}$ (*d*) $\begin{aligned} kx_1 - x_2 &= 1 \\ x_1 + kx_2 &= 2 \end{aligned}$

(*e*) $\begin{aligned} kx_1 - x_2 &= 1 \\ x_1 + kx_2 &= i \end{aligned}$ (*f*) $\begin{aligned} kx + |k|y &= 1 \\ |k|x - ky &= -1 \end{aligned}$

7. Use gaussian elimination to solve

$$\begin{aligned} ax_1 + bx_2 &= r \\ cx_1 + dx_2 &= s \end{aligned}$$

where a, b, c, and d are nonzero real numbers. State a condition on $ad - bc$ which guarantees a unique solution.

8. If $ad - bc = 0$ in

$$\begin{aligned} ax + by &= r \\ cx + dy &= s \end{aligned} \qquad a \neq 0$$

what conditions on r and s will guarantee no solutions? An infinite number of solutions?

9. Use gaussian elimination to solve

$$\begin{aligned} ax_1 + bx_2 \qquad\qquad &= r \qquad a \neq 0 \\ cx_1 + dx_2 \qquad\qquad &= s \qquad e \neq 0 \\ ex_3 + fx_4 &= t \\ gx_3 + hx_4 &= u \end{aligned}$$

State conditions on a, b, c, d, e, f, g, and h which will guarantee a unique solution.

10. Show that if $ad - bc \neq 0$, then the reduced row echelon form of

$$\begin{pmatrix} a & b \\ c & d \end{pmatrix} \quad \text{is} \quad \begin{pmatrix} 1 & 0 \\ 0 & 1 \end{pmatrix}$$

11. If $ad - bc = 0$, what can you say about the reduced row echelon form of the following matrix?

$$\begin{pmatrix} a & b \\ c & d \end{pmatrix}$$

12. Solve the following homogeneous systems.

(*a*) $\begin{aligned} 3x - 4y &= 0 \\ 2x + y &= 0 \end{aligned}$

(*b*) $\begin{aligned} x - y &= 0 \\ -3x + 3y &= 0 \\ x + y &= 0 \end{aligned}$

(*c*) $\begin{aligned} 2x - 3y + 4z &= 0 \\ x + 2y - z &= 0 \\ 4x + y + 2z &= 0 \end{aligned}$

(*d*) $\begin{aligned} x - y + z &= 0 \\ 3x + y + z &= 0 \\ -x - 2y - z &= 0 \end{aligned}$

(*e*) $\begin{aligned} x_1 - 2x_2 - x_3 + x_4 &= 0 \\ 3x_1 + x_2 - 5x_3 - x_4 &= 0 \\ x_1 \qquad - x_3 + 2x_4 &= 0 \\ x_1 - x_2 + 7x_3 \qquad &= 0 \end{aligned}$

(*f*) $\begin{aligned} ix + 2y &= 0 \\ 0.5x - iy &= 0 \end{aligned}$

13. Let $x = r$, $y = s$ be a solution of

$$\begin{aligned} ax + by &= 0 \\ cx + dy &= 0 \end{aligned}$$

and $x = u$, $y = v$ be a solution of

$$\begin{aligned} ax + by &= m \\ cx + by &= n \end{aligned}$$

Show that $x = u + r$, $y = v + s$ is also a solution of the second set of equations.

14. Let $x = r$, $y = s$ and $x = u$, $y = v$ be solutions of

$$\begin{aligned} ax + by &= 0 \\ cx + dy &= 0 \end{aligned}$$

Show that (*a*) $x = r + u$, $y = s + v$ is a solution; (*b*) $x = r - u$, $y = s - v$ is a solution; (*c*) $x = kr$, $y = ks$, where k is a constant, is a solution. (*d*) Use an example to show that $x = r + u$, $y = s - v$ is not necessarily a solution.

15. Solve the equations in Example 13 with $w = 2 - \sqrt{2}$.

16. Show that if matrix A is row-equivalent to B, then B is row-equivalent to A. Show that if A is row-equivalent to B and B is row-equivalent to C, then A is row-equivalent to C.

17. It is true in general that a matrix with m rows and n columns is row-equivalent to only one reduced row echelon form. Show the truth of this statement for matrices with two rows and two columns by completing (*a*), (*b*), and (*c*).

(*a*) Show that a matrix with two rows and two columns can have only one of the following reduced row echelon forms:

$$\begin{pmatrix} 1 & 0 \\ 0 & 1 \end{pmatrix} \quad \begin{pmatrix} 0 & 1 \\ 0 & 0 \end{pmatrix} \quad \begin{pmatrix} 1 & r \\ 0 & 0 \end{pmatrix} \quad \begin{pmatrix} 0 & 0 \\ 0 & 0 \end{pmatrix}$$

(*b*) Show that none of the four matrices in part (*a*) is row-equivalent.

(*c*) Using (*a*) and (*b*), argue that there can be only one reduced row echelon form for the matrices under consideration.

18. The reduced row echelon form of

$$\begin{pmatrix} 1 & 2 & 3 \\ 4 & 5 & 6 \\ 7 & 8 & 9 \end{pmatrix}$$

has all zeros in the last row; show that the same is true for

$$\begin{pmatrix} 1 & 2 & 3 & 4 \\ 5 & 6 & 7 & 8 \\ 9 & 10 & 11 & 12 \\ 13 & 14 & 15 & 16 \end{pmatrix}$$

19. What can you say about the last row of the reduced row echelon form of the following?

$$\begin{pmatrix} 1 & 2 & 3 & \cdots & n \\ n+1 & n+2 & n+3 & \cdots & 2n \\ 2n+1 & 2n+2 & 2n+3 & \cdots & 3n \\ \cdots & \cdots & \cdots & \cdots & \cdots \\ n^2-n+1 & n^2-n+2 & n^2-n+3 & \cdots & n^2 \end{pmatrix}$$

20. Consider the system of equations

$$x - \bar{y} = -2 + 2i$$
$$\bar{x} + y = 2$$

Solve this system by taking the complex conjugate of both sides of the second equation and adding to solve for x; then substitute to find y.

21. Solve

$$\bar{x} + iy = 2$$
$$ix + \bar{y} = 2i$$

(Multiply the second equation by i, then take conjugates.)

22. Solve Probs. 20 and 21 by solving the associated real systems.

23. Prove parts (*a*) and (*b*) of Theorem 1.2.1.

1.3 COMPUTERS, ASSOCIATED ERRORS, AND STRATEGY IN GAUSSIAN ELIMINATION

In practice, systems of equations are not solved by hand; computers are used. Because of this we must consider some aspects of machine computation. Computers are finite machines. Although a human being can think of a number such as $\frac{7}{9}$ (and the concept behind it), a computer which stores six digits of a number would store

$$.777777$$

or

$$.777778$$

depending on whether truncation (cutting off) or rounding was used to convert $.7777777\cdots$ to a six-digit number. Both representations of $\frac{7}{9}$ are not exact, and any time these numbers enter a calculation in place of $\frac{7}{9}$, error is introduced. For example, $9 \times \frac{7}{9} = 7$, but $9 \times .777777 = 6.99999$ (using six digits and truncation). In problems involving thousands of calculations, *error accumulation* can be significant.

To solve a mathematical problem numerically with a computer involves several steps, one of which is concerned with the errors just mentioned. The steps are as follows:

1. Formulate the *mathematical problem.*
2. Develop a step-by-step numerical method, called an *algorithm*, for approximating the solution to the problem.
3. Write a *computer program* which puts the algorithm in a form that the computer can understand and carry out.
4. Conduct an *error analysis* to evaluate the accuracy of the computer's solution to the problem.

Numerical analysis is the branch of mathematics concerned with items 2 through

4 above. We concentrate on step 2. Before doing this, we discuss in greater detail how computers store numbers and how errors arise.

Computers store floating point (real) numbers in the form

$$(+ \text{ or } -) \times (.a_1 a_2 \cdots a_n) \times B^k \qquad n \text{ a positive integer} \qquad (1.3.1)$$

where $.a_1 a_2 \cdots a_n$ is called the *mantissa*, B is called the *base*, and k is called the *exponent*. The number n is called the *number of significant digits* of the machine. Since all numbers used in computers are put in the form of (1.3.1), rounding or truncation must take place. Usually $B = 2$, 8, 10, or 16. We use $B = 10$ for ease of understanding.

EXAMPLE 1 Suppose a computer stores numbers in the form $\pm .a_1 a_2 a_3 \times 10^n$. That is, the computer stores three significant digits. For the following numbers, write their stored versions, assuming rounding and truncation: $-\frac{2}{3}$, 3678, .2224, 4.278, $-.0667$.

Solution

NUMBER	ROUNDED VERSION	TRUNCATED VERSION
$-\frac{2}{3}$	$-.667 \times 10^0$	$-.666 \times 10^0$
3678	$.368 \times 10^4$	$.367 \times 10^4$
.2224	$.222 \times 10^0$	$.222 \times 10^0$
4.278	$.428 \times 10^1$	$.427 \times 10^1$
$-.0667$	$-.667 \times 10^{-1}$	$-.667 \times 10^{-1}$

EXAMPLE 2 Suppose the computer from Example 1 must add 23.4 and .213. The numbers are stored in the form $.234 \times 10^2$ and $.213 \times 10^0$. To be added, the exponents must agree, so we change $.213 \times 10^0$ to $.00213 \times 10^2$. Adding gives $(.234 + .00213) \times 10^2 = .23613 \times 10^2$ which must be rounded to $.236 \times 10^2$ to be stored. So the machine reports the sum as 23.6. Since the actual sum is 23.613, a roundoff error of .013 has been introduced.

EXAMPLE 3 Suppose that the computer in Example 2, after adding 23.4 and .213, must multiply that sum by 100,000 at a later time. The calculations would proceed as follows, since 100,000 is stored as $.1 \times 10^6$:

$$(.1 \times 10^6)(.236 \times 10^2) = .0236 \times 10^8 \qquad \text{stored as } .024 \times 10^8$$

The answer would be reported as 240,000. The actual product of 100,000(23.4 + .213) is 236,130. The roundoff error from Example 3 has accumulated because of its entry in another calculation.

Example 3 illustrates the fact that multiplication by a large number (or division by a small number) can cause a large roundoff error. Other causes of large roundoff error are adding a large number to a small number or subtracting two numbers which are almost equal. This is shown in Example 4.

EXAMPLE 4 Using the same computer as in the previous examples calculate (*a*) 3288 − 3286 and (*b*) 2460 + 2.

Solution In the calculations we always use the stored forms of the numbers.

(*a*) Difference $= .329 \times 10^4 - .329 \times 10^4 = .000 \times 10^4$. The difference would be reported as 0.

(*b*)
$$\text{Sum} = .246 \times 10^4 + .2 \times 10^1 = .246 \times 10^4 + .0002 \times 10^4$$
$$= .2462 \times 10^4 \underset{\substack{\uparrow \\ \text{Round off}}}{=} .246 \times 10^4$$

The sum would be reported as 2460.

In both cases the error had magnitude 2.

We have shown several examples of how error can be introduced and propagated in machine computations. Definitions of errors and measures of error are now given.

DEFINITION 1.3.1 Any error due to roundoff or truncation is called *roundoff error*.

DEFINITION 1.3.2 Let N be a number and N_{calc} be the calculated approximation of N. Then the *absolute error* due to using N_{calc} is

$$N_{\text{calc}} - N$$

The *relative error* is

$$\frac{N_{\text{calc}} - N}{N} \qquad N \neq 0$$

EXAMPLE 5 Determine the absolute and relative errors in Example 4*a*.

Solution Since $N = 2$ and $N_{\text{calc}} = 0$, the absolute error is -2 and the relative error is $-\frac{2}{2} = -100$ percent.

These examples should not lead us to undue pessimism regarding machine calculations. Rather, they keep us alert to the problem of error accumulation. Clearly with modern computing machines we can easily obtain 12 significant digits of accuracy, and even more if we are willing to pay the cost. By controlling error propagation we can expect results which, while not error-free, are fairly accurate. One way is to avoid division by small numbers or (equivalently) multiplication by large numbers (see Example 3).

Implementation of gaussian elimination on a computer entails minimizing the introduction of roundoff error. Row reduction involves division of a row of a matrix by a number. Therefore, we always want this division to be by as large a number as possible. Remember, division by small numbers increases roundoff error. We now illustrate a strategy for avoiding such operations. The method is called *partial pivoting*.

EXAMPLE 6 Suppose we solve the system

$$\begin{aligned} 0.0001x_1 + x_2 &= 1.000 \\ x_1 - x_2 &= 0 \end{aligned}$$

(which has exact solution $x_1 = x_2 = 0.9999000\overline{09999000}$) always rounding all calculations to four significant digits. If we do not rearrange the equations, we have

$$\left(\begin{array}{cc|c} 0.0001 & 1.000 & 1.000 \\ 1.000 & -1.000 & 0 \end{array}\right) \xrightarrow{\frac{1}{0.0001}\text{R1}} \left(\begin{array}{cc|c} 1 & 10{,}000 & 10{,}000 \\ 1 & -1 & 0 \end{array}\right)$$

$$\xrightarrow{-\text{R1}+\text{R2}} \left(\begin{array}{cc|c} 1 & 10{,}000 & 10{,}000 \\ 0 & -10{,}000 & -10{,}000 \end{array}\right)$$

which has solution $x_1 = 0$, $x_2 = 1$. The calculated solution for x_1 is not close to the true solution. However, if we switch rows so that the first coefficient in the first row is the largest number in the first column, we do better:

$$\left(\begin{array}{cc|c} 1 & -1 & 0 \\ 0.0001 & 1 & 1 \end{array}\right) \xrightarrow{-0.0001\text{R1}+\text{R2}} \left(\begin{array}{cc|c} 1 & -1 & 0 \\ 0 & 1 & 1 \end{array}\right)$$

$$x_1 = x_2 = 1$$

The solution calculated in the second way is very close to the actual solution.

The strategy of rearranging rows so that the pivot element is the largest (in magnitude) in its column is called *partial pivoting*. This strategy is illustrated in Example 7. (*Total pivoting* is explained a bit later.)

EXAMPLE 7 Solve

$$\begin{aligned} x - 2y + z &= 0 \\ 2x + y - 3z &= -5 \\ 4x - y + 5z &= 17 \end{aligned}$$

by using partial pivoting.

Solution The system has an augmented matrix

$$\left(\begin{array}{ccc|c} 1 & -2 & 1 & 0 \\ 2 & 1 & -3 & -5 \\ 4 & -1 & 5 & 17 \end{array}\right)$$

Beginning with the leftmost column, we see that 4 is the largest entry, so we interchange rows 1 and 3 and proceed;

$$\left(\begin{array}{ccc|c} 4 & -1 & 5 & 17 \\ 2 & 1 & -3 & -5 \\ 1 & -2 & 1 & 0 \end{array}\right) \xrightarrow{\frac{1}{4}R1} \left(\begin{array}{ccc|c} 1 & -\frac{1}{4} & \frac{5}{4} & \frac{17}{4} \\ 2 & 1 & -3 & -5 \\ 1 & -2 & 1 & 0 \end{array}\right)$$

$$\xrightarrow[-R1 + R2]{-2R1 + R2} \left(\begin{array}{ccc|c} 1 & -\frac{1}{4} & \frac{5}{4} & \frac{17}{4} \\ 0 & \frac{3}{2} & -\frac{11}{2} & -\frac{27}{2} \\ 0 & -\frac{7}{4} & -\frac{1}{4} & -\frac{17}{4} \end{array}\right)$$

Now we turn our attention to the submatrix

$$\left(\begin{array}{cc|c} \frac{3}{2} & -\frac{11}{2} & -\frac{27}{2} \\ -\frac{7}{4} & -\frac{1}{4} & -\frac{17}{4} \end{array}\right)$$

and notice that $\left|-\frac{7}{4}\right| > \left|\frac{3}{2}\right|$, so we interchange rows 2 and 3. Continuing the elimination process, we find

$$\left(\begin{array}{ccc|c} 1 & -\frac{1}{4} & \frac{5}{4} & \frac{17}{4} \\ 0 & -\frac{7}{4} & -\frac{1}{4} & -\frac{17}{4} \\ 0 & \frac{3}{2} & -\frac{11}{2} & -\frac{27}{2} \end{array}\right) \xrightarrow{-\frac{4}{7}R2} \left(\begin{array}{ccc|c} 1 & -\frac{1}{4} & \frac{5}{4} & \frac{17}{4} \\ 0 & 1 & \frac{1}{7} & \frac{17}{7} \\ 0 & \frac{3}{2} & -\frac{11}{2} & -\frac{27}{2} \end{array}\right)$$

$$\xrightarrow{-\frac{3}{2}R2 + R3} \left(\begin{array}{ccc|c} 1 & -\frac{1}{4} & \frac{5}{4} & \frac{17}{4} \\ 0 & 1 & \frac{1}{7} & \frac{17}{7} \\ 0 & 0 & -\frac{80}{14} & -\frac{120}{7} \end{array}\right)$$

$$\xrightarrow{-\frac{14}{80}R3} \left(\begin{array}{ccc|c} 1 & -\frac{1}{4} & \frac{5}{4} & \frac{17}{4} \\ 0 & 1 & \frac{1}{7} & \frac{17}{7} \\ 0 & 0 & 1 & 3 \end{array}\right)$$

The solution is $x_1 = 1$, $x_2 = 2$, $x_3 = 3$.

The method just illustrated is known as *partial pivoting*, to distinguish it from total pivoting. In partial pivoting, we search only one column at each step for the pivot element largest in magnitude. In total pivoting we search the entire coefficient matrix (or submatrix) for the largest element to choose as a pivot element. To show how total pivoting works, we rework Example 7.

EXAMPLE 8 Solve the system of Example 7 with total pivoting.

Solution The system has an augmented matrix

$$\left(\begin{array}{rrr|r} 1 & -2 & 1 & 0 \\ 2 & 1 & -3 & -5 \\ 4 & -1 & \mathbf{5} & 17 \end{array}\right)$$

We search the coefficient matrix and find 5 as the element largest in magnitude. Interchanging *columns* 1 and 3 and *then rows* 1 and 3, we have

$$\left(\begin{array}{rrr|r} \mathbf{5} & -1 & 4 & 17 \\ -3 & 1 & 2 & -5 \\ 1 & -2 & 1 & 0 \end{array}\right)$$

but we must remember that the column corresponding to x is now column 3 and the column corresponding to z is now column 1. To emphasize this, we write the unknowns below their columns. Proceeding with the elimination process, we have

$$\begin{array}{c}\left(\begin{array}{rrr|r} 5 & -1 & 4 & 17 \\ -3 & 1 & 2 & -5 \\ 1 & -2 & 1 & 0 \end{array}\right) \\ \begin{array}{ccc} z & y & x \end{array}\end{array} \xrightarrow{\frac{1}{5}R1} \begin{array}{c}\left(\begin{array}{rrr|r} 1 & -\frac{1}{5} & \frac{4}{5} & \frac{17}{5} \\ -3 & 1 & 2 & -5 \\ 1 & -2 & 1 & 0 \end{array}\right) \\ \begin{array}{ccc} z & y & x \end{array}\end{array}$$

$$\xrightarrow[-R1+R2]{3R1+R2} \begin{array}{c}\left(\begin{array}{rrr|r} 1 & -\frac{1}{5} & \frac{4}{5} & \frac{17}{5} \\ 0 & \frac{2}{5} & \mathbf{\frac{22}{5}} & \frac{26}{5} \\ 0 & -\frac{9}{5} & \frac{1}{5} & -\frac{17}{5} \end{array}\right) \\ \begin{array}{ccc} z & y & x \end{array}\end{array}$$

Now in the coefficient part of the submatrix $\frac{22}{5}$ is the element largest in magnitude. To put it in the pivot position, columns 2 and 3 must be interchanged. *Notice that this interchange affects entries outside the submatrix, too; this is a difference between total and partial pivoting.* Continuing with elimination gives us

$$\begin{array}{c}\left(\begin{array}{rrr|r} 1 & -\frac{1}{5} & \frac{4}{5} & \frac{17}{5} \\ 0 & \frac{2}{5} & \mathbf{\frac{22}{5}} & \frac{26}{5} \\ 0 & -\frac{9}{5} & \frac{1}{5} & -\frac{17}{5} \end{array}\right) \\ \begin{array}{ccc} z & y & x \end{array}\end{array} \xrightarrow{C2 \leftrightarrow C3} \begin{array}{c}\left(\begin{array}{rrr|r} 1 & \frac{4}{5} & -\frac{1}{5} & \frac{17}{5} \\ 0 & \mathbf{\frac{22}{5}} & \frac{2}{5} & \frac{26}{5} \\ 0 & \frac{1}{5} & -\frac{9}{5} & -\frac{17}{5} \end{array}\right) \\ \begin{array}{ccc} z & x & y \end{array}\end{array}$$

$$\xrightarrow{\frac{5}{22}R2} \left(\begin{array}{ccc|c} 1 & \frac{4}{5} & -\frac{1}{5} & \frac{17}{5} \\ 0 & 1 & \frac{1}{11} & \frac{13}{11} \\ 0 & \frac{1}{5} & -\frac{9}{5} & -\frac{17}{5} \end{array}\right)$$
$$\quad z \quad x \quad y$$

$$\xrightarrow{-\frac{1}{5}R2 + R3} \left(\begin{array}{ccc|c} 1 & \frac{4}{5} & -\frac{1}{5} & \frac{17}{5} \\ 0 & 1 & \frac{1}{11} & \frac{13}{11} \\ 0 & \frac{1}{5} & -\frac{100}{55} & -\frac{200}{55} \end{array}\right)$$
$$\quad z \quad x \quad y$$

And again $z = 3$, $x = 1$, $y = 2$.

Example 8 shows that total pivoting requires more bookkeeping than partial pivoting since columns may be interchanged. Of course, if a computer is doing the bookkeeping, this causes us no problem. What *is* significant is the extra time the computer needs to make needed comparisons. For large systems the time difference can be significant. Considering Fig. 1.3.1, we can see that the total comparisons for a 50×50 system are 42,875 for total pivoting and only

Figure 1.3.1 Number of comparisons in pivoting methods.

Total pivoting		Partial pivoting	
Comparison 1; entire matrix	$n^2 - 1$	Comparison 1; column 1	$n - 1$
Comparison 2; first submatrix	$(n - 1)^2 - 1$	Comparison 2; column 2	$n - 2$
Comparison 3; second submatrix	$(n - 2)^2 - 1$	Comparison 3; column 3	$n - 3$
⋮			
Last comparison 2×2 matrix	$2^2 - 1$	Last comparison	1
Total: $n^2 + (n - 1)^2 + (n - 2)^2 + \cdots + 2^2 - (n - 1) = \frac{(2n + 1)(n)(n + 1)}{6} - n$		Total: $(n - 1) + (n - 2) + \cdots + 2 + 1 = \frac{n(n - 1)}{2}$	

1225 for partial pivoting. For a 50×50 system, total pivoting requires about 35 times as much computer time. Because computer time costs money, partial pivoting is preferable to total pivoting, even though total pivoting is superior with regard to error control. The error control gained by partial pivoting is sufficient for most problems.

Other pivoting strategies can be devised. In general, there is no way to pick a "best" pivoting strategy.

PROBLEMS 1.3

1. Round the following numbers to four significant digits.

(*a*) .000786 (*b*) 5,479,829.3 (*c*) .793029847 (*d*) 5000.004

2. Consider the list of numbers

$$\begin{array}{r} .1246 \\ .3672 \\ 4.067 \\ 7.125 \\ 35.71 \\ 64.89 \\ 281.6 \\ 419.8 \\ 3657. \\ 8785. \end{array}$$

(*a*) Add from top to bottom, rounding to four significant digits with each addition.

(*b*) Add from bottom to top, rounding to four significant digits with each addition.

(*c*) Add the numbers without rounding.

(*d*) Compare answers in (*a*), (*b*), and (*c*).

3. In algebra we have the property for real numbers a, b, c, and d that

$$\frac{a/b}{c/d} = \frac{ad}{bc}$$

Using four significant digits and rounding, show that if $a = -3$, $b = 3$, $c = .0003$, and $d = 2$, then this equation does not hold for computed results.

4. Choose values for a, b, and c which when used in

$$a - (b + c) = (a - b) - c$$

with four significant digits and rounding make the equation false.

In Probs. 5 to 8, systems of equations are given. Solve each system (*a*) without pivoting, (*b*) with partial pivoting, (*c*) with total pivoting. Round after each computation.

5. $.0001x_1 + 3.172x_2 = 3.173$
$.6721x_1 + 4.227x_2 = 2.494$
(Keep only four significant digits.)

6. $3x_1 - 2x_2 + 4x_3 = 71$
$-x_1 + \frac{1}{10}x_2 - x_3 = -3.8$
$\frac{1}{2}x_1 - x_3 = 2.5$
(Keep only two significant digits.)

7. $.01x_1 - x_2 = 0$
$-x_1 + x_2 - x_3 = -99.1$
$- x_2 + 10x_3 = 0$
(Keep only three significant digits.)

8. $.01x_1 + x_2 = 1$
$x_1 - x_2 = 0$
(Keep only two significant digits.)

9. Solve

$$x_1 + \tfrac{1}{2}x_2 + \tfrac{1}{3}x_3 = \tfrac{11}{6}$$
$$\tfrac{1}{2}x_1 + \tfrac{1}{3}x_2 + \tfrac{1}{4}x_3 = \tfrac{13}{12}$$
$$\tfrac{1}{3}x_1 + \tfrac{1}{4}x_2 + \tfrac{1}{5}x_3 = \tfrac{47}{60}$$

(*a*) using exact arithmetic and (*b*) keeping only three significant digits.

10. Using five significant digits and truncation (that is, .011237 = .01123 and so on), calculate the solutions of $x^2 + 205.16x + 2.1123 = 0$ by using the quadratic formula

$$r = \frac{-b \pm \sqrt{b^2 - 4ac}}{2a}$$

Actual solutions are $-.010296\cdots$ and $-205.1497\cdots$. Why is the relative error in your smaller solution (smallest absolute value) nearly 50 percent? Use

$$r = \frac{-2c}{b + \sqrt{b^2 - 4ac}}$$

to calculate the smaller solution, again using five significant digits and truncation. Why is your answer more accurate this time?

1.4 MATRICES

In previous sections we used gaussian elimination to solve systems of linear equations. Augmented matrices were used to list the coefficients of the unknowns and the right-hand-side values. Matrices are useful in many areas of science and engineering, not just for solving linear equations which come up in those disciplines.

EXAMPLE 1 (Chemical engineering) Matrices arise in chemical engineering in the analysis of staged absorbers.[8] Roughly speaking, liquid and gaseous phases of a chemical pass through a column of n plates, each of which absorbs some of the chemical. To determine the concentration at each stage, the matrix (with n rows and n columns)

$$\begin{pmatrix} -(L+GK) & GK & 0 & \cdots & 0 \\ L & -(L+GK) & GK & \ddots & \vdots \\ 0 & L & -(L+GK) & \ddots & 0 \\ \vdots & \ddots & \ddots & -(L+GK) & GK \\ 0 & \cdots & 0 & L & -(L+GK) \end{pmatrix}$$

is part of a mathematical model; K, L, G are constants related to the particular chemical species and the types of plates.

We can locate a number in a matrix by specifying its row and column. For example, in matrix A (capital letters denote matrices)

$$A = \begin{pmatrix} 1 & -1 & 2 \\ 3+i & 6 & -4 \\ 0 & 5i & -2 \end{pmatrix}$$

the number $5i$ is in row 3 and column 2. Another way to say this is, "$5i$ is the *entry* of A in the 3,2 position." It is this convention of identifying position by row and column which is used to store a matrix in a computer.

EXAMPLE 2 (Graph theory) Another area in which matrices arise is graph theory, which can be applied to theoretical physics and electrical networks, among other fields. A *graph* is a collection of vertices, some of which are joined by lines and edges. An example of a graph is shown in Fig. 1.4.1, where letters label the *vertices* and numbers label the edges. The *incidence matrix* $\mathcal{I}$ of the graph is a matrix with entries 0 and 1 as follows:

$$a_{ij} = \begin{cases} 1 & \text{if edge } j \text{ touches vertex } i \\ 0 & \text{otherwise} \end{cases}$$

For the graph in Fig. 1.4.1,

[8] N. R. Amundson, *Mathematical Methods in Chemical Engineering*, Prentice-Hall, Englewood Cliffs, N. J., 1966.

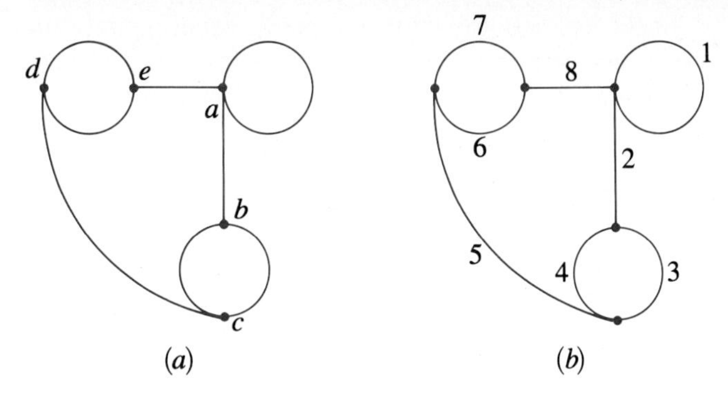

Figure 1.4.1 (*a*) Graph with vertices labeled. (*b*) Graph with edges labeled.

$$\mathscr{I} = \begin{array}{c} \\ a \\ b \\ c \\ d \\ e \end{array} \begin{array}{c} \begin{array}{cccccccc} 1 & 2 & 3 & 4 & 5 & 6 & 7 & 8 \end{array} \\ \begin{pmatrix} 1 & 1 & 0 & 0 & 0 & 0 & 0 & 1 \\ 0 & 1 & 1 & 1 & 0 & 0 & 0 & 0 \\ 0 & 0 & 1 & 1 & 1 & 0 & 0 & 0 \\ 0 & 0 & 0 & 0 & 1 & 1 & 1 & 0 \\ 0 & 0 & 0 & 0 & 0 & 1 & 1 & 1 \end{pmatrix} \end{array}$$

DEFINITION 1.4.1 A matrix is called $m \times n$ (said m by n) or *of order* $m \times n$ if it has m rows and n columns. If we want to indicate the size of an $m \times n$ matrix B, we can write $B_{m \times n}$. A matrix is *square* if it is $n \times n$.

Let A and B be these $m \times n$ matrices:

$$A = \begin{pmatrix} a_{11} & a_{12} & a_{13} & \cdots & a_{1n} \\ a_{21} & a_{22} & a_{23} & \cdots & a_{2n} \\ a_{31} & a_{32} & a_{33} & \cdots & a_{3n} \\ \cdots & \cdots & \cdots & \cdots & \cdots \\ a_{m1} & a_{m2} & a_{m3} & \cdots & a_{mn} \end{pmatrix} \qquad B = \begin{pmatrix} b_{11} & b_{12} & b_{13} & \cdots & b_{1n} \\ b_{21} & b_{22} & b_{23} & \cdots & b_{2n} \\ b_{31} & b_{32} & b_{33} & \cdots & b_{3n} \\ \cdots & \cdots & \cdots & \cdots & \cdots \\ b_{m1} & b_{m2} & b_{m3} & \cdots & b_{mn} \end{pmatrix}$$

We say A and B are *equal* if $a_{ij} = b_{ij}$ for all $i = 1, 2, \ldots, m$ and $j = 1, 2, \ldots, n$. That is, the entries of A and B in the same position are equal. Thus

$$\begin{pmatrix} 1 & 0 \\ 5 & 6 \end{pmatrix} = \begin{pmatrix} 1 & 0 \\ 5 & 6 \end{pmatrix}$$

but

$$\begin{pmatrix} 1 & 0 \\ 5 & 6 \end{pmatrix} \neq \begin{pmatrix} 1 & 1 \\ 5 & 6 \end{pmatrix}$$

and

$$\begin{pmatrix} 1 & 0 \\ 5 & 6 \end{pmatrix} \neq \begin{pmatrix} 1 & 0 & 0 \\ 5 & 6 & 0 \end{pmatrix}$$

In the last case the matrices are not equal because they are not the same size.

There are three basic operations involving matrices, and all are related to solving systems of linear equations. The operations are matrix addition, scalar multiplication, and matrix multiplication.

DEFINITION 1.4.2 The *scalar multiple* cA of matrix A, where c is a complex number (scalar), is a matrix defined by

$$c\begin{pmatrix} a_{11} & & \cdots & & a_{1n} \\ a_{i1} & \cdots & a_{ij} & \cdots & a_{in} \\ a_{m1} & & \cdots & & a_{mn} \end{pmatrix} = \begin{pmatrix} ca_{11} & & \cdots & & ca_{1n} \\ ca_{i1} & \cdots & ca_{ij} & \cdots & ca_{in} \\ ca_{m1} & & \cdots & & ca_{mn} \end{pmatrix}$$

or $cA = c(a_{ij})_{m \times n} = (ca_{ij})_{m \times n}$. That is, each entry of A is multiplied by c.

EXAMPLE 3 Let

$$A = \begin{pmatrix} 1 & -2 & 3 \\ 7 & 4 & -6 \end{pmatrix}$$

Calculate $2A$, $-3A$, $0A$, $1A$, and iA.

Solution Using the definition yields

$$2A = \begin{pmatrix} 2(1) & 2(-2) & 2(3) \\ 2(7) & 2(4) & 2(-6) \end{pmatrix} = \begin{pmatrix} 2 & -4 & 6 \\ 14 & 8 & -12 \end{pmatrix}$$

In the same way

$$-3A = \begin{pmatrix} -3 & 6 & -9 \\ -21 & -12 & 18 \end{pmatrix} \qquad 1A = \begin{pmatrix} 1 & -2 & 3 \\ 7 & 4 & -6 \end{pmatrix}$$

$$0A = \begin{pmatrix} 0 & 0 & 0 \\ 0 & 0 & 0 \end{pmatrix} \qquad iA = \begin{pmatrix} i & -2i & 3i \\ 7i & 4i & -6i \end{pmatrix}$$

Notice that $1A = A$.

The operation of multiplication by scalars corresponds to multiplying an entire system of equations by a constant. For example, if we were solving

$$\begin{aligned} \tfrac{1}{8}x - \tfrac{1}{4}y &= 2 \\ x + \tfrac{1}{2}y &= \tfrac{1}{8} \end{aligned} \tag{1.4.1}$$

by hand, we might multiply both equations by 8, in order to "clear the system of fractions," to obtain

$$\begin{aligned} x - 2y &= 16 \\ 8x + 4y &= 1 \end{aligned}$$

The corresponding operation for the augmented matrix for (1.4.1) would be scalar multiplication by 8:

$$8\left(\begin{array}{cc|c} \frac{1}{8} & -\frac{1}{4} & 2 \\ 1 & \frac{1}{2} & \frac{1}{8} \end{array}\right) = \left(\begin{array}{cc|c} 1 & -2 & 16 \\ 8 & 4 & 1 \end{array}\right)$$

DEFINITION 1.4.3 If A and B are both $m \times n$, then they are *conformable for addition* and the *sum* is defined by

$$A + B = \begin{pmatrix} a_{11} + b_{11} & \cdots & a_{1n} + b_{1n} \\ \cdots & a_{ij} + b_{ij} & \cdots \\ a_{n1} + b_{n1} & \cdots & a_{mn} + b_{mn} \end{pmatrix}$$

That is, the addition is performed entry by entry.

The difference $A - B$ of A and B is defined in virtually the same way as the sum, with plus replaced by minus.

EXAMPLE 4 For the following pairs of matrices, determine the sum and difference, if they exist.

(*a*) $A = \begin{pmatrix} 1 & -1 & 2 \\ 0 & 1 & 3 \end{pmatrix}$ $\quad B = \begin{pmatrix} 2 & 1.5 & 6 \\ -3 & 2+i & 0 \end{pmatrix}$

(*b*) $A = \begin{pmatrix} 1 & 0 \\ 3 & -4 \end{pmatrix}$ $\quad B = \begin{pmatrix} 1 & 1 & 2 \\ 0 & -2 & 0 \end{pmatrix}$

(*c*) $A = \begin{pmatrix} 2 \\ 3 \end{pmatrix}$ $\quad B = \begin{pmatrix} -2 \\ 1.7 \end{pmatrix}$

(*d*) $A = (\frac{3}{2}, \; -\frac{5}{8})$ $\quad B = \begin{pmatrix} 1 \\ 0 \end{pmatrix}$

Solution (*a*) Matrices A and B are 2×3 and conformable for addition and subtraction.

$$A + B = \begin{pmatrix} 1+2 & -1+1.5 & 2+6 \\ 0+-3 & 1+2+i & 3+0 \end{pmatrix} = \begin{pmatrix} 3 & 0.5 & 8 \\ -3 & 3+i & 3 \end{pmatrix}$$

$$A - B = \begin{pmatrix} 1-2 & -1-1.5 & 2-6 \\ 0-(-3) & 1-(2+i) & 3-0 \end{pmatrix} = \begin{pmatrix} -1 & -2.5 & -4 \\ 3 & -1-i & 3 \end{pmatrix}$$

(*b*) Matrix A is 2×2, and B is 2×3. Since A and B are not the same size, they are not conformable for addition or subtraction.

(*c*) Matrices A and B are conformable for addition and subtraction since they are the same size.

$$A + B = \begin{pmatrix} 2 + (-2) \\ 3 + 1.7 \end{pmatrix} = \begin{pmatrix} 0 \\ 4.7 \end{pmatrix}$$

$$A - B = \begin{pmatrix} 2 - (-2) \\ 3 - 1.7 \end{pmatrix} = \begin{pmatrix} 4 \\ 1.3 \end{pmatrix}$$

(*d*) Matrices A and B are not conformable for addition or subtraction since A is 1×2 and B is 2×1.

The operation of addition corresponds to adding systems of equations. For example, if we want to solve the system

$$\begin{aligned} x + y &= 4 \\ x - y &= 2 \end{aligned} \tag{1.4.2}$$

we could add the first equation to the second and keep the result. This corresponds to adding a system

$$\begin{aligned} 0x + 0y &= 0 \\ x + y &= 4 \end{aligned} \tag{1.4.3}$$

to the system (1.4.2). Adding the augmented matrices

$$\left(\begin{array}{cc|c} 1 & 1 & 4 \\ 1 & -1 & 2 \end{array}\right) + \left(\begin{array}{cc|c} 0 & 0 & 0 \\ 1 & 1 & 4 \end{array}\right) = \left(\begin{array}{cc|c} 1 & 1 & 4 \\ 2 & 0 & 6 \end{array}\right)$$

we see that $2x = 6$, or $x = 3$. Then $y = 1$ is found by using the first equation.

The most important basic operation on two matrices is the product. It can be used in electrical engineering to describe the behavior of certain black boxes which contain electrical parts and circuits. In particular, a four-terminal black-box is shown in Fig. 1.4.2. We see that the box has four terminals and that four quantities—two voltages (v_1 and v_2) and two currents (I_1 and I_2)—are associated with the box as inputs (v_1 and I_1) and outputs (v_2 and I_2). When the box has what is called *linear operation*, there exist constants a_{11}, a_{12}, a_{21} and a_{22} (which depend on what is inside the box) such that

$$\begin{aligned} v_1 &= a_{11}v_2 + a_{12}I_2 \\ I_1 &= a_{21}v_2 + a_{22}I_2 \end{aligned}$$

We can think of the right-hand sides of these equations as being generated in

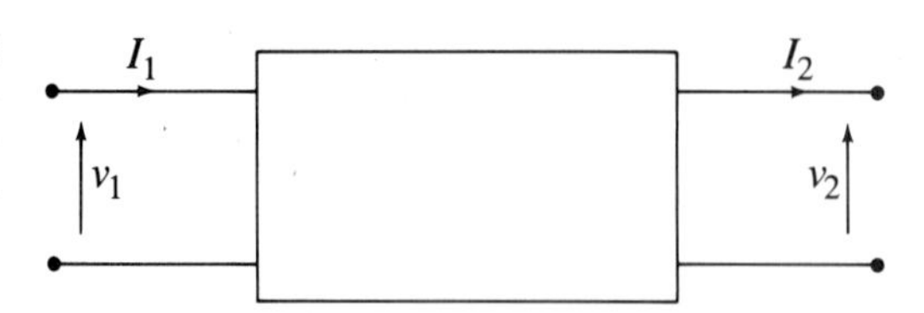

Figure 1.4.2 Black box with inputs on the left and outputs on the right.

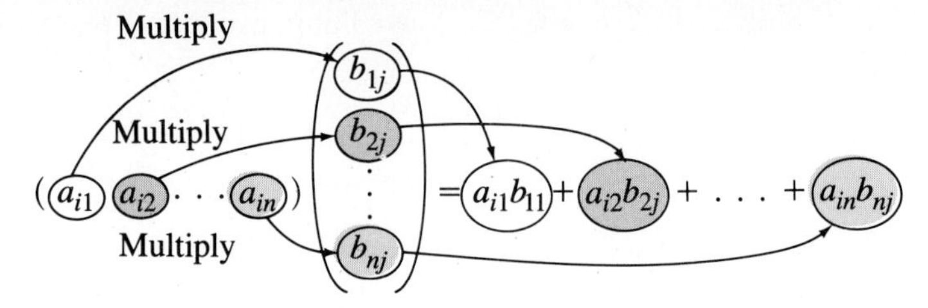

Figure 1.4.3
Row column product.

the following way:

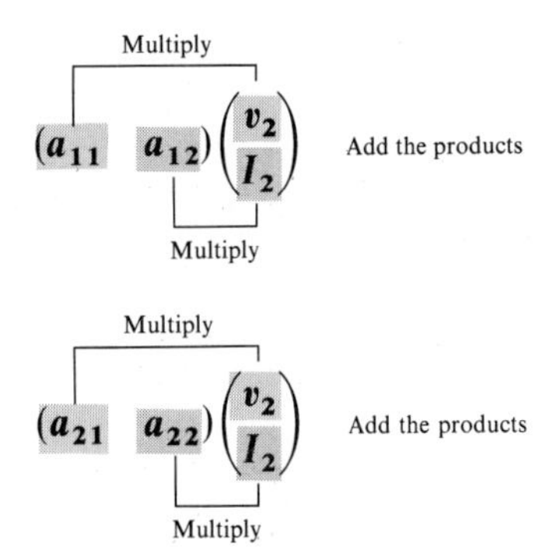

This brings us to the idea of a row-column product.

DEFINITION 1.4.4 Let A be $m \times n$ and B be $n \times p$. The *row-column product* of row i of A [which looks like $(a_{i1} a_{i2} \cdots a_{in})$] with column j of B which looks like

$$\begin{pmatrix} b_{1j} \\ b_{2j} \\ \vdots \\ b_{nj} \end{pmatrix}$$

is denoted $(AB)_{ij}$ and is defined as

$$(AB)_{ij} = a_{i1}b_{1j} + a_{i2}b_{2j} + \cdots + a_{in}b_{nj}$$

(See Fig. 1.4.3.)

EXAMPLE 5 Let

$$A = \begin{pmatrix} 1 & -1 & 2 \\ 0 & 1 & 3 \end{pmatrix} \quad \text{and} \quad B = \begin{pmatrix} 1 & -1 & -3 \\ 1 & 2 & 4 \\ 1 & 3 & 6 \end{pmatrix}$$

Calculate $(AB)_{12}$ and $(AB)_{23}$.

Solution Now, $(AB)_{12}$ denotes the row-column product of row 1 or A and column 2 of B:

$$(1 \quad \mathbf{-1} \quad 2)\begin{pmatrix} -1 \\ \mathbf{2} \\ 3 \end{pmatrix} = 1(-1) + \mathbf{(-1)2} + 2(3) = -1 - 2 + 6 = 3$$

$$(AB)_{23} = (0 \quad \mathbf{1} \quad 3)\begin{pmatrix} -3 \\ \mathbf{4} \\ 6 \end{pmatrix} = 0(-3) + \mathbf{1(4)} + 3(6) = 22$$

Now we see that the type of blackbox we discussed can be drawn as in Fig. 1.4.4. Two four-terminal boxes can be connected as in Fig. 1.4.5. If we want to write the combined action of the two boxes, we can first calculate the action of box B and then calculate the action of box A. Doing this, we have (refer to Fig. 1.4.5)

$$v_2 = b_{11}v_3 + b_{12}I_3 \qquad v_1 = a_{11}v_2 + a_{12}I_2$$
$$I_2 = b_{21}v_3 + b_{22}I_3 \qquad I_1 = a_{21}v_2 + a_{22}I_2$$

Substituting the expressions for v_2 and I_2 into the last two equations, we obtain

$$v_1 = (a_{11}b_{11} + a_{12}b_{21})v_3 + (a_{11}b_{12} + a_{12}b_{22})I_3$$
$$I_1 = (a_{21}b_{11} + a_{22}b_{21})v_3 + (a_{21}b_{12} + a_{22}b_{22})I_3$$

Therefore, the combined box is characterized by the four coefficients on the right-hand side of the last two equations. To compute these coefficients in a less painful way, we define the matrix product.

DEFINITION 1.4.5 Let A be $m \times n$ and B be $n \times p$. Then A and B are *conformable* for the matrix product AB. The *product* is $m \times p$, and the entry in row i and column j of AB

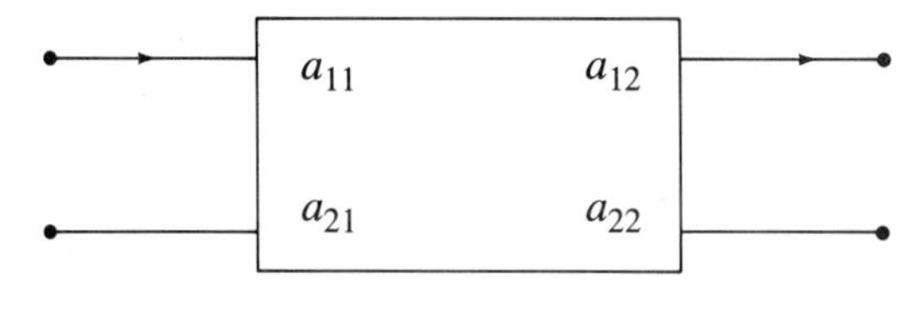

Figure 1.4.4 Black box with coefficients. The box can be referred to by the matrix

$$A = \begin{pmatrix} a_{11} & a_{12} \\ a_{21} & a_{22} \end{pmatrix}$$

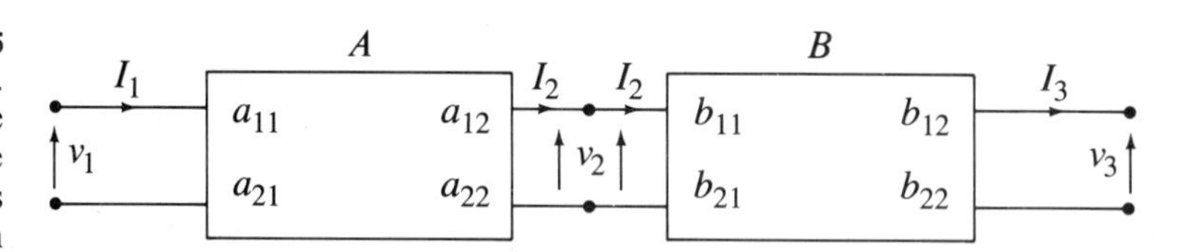

Figure 1.4.5 Black boxes connected. Together they can be considered as one black box with inputs v_1 and I_1 and with outputs v_3 and I_3.

is $(AB)_{ij}$. If the product is represented by $C_{m \times p}$, then

$$c_{ij} = \sum_{k=1}^{n} a_{ik} b_{kj}$$

EXAMPLE 6 Let

$$A = \begin{pmatrix} 1 & -1 & 2 \\ 0 & 1 & 3 \end{pmatrix} \quad \text{and} \quad B = \begin{pmatrix} 1 & -1 & -3 \\ 1 & 2 & 4 \\ 1 & 3 & 6 \end{pmatrix}$$

Calculate AB. Does the product BA exist?

Solution

$$AB = \begin{pmatrix} (1 \quad -1 \quad 2)\begin{pmatrix} 1 \\ 1 \\ 1 \end{pmatrix} & (1 \quad -1 \quad 2)\begin{pmatrix} -1 \\ 2 \\ 3 \end{pmatrix} & (1 \quad -1 \quad 2)\begin{pmatrix} -3 \\ 4 \\ 6 \end{pmatrix} \\ (0 \quad 1 \quad 3)\begin{pmatrix} 1 \\ 1 \\ 1 \end{pmatrix} & (0 \quad 1 \quad 3)\begin{pmatrix} -1 \\ 2 \\ 3 \end{pmatrix} & (0 \quad 1 \quad 3)\begin{pmatrix} -3 \\ 4 \\ 6 \end{pmatrix} \end{pmatrix}$$

$$= \begin{pmatrix} 2 & 3 & 5 \\ 4 & 11 & 22 \end{pmatrix}$$

The product BA is not defined:

$$B_{3 \times 3} \qquad A_{2 \times 3}$$

Not equal, so not conformable

EXAMPLE 7 Let

$$A = \begin{pmatrix} 3 & 1 \\ 0 & 2 \end{pmatrix} \quad \text{and} \quad B = \begin{pmatrix} 1 & 0 \\ 3 & 1 \end{pmatrix}$$

Calculate AB and BA.

Solution

$$AB = \begin{pmatrix} 3 & 1 \\ 0 & 2 \end{pmatrix}\begin{pmatrix} 1 & 0 \\ 3 & 1 \end{pmatrix} = \begin{pmatrix} 6 & 1 \\ 6 & 2 \end{pmatrix}$$

$$BA = \begin{pmatrix} 1 & 0 \\ 3 & 1 \end{pmatrix}\begin{pmatrix} 3 & 1 \\ 0 & 2 \end{pmatrix} = \begin{pmatrix} 3 & 1 \\ 9 & 5 \end{pmatrix}$$

Note that although AB and BA are both defined, $AB \neq BA$. Whether $AB = BA$ depends on matrices A and B.

Given the definition of the matrix product, we see that the action of the pair of blackboxes in Fig. 1.4.5 is expressed in terms of the product AB of the matrices which represent the individual boxes. This use of matrix products helps us see why, in general, AB may not be equal to BA: If the blackboxes in a circuit are reversed, the action of the circuit may not be the same. One case in which AB is equal to BA is when $A = B$; in terms of blackboxes, this means that if we reverse adjacent identical blackboxes, we do not change the action of the entire circuit.

Another important application of the matrix product is to the writing of systems of linear equations in compact form. If the matrix of coefficients of the system is A and the column of the right-hand sides is B, then A will be $m \times n$ if there are m equations and n unknowns and B will be $m \times 1$. If we write the unknowns as

$$X = \begin{pmatrix} x_1 \\ x_2 \\ \vdots \\ x_n \end{pmatrix}$$

then the system can be written

$$AX = B$$

With this convention, Theorem 1.2.3 can be restated by letting

$$0 = \begin{pmatrix} 0 \\ \vdots \\ 0 \end{pmatrix}_{m \times 1}$$

THEOREM 1.4.1 If $m < n$, then the system

$$A_{m \times n} X_{n \times 1} = 0$$

has a nontrivial solution.

After a discussion of matrix algebra, we will be able to state concisely, and prove other theorems about, systems of linear equations.

PROBLEMS 1.4

1. A matrix and an entry of the matrix are given. State the entry's row and column.

(a) $\begin{pmatrix} 1 & -3 \\ 2 & 1 \end{pmatrix}; -3$ (b) $\begin{pmatrix} 1-i & -1 & 2 \\ 0 & 1 & 1 \\ 3i & -6+i & 7 \end{pmatrix}; -6+i$

(c) $\begin{pmatrix} -3 & 2 & 1 \\ 4 & 6 & 7 \end{pmatrix}; 4$ (d) $\begin{pmatrix} 3 & 2 \\ 4 & 1 \\ 6 & 7 \\ 5 & 9 \end{pmatrix}; 5$

(e) $\begin{pmatrix} a & b & c & d \\ e & f & g & h \end{pmatrix}$; g (f) $\begin{pmatrix} 1 \\ 1 \\ 1 \\ 3 \\ 3 \\ 2 \\ 7 \end{pmatrix}$; 2

2. A matrix as well as a row number and a column number are given in an ordered pair. State the corresponding entry.

(a) $\begin{pmatrix} 3 & 6 \\ a & 5 \end{pmatrix}$; (2, 1) (b) $\begin{pmatrix} 3+2i & 6 & -2 & i\pi \\ 4-i & -6 & 2 & 5 \end{pmatrix}$; (2, 3)

(c) $\begin{pmatrix} a & b \\ c & d \\ e & f \\ g & h \end{pmatrix}$; (3, 2) (d) $\begin{pmatrix} 1 & 2 & -1 & 6 & 4 \\ 3 & -1 & 2 & 7 & 6 \\ 4 & 4 & 2 & 6 & 5 \end{pmatrix}$; (1, 4)

3. Write the augmented matrices for the systems in Prob. 1 of Sec. 1.2. State the size, $m \times n$, of the matrix.

4. Let

$$A = \begin{pmatrix} -1 & 2 \\ 2 & 3 \end{pmatrix} \quad B = \begin{pmatrix} 1 & 0 \\ 1 & 0 \end{pmatrix} \quad C = \begin{pmatrix} 1 & 0 & -1 \\ 2 & 1 & 3 \end{pmatrix} \quad D = \begin{pmatrix} 2 & 0 \\ 1 & 2 \\ 1 & 3 \end{pmatrix}$$

$$E = \begin{pmatrix} i & 1 \\ -1 & i \end{pmatrix}$$

For each part of the problem, perform the matrix calculation shown. If a calculation cannot be performed, say so.

(a) $3A$ (b) $A + C$ (c) CD

(d) $A - B$ (e) $3A + 4B$ (f) $AB + DC$

(g) $AB + CD$ (h) $(2 + i)A$ (i) iE

(j) $iA + E$ (k) AE (l) $A(B + C)$

(m) $AB + AC$ (n) $A + B$ (o) $B + A$

(p) AA (q) EE (r) $(AB)C$

5. Give an example of $A_{3\times 3}$ and $B_{3\times 3}$ such that $AB = BA$.

6. Give an example of $A_{3\times 3}$ and $B_{3\times 3}$ such that $AB \neq BA$.

7. Let

$$A = \begin{pmatrix} 1 & 1 \\ 1 & 1 \end{pmatrix} \quad B = \begin{pmatrix} -2 & -2 \\ 4 & 4 \end{pmatrix} \quad C = \begin{pmatrix} -1 & -1 \\ 3 & 3 \end{pmatrix}$$

Show that $AB = AC$ even though $B \neq C$.

8. (*a*) Let A be a matrix. Show that $1A = A$.

(*b*) Let A be a matrix. Show that $0A$ is a matrix with all entries equal to zero.

9. Must the product of conformable matrices with nonreal entries be a matrix with nonreal entries? If your answer is no, give an example to illustrate.

10. For nonzero real numbers a, $a \cdot a \neq 0$. If a matrix A has nonzero real entries, must AA have positive entries? If your answer is no, illustrate with an example.

11. Show that the matrix in Example 1 can be written in the form

$$\begin{pmatrix} -(a+b) & a & 0 & 0 & \cdots & 0 \\ b & -(a+b) & a & 0 & \cdots & 0 \\ 0 & b & -(a+b) & a & \cdots & 0 \\ \vdots & & \ddots & \ddots & \ddots & \vdots \\ 0 & & & & & a \\ & & \cdots & & b & -(a+b) \end{pmatrix}$$

12. Find incidence matrices for the following graphs (see Example 2).

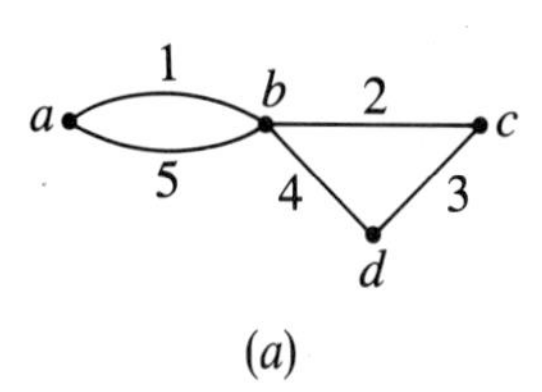

(*a*)

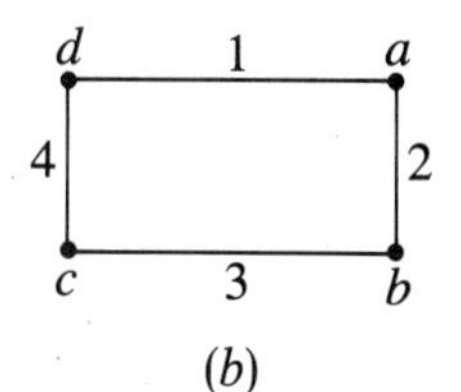

(*b*)

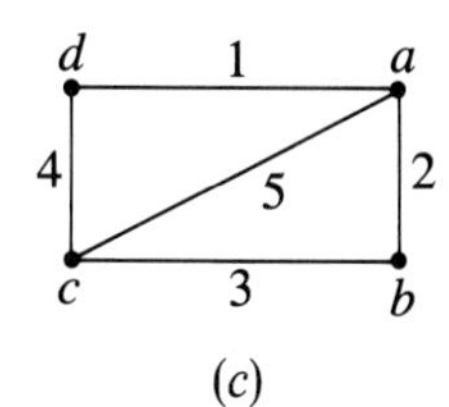

(*c*)

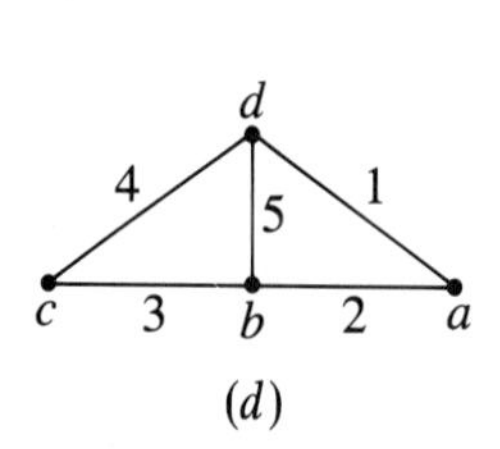

(*d*)

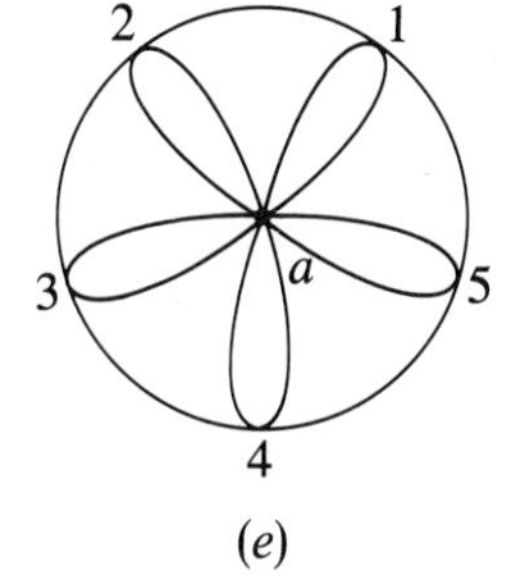

(*e*)

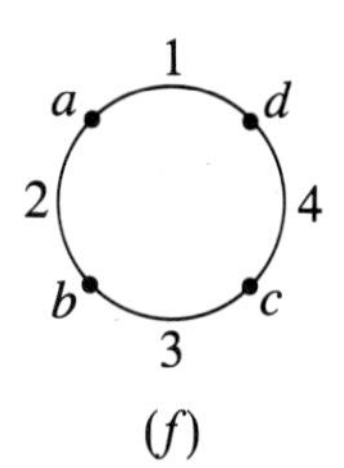

(*f*)

13. Let A and B be conformable for addition. If $A + B = A$, what can you say about B?

14. Let A and B be conformable for the product AB. If each entry of B is zero, what can you say about the entries of AB?

15. If A, B, and C are all $n \times n$ and $AB = BA$ and $BC = CB$, must $AC = CA$? If not, illustrate with an example.

16. In some applications, different kinds of matrix products are useful. For example, a bank may have made commercial loans and personal loans at different interest rates, and the information may be stored in matrices (let p and q represent the principals):

Commercial	p_1 p_2 p_3			Corresponding	i_1 i_2 i_3		
Personal	q_1 q_2 q_3			Interest rates	I_1 I_2 I_3		

To determine the total amount due to interest at the end of the loans, we calculate

$$p_1 i_1 + p_2 i_2 + p_3 i_3 + q_1 I_1 + q_2 I_2 + q_3 I_3$$

This corresponds to multiplying the matrices entrywise and adding the results. With this product then, we multiply two matrices and obtain a real number, as in the example:

$$\begin{pmatrix} 1 & 2 & -1 \\ 3 & 0 & 2 \end{pmatrix}\begin{pmatrix} 5 & 0 & 3 \\ -1 & -1 & 6 \end{pmatrix} = 5 + 0 - 3 - 3 - 0 + 12 = 11$$

Using this product, multiply the following pairs of matrices, if possible.

(*a*) $\begin{pmatrix} 1 & 0 \\ 0 & 1 \end{pmatrix}\begin{pmatrix} 0 & 1 \\ 1 & 0 \end{pmatrix}$ (*b*) $\begin{pmatrix} 2 & 3 \\ 6 & 5 \end{pmatrix}\begin{pmatrix} 4 & 2 & 1 \\ 3 & 5 & 2 \end{pmatrix}$

(*c*) $\begin{pmatrix} 1 & 1 \\ 1 & 1 \end{pmatrix}\begin{pmatrix} 1 & -1 \\ -1 & 1 \end{pmatrix}$ (*d*) $\begin{pmatrix} a & b \\ c & d \end{pmatrix}\begin{pmatrix} 0 & 0 \\ 0 & 0 \end{pmatrix}$

17. Regarding Prob. 16, if the bank wanted to know the amount due to interest on *each* loan, the products $p_1 i_1$, $p_2 i_2$, $p_3 i_3$, $q_1 I_1$, $q_2 I_2$, and $q_3 I_3$ would not be added, but placed in a third matrix:

$$\begin{pmatrix} p_1 & p_2 & p_3 \\ q_1 & q_2 & q_3 \end{pmatrix}\begin{pmatrix} i_1 & i_2 & i_3 \\ I_1 & I_2 & I_3 \end{pmatrix} = \begin{pmatrix} p_1 i_1 & p_2 i_2 & p_3 i_3 \\ q_1 I_1 & q_2 I_2 & q_3 I_3 \end{pmatrix}$$

So, for this kind of product,

$$\begin{pmatrix} 1 & 3 & 0 \\ 4 & 2 & 1 \end{pmatrix}\begin{pmatrix} 1 & 0 & 1 \\ -2 & 3 & 5 \end{pmatrix} = \begin{pmatrix} 1 & 0 & 0 \\ -8 & 6 & 5 \end{pmatrix}$$

State in words the rule for this product. Using this product, multiply the following pairs of matrices, if possible.

(*a*) $\begin{pmatrix} 1 & 1 \\ 0 & 2 \\ 1 & 1 \end{pmatrix}\begin{pmatrix} 3 & 1 \\ 0 & 2 \end{pmatrix}$ (*b*) $\begin{pmatrix} 3 & 6 & 9 \\ 1 & 2 & 3 \end{pmatrix}\begin{pmatrix} 1 & 0 & 1 \\ -2 & 3 & 0 \end{pmatrix}$

(c) $\begin{pmatrix} 5 & 10 \\ -1 & 7 \end{pmatrix}\begin{pmatrix} 3 & 6 \\ 4 & 7 \end{pmatrix}$ (d) $\begin{pmatrix} 5 \\ 3 \\ 6 \\ 1 \end{pmatrix}\begin{pmatrix} -1 \\ 0 \\ 2 \\ 3 \end{pmatrix}$

18. The Pauli spin matrices

$$A = \begin{pmatrix} 0 & 1 \\ 1 & 0 \end{pmatrix} \quad B = \begin{pmatrix} 0 & -i \\ i & 0 \end{pmatrix} \quad C = \begin{pmatrix} 1 & 0 \\ 0 & -1 \end{pmatrix}$$

are used in the study of electron spin. Calculate AB, BA, BB, AC, CA, and $A + iB$.

1.5 MATRIX ALGEBRA

In the last section, we defined the sum, difference, scalar multiple, and product of matrices. Calculations involving matrices and these operations are governed by certain rules, which are discussed in this section. The resulting structure is called *matrix algebra.*

Many of the properties we derive look similar to properties of the real numbers. However, some definitions and concepts will be new. For instance, the first property we state is the commutative property for addition. That is, $A + B = B + A$. This looks exactly like the commutative property of addition for the real numbers, except the letters A and B stand for matrices instead of single real numbers. However, as we will see, the important concept of the transpose of a matrix when reduced to the real number case gives nothing new.

Basic rules for matrix algebra are contained in the following theorem.

THEOREM 1.5.1 In each statement (*a*) through (*h*), assume that the matrices involved are the correct order for the operations to be defined, and let r and s be complex numbers. Statements (*a*) through (*h*) are true.

(*a*) $A + B = B + A$ — Addition is commutative.

(*b*) $A + (B + C) = (A + B) + C$ — Addition is associative.

(*c*) $A(B + C) = AB + AC$

(*d*) $(B + C)A = BA + CA$

(*c*)–(*d*): Multiplication distributes over addition.

(*e*) $(AB)C = A(BC)$ — Multiplication is associative.

(*f*) $r(A + B) = rA + rB$

(*g*) $(r + s)A = rA + sA$

(*h*) $(rs)A = r(sA) = s(rA) = (sr)A$

(*f*)–(*h*): Scalar multiplication distributes over addition.

Before we prove this theorem, a few words about proofs are in order. The reader may be perfectly willing to accept the truth of a stated theorem, even though a proof might not be supplied. But a study of the proof of a theorem gives additional insight to the concept, topic, or method being considered. Moreover, study of several proofs may reveal a common technique or method which helps to solve problems or to decide what is true in new or more general situations. Of course, a proof should be convincing; yet it must be correct. In a textbook the degree of detail is necessarily limited. A balance of technical computations, reference to previous results, hand waving, and "leaving the rest to the reader" must necessarily result.

Many theorems in linear algebra are propositions involving the positive integers and can be proved by using the principle of mathematical induction (see App. II). However, such proofs can mask what makes things work. For example, consider this proposition:

For $n = 2, 3, \ldots$ the sum of the first n positive integers is $n(n + 1)/2$.

Another way to write this is

$$1 + 2 + \cdots + n = \frac{n(n + 1)}{2} \qquad n = 2, 3, \ldots$$

Let us give two proofs.

Proof 1 *(by Induction)* The proposition is true for $n = 2$ because

$$3 = 1 + 2 = \frac{2(2 + 1)}{2}$$

Assume that [this is the induction hypothesis (IH)]

$$1 + 2 + \cdots + k = \frac{k(k + 1)}{2} \qquad \text{(IH)}$$

and prove that

$$1 + 2 + \cdots + k + (k + 1) = \frac{(k + 1)(k + 2)}{2}$$

Add $k + 1$ to both sides of the IH to obtain

$$1 + 2 + \cdots + k + (k + 1) = \frac{k(k + 1)}{2} + (k + 1) = \frac{k(k + 1) + 2(k + 1)}{2}$$

$$= \frac{(k + 2)(k + 1)}{2}$$

Therefore, by the principle of mathematical induction, the proposition is true.

Proof 2 Suppose n is even. The terms in the sum can be grouped

$$1 + 2 + 3 + \cdots + (n - 2) + (n - 1) + n$$

Sums are $n + 1$

so that we have $n/2$ pairs of numbers, each pair adding to $n + 1$. Therefore the total sum is $n(n + 1)/2$. If n is odd, group as follows:

$$1 + 2 + 3 + \cdots + (n - 3) + (n - 2) + (n - 1) + n$$

Since $n - 1$ is even, the sum is

$$\frac{(n - 1)(n)}{2} + n = \frac{(n + 1)n}{2}$$

The second proof might be rejected by some, yet it shows why the proposition is true. Sometimes we give proofs of this sort, and other times we use induction. Our purpose is to make as many proofs as illuminating as possible. Now to a proof of Theorem 1.5.1.

Proof of Theorem 1.5.1. The proof of each statement reduces to properties of complex numbers by looking at the general elements of the matrices. For example, to prove Theorem 1.5.1f, suppose $A = (a_{ij})_{m \times n}$ and $B = (b_{ij})_{m \times n}$. Then the left-hand side of (f) is

$$\begin{aligned} r(A + B) &= r((a_{ij}) + (b_{ij})) = r((a_{ij} + b_{ij})) \\ &= (r(a_{ij} + b_{ij})) = (ra_{ij} + rb_{ij}) \end{aligned}$$

By distributive property of complex numbers

The right-hand side is

$$\begin{aligned} rA + rB &= r(a_{ij}) + r(b_{ij}) = (ra_{ij}) + (rb_{ij}) \\ &= (ra_{ij} + rb_{ij}) \end{aligned}$$

Since both sides of (f) reduce to the same matrix, they are equal.

The other parts work the same way. For part (a), we can look at both sides again:

$$A + B = (a_{ij}) + (b_{ij}) = (a_{ij} + b_{ij})$$

$$B + A = (b_{ij}) + (a_{ij}) = (b_{ij} + a_{ij})$$

Since a_{ij} and b_{ij} [illegible] complex numbers, $a_{ij} + b_{ij} = b_{ij} + a_{ij}$, and so $A + B = B + A$.

EXAMPLE 1 Give an example to illustrate Theorem 1.5.1c.

Solution To illustrate $A(B + C) = AB + AC$, we should use small matrices to cut down computations; 2×2 matrices will be fine. Let

$$A = \begin{pmatrix} 1 & -1 \\ 2 & 3 \end{pmatrix} \qquad B = \begin{pmatrix} 0 & -2 \\ -4 & 1 \end{pmatrix} \qquad C = \begin{pmatrix} -4 & 2 \\ 1 & 7 \end{pmatrix}$$

We have

$$A(B + C) = \begin{pmatrix} 1 & -1 \\ 2 & 3 \end{pmatrix}\left(\begin{pmatrix} 0 & -2 \\ -4 & 1 \end{pmatrix} + \begin{pmatrix} -4 & 2 \\ 1 & 7 \end{pmatrix}\right) = \begin{pmatrix} 1 & -1 \\ 2 & 3 \end{pmatrix}\begin{pmatrix} -4 & 0 \\ -3 & 8 \end{pmatrix}$$
$$= \begin{pmatrix} -1 & -8 \\ -17 & 24 \end{pmatrix}$$

and

$$AB + AC = \begin{pmatrix} 1 & -1 \\ 2 & 3 \end{pmatrix}\begin{pmatrix} 0 & -2 \\ -4 & 1 \end{pmatrix} + \begin{pmatrix} 1 & -1 \\ 2 & 3 \end{pmatrix}\begin{pmatrix} -4 & 2 \\ 1 & 7 \end{pmatrix}$$
$$= \begin{pmatrix} 4 & -3 \\ -12 & -1 \end{pmatrix} + \begin{pmatrix} -5 & -5 \\ -5 & 25 \end{pmatrix}$$
$$= \begin{pmatrix} -1 & -8 \\ -17 & 24 \end{pmatrix}$$

The set of complex numbers contains some numbers such as 0 and 1 which have special properties. For example, $a \cdot 0 = 0$ for all a, $a + 0 = a$ for all a, and $a \cdot 1 = a$ for all a. Accordingly, we also have some special matrices.

The *zero matrix* $0_{m \times n}$ has all entries zero:

$$0_{m \times n} = \begin{pmatrix} 0 & \cdots & 0 \\ 0 & \cdots & 0 \\ \cdots & \cdots & \cdots \\ 0 & \cdots & 0 \end{pmatrix}_{m \times n}$$

The *identity matrix* $I_{n \times n}$ has $a_{ii} = 1$, $i = 1, 2, \ldots, n$, and all other entries zero:

$$I_n = \begin{pmatrix} 1 & 0 & 0 & \cdots & 0 \\ 0 & 1 & 0 & \cdots & 0 \\ 0 & 0 & 1 & \cdots & 0 \\ \cdots & \cdots & \cdots & \cdots & \cdots \\ 0 & 0 & 0 & \cdots & 1 \end{pmatrix}_{n \times n}$$

Another way to state this is to use the *Kronecker delta* symbol δ_{ij} which equals 1 if $i = j$ and equals 0 if $i \neq j$. Thus $I_n = (\delta_{ij})_{n \times n}$.

The *scalar matrix* $rI_{n \times n}$ has $a_{ii} = r$, $i = 1, 2, \ldots, n$, and all other entries zero:

$$rI_n = \begin{pmatrix} r & 0 & 0 & \cdots & 0 \\ 0 & r & 0 & \cdots & 0 \\ 0 & 0 & r & \cdots & 0 \\ \cdots & \cdots & \cdots & \cdots & \cdots \\ 0 & 0 & 0 & \cdots & r \end{pmatrix} = (r\delta_{ij})_{n \times n}$$

Some of the properties of these special matrices are contained in Theorem 1.5.2.

THEOREM 1.5.2 In each statement (a) through (h), assume that the matrices involved are the correct order for the operations to be defined, and let r be a number. The following statements are true.

(a) $A + 0 = A, 0 + A = A$ (b) $A - A = 0$

(c) $A0 = 0$ (d) $0A = 0$

(e) $AI = A$ (f) $IA = A$

(g) $rIA = rA$ (h) $A(rI) = rA$

Proof We will prove (e) and leave the rest to the reader. Let $A = (a_{ij})_{m \times n}$. Then

$$AI = \begin{pmatrix} a_{11} & a_{12} & \cdots & a_{1n} \\ a_{21} & a_{22} & \cdots & a_{2n} \\ \cdots & \cdots & \cdots & \cdots \\ a_{m1} & a_{m2} & \cdots & a_{mn} \end{pmatrix} \begin{pmatrix} 1 & 0 & \cdots & 0 \\ 0 & 1 & \cdots & 0 \\ \cdots & \cdots & \cdots & \cdots \\ 0 & 0 & \cdots & 1 \end{pmatrix}_{n \times n}$$

The ij entry of AI is the ith row of A times the jth column of I, that is,

$$\begin{bmatrix} a_{i1} & a_{i2} & \cdots & a_{ij} & \cdots & a_{in} \end{bmatrix} \begin{bmatrix} 0 \\ 0 \\ \vdots \\ 1 \\ \vdots \\ 0 \end{bmatrix} \leftarrow \text{Position } j$$

So the ij entry of AI is a_{ij}, the ij entry of A. Thus $AI = A$.

EXAMPLE 2 Illustrations of Theorem 1.5.2a, b, c, and e.

(a) Let

$$A = \begin{pmatrix} 1+i & -1 & 2 \\ 0 & 3 & 7 \end{pmatrix}$$

Then

$$\begin{aligned} A + 0 &= \begin{pmatrix} 1+i & -1 & 2 \\ 0 & 3 & 7 \end{pmatrix} + \begin{pmatrix} 0 & 0 & 0 \\ 0 & 0 & 0 \end{pmatrix} \\ &= \begin{pmatrix} 1+i+0 & -1+0 & 2+0 \\ 0+0 & 3+0 & 7+0 \end{pmatrix} \\ &= \begin{pmatrix} 1+i & -1 & 2 \\ 0 & 3 & 7 \end{pmatrix} = A \end{aligned}$$

(*b*) Let

$$A = \begin{pmatrix} 1 & -1 \\ 2-3i & 4 \end{pmatrix}$$

Then

$$A - A = \begin{pmatrix} 1 & -1 \\ 2-3i & 4 \end{pmatrix} - \begin{pmatrix} 1 & -1 \\ 2-3i & 4 \end{pmatrix}$$
$$= \begin{pmatrix} 1-1 & -1-(-1) \\ 2-2-(3-3)i & 4-4 \end{pmatrix} = \begin{pmatrix} 0 & 0 \\ 0 & 0 \end{pmatrix}$$

(*c*) Let

$$A = \begin{pmatrix} 1 & 3 \\ 2 & -1 \\ 4 & 6 \end{pmatrix} \quad \text{and} \quad 0 = \begin{pmatrix} 0 & 0 \\ 0 & 0 \end{pmatrix}$$

Then

$$A0 = \begin{pmatrix} 1 & 3 \\ 2 & -1 \\ 4 & 6 \end{pmatrix}\begin{pmatrix} 0 & 0 \\ 0 & 0 \end{pmatrix} = \begin{pmatrix} 1\cdot 0 + 3\cdot 0 & 1\cdot 0 + 3\cdot 0 \\ 2\cdot 0 - 1\cdot 0 & 2\cdot 0 - 1\cdot 0 \\ 4\cdot 0 + 6\cdot 0 & 4\cdot 0 + 6\cdot 0 \end{pmatrix} = \begin{pmatrix} 0 & 0 \\ 0 & 0 \\ 0 & 0 \end{pmatrix} = 0$$

Note that $0A$ is not defined in this example, because that product is not defined.

(*e*) Let

$$A = \begin{pmatrix} 1 & -1 & 2 \\ 4 & -3 & 6 \end{pmatrix} \quad \text{and} \quad I = \begin{pmatrix} 1 & 0 & 0 \\ 0 & 1 & 0 \\ 0 & 0 & 1 \end{pmatrix}$$

Then

$$AI = \begin{pmatrix} 1 & -1 & 2 \\ 4 & -3 & 6 \end{pmatrix}\begin{pmatrix} 1 & 0 & 0 \\ 0 & 1 & 0 \\ 0 & 0 & 1 \end{pmatrix} = \begin{pmatrix} 1+0+0 & 0-1+0 & 0+0+2 \\ 4+0+0 & 0-3+0 & 0+0+6 \end{pmatrix}$$
$$= \begin{pmatrix} 1 & -1 & 2 \\ 4 & -3 & 6 \end{pmatrix} = A$$

Note that IA is not defined in this example, because that product is not defined.

Other special matrices important in applications are diagonal, tridiagonal, triangular, and symmetric matrices.

DEFINITION 1.5.1 A *diagonal matrix* is a square matrix (a_{ij}) with $a_{ij} = 0$ if $i \neq j$. That is, the elements which are not on the *main diagonal* of A must be zero.

EXAMPLE 3 The following are examples of diagonal matrices.

$$\begin{pmatrix} 1 & 0 & 0 \\ 0 & -1 & 0 \\ 0 & 0 & 3 \end{pmatrix} \quad \begin{pmatrix} 3+i & 0 \\ 0 & 2 \end{pmatrix} \quad \begin{pmatrix} 17 & 0 & 0 & 0 \\ 0 & -8 & 0 & 0 \\ 0 & 0 & 4 & 0 \\ 0 & 0 & 0 & \pi \end{pmatrix} \quad \begin{pmatrix} 1 & 0 & 0 \\ 0 & i & 0 \\ 0 & 0 & 0 \end{pmatrix}$$

The main diagonal is the diagonal which runs from the upper left-hand corner to the lower right-hand corner of each matrix.

DEFINITION 1.5.2 A square matrix $A_{n \times n}$, $n \geq 3$, is *tridiagonal* if $a_{ij} = 0$ when $i > j + 1$ or $j > i + 1$. That is, if the row number and column number of an entry differ by 2 or more, the entry must be zero. The other entries are unrestricted.

EXAMPLE 4 According to the definition, in a 4×4 tridiagonal matrix

$$\begin{pmatrix} a_{11} & a_{12} & a_{13} & a_{14} \\ a_{21} & a_{22} & a_{23} & a_{24} \\ a_{31} & a_{32} & a_{33} & a_{34} \\ a_{41} & a_{42} & a_{43} & a_{44} \end{pmatrix}$$

the entries $a_{13}, a_{14}, a_{24}, a_{31}, a_{41}$, and a_{42} must all be zero. Some 4×4 tridiagonal matrices are

$$\begin{pmatrix} 3 & 4 & 0 & 0 \\ 6 & -1 & 2 & 0 \\ 0 & -17 & 2 & 0 \\ 0 & 0 & -1 & 0 \end{pmatrix} \quad \begin{pmatrix} 1 & 0 & 0 & 0 \\ 1 & 1 & 7-2i & 0 \\ 0 & i & 0 & 1 \\ 0 & 0 & 1+i & 1 \end{pmatrix}$$

All diagonal matrices are also tridiagonal.

An example of a tridiagonal matrix arising in chemical engineering was given in Example 1 of the last section. The matrix was of the form

$$\begin{pmatrix} -(a+b) & a & 0 & 0 & \cdots & 0 \\ b & -(a+b) & a & 0 & \cdots & 0 \\ 0 & b & -(a+b) & \ddots & \cdots & 0 \\ \vdots & & \ddots & \ddots & \ddots & \vdots \\ & & & & -(a+b) & a \\ 0 & 0 & 0 & \cdots & b & -(a+b) \end{pmatrix}$$

Such tridiagonal matrices are called *Jacobi matrices*.

DEFINITION 1.5.3 A square matrix (a_{ij}) is *upper triangular* if $a_{ij} = 0$ whenever $i > j$. That is, if a_{ij} is below the main diagonal, $a_{ij} = 0$. A square matrix (a_{ij}) is *lower triangular* if $a_{ij} = 0$ whenever $i < j$.

EXAMPLE 5 Some upper triangular matrices are

$$\begin{pmatrix} 0 & 3 \\ 0 & 0 \end{pmatrix} \quad \begin{pmatrix} 1 & 4 \\ 0 & 2 \end{pmatrix} \quad \begin{pmatrix} 1 & 7 & 6 \\ 0 & 2 & 0 \\ 0 & 0 & 3 \end{pmatrix} \quad \begin{pmatrix} 5 & \pi & -3 & 6+i \\ 0 & 0 & 4 & 2 \\ 0 & 0 & -6 & 2i \\ 0 & 0 & 0 & 2 \end{pmatrix}$$

The following matrices are lower triangular:

$$\begin{pmatrix} -1 & 0 \\ -3 & 2 \end{pmatrix} \quad \begin{pmatrix} 0 & 0 & 0 \\ 2-i & 1 & 0 \\ 0 & 2 & -8 \end{pmatrix} \quad \begin{pmatrix} 1 & 0 & 0 & 0 \\ 7 & 12 & 0 & 0 \\ 1 & -6 & 3 & 0 \\ 7i & 2 & 5 & 2 \end{pmatrix}$$

One reason why triangular matrices are important is that following gaussian elimination for an $n \times n$ system the resulting system has an upper triangular matrix of coefficients. In Gauss-Jordan elimination, the resulting system sometimes has a diagonal matrix of coefficients.

An extremely important set of matrices is the set of square matrices which are unchanged when rows and columns are interchanged. These are called *symmetric* matrices; to define this set, we need the idea of the transpose of a matrix.

DEFINITION 1.5.4 The *transpose* of an $m \times n$ matrix A is the $n \times m$ matrix A^T such that the ij entry of A^T is the ji entry of A.

EXAMPLE 6 Calculate A^T for the following matrices.

(*a*) $A = \begin{pmatrix} 1 & 2i \\ 0 & 6 \end{pmatrix}$

(*b*) $A = \begin{pmatrix} 1 & -2 & 2 \\ 3 & 0 & 4 \\ -6 & \pi & 2 \end{pmatrix}$

(*c*) $A = \begin{pmatrix} 3 & 1 & 1 \\ 0 & 2 & 5 \end{pmatrix}$

Solution We calculate A^T by making the first row of A the first column of A^T, the second row of A the second column of A^T, and so on. For example, in part (*a*),

$$A = \begin{pmatrix} \mathbf{1} & \mathbf{2i} \\ \mathbf{0} & \mathbf{6} \end{pmatrix} \quad \text{so} \quad A^T = \begin{pmatrix} \mathbf{1} & \mathbf{0} \\ \mathbf{2i} & \mathbf{6} \end{pmatrix}$$

(*b*)

$$A^T = \begin{pmatrix} 1 & 3 & -6 \\ -2 & 0 & \pi \\ 2 & 4 & 2 \end{pmatrix}$$

1st row of A 2d row of A 3d row of A

(*c*)

$$A^T = \begin{pmatrix} 3 & 0 \\ 1 & 2 \\ 1 & 5 \end{pmatrix}$$

Note that A was 2×3 and A^T is 3×2.

DEFINITION 1.5.5 A square matrix is *symmetric* if $A^T = A$. That is, $A = (a_{ij})$ is symmetric if and only if $a_{ij} = a_{ji}$ for all i and j.

EXAMPLE 7 Show that

$$A = \begin{pmatrix} 1 & -1 & 17 \\ -1 & 2 & 4 \\ 17 & 4 & 5 \end{pmatrix}$$

is symmetric.

Solution We need to show that $A^T = A$. Now

$$A^T = \begin{pmatrix} 1 & -1 & 17 \\ -1 & 2 & 4 \\ 17 & 4 & 5 \end{pmatrix}$$

1st row of A 2d row of A 3d row of A

which equals A. Notice that A has a "mirror image" property: If the matrix is folded along the main diagonal, the numbers on top of each other are equal.

$$\begin{pmatrix} 1 & -1 & 17 \\ -1 & 2 & 4 \\ 17 & 4 & 5 \end{pmatrix}$$

Fold

Symmetric matrices are used in the method of linear least squares (see App. III), which is used in turn to forecast trends and estimate parameters in engineering problems.

Important properties of the transpose operation and triangular and symmetric matrices are contained in Theorem 1.5.3.

THEOREM 1.5.3 Let A and B be matrices of the correct size so that (*a*) through (*h*) have meaning. The following statements are true:

(*a*) $(A^T)^T = A$

(*b*) $(A + B)^T = A^T + B^T$

(*c*) $(rA)^T = rA^T$

(*d*) $(AB)^T = B^T A^T$

(The transpose of a product is the product of the transposes in reverse order.)

(*e*) If A is upper (lower) triangular, A^T is lower (upper) triangular.

(*f*) If A is diagonal (tridiagonal), A^T is diagonal (tridiagonal).

(*g*) If A and B are symmetric and $AB = BA$, then AB is symmetric.

(*h*) If A and B are upper (lower) triangular, then AB is upper (lower) triangular.

Before proving parts of this theorem, we give examples illustrating parts (*b*) and (*d*).

EXAMPLE 8 Illustrate Theorem 1.5.3*b* and *d*; that is $(A + B)^T = A^T + B^T$ and $(AB)^T = B^T A^T$.

Solution (*b*) Let

$$A = \begin{pmatrix} 1 & 0 \\ -1 & 2 \end{pmatrix} \qquad B = \begin{pmatrix} 3 & 7 \\ 6 & 4 \end{pmatrix}$$

Then

$$(A + B)^T = \begin{pmatrix} 4 & 7 \\ 5 & 6 \end{pmatrix}^T = \begin{pmatrix} 4 & 5 \\ 7 & 6 \end{pmatrix}$$

$$A^T + B^T = \begin{pmatrix} 1 & -1 \\ 0 & 2 \end{pmatrix} + \begin{pmatrix} 3 & 6 \\ 7 & 4 \end{pmatrix} = \begin{pmatrix} 4 & 5 \\ 7 & 6 \end{pmatrix}$$

(*d*) Let

$$A = \begin{pmatrix} 1 & 0 \\ -1 & 2 \end{pmatrix} \qquad B = \begin{pmatrix} 3 & 7 \\ 6 & 4 \end{pmatrix}$$

Then

$$(AB)^T = \begin{pmatrix} 3 & 7 \\ 9 & 1 \end{pmatrix}^T = \begin{pmatrix} 3 & 9 \\ 7 & 1 \end{pmatrix}$$

$$B^T A^T = \begin{pmatrix} 3 & 6 \\ 7 & 4 \end{pmatrix} \begin{pmatrix} 1 & -1 \\ 0 & 2 \end{pmatrix} = \begin{pmatrix} 3 & 9 \\ 7 & 1 \end{pmatrix}$$

Proof of Theorem 1.5.3b and d For part (*b*), let A and B be $m \times n$. The *ij* entry of $(A + B)^T =$ *ji* entry of $A + B =$ *ji* entry of $A +$ *ji* entry of $B =$ *ij* entry of $A^T +$ *ij* entry of $B^T =$ *ij* entry of $A^T + B^T$.

For (d), let A be $m \times n$ and let B be $n \times p$.

$$\begin{aligned} ij \text{ entry of } (AB)^T &= ji \text{ entry of } AB = \sum_{k=1}^{n} a_{jk}b_{ki} \\ &= \sum_{k=1}^{n} b_{ki}a_{jk} = \sum_{k=1}^{n} [(ik \text{ entry of } B^T)(kj \text{ entry of } A^T)] \\ &= ij \text{ entry of } B^T A^T \end{aligned}$$

The proofs of the other parts of the theorem are left to the problems.

EXAMPLE 9 Give an example of two symmetric matrices A and B for which AB is not symmetric.

Solution Let

$$A = \begin{pmatrix} 1 & 2 \\ 2 & 4 \end{pmatrix} \qquad B = \begin{pmatrix} -1 & 6 \\ 6 & 3 \end{pmatrix}$$

Then

$$AB = \begin{pmatrix} 11 & 12 \\ 22 & 24 \end{pmatrix}$$

Note that $AB \neq BA$.

In Example 9 we saw matrices A and B for which AB and BA were defined but $AB \neq BA$. This shows that matrix multiplication is, in general, not commutative. This is an important difference between multiplication of complex numbers (which is commutative) and multiplication of matrices.

The power of a square matrix, which is important in the subject of Markov chains (which we mentioned briefly in Sec. 1.1) is now defined.

DEFINITION 1.5.6 The nth power of a square matrix A (where n is a positive integer) is written A^n and is defined as follows:

If $n = 1$, $A^1 = A$; for $n \geq 2$, $A^n = A(A^{n-1})$.

EXAMPLE 10 Let

$$A = \begin{pmatrix} 1 & 2 \\ 0 & 4 \end{pmatrix}$$

Calculate A^2, A^3, and A^6.

Solution

$$A^2 = \begin{pmatrix} 1 & 2 \\ 0 & 4 \end{pmatrix}\begin{pmatrix} 1 & 2 \\ 0 & 4 \end{pmatrix} = \begin{pmatrix} 1 & 10 \\ 0 & 16 \end{pmatrix}$$

$$A^3 = \begin{pmatrix} 1 & 2 \\ 0 & 4 \end{pmatrix}\begin{pmatrix} 1 & 10 \\ 0 & 16 \end{pmatrix} = \begin{pmatrix} 1 & 42 \\ 0 & 64 \end{pmatrix}$$

To calculate A^6, we try A^3A^3, hoping that laws of exponents hold for powers of matrices (they do, as we shall see):

$$A^6 = A^3A^3 = \begin{pmatrix} 1 & 42 \\ 0 & 64 \end{pmatrix}\begin{pmatrix} 1 & 42 \\ 0 & 64 \end{pmatrix} = \begin{pmatrix} 1 & 2730 \\ 0 & 4096 \end{pmatrix}$$

Notice that A is upper triangular and the powers we calculated are all upper triangular.

An important property of the power of a matrix is that

$$A^mA^n = A^{m+n} \qquad \text{for all positive integers } m \text{ and } n$$

This can be seen by using the associativity of multiplication and writing

$$A^mA^n = \underbrace{A\cdots A}_{m \text{ times}}\underbrace{A\cdots A}_{n \text{ times}} = \underbrace{A\cdots A\,A\cdots A}_{m+n \text{ times}} = A^{m+n}$$

That $(A^m)^n = A^{mn}$ for positive integers m and n is shown similarly.

If a matrix A has nonreal entries, taking the conjugate of those entries will result in a new matrix, called the *conjugate of A*.

DEFINITION 1.5.7 Let $A = (a_{ij})_{m\times n}$. The *conjugate* of A is written $\bar{A}$ and defined by

$$\bar{A} = (\bar{a}_{ij})_{m\times n}$$

where $\overline{c + id} = c - id$. The *conjugate transpose* of A is written A^* (read "A star") and is defined by

$$A^* = \bar{A}^T$$

EXAMPLE 11 For the following matrices, calculate the conjugate and conjugate transpose.

$$A = \begin{pmatrix} 2 & 1+i \\ 1-i & 3 \end{pmatrix} \qquad B = \begin{pmatrix} 1 & -2 & 0 \\ 7 & -2 & 6 \\ 1 & 5 & 0 \end{pmatrix}$$

Solution

$$\bar{A} = \begin{pmatrix} \bar{2} & \overline{1+i} \\ \overline{1-i} & \bar{3} \end{pmatrix} = \begin{pmatrix} 2 & 1-i \\ 1+i & 3 \end{pmatrix} \qquad A^* = \bar{A}^T = \begin{pmatrix} 2 & 1+i \\ 1-i & 3 \end{pmatrix}$$

$$\bar{B} = B \qquad B^* = \bar{B}^T = B^T = \begin{pmatrix} 1 & 7 & 1 \\ -2 & -2 & 5 \\ 0 & 6 & 0 \end{pmatrix}$$

In Example 11, notice that $A^* = A$. Matrices for which this is true are called *hermitian*. Note that in a hermitian matrix, $\bar{a}_{ij} = a_{ji}$.

EXAMPLE 12 (Mechanics) Matrix algebra can help in representing the *stress tensor* in problems from mechanics. The stress tensor is representable by a 3×3 matrix

$$\sigma = \begin{pmatrix} \sigma_{11} & \sigma_{12} & \sigma_{13} \\ \sigma_{21} & \sigma_{22} & \sigma_{23} \\ \sigma_{31} & \sigma_{32} & \sigma_{33} \end{pmatrix}$$

and occasionally we need to calculate a matrix

$$\sum = (\sum_{ij})$$

where

$$\begin{aligned} \sum_{ij} = \sigma_{11} l_{i1} l_{j1} + \sigma_{12} l_{i1} l_{j2} + \sigma_{13} l_{i1} l_{j3} \\ + \sigma_{21} l_{i2} l_{j1} + \sigma_{22} l_{i2} l_{j2} + \sigma_{23} l_{i2} l_{j3} \\ + \sigma_{31} l_{i3} l_{j1} + \sigma_{32} l_{i3} l_{j2} + \sigma_{33} l_{i3} l_{j3} \end{aligned}$$

and the l_{ij} come from thc matrices

$$L_i = \begin{pmatrix} l_{i1} \\ l_{i2} \\ l_{i3} \end{pmatrix} \qquad i = 1, 2, 3$$

The formula for $\sum_{ij}$ is hard to remember as it stands. However, using matrix products, we have

$$\sum_{ij} = L_i^T \sigma L_j$$

which is easier to remember and program for a computer.

We can now state more theorems about the theory of equations. As usual, we let X be the column matrix of unknowns.

THEOREM 1.5.4 If A is $n \times n$ and the system $AX = 0$ has no nontrivial solution, then A is row-equivalent to I.

Proof Let B be the reduced row echelon form of A. Then $BX = 0$ has no nontrivial solution. Therefore, by Theorem 1.4.1, if r is the number of nonzero rows of B, we must have $r \geq n$; on the other hand, since B is $n \times n$, we know that $r \leq n$. Thus $r = n$ and $B = I$ (see Prob. 35).

THEOREM 1.5.5 If $AX = B$ and $AY = 0$, then $A(X + Y) = B$. That is, if X is a solution of the nonhomogeneous equation and Y is a solution of the homogeneous equation, their sum is a solution of the nonhomogeneous equation. Moreover every solution of $AX = B$ is of this form.

Proof To show that $X + Y$ is a solution, calculate

$$A(X + Y) = AX + AY = B + 0 = B$$

Now let Z be any solution of the nonhomogeneous equation, that is, $AZ = B$. We can write $Z = (Z - Y) + Y$, and $Z - Y$ is a solution of the nonhomogeneous problem because $A(Z - Y) = AZ - AY = B - 0 = B$.

THEOREM 1.5.6 If A is $n \times n$ and $AX = 0$ has no nontrivial solution, then $AX = B$ has a unique solution.

Proof By Theorem 1.5.4, A is row-equivalent to I. That is, the augmented matrix is row-equivalent to

$$\left(\begin{array}{cccc|c} 1 & 0 & \cdots & 0 & s_1 \\ 0 & 1 & \cdots & 0 & s_2 \\ \vdots & & \ddots & \vdots & \vdots \\ 0 & 0 & \cdots\ 0 & 1 & s_n \end{array}\right)$$

and the system $IX = S$ is equivalent to $AX = B$. Therefore, the unique solution to $AX = B$ is $x_1 = s_1, x_2 = s_2, \ldots, x_n = s_n$.

EXAMPLE 13 (Illustration of Theorem 1.5.4) We saw in Example 5 of Sec. 1.2 that the system

$$AX = \begin{pmatrix} 1 & -2 & 1 \\ 3 & -1 & 1 \\ -1 & 4 & -1 \end{pmatrix} \begin{pmatrix} x_1 \\ x_2 \\ x_3 \end{pmatrix} = \begin{pmatrix} 0 \\ 0 \\ 0 \end{pmatrix}$$

has no nontrivial solution. Following the row reduction there with the row operations $\frac{2}{5}$R3 + R2, $-$R3 + R1, 2R2 + R1, we see that A has I for its reduced row echelon form.

EXAMPLE 14 (Illustration of Theorem 1.5.5) Consider the system

$$AX = \begin{pmatrix} 1 & 2 & 3 \\ 4 & 5 & 6 \\ 7 & 8 & 9 \end{pmatrix} \begin{pmatrix} x_1 \\ x_2 \\ x_3 \end{pmatrix} = \begin{pmatrix} 4 \\ 7 \\ 10 \end{pmatrix} = B$$

from Example 2 of Sec. 1.2. We found the solution

$$\begin{pmatrix} -2 + k \\ 3 - 2k \\ k \end{pmatrix}$$

where k is arbitrary. This can be written as

$$\begin{pmatrix} -2 \\ 3 \\ 0 \end{pmatrix} + \begin{pmatrix} k \\ -2k \\ k \end{pmatrix}$$

where the second term in the sum solves $AX = 0$ and the first term solves $AX = B$. This "split" of the solution is not done by magic; $AX = 0$ is solved by elimination (see Example 4, Sec. 1.2).

PROBLEMS 1.5

Note to the student: Some of these problems do not "have the answer in them." For example, Prob. 4 asks, Is the sum of tridiagonal matrices also tridiagonal? If the answer is no, an example should be given to illustrate. Regardless, one way to start searching for the answer is to try examples—if for several examples the answer is yes, try to develop a proof or demonstration of the fact. For early examples use special matrices that fit the requirements of the problem: Use I, zero, scalar, or diagonal matrices. If one of your examples shows the answer to be no, use that example in your answer.

1. Let

$$A = \begin{pmatrix} 1 & 2 \\ -3 & 2 \end{pmatrix} \quad B = \begin{pmatrix} 0 & 4 \\ 2 & 5 \end{pmatrix} \quad C = \begin{pmatrix} 1 & 0 \\ 5 & 2 \end{pmatrix} \quad r = 4 \quad s = -7$$

Use these matrices and real numbers to illustrate Theorem 1.5.1*a* to *h*.

2. Prove Theorem 1.5.2*a* and *b*.

3. Is the sum of diagonal matrices also diagonal?

4. Is the sum of tridiagonal matrices also tridiagonal?

5. Is the sum of symmetric matrices also symmetric?

6. Is the sum of upper triangular matrices also upper triangular?

7. Answer Probs. 3 through 6 with *sum* replaced by *product*.

8. Prove Theorem 1.5.3*a*, *c*, *e*, *f*, and *h*.

9. Using A and B from Prob. 1, calculate (*a*) $(A + B)^2$ and (*b*) $A^2 + 2AB + B^2$.

10. (*a*) Show that if A and B are $n \times n$ matrices, then

$$(A + B)^2 = A^2 + AB + BA + B^2$$

(*b*) Deduce that $(A + B)^2 = A^2 + 2AB + B^2$ if and only if $AB = BA$.

11. A matrix is called *skew-symmetric* (or antisymmetric) if $A^T = -A$.
(*a*) Give an example of a skew-symmetric matrix.
(*b*) Is the sum of skew-symmetric matrices also skew-symmetric?
(*c*) Is the square of a skew-symmetric matrix also skew-symmetric?

(*d*) Is the cube of a skew-symmetric matrix also skew-symmetric?
(*e*) What can you say about the product of a skew-symmetric and a symmetric matrix?
(*f*) What can you say about the main diagonal of a skew-symmetric matrix?

12. Prove Theorem 1.5.2*c*, *d*, *f*, *g*, and *h*.

13. Does $(A + B)(A - B)$ equal $A^2 - B^2$?

14. Show that $(A^n)^T = (A^T)^n$ for n a positive integer.

15. Let

$$A = \begin{pmatrix} 1 & -1 \\ 1 & -1 \end{pmatrix} \qquad B = \begin{pmatrix} 1 & 0 \\ 1 & 0 \end{pmatrix}$$

Show that $BA = A$, even though $B \neq I$.

16. If A and B are $n \times n$ diagonal matrices, must AB be equal to BA?

17. If A is $n \times n$ and D is an $n \times n$ diagonal matrix, must AD be equal to DA? What if D is a scalar matrix?

18. Calculate $\bar{A}$ and A^*.

(*a*) $A = \begin{pmatrix} i & -i \\ -i & i \end{pmatrix}$ (*b*) $A = \begin{pmatrix} 1 & 1-i \\ 2+3i & 1 \end{pmatrix}$

(*c*) $A = \begin{pmatrix} 1 & i & -i \\ 1 & 2 & 7i \\ -2 & 6 & -4 \end{pmatrix}$ (*d*) $A = \begin{pmatrix} a & z \\ \bar{z} & b \end{pmatrix}$

19. Show that if all the entries of a matrix are real, then $\bar{A} = A$.

20. *Decomposition of complex systems.* Suppose we are asked to solve $CZ = D$, where $C_{n \times n}$ and $D_{n \times 1}$ are complex matrices. Show that by writing $Z = X + iY$ and equating real and imaginary parts of equations, the complex $n \times n$ system is equivalent to the $2n \times 2n$ real system

$$\begin{aligned} AX - BY &= R \\ BX + AY &= S \end{aligned}$$

where $D = R + iS$. Show further that the system can be written

$$\left(\begin{array}{c|c} A & -B \\ \hline B & A \end{array}\right)\left(\begin{array}{c} X \\ \hline Y \end{array}\right) = \left(\begin{array}{c} R \\ \hline S \end{array}\right)$$

These last matrices are called *partitioned matrices* (see Prob. 32).

21. Recall Example 10 from Sec. 1.2.

$$\begin{aligned} (3 + i)z_1 + (1 + i)z_2 &= 4 + 4i \\ z_1 - \qquad z_2 &= 2i \end{aligned}$$

Use the result of Prob. 20 to obtain the real system

$$\left(\begin{array}{cc|cc} 3 & 1 & -1 & -1 \\ 1 & -1 & 0 & 0 \\ \hline 1 & 1 & 3 & 1 \\ 0 & 0 & 1 & -1 \end{array}\right)\begin{pmatrix} x_1 \\ x_2 \\ y_1 \\ y_2 \end{pmatrix} = \begin{pmatrix} 4 \\ 0 \\ 4 \\ 2 \end{pmatrix}$$

22. Show that

(*a*) $A = \bar{\bar{A}}$ (*b*) $\overline{(A + B)} = \bar{A} + \bar{B}$
(*c*) $\overline{rA} = rA$ (*d*) $\overline{AB} = \bar{A}\bar{B}$

23. Show that

(*a*) $A^{**} = A$ (*b*) $(A + B)^* = A^* + B^*$ (*c*) $(AB)^* = B^*A^*$

24. Show that $\bar{A}^T = (\overline{A^T})$.

25. Show that A^*A is hermitian.

26. Show that if A and B are $n \times n$, and AB is symmetric, and A and B are symmetric, then $AB = BA$.

27. Let A be square. Show that $A + A^T$ is symmetric.

28. Let A be square. Show that $A - A^T$ is skew-symmetric.

29. Show that any square matrix can be written as the sum of a symmetric and a skew-symmetric matrix.

30. What can you say about the diagonal elements of a hermitian matrix? If A is hermitian, can iA be hermitian?

31. (*a*) Let

$$A = \begin{pmatrix} 0 & 1 \\ 0 & 0 \end{pmatrix}$$

Show that $A^2 = 0$.

(*b*) Let

$$A = \begin{pmatrix} 0 & 1 & 0 \\ 0 & 0 & 1 \\ 0 & 0 & 0 \end{pmatrix}$$

Show that $A^3 = 0$.

(*c*) Let A be $n \times n$ with $a_{ij} = 1$ if $j = i + 1$ and $a_{ij} = 0$ if $j \neq i + 1$. What can you say about A^n?

32. *Partitioned matrices.* A matrix which has blocks separated by dotted or solid lines is called a *partitioned matrix*. Augmented matrices from Sec. 1.2 are partitioned matrices. The matrix

$$\left(\begin{array}{cc|c} 1 & -1 & 0 \\ 2 & 3 & 0 \\ \hline 0 & 0 & 1 \\ 0 & 0 & 6 \end{array}\right)$$

is a partitioned matrix; it is partitioned to set the blocks of 0s apart. Products of partitioned matrices can be computed by ordinary matrix multiplication, treating blocks as entries, as long as the blocks which must be multiplied are conformable. Letting the matrix above be

$$A = \left(\begin{array}{c|c} B_{2\times 2} & 0_{2\times 1} \\ \hline 0_{2\times 2} & C_{2\times 1} \end{array}\right)$$

and letting

$$D = \left(\begin{array}{c} E_{2\times 3} \\ \hline F_{1\times 3} \end{array}\right) = \left(\begin{array}{ccc} 1 & -1 & 2 \\ 3 & 2 & 1 \\ \hline 1 & 1 & 0 \end{array}\right)$$

we can write

$$AD = \left(\begin{array}{c|c} B & 0 \\ \hline 0 & C \end{array}\right)\left(\begin{array}{c} E \\ \hline F \end{array}\right) = \left(\begin{array}{c} BE + 0F \\ \hline 0E + CF \end{array}\right) = \left(\begin{array}{c} BE \\ \hline CF \end{array}\right) = \left(\begin{array}{ccc} -2 & -3 & 1 \\ 11 & 4 & 7 \\ \hline 1 & 1 & 0 \\ 6 & 6 & 0 \end{array}\right)$$

Multiply the following matrices, by partitioning first.

$$\begin{pmatrix} 3 & 2 & 0 & 0 \\ 1 & -1 & 0 & 0 \\ 0 & 0 & 1 & 2 \\ 0 & 0 & 2 & 6 \end{pmatrix} \quad \begin{pmatrix} 0 & 1 & 1 & 0 \\ 1 & 1 & 0 & 1 \\ 2 & 7 & 0 & 0 \\ -3 & 6 & 0 & 0 \end{pmatrix}$$

33. A Seifert matrix, which is used in the mathematical theory of knots, is a square matrix V which satisfies

$$V - V^T = F$$

where F is a given matrix. For the following F's find V (if it exists).

$$(a)\ F = \begin{pmatrix} 0 & 1 \\ 1 & 0 \end{pmatrix} \qquad (b)\ F = \begin{pmatrix} 0 & -1 \\ 1 & 0 \end{pmatrix} \qquad (c)\ F = 0_{n\times n}$$

34. In order for the matrix equation $V - V^T = F$ to have a solution, what must be true for the diagonal entries of F? What must be true regarding the symmetry of F?

35. Show that if an $n \times n$ matrix is in reduced row echelon form and has no zero rows, then the matrix is I_n.

36. Make a reasonable definition for "skew-hermitian matrix."

37. The Pauli spin matrices, used to study electron spin, were given in Prob. 18 of Sec. 1.4. Which of the spin matrices are symmetric, skew-symmetric, or hermitian?

38. Show that if r is a scalar, and A is a square matrix, then $(rA)^n = r^n A^n$, for n a positive integer.

39. Show that if r is a scalar and A and B are square matrices with $AB = BA$, then $(AB)^n = A^n B^n = B^n A^n$ and $A^m B^n = B^n A^m$, where m and n are positive integers.

1.6 DETERMINANTS AND CRAMER'S RULE FOR LINEAR EQUATIONS

The *determinant* of a square matrix is a number which is quite useful in the theory of equations and can be computed in a straightforward manner.

DEFINITION 1.6.1 The *determinant* of the 2×2 matrix

$$A = \begin{pmatrix} a_{11} & a_{12} \\ a_{21} & a_{22} \end{pmatrix}$$

is the number $a_{11}a_{22} - a_{12}a_{21}$. The notation for the determinant of A is det A.

EXAMPLE 1 Calculate the determinant of the following matrices.

(*a*) $\begin{pmatrix} 1 & 2 \\ 0 & 3 \end{pmatrix}$ (*b*) $\begin{pmatrix} -1 & 2 \\ -1 & 7 \end{pmatrix}$ (*c*) $\begin{pmatrix} 1 & i \\ -3 & -3 \end{pmatrix}$

Solution (*a*) $\det \begin{pmatrix} 1 & 2 \\ 0 & 3 \end{pmatrix} = 3 \cdot 1 - 2 \cdot 0 = 3$

(*b*) $\det \begin{pmatrix} -1 & 2 \\ -1 & 7 \end{pmatrix} = (-1)7 - (-1)2 = -5$

(*c*) $\det \begin{pmatrix} 1 & i \\ -3 & -3 \end{pmatrix} = 1(-3) - i(-3) = -3 + 3i$

Our definition of det A for a 2×2 matrix allows us to calculate det $A_{n \times n}$ by reducing the problem to several determinants of 2×2 matrices. In the definition, we show an example alongside.

DEFINITION 1.6.2 For an $n \times n$ matrix A, det A is the number which can be calculated in the following way:

Consider $A = \begin{pmatrix} 1 & -2 & 2 \\ 4 & 1 & -3 \\ 2 & 1 & 1 \end{pmatrix}$

1. Choose any row of A, say, $a_{i1}, a_{i2}, a_{i3}, \ldots, a_{in}$ (row i, where i is arbitrarily chosen).

1. Choose row 2: $\begin{pmatrix} 1 & -2 & 2 \\ \mathbf{4} & \mathbf{1} & \mathbf{-3} \\ 2 & 1 & 1 \end{pmatrix}$

2. Let M_{ij} be the $(n-1) \times (n-1)$ matrix obtained from A by crossing out the ith row and jth column of A.

2. $M_{21} = \begin{pmatrix} -2 & 2 \\ 1 & 1 \end{pmatrix}$

$M_{22} = \begin{pmatrix} 1 & 2 \\ 2 & 1 \end{pmatrix}$

$M_{23} = \begin{pmatrix} 1 & -2 \\ 2 & 1 \end{pmatrix}$

3. $$\det A = (-1)^{i+1}a_{i1} \det M_{i1} + (-1)^{i+2}a_{i2} \det M_{i2} + \cdots + (-1)^{i+n}a_{in} \det M_{in}$$

3. $$\det A = (-1)^3\mathbf{4}(-2-2) + (-1)^4\mathbf{1}(1-4) + (-1)^5(\mathbf{-3})[1-(-4)] = 16 - 3 + 15 = 28$$

or the number can be calculated in an alternative way:

1. Choose any column of A, say,

$$\begin{pmatrix} a_{1j} \\ a_{2j} \\ \vdots \\ a_{nj} \end{pmatrix}$$

(column j, j arbitrary).

1. Choose column 3: $\begin{pmatrix} 1 & -2 & \mathbf{2} \\ 4 & 1 & \mathbf{-3} \\ 2 & 1 & \mathbf{1} \end{pmatrix}$

2. Let M_{ij} be as in 2 above.

2. $M_{13} = \begin{pmatrix} 4 & 1 \\ 2 & 1 \end{pmatrix}$

$M_{23} = \begin{pmatrix} 1 & -2 \\ 2 & 1 \end{pmatrix}$

$M_{33} = \begin{pmatrix} 1 & -2 \\ 4 & 1 \end{pmatrix}$

3. $$\det A = (-1)^{1+j}a_{1j} \det M_{1j} + (-1)^{2+j}a_{2j} \det M_{2j} + \cdots + (-1)^{n+j}a_{nj} \det M_{nj}$$

3. $$\det A = (-1)^4\mathbf{2}(4-2) + (-1)^5(\mathbf{-3})[1-(-4)] + (-1)^6\mathbf{1}[1-(-8)] = 4 + 15 + 9 = 28$$

Note that the determinant of an $n \times n$ matrix is defined as a sum of ± 1 times determinants of $(n-1) \times (n-1)$ matrices. Each of those determinants is calculated as a sum of ± 1 times determinants of $(n-2) \times (n-2)$ matrices. Continuing this process, we work down to determinants of 2×2 matrices, which we know how to compute. It is not at all obvious (but true) that det A is independent of the choice of row or column for calculation. The determinant is not defined for nonsquare matrices.

EXAMPLE 2 Calculate

$$\det\begin{pmatrix} 1 & 2 & -1 \\ 4 & 1 & 0 \\ 3 & -1 & 0 \end{pmatrix}$$

(*a*) by choosing a row and (*b*) by choosing a column.

Solution First we show a simple way to remember whether to write $+1$ or -1 in using the definition. For a_{11}, $(-1)^{1+1} = +1$; when we move to the right or down, we have a sign of -1 since $(-1)^{2+1} = -1$ and $(-1)^{1+2} = -1$. That is, the sign changes as we move one row up or down, or one column left or right. So for a 3×3 matrix the signs are

$$\begin{pmatrix} + & - & + \\ - & + & - \\ + & - & + \end{pmatrix}$$

(*a*) Use the second row. This is preferable to using row 1 because the zero entry will be multiplied by another determinant, which will give a zero. Using the second row (see Fig. 1.6.1) gives

$$\begin{aligned} \det\begin{pmatrix} 1 & 2 & -1 \\ \mathbf{4} & \mathbf{1} & \mathbf{0} \\ 3 & -1 & 0 \end{pmatrix} &= -\mathbf{4}\det\begin{pmatrix} 2 & -1 \\ -1 & 0 \end{pmatrix} + \mathbf{1}\det\begin{pmatrix} 1 & -1 \\ 3 & 0 \end{pmatrix} \\ &\quad - \mathbf{0}\det\begin{pmatrix} 1 & 2 \\ 3 & -1 \end{pmatrix} \\ &= -4(-1) + 1(3) - 0 \\ &= 7 \end{aligned}$$

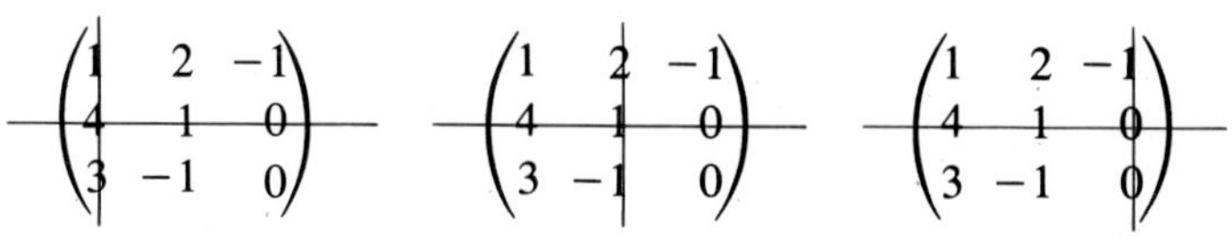

Figure 1.6.1
Crossing out rows and columns to find submatrices.

$$\begin{pmatrix} 1 & 2 & -1 \\ 4 & 1 & 0 \\ 3 & -1 & 0 \end{pmatrix} \quad \begin{pmatrix} 1 & 2 & -1 \\ 4 & 1 & 0 \\ 3 & -1 & 0 \end{pmatrix} \quad \begin{pmatrix} 1 & 2 & -1 \\ 4 & 1 & 0 \\ 3 & -1 & 0 \end{pmatrix}$$

(*b*) Use column 3 since it has the most zeros.

$$\det\begin{pmatrix}1 & 2 & \mathbf{-1}\\ 4 & 1 & \mathbf{0}\\ 3 & -1 & \mathbf{0}\end{pmatrix} = +(\mathbf{-1})\det\begin{pmatrix}4 & 1\\ 3 & -1\end{pmatrix}$$

$$-\mathbf{0}\det\begin{pmatrix}1 & 2\\ 3 & -1\end{pmatrix} + \mathbf{0}\det\begin{pmatrix}1 & 2\\ 4 & 1\end{pmatrix}$$

$$= -1(-7) - 0 + 0 = 7$$

EXAMPLE 3 Calculate

$$\det\begin{pmatrix}1 & -2 & 4 & 0\\ 7 & 3 & 0 & 3\\ -1 & 1 & -4 & 0\\ 0 & 3 & 2 & 1\end{pmatrix}$$

Solution Choose column 4 since it has the most zeros. The signs are

$$\begin{pmatrix}+ & - & + & -\\ - & + & - & +\\ + & - & + & -\\ - & + & - & +\end{pmatrix}$$

So

$$\det\begin{pmatrix}1 & -2 & 4 & \mathbf{0}\\ 7 & 3 & 0 & \mathbf{3}\\ -1 & 1 & -4 & \mathbf{0}\\ 0 & 3 & 2 & \mathbf{1}\end{pmatrix} = \mathbf{+3}\underbrace{\det\begin{pmatrix}1 & -2 & 4\\ -1 & 1 & -4\\ 0 & 3 & 2\end{pmatrix}}_{\text{Use column 1}}$$

$$+\mathbf{1}\underbrace{\det\begin{pmatrix}1 & -2 & 4\\ 7 & 3 & 0\\ -1 & 1 & -4\end{pmatrix}}_{\text{Use column 3}}$$

$$= \mathbf{3}\left[\mathbf{1}\det\begin{pmatrix}1 & -4\\ 3 & 2\end{pmatrix} - (\mathbf{-1})\det\begin{pmatrix}-2 & 4\\ 3 & 2\end{pmatrix}\right]$$

$$+\mathbf{1}\left[\mathbf{4}\det\begin{pmatrix}7 & 3\\ -1 & 1\end{pmatrix} + (\mathbf{-4})\det\begin{pmatrix}1 & -2\\ 7 & 3\end{pmatrix}\right]$$

$$= 3[14 + (-16)] + [4(10) - 4(17)] = -34$$

EXAMPLE 4 Calculate

$$\det\begin{pmatrix} 2 & -1 & 4 & 6 \\ 3 & -2 & 7 & 17 \\ 0 & 0 & 0 & 0 \\ -4 & 3 & -6 & 12 \end{pmatrix}$$

Solution Choose row 3, since it has the most zeros.

$$\det\begin{pmatrix} 2 & -1 & 4 & 6 \\ 3 & -2 & 7 & 17 \\ 0 & 0 & 0 & 0 \\ -4 & 3 & -6 & 12 \end{pmatrix} = 0 \cdot \det(\ ) + 0 \cdot \det(\ ) + 0 \cdot \det(\ ) + 0 \cdot \det(\ )$$
$$= 0$$

This example illustrates the fact that if a matrix has a row (or column) containing all zeros, the determinant is zero.

EXAMPLE 5 Calculate

$$\det\begin{pmatrix} 2 & 16 & 17 & 4 \\ 0 & -3 & 22 & -3 \\ 0 & 0 & 6 & 0 \\ 0 & 0 & 0 & 1 \end{pmatrix}$$

Solution Choose column 1 since it has the most zeros.

$$\det\begin{pmatrix} 2 & 16 & 17 & 4 \\ 0 & -3 & 22 & -3 \\ 0 & 0 & 6 & 0 \\ 0 & 0 & 0 & 1 \end{pmatrix} = 2\det\underbrace{\begin{pmatrix} -3 & 22 & -3 \\ 0 & 6 & 0 \\ 0 & 0 & 1 \end{pmatrix}}_{\text{Use column 1}} = +2(-3)\det\begin{pmatrix} 6 & 0 \\ 0 & 1 \end{pmatrix}$$
$$= +2(-3)(6) = -36$$

Another look at Example 5 shows us that the determinant of the given matrix was the product of the diagonal elements. Although this does not happen for all matrices, it does if the matrix is upper or lower triangular.

THEOREM 1.6.1 If $A_{n \times n}$ is upper (or lower) triangular, then $\det A = a_{11}a_{22}\cdots a_{nn}$.

Proof Let us use the principle of mathematical induction. The proposition $\mathscr{P}(n)$ is as follows: An $n \times n$ upper triangular matrix A has determinant $a_{11}a_{22}\cdots a_{nn}$.

First, we check $\mathscr{P}(2)$. When $n = 2$,

$$A = \begin{pmatrix} a_{11} & a_{12} \\ 0 & a_{22} \end{pmatrix}$$

and by definition $\det A = a_{11}a_{22}$. The proposition is true for $n = 2$.

For the induction hypothesis we suppose that $\mathscr{P}(k)$ is true. That is, suppose that if $A_{k \times k}$ is upper triangular, then

$$\det A_{k \times k} = a_{11}a_{22} \cdots a_{kk}$$

To complete the proof, we must show that $\det A_{k+1 \times k+1} = a_{11}a_{22} \cdots a_{k+1,k+1}$. Writing $A_{k+1 \times k+1}$, we have

$$\begin{pmatrix} a_{11} & a_{12} & \cdots & & a_{1k} & a_{1,k+1} \\ 0 & a_{22} & \ddots & & a_{2k} & a_{2,k+1} \\ \vdots & & & \ddots & & \vdots \\ 0 & 0 & & & a_{kk} & a_{k,k+1} \\ 0 & 0 & 0 & \cdots\ 0 & 0 & a_{k+1,k+1} \end{pmatrix}$$

We compute $\det A_{k+1 \times k+1}$ by using row $k + 1$ to find

$$\begin{aligned} \det A_{k+1 \times k+1} &= (-1)^{2k+2} a_{k+1,k+1} \det A_{k \times k} \\ &\overset{\text{By induction hypothesis}}{=} a_{k+1,k+1}(a_{11} \cdots a_{kk}) \\ &= a_{11} \cdots a_{kk} a_{k+1,k+1} \end{aligned}$$

Thus by the principle of mathematical induction, the proposition is true for all n.

(Alternative) Proof Let

$$A = \begin{pmatrix} a_{11} & & & \\ 0 & a_{22} & & \\ 0 & 0 & a_{33} & \boldsymbol{a_{ij}}\text{'s} \\ \vdots & 0\text{'s} & \vdots & \ddots \\ 0 & \cdots & 0 & \cdots\ a_{nn} \end{pmatrix}$$

Use the first column to calculate:

$$\det A = a_{11} \det \begin{pmatrix} a_{22} & & \\ 0 & a_{33} & \boldsymbol{a_{ij}}\text{'s} \\ \vdots & 0\text{'s} & \ddots \\ 0 & \cdots\ 0 & \cdots\ a_{nn} \end{pmatrix}$$

Use the first column again:

$$\det A = a_{11}a_{22} \det \begin{pmatrix} a_{33} & \boldsymbol{a_{ij}}\text{'s} \\ \boldsymbol{0}\text{'s} & \ddots \\ & a_{nn} \end{pmatrix}$$

Figure 1.6.2 Calculating det A by row reduction.

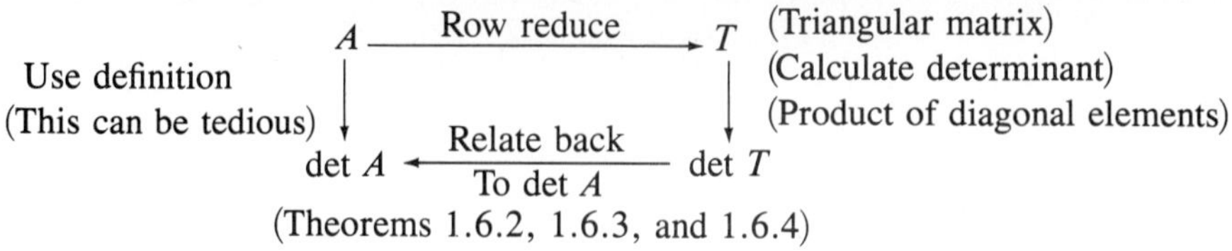

$$\begin{array}{ccc} A & \xrightarrow{\text{Row reduce}} & T \text{ (Triangular matrix)} \\ \text{Use definition (This can be tedious)} \downarrow & & \downarrow \text{(Calculate determinant) (Product of diagonal elements)} \\ \det A & \xleftarrow[\text{To det } A]{\text{Relate back}} & \det T \end{array}$$

(Theorems 1.6.2, 1.6.3, and 1.6.4)

Continuing to always use the first column gives

$$\det A = a_{11}a_{22}\cdots a_{n-2,n-2}\det\begin{pmatrix} a_{n-1,n-1} & a_{n-1,n} \\ 0 & a_{nn}\end{pmatrix}$$
$$= a_{11}a_{22}\cdots a_{nn}$$

Both proofs are almost the same for lower triangular matrices. This is left to the problems.

So, if A is upper or lower triangular, the determinant is easy to calculate. To use this fact, we can row-reduce a matrix to upper or lower triangular form, calculate the determinant of the resulting matrix, and then relate that determinant to the determinant of the original matrix. This process is diagrammed in Fig. 1.6.2.

The relationship between the determinant of a matrix and a row-reduced version is contained in the following theorems. These theorems are stated in terms of row operations—they are also true for column operations.

THEOREM 1.6.2 Let A be an $n \times n$ matrix. If two adjacent rows of A are interchanged, the determinant of the new matrix is $-\det A$.

Proof Let B be the result of interchanging two adjacent rows of A, so that

$$A = \begin{pmatrix} \cdots & \cdots & \cdots & \cdots \\ a_{i,1} & a_{i,2} & \cdots & a_{i,n} \\ a_{i+1,1} & a_{i+1,2} & \cdots & a_{i+1,n} \\ \cdots & \cdots & \cdots & \cdots \end{pmatrix}$$

$$B = \begin{pmatrix} \cdots & \cdots & \cdots & \cdots \\ a_{i+1,1} & a_{i+1,2} & \cdots & a_{i+1,n} \\ a_{i,1} & a_{i,2} & \cdots & a_{i,n} \\ \cdots & \cdots & \cdots & \cdots \end{pmatrix} \begin{matrix} \\ \leftarrow R_i \\ \leftarrow R_{i+1} \\ \\ \end{matrix}$$

Now calculate det A by using row i, and calculate det B by using row $i + 1$:

$$\det A = (-1)^{i+1}a_{i1}\det M_{i1} + \cdots + (-1)^{i+n}a_{i,n}\det M_{in}$$
$$\det B = (-1)^{i+2}a_{i1}\det M_{i1} + \cdots + (-1)^{i+1+n}a_{i,n}\det M_{in}$$

Notice that the determinants of the submatrices (or *subdeterminants*) are the same. The only differences between det A and det B are the signs. In particular,

$$\det A = -\det B \qquad \text{or} \qquad \det B = -\det A$$

THEOREM 1.6.3 If any two rows of A are interchanged, the determinant of the new matrix equals $-\det A$.

Proof The interchange of two rows can always be accomplished with an odd number of adjacent row interchanges. Suppose we wish to interchange row i with row $i + m$. Moving row i down to row $i + m$ requires m adjacent interchanges. Then to move row $i + m$ (which is now one row higher than it was) up to row m requires $m - 1$ adjacent interchanges. All other rows are in their original positions. The total number of interchanges is $m + m - 1 = 2m - 1$, which is an odd number. Each adjacent interchange changes the sign of the determinant. An odd number of sign changes alters the sign of the original determinant. (The fact that the interchange of two rows can be accomplished with only an odd number of adjacent row interchanges follows from the study of parity of a permutation, which is not covered in this text.)

COROLLARY If two rows of A are identical, then det $A = 0$.

Proof If we interchange the identical rows, the determinant of the new matrix is $-\det A$. But the new matrix is equal to A. So

$$\det A = -\det A$$

This can happen only if det $A = 0$.

THEOREM 1.6.4 Let A be an $n \times n$ matrix.

(*a*) If a row of A is multiplied by a number k, the determinant of the new matrix is k det A.

(*b*) If the result of multiplying a row of A by a constant is added to another row of A, the determinant of the new matrix is equal to det A.

Proof (*a*) Let

$$B = \begin{pmatrix} a_{11} & a_{12} & \cdots & a_{1n} \\ \cdots & \cdots & \cdots & \cdots \\ \mathbf{ka_{i1}} & \mathbf{ka_{i2}} & \cdots & \mathbf{ka_{in}} \\ \cdots & \cdots & \cdots & \cdots \\ a_{n1} & a_{n2} & \cdots & a_{nn} \end{pmatrix}$$

which is the matrix A with the row i multiplied by k. Calculate det B by using row i.

$$\begin{aligned}\det B &= (-1)^{i+1}ka_{i1} \det M_{i1} + \cdots + (-1)^{i+n}ka_{in} \det M_{in} \\ &= k((-1)^{i+1}a_{i1} \det M_{i1} + \cdots + (-1)^{i+n}a_{in} \det M_{in}) \\ &= k \det A\end{aligned}$$

(*b*) Suppose we have added r times row i of A to row m, to get a matrix B;

$$B = \begin{pmatrix} a_{11} & \cdots & a_{1n} \\ \cdots & \cdots & \cdots \\ a_{i1} & \cdots & a_{in} \\ \cdots & \cdots & \cdots \\ \boldsymbol{ra_{i1} + a_{m1}} & \cdots & \boldsymbol{ra_{in} + a_{mn}} \\ \cdots & \cdots & \cdots \\ a_{n1} & \cdots & a_{nn} \end{pmatrix}$$

Use row m to calculate det B:

$$\begin{aligned}\det B &= (-1)^{m+1}(\boldsymbol{ra_{i1} + a_{m1}}) \det M_{m1} \\ &\quad + \cdots + (-1)^{m+n}(\boldsymbol{ra_{in} + a_{mn}}) \det M_{mn} \\ &= \boldsymbol{r}[(-1)^{m+1}\, \boldsymbol{a_{i1}} \det M_{m1} + \cdots + (-1)^{m+n}\, \boldsymbol{a_{in}}\, M_{mn}] \\ &\quad + [(-1)^{m+1}\, \boldsymbol{a_{m1}} \det M_{m1} + \cdots + (-1)^{m+n}\, \boldsymbol{a_{mn}} \det M_{mn}] \\ &= \boldsymbol{r} \det \begin{pmatrix} a_{11} & \cdots & a_{1n} \\ \cdots & \cdots & \cdots \\ a_{i1} & \cdots & a_{in} \\ \cdots & \cdots & \cdots \\ \boldsymbol{a_{i1}} & \cdots & \boldsymbol{a_{in}} \\ \cdots & \cdots & \cdots \\ a_{n1} & \cdots & a_{nn} \end{pmatrix} + \det A\end{aligned}$$

By the corollary the first determinant is zero because two rows are identical. Therefore, det B = det A.

Now we use these results to calculate some determinants.

EXAMPLE 6 Let

$$A = \begin{pmatrix} 1 & -1 & 2 & 7 \\ 4 & 3 & 1 & 2 \\ -1 & 8 & 6 & 2 \\ 2 & -2 & 4 & -3 \end{pmatrix}$$

Calculate det A.

Solution We row-reduce

$$\begin{pmatrix} 1 & -1 & 2 & 7 \\ 4 & 3 & 1 & 2 \\ -1 & 8 & 6 & 2 \\ 2 & -2 & 4 & -3 \end{pmatrix} \xrightarrow[\substack{R1+R3 \\ -2R1+R4}]{-4R1+R2} \begin{pmatrix} 1 & -1 & 2 & 7 \\ 0 & 7 & -7 & -26 \\ 0 & 7 & 8 & 9 \\ 0 & 0 & 0 & -17 \end{pmatrix}$$

$$\xrightarrow{-R2+R3} \begin{pmatrix} 1 & -1 & 2 & 7 \\ 0 & 7 & -7 & -26 \\ 0 & 0 & 15 & 35 \\ 0 & 0 & 0 & -17 \end{pmatrix}$$

By Theorems 1.6.4*b* and 1.6.1,

$$\det A = \det \begin{pmatrix} 1 & -1 & 2 & 7 \\ 0 & 7 & -7 & -26 \\ 0 & 0 & 15 & 35 \\ 0 & 0 & 0 & -17 \end{pmatrix} = (1)(7)(15)(-17) = -1785$$

EXAMPLE 7 Let

$$A = \begin{pmatrix} 1 & 7 & -2 & 3 \\ -1 & -11 & 2 & 2 \\ 2 & 4 & -4 & -7 \\ 3 & 2 & -6 & -5 \end{pmatrix}$$

Calculate det A.

Solution We notice that column 3 is just -2 times column 1. So

$$A \xrightarrow{2C1+C3} \begin{pmatrix} 1 & 7 & 0 & 3 \\ -1 & -11 & 0 & 2 \\ 2 & 4 & 0 & -7 \\ 3 & 2 & 0 & -5 \end{pmatrix}$$

Since the last matrix has a column of all zeros, its determinant is zero by definition. Remember, our theorems hold true with "row" replaced by "column."

Cramer's rule is an interesting application of determinants, which we state now. This rule can be used to calculate solutions of

$$AX = B$$

when A is square. For calculating solutions it is primarily useful only when A is of small order. We do not prove Cramer's rule.

Let A be an $n \times n$ matrix. The system of equations

$$AX = B$$

has a unique solution if and only if $\det A \neq 0$. The solution can be calculated in that case as follows. Let A_k be the matrix obtained by replacing column k of A by the column matrix B. Then

$$\begin{aligned} x_1 &= \frac{\det A_1}{\det A} \\ x_2 &= \frac{\det A_2}{\det A} \\ &\vdots \\ x_n &= \frac{\det A_n}{\det A} \end{aligned}$$

EXAMPLE 8 Solve

$$\begin{aligned} 3x_1 - x_2 + x_3 &= 4 \\ x_1 + x_2 + x_3 &= 6 \\ x_1 - x_2 - x_3 &= -4 \end{aligned}$$

by using Cramer's rule.

Solution First we calculate $\det A$:

$$\det A = \det \begin{pmatrix} 3 & -1 & 1 \\ 1 & 1 & 1 \\ 1 & -1 & -1 \end{pmatrix} \underset{\substack{\uparrow \\ R2 + R3}}{=} \det \begin{pmatrix} 3 & -1 & 1 \\ 1 & 1 & 1 \\ 2 & 0 & 0 \end{pmatrix} = 2 \det \begin{pmatrix} -1 & 1 \\ 1 & 1 \end{pmatrix} = -4$$

Now substitute

$$B = \begin{pmatrix} 4 \\ 6 \\ -4 \end{pmatrix}$$

for column 1 and calculate

$$x_1 = \frac{\det \begin{pmatrix} 4 & -1 & 1 \\ 6 & 1 & 1 \\ -4 & -1 & -1 \end{pmatrix}}{-4} \underset{\substack{\uparrow \\ -C2 + C3}}{=} \frac{\det \begin{pmatrix} 4 & -1 & 2 \\ 6 & 1 & 0 \\ -4 & -1 & 0 \end{pmatrix}}{-4} = \frac{-4}{-4} = 1$$

Similarly,

$$x_2 = \frac{\det\begin{pmatrix} 3 & 4 & 1 \\ 1 & 6 & 1 \\ 1 & -4 & -1 \end{pmatrix}}{-4} \underset{\substack{\uparrow \\ -4C3 + C2}}{=} \frac{\det\begin{pmatrix} 3 & 0 & 1 \\ 1 & 2 & 1 \\ 1 & 0 & -1 \end{pmatrix}}{-4} = \frac{-8}{-4} = 2$$

$$x_3 = \frac{\det\begin{pmatrix} 3 & -1 & 4 \\ 1 & 1 & 6 \\ 1 & -1 & -4 \end{pmatrix}}{-4} \underset{\substack{\uparrow \\ R2 + R3 \\ R1 + R2}}{=} \frac{\det\begin{pmatrix} 3 & -1 & 4 \\ 4 & 0 & 10 \\ 2 & 0 & 2 \end{pmatrix}}{-4} = \frac{-12}{-4} = 3$$

The answer checks, by direct substitution.

This theorem will be useful later.

THEOREM 1.6.5 For all $n \times n$ matrices A and B, (*a*) $\det A^T = \det A$ and (*b*) $\det (AB) = \det A \det B$.

The proof of part (*a*) is left to the problems; the proof of (*b*) is delayed until we discuss elementary matrices in the next section.

EXAMPLE 9 Using the matrices in Examples 3 and 5, illustrate Theorem 1.6.5*b*.

Solution Let

$$A = \begin{pmatrix} 1 & -2 & 4 & 0 \\ 7 & 3 & 0 & 3 \\ -1 & 1 & -4 & 0 \\ 0 & 3 & 2 & 1 \end{pmatrix} \quad \text{and} \quad B = \begin{pmatrix} 2 & 16 & 17 & 4 \\ 0 & -3 & 22 & -3 \\ 0 & 0 & 6 & 0 \\ 0 & 0 & 0 & 1 \end{pmatrix}$$

Then

$$AB = \begin{pmatrix} 2 & 22 & -3 & 10 \\ 14 & 103 & 185 & 22 \\ -2 & -19 & -19 & -7 \\ 0 & -9 & 78 & -8 \end{pmatrix}$$

The operations 7R3 + R2 and R3 + R1 show that

$$\det(AB) = \det\begin{pmatrix} 0 & 3 & -22 & 3 \\ 0 & -30 & 52 & -27 \\ -2 & -19 & -19 & -7 \\ 0 & -9 & 78 & -8 \end{pmatrix}$$

$$= -2\det\begin{pmatrix} 3 & -22 & 3 \\ -30 & 52 & -27 \\ -9 & 78 & -8 \end{pmatrix}$$

Now using $-C1 + C3$, $10C3 + C1$, and $-3R3 + R1$, in that order, we have

$$\det AB = -2\det\begin{pmatrix} 0 & -256 & -3 \\ 0 & 52 & 3 \\ 1 & 78 & 1 \end{pmatrix}$$

$$= -2\det\begin{pmatrix} -256 & -3 \\ 52 & 3 \end{pmatrix} = 1224$$

On the other hand, $(\det A)(\det B) = (-34)(-36) = 1224$.

PROBLEMS 1.6

1. Calculate the determinants of the following matrices by using the definition.

(*a*) $\begin{pmatrix} 1 & 2 \\ 3 & -1 \end{pmatrix}$ (*b*) $\begin{pmatrix} 2 & 0 & 1 \\ 5 & -2 & 3 \\ 6 & 1 & 2 \end{pmatrix}$ (*c*) $\begin{pmatrix} 3 & 5 & 2 & 1 \\ 0 & 2 & 0 & 5 \\ 6 & -7 & 1 & 0 \\ 2 & 0 & 3 & 0 \end{pmatrix}$

2. Calculate the determinants of the following matrices by using row reduction.

(*a*) $\begin{pmatrix} 1 & -2 & 1 \\ 0 & 3 & 6 \\ 1 & 4 & 7 \end{pmatrix}$ (*b*) $\begin{pmatrix} 2 & 1 & 3 \\ 5 & 4 & 6 \\ 8 & 7 & 9 \end{pmatrix}$ (*c*) $\begin{pmatrix} 1 & -2 & 0 & 3 \\ 2 & 4 & 1 & 6 \\ -1 & 3 & 4 & 1 \\ 2 & 5 & 5 & 10 \end{pmatrix}$

3. Evaluate the determinants of the following matrices by using theorems or corollaries or special structure of the matrix.

(*a*) $\begin{pmatrix} 1 & -1 & 2 \\ 3 & 6 & 10 \\ 1 & -1 & 2 \end{pmatrix}$ (*b*) $\begin{pmatrix} 1 & 0 & 0 & 0 \\ 2 & 4 & 0 & 0 \\ 7 & 3 & 2 & 0 \\ 8 & 6 & 5 & 7 \end{pmatrix}$ (*c*) $\begin{pmatrix} 0 & 1 & -3 \\ 0 & 4 & -6 \\ 0 & 7 & 13 \end{pmatrix}$

4. A rule for calculating the determinant of a 3×3 matrix

$$\begin{pmatrix} a & b & c \\ d & e & f \\ g & h & i \end{pmatrix}$$

can be illustrated as

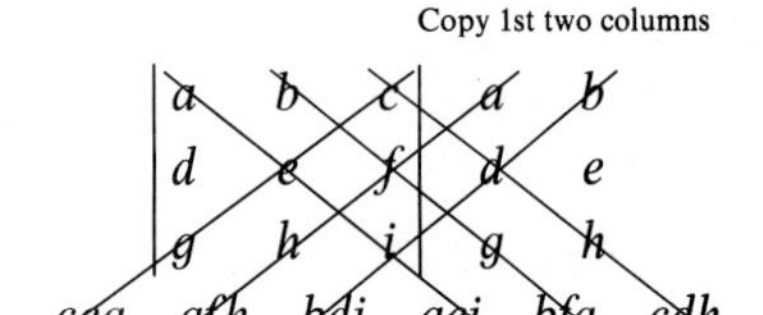

$$\det\begin{pmatrix} a & b & c \\ d & e & f \\ g & h & i \end{pmatrix} = aei + bfg + cdh - ceg - afh - bdi$$

So, to calculate

$$\det\begin{pmatrix} 1 & -1 & 2 \\ 0 & 3 & 6 \\ 4 & -7 & 2 \end{pmatrix}$$

we write

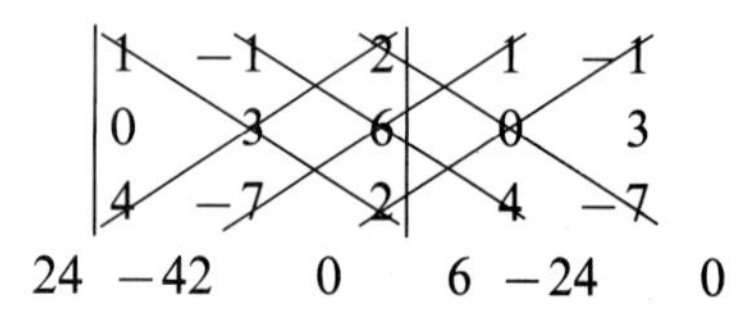

and have

$$\begin{aligned} \det &= 6 + (-24) + 0 - 24 - (-42) - 0 \\ &= 0 \end{aligned}$$

Use this method to calculate the determinants in Probs. 1*b*, 2*a*, 2*b*, and 3*c*.

5. Use Cramer's rule to solve the following systems.

(*a*) $\begin{aligned} 3x - 2y &= 7i \\ 4x + y &= 8 \end{aligned}$ (*b*) $\begin{aligned} x - 2y + z &= 3 \\ -x - y - z &= 5 \\ 3x + y - 7z &= -3 \end{aligned}$

6. Prove Theorem 1.6.1 for the case of lower triangular matrices.

7. Use Theorem 1.6.5 to show that $\det AB = \det BA$.

8. Show for

$$A = \begin{pmatrix} 1 & -1 & 2 \\ -3 & 1 & 4 \\ 1 & 2 & 3 \end{pmatrix}$$

that $\det A = \det A^T$.

9. Using

$$A = \begin{pmatrix} 1 & 2 \\ -6 & 3 \end{pmatrix} \quad \text{and} \quad B = \begin{pmatrix} 2 & -3 \\ 7 & 11 \end{pmatrix}$$

illustrate

$$\det AB = \det A \det B$$

10. Find values for k which make the determinant of the given matrix equal to zero.

(a) $\begin{pmatrix} k-1 & 2 \\ 1 & k \end{pmatrix}$ (b) $\begin{pmatrix} k-2 & 5 \\ 10 & k+3 \end{pmatrix}$ (c) $\begin{pmatrix} k & 1 \\ -1 & k \end{pmatrix}$

11. Prove Theorem 1.6.5*a*. (*Hint*: Use rows for det A, columns for det A^T.)

12. Show that $\det \bar{A} = \overline{\det A}$.

13. Show that $\det A^* = \overline{\det A}$.

14. Show that $\det I_n = 1$.

15. Show that $\det cI_n = c^n$.

16. Show that if $A_{n\times n}B_{n\times n} = I$, then $\det(ACB) = \det C$ (C is $n \times n$).

17. If $A^2 = A$, then A is called *idempotent*. Show that if A is idempotent, then the determinant of A is either 1 or 0.

18. If $A^2 = I$, then A is called *involutory*. Show that if A is involutory, then the absolute value of the determinant of A equals 1.

19. If $A^n = 0$ for some positive integer n, then A is called *nilpotent* of order (or exponent) n. For such a matrix, what can you say about its determinant?

20. Is it possible for an $n \times n$ matrix A with real entries to satisfy the equation $A^2 = -I$ when n is odd? What if n is even?

21. If $\det A = \det B$, must A be equal to B?

22. Show that $\det cA = c^n \det A$.

23. If A is skew-symmetric, what can you say about $\det A$?

24. Use Prob. 13 to show that if A is hermitian, then $\det A$ is real.

25. Let $H_n = (h_{ij})_{n\times n}$, where $h_{ij} = 1/(i+j-1)$. Calculate $\det H_n$ for $n = 2, 3, 4$.

1.7 METHOD OF MATRIX INVERSES

To solve the equation

$$ax = b \tag{1.7.1}$$

where a and b are real or complex numbers, we can simply divide both sides of (1.7.1) by a (if $a \neq 0$) to find $x = b/a$. There is a similar method for solving

$$AX = B \tag{1.7.2}$$

where A and B are matrices. This method is called the *method of matrix inversion.* Although this method is not usually employed for numerical calculations, it has theoretical use in developing other numerical methods, as we will see in the next section.

We cannot "divide" both sides of (1.7.2) by A. To see what is needed, we recall the multiplicative inverse a^{-1} of a real number a. If we multiply both sides of (1.7.1) by a^{-1}, we have

$$a^{-1}ax = a^{-1}b$$
$$1x = a^{-1}b$$
$$x = a^{-1}b$$

To pattern this method for $AX = B$, we need to define the multiplicative inverse of a square matrix A.

DEFINITION 1.7.1 A *multiplicative inverse* of $A_{n \times n}$ is any matrix $M_{n \times n}$ for which

$$MA = AM = I_n$$

EXAMPLE 1 Show that

$$\begin{pmatrix} 4 & -1 \\ -3 & 1 \end{pmatrix}$$

is a multiplicative inverse of

$$\begin{pmatrix} 1 & 1 \\ 3 & 4 \end{pmatrix}$$

Solution Since

$$\begin{pmatrix} 4 & -1 \\ -3 & 1 \end{pmatrix}\begin{pmatrix} 1 & 1 \\ 3 & 4 \end{pmatrix} = \begin{pmatrix} 1 & 0 \\ 0 & 1 \end{pmatrix} = I_2$$

and

$$\begin{pmatrix} 1 & 1 \\ 3 & 4 \end{pmatrix}\begin{pmatrix} 4 & -1 \\ -3 & 1 \end{pmatrix} = \begin{pmatrix} 1 & 0 \\ 0 & 1 \end{pmatrix} = I_2$$

we know that

$$\begin{pmatrix} 4 & -1 \\ -3 & 1 \end{pmatrix}$$

is a multiplicative inverse of

$$\begin{pmatrix} 1 & 1 \\ 3 & 4 \end{pmatrix}$$

A square matrix need not have a multiplicative inverse. For example, consider

$$A = \begin{pmatrix} 2 & 0 \\ 0 & 0 \end{pmatrix}$$

and try to find a multiplicative inverse

$$M = \begin{pmatrix} a & b \\ c & d \end{pmatrix}$$

Since AM has to be equal to I, we must have

$$\begin{pmatrix} 2 & 0 \\ 0 & 0 \end{pmatrix}\begin{pmatrix} a & b \\ c & d \end{pmatrix} = \begin{pmatrix} 1 & 0 \\ 0 & 1 \end{pmatrix}$$

or

$$\begin{pmatrix} 2a & 2b \\ 0 & 0 \end{pmatrix} = \begin{pmatrix} 1 & 0 \\ 0 & 1 \end{pmatrix}$$

or

$$2a = 1 \qquad 2b = 0 \qquad 0 = 0 \qquad 0 = 1$$

Since this last equation can never hold, there is no choice for a, b, c, and d for which $AM = I$.

So we see that some matrices have multiplicative inverses and others do not. We now show that *if A has a multiplicative inverse, the inverse is unique.*

THEOREM 1.7.1 If $A_{n \times n}$ has a multiplicative inverse, the inverse is unique.

Proof We will show that if two matrices are inverses of A, then they are actually the same. Suppose M_1 and M_2 are inverses of A, so that $M_1A = AM_1 = I$ and $M_2A = AM_2 = I$. Start with

$$M_1A = I$$

and multiply on the right by M_2 to find

$$(M_1A)M_2 = IM_2 = M_2$$

which by the associative law is

$$M_1(AM_2) = M_2$$

Now $AM_2 = I$, so

$$M_1I = M_2$$

and

$$M_1 = M_2$$

Since the multiplicative inverse of A is unique, we will use a special symbol to represent it: A^{-1}. Thus we have

$$A^{-1}A = AA^{-1} = I$$

Now to solve $AX = B$, one can use the matrix inverse, if it exists:

1. Find A^{-1} (if it exists).
2. Multiply on left by A^{-1} to get $A^{-1}AX = A^{-1}B$.
3. The solution is $X = A^{-1}B$.

EXAMPLE 2 Solve

$$\begin{aligned} x_1 + x_2 &= 2 \\ 3x_1 + 4x_2 &= 1 \end{aligned}$$

by the matrix inverse method.

Solution Since

$$A = \begin{pmatrix} 1 & 1 \\ 3 & 4 \end{pmatrix}$$

we know that

$$A^{-1} = \begin{pmatrix} 4 & -1 \\ -3 & 1 \end{pmatrix}$$

(See Example 1.) Therefore

$$\begin{pmatrix} x_1 \\ x_2 \end{pmatrix} = A^{-1}\begin{pmatrix} 2 \\ 1 \end{pmatrix} = \begin{pmatrix} 4 & -1 \\ -3 & 1 \end{pmatrix}\begin{pmatrix} 2 \\ 1 \end{pmatrix} = \begin{pmatrix} 7 \\ -5 \end{pmatrix}$$

and $x_1 = 7$ and $x_2 = -5$.

Although we have seen the inverse method work, two very important questions are still unanswered:

1. Given A, can we determine whether A^{-1} exists?
2. If A^{-1} exists, how do we find it?

To get an idea as to how to answer these questions, we can look at the case of 2×2 matrices. Let

$$A = \begin{pmatrix} a_{11} & a_{12} \\ a_{21} & a_{22} \end{pmatrix}$$

be an arbitrary 2×2 matrix and try to find a 2×2 matrix

$$\begin{pmatrix} x & y \\ z & w \end{pmatrix}$$

so that

$$\begin{pmatrix} a_{11} & a_{12} \\ a_{21} & a_{22} \end{pmatrix}\begin{pmatrix} x & y \\ z & w \end{pmatrix} = \begin{pmatrix} 1 & 0 \\ 0 & 1 \end{pmatrix} \tag{1.7.3}$$

If we can do this, then

$$\begin{pmatrix} x & y \\ z & w \end{pmatrix}$$

is A^{-1}. Writing out the left-hand side of (1.7.3) and equating elements of matrices, we have

$$\begin{aligned} a_{11}x + a_{12}z + \qquad\qquad\qquad &= 1 \\ a_{11}y + a_{12}w &= 0 \\ a_{21}x + a_{22}z \qquad\qquad\qquad &= 0 \\ a_{21}y + a_{22}w &= 1 \end{aligned}$$

We can rewrite this as two systems:

$$\begin{aligned} a_{11}x + a_{12}z &= 1 \qquad & a_{11}y + a_{12}w &= 0 \\ a_{21}x + a_{22}x &= 0 \qquad & a_{21}y + a_{22}w &= 1 \end{aligned} \tag{1.7.4}$$

From Cramer's rule we can solve uniquely for x, z, y, and w if and only if

$$\det \begin{pmatrix} a_{11} & a_{12} \\ a_{21} & a_{22} \end{pmatrix} \neq 0$$

that is, $\det A \neq 0$. So for a 2×2 matrix A, A^{-1} exists if and only if $\det A \neq 0$. In fact, this is true for $n \times n$ matrices, and the proof involves writing out sets of equations as above and using Cramer's rule.

THEOREM 1.7.2 Matrix A has a multiplicative inverse if and only if $\det A \neq 0$.

EXAMPLE 3 Which of the following matrices has a multiplicative inverse?

(*a*) $\begin{pmatrix} 0 & 1 \\ 1 & 0 \end{pmatrix}$ (*b*) $\begin{pmatrix} 1 & 2 & 3 \\ 7 & 8 & 9 \\ 4 & 5 & 6 \end{pmatrix}$ (*c*) $\begin{pmatrix} 5 & 0 & 2 \\ 0 & -1 & 7 \\ 0 & 0 & 2 \end{pmatrix}$ (*d*) $\begin{pmatrix} 3i & 2 \\ 1 & 1 + i \end{pmatrix}$

Solution (*a*) The determinant is -1, which is nonzero; the matrix has an inverse.

(*b*) The determinant is 0. The matrix does not have an inverse.

(c) The determinant is -10, which is nonzero; the matrix has an inverse.

(d) The determinant is $-5 + 3i \neq 0$; the matrix has an inverse.

Now we show how to find A^{-1}, by solving the equations in (1.7.4) for x, y, z, and w. In augmented form we have

$$\underset{\text{(for } x, z)}{\left(\begin{array}{cc|c} a_{11} & a_{12} & 1 \\ a_{21} & a_{22} & 0 \end{array}\right)} \qquad \underset{\text{(for } y, w)}{\left(\begin{array}{cc|c} a_{11} & a_{12} & 0 \\ a_{21} & a_{22} & 1 \end{array}\right)} \tag{1.7.5}$$

So if we eliminate variables by using Gauss-Jordan elimination, we find x, y, z, and w by reading off the right-hand sides. For example, we can consider

$$A = \begin{pmatrix} 1 & 1 \\ 3 & 4 \end{pmatrix}$$

Equations (1.7.5) are

$$\begin{array}{ll} \left(\begin{array}{cc|c} 1 & 1 & \mathbf{1} \\ 3 & 4 & \mathbf{0} \end{array}\right) & \left(\begin{array}{cc|c} 1 & 1 & \mathbf{0} \\ 3 & 4 & \mathbf{1} \end{array}\right) \\ \xrightarrow{-3R1 + R2} \left(\begin{array}{cc|c} 1 & 1 & 1 \\ 0 & 1 & -3 \end{array}\right) & \xrightarrow{-3R1 + R2} \left(\begin{array}{cc|c} 1 & 1 & 0 \\ 0 & 1 & 1 \end{array}\right) \\ \xrightarrow{-R2 + R1} \left(\begin{array}{cc|c} 1 & 0 & 4 \\ 0 & 1 & -3 \end{array}\right) & \xrightarrow{-R2 + R1} \left(\begin{array}{cc|c} 1 & 0 & -1 \\ 0 & 1 & 1 \end{array}\right) \\ x = 4 \quad z = -3 & y = -1 \quad w = 1 \end{array} \tag{1.7.6}$$

So

$$A^{-1} = \begin{pmatrix} 4 & -1 \\ -3 & 1 \end{pmatrix}$$

Since we used the same row operations in both cases in (1.7.6), we can save time and space by joining the augmented matrices to obtain

$$\left(\begin{array}{cc|cc} 1 & 1 & \mathbf{1} & \mathbf{0} \\ 3 & 4 & \mathbf{0} & \mathbf{1} \end{array}\right) \xrightarrow{-3R1 + R2} \left(\begin{array}{cc|cc} 1 & 1 & 1 & 0 \\ 0 & 1 & -3 & 1 \end{array}\right) \xrightarrow{-R2 + R1} \left(\begin{array}{cc|cc} 1 & 0 & 4 & -1 \\ 0 & 1 & -3 & 1 \end{array}\right)$$

Notice that when A has been reduced to I_2, then I_2 has been changed to A^{-1}. In fact, this happens for square matrices of any order.

THEOREM 1.7.3 Let A be an $n \times n$ matrix which has a multiplicative inverse. Then A^{-1} can be found by the process

$$(A \mid I) \xrightarrow[\text{reduce } A \text{ to } I]{\text{row operations to}} (I \mid A^{-1})$$

A proof of this theorem can be given by using *elementary matrices*, which we discuss later in the section.

EXAMPLE 4 Solve

$$\begin{aligned} x - \ y &= 2 \\ 2x + 4y &= 4 \end{aligned}$$

by matrix inversion.

Solution

$$A = \begin{pmatrix} 1 & -1 \\ 2 & 4 \end{pmatrix} \qquad \text{and} \qquad \det A = 6 \neq 0$$

Therefore, A^{-1} exists. To find it, do the row operations

$$\left(\begin{array}{cc|cc} 1 & -1 & 1 & 0 \\ 2 & 4 & 0 & 1 \end{array}\right) \xrightarrow{-2R1 + R2} \left(\begin{array}{cc|cc} 1 & -1 & 1 & 0 \\ 0 & 6 & -2 & 1 \end{array}\right)$$

$$\xrightarrow{\frac{1}{6}R2} \left(\begin{array}{cc|cc} 1 & -1 & 1 & 0 \\ 0 & 1 & -\frac{1}{3} & \frac{1}{6} \end{array}\right)$$

$$\xrightarrow{R2 + R1} \left(\begin{array}{cc|cc} 1 & 0 & \frac{2}{3} & \frac{1}{6} \\ 0 & 1 & -\frac{1}{3} & \frac{1}{6} \end{array}\right)$$

So

$$A^{-1} = \begin{pmatrix} \frac{2}{3} & \frac{1}{6} \\ -\frac{1}{3} & \frac{1}{6} \end{pmatrix} \qquad \text{and} \qquad \begin{pmatrix} x \\ y \end{pmatrix} = \begin{pmatrix} \frac{2}{3} & \frac{1}{6} \\ -\frac{1}{3} & \frac{1}{6} \end{pmatrix} \begin{pmatrix} 2 \\ 4 \end{pmatrix} = \begin{pmatrix} 2 \\ 0 \end{pmatrix}$$

If a square matrix A has a multiplicative inverse, A is called *invertible*, or *nonsingular*. If A^{-1} does not exist, then A is called *singular*. When we try to find A^{-1} for a given matrix A, we need not calculate $\det A$ before beginning row operations on $(A \mid I)$. If in the sequence of row operations we find a row of all zeros in the row-reduced form of A, we know that $\det A = 0$ and A^{-1} does not exist. We know $\det A = 0$ because the determinant of a row reduction of A differs from $\det A$ only by a nonzero multiple.

Some properties of A^{-1} which we will need later are contained in the next theorem.

THEOREM 1.7.4 Let A and B be $n \times n$ invertible matrices. Then the following are true.

(*a*) Matrix AB is invertible, and $(AB)^{-1} = B^{-1}A^{-1}$.

(*b*) For $k = 1, 2, 3, \ldots, (A^k)^{-1} = (A^{-1})^k$.

(c) $\det A^{-1} = \dfrac{1}{\det A}$

(d) Matrix A^T is invertible, and $(A^T)^{-1} = (A^{-1})^T$.

(e) If r is a nonzero scalar, then rA is invertible and $(rA)^{-1} = (1/r)A^{-1}$.

(f) Matrix A^{-1} is invertible, and $(A^{-1})^{-1} = A$.

Proof (a) This part is shown by actual multiplication by the candidate $B^{-1}A^{-1}$ for an inverse of AB.

$$(AB)(B^{-1}A^{-1}) = A(BB^{-1})A^{-1} = AIA^{-1} = AA^{-1} = I$$
$$(B^{-1}A^{-1})(AB) = B^{-1}(A^{-1}A)B = B^{-1}IB = B^{-1}B = I$$

Since these products equal I, matrix AB is invertible, and $B^{-1}A^{-1}$ is the inverse of AB.

(b) $$(A^n)^{-1} = (\underbrace{AA\cdots A}_{n \text{ copies}})^{-1} = [A(\underbrace{A\cdots A}_{n-1 \text{ copies}})]^{-1}$$

$$\underset{\text{By part (a)}}{=} (\underbrace{A\cdots A}_{n-1 \text{ copies}})^{-1}A^{-1} = [A(\underbrace{A\cdots A}_{n-2 \text{ copies}})]^{-1}A^{-1}$$

$$= (\underbrace{A\cdots A}_{n-2 \text{ copies}})^{-1}A^{-1}A^{-1}$$

Continuing this process, we finally obtain

$$\underbrace{A^{-1}A^{-1}\cdots A^{-1}}_{n \text{ copies}} = (A^{-1})^n$$

(c) Since $AA^{-1} = I$, $\det(AA^{-1}) = \det I$. This means that $(\det A)(\det A^{-1}) = 1$. The invertibility of A implies that $\det A \neq 0$, so we can divide by $\det A$ to find

$$\det A^{-1} = \frac{1}{\det A}$$

Proofs of (d), (e), and (f) are left to the problems.

Now to our discussion of elementary matrices. An $n \times n$ *elementary matrix* is any matrix obtained from I_n by an elementary row operation. So

$$\begin{pmatrix} 1 & 0 & 0 \\ 0 & 1 & 0 \\ 2 & 0 & 1 \end{pmatrix} \quad \begin{pmatrix} 1 & 0 & 0 \\ 0 & -3 & 0 \\ 0 & 0 & 1 \end{pmatrix} \quad \begin{pmatrix} 0 & 1 & 0 \\ 1 & 0 & 0 \\ 0 & 0 & 1 \end{pmatrix}$$

are elementary matrices resulting from the operations $2\text{R}1 + \text{R}3$, $-3\text{R}2$, and $\text{R}1 \leftrightarrow \text{R}2$ on I_3.

The importance of elementary matrices is that the result of a row operation of a matrix A can be obtained by multiplying A by the appropriate elementary matrix.

Let A be an $n \times n$ matrix and E the elementary matrix corresponding to a row operation. If the row operation on A results in B, then

$$EA = B$$

For example, in the 3×3 case, let the row operation be $-3\text{R1} + \text{R3}$. The associated elementary matrix is

$$E = \begin{pmatrix} 1 & 0 & 0 \\ 0 & 1 & 0 \\ -3 & 0 & 1 \end{pmatrix}$$

Now if

$$A = \begin{pmatrix} 1 & 2 & 1 \\ -2 & 6 & 1 \\ 4 & 2 & 3 \end{pmatrix}$$

then

$$A \xrightarrow{-3R1 + R3} \begin{pmatrix} 1 & 2 & 1 \\ -2 & 6 & 1 \\ 1 & -4 & 0 \end{pmatrix}$$

and

$$EA = \begin{pmatrix} 1 & 0 & 0 \\ 0 & 1 & 0 \\ -3 & 0 & 1 \end{pmatrix}\begin{pmatrix} 1 & 2 & 1 \\ -2 & 6 & 1 \\ 4 & 2 & 3 \end{pmatrix} = \begin{pmatrix} 1 & 2 & 1 \\ -2 & 6 & 1 \\ 1 & -4 & 0 \end{pmatrix}$$

Of course, a sequence of m row operations leads to a sequence of elementary matrices $E_1, E_2, E_3, \ldots, E_m$, and the result of the row operations on $A_{n \times n}$ is

$$(E_m \cdots (E_3(E_2(E_1 A))) \cdots) = E_m E_{m-1} \cdots E_3 E_2 E_1 A$$

For example, if

$$A = \begin{pmatrix} 1 & 2 \\ 3 & 4 \end{pmatrix}$$

and we perform the row operations

$$\begin{pmatrix} 1 & 2 \\ 3 & 4 \end{pmatrix} \xrightarrow{-3\text{R1} + \text{R2}} \begin{pmatrix} 1 & 2 \\ 0 & -2 \end{pmatrix} \xrightarrow{\text{R2} + \text{R1}} \begin{pmatrix} 1 & 0 \\ 0 & -2 \end{pmatrix} \xrightarrow{-\frac{1}{2}\text{R2}} \begin{pmatrix} 1 & 0 \\ 0 & 1 \end{pmatrix}$$

we generate

$$E_1 = \begin{pmatrix} 1 & 0 \\ -3 & 1 \end{pmatrix} \qquad E_2 = \begin{pmatrix} 1 & 1 \\ 0 & 1 \end{pmatrix} \qquad E_3 = \begin{pmatrix} 1 & 0 \\ 0 & -\frac{1}{2} \end{pmatrix}$$

Multiplying, we have

$$
\begin{aligned}
E_3E_2E_1A &= \begin{pmatrix} 1 & 0 \\ 0 & -\frac{1}{2} \end{pmatrix}\begin{pmatrix} 1 & 1 \\ 0 & 1 \end{pmatrix}\begin{pmatrix} 1 & 0 \\ -3 & 1 \end{pmatrix}\begin{pmatrix} 1 & 2 \\ 3 & 4 \end{pmatrix} \\
&= \begin{pmatrix} 1 & 0 \\ 0 & -\frac{1}{2} \end{pmatrix}\begin{pmatrix} 1 & 1 \\ 0 & 1 \end{pmatrix}\begin{pmatrix} 1 & 2 \\ 0 & -2 \end{pmatrix} \\
&= \begin{pmatrix} 1 & 0 \\ 0 & -\frac{1}{2} \end{pmatrix}\begin{pmatrix} 1 & 0 \\ 0 & -2 \end{pmatrix} \\
&= \begin{pmatrix} 1 & 0 \\ 0 & 1 \end{pmatrix}
\end{aligned}
$$

The most important fact about elementary matrices is that they are invertible. This follows because

$rR_i + R_j$	is inverted by	$-rR_i + R_j$	
rR_i	is inverted by	$\frac{1}{r}R_i$	(remember, $r \neq 0$)
$R_i \leftrightarrow R_j$	is inverted by	$R_j \leftrightarrow R_i$	

So, for example, in the 3×3 case $-3\text{R}1 + \text{R}3$ results in

$$
\begin{pmatrix} 1 & 0 & 0 \\ 0 & 1 & 0 \\ -3 & 0 & 1 \end{pmatrix}
$$

and $3\text{R}1 + \text{R}3$ results in

$$
\begin{pmatrix} 1 & 0 & 0 \\ 0 & 1 & 0 \\ 3 & 0 & 1 \end{pmatrix}
$$

The product is

$$
\begin{pmatrix} 1 & 0 & 0 \\ 0 & 1 & 0 \\ -3 & 0 & 1 \end{pmatrix}\begin{pmatrix} 1 & 0 & 0 \\ 0 & 1 & 0 \\ 3 & 0 & 1 \end{pmatrix} = \begin{pmatrix} 1 & 0 & 0 \\ 0 & 1 & 0 \\ -3+3 & 0 & 1 \end{pmatrix} = I_3
$$

If A is invertible, its reduced row echelon form will be I, because by properties of determinants det A will just be a constant times the determinant of the reduced row echelon form of A and det $A \neq 0$. Therefore, if the elementary matrices corresponding to the row operations which reduce A to I are denoted by $E_1, E_2, \ldots, E_m$, we have

$$
E_m \cdots E_2E_1A = I
$$

and

$$
A = E_1^{-1}E_2^{-1} \cdots E_m^{-1}I
$$

Now the inverse of a product is the product of the inverses in the reverse order, so

$$\begin{aligned} A^{-1} &= I^{-1}(E_m^{-1})^{-1} \cdots (E_2^{-1})^{-1}(E_1^{-1})^{-1} \\ &= E_m \cdots E_2 E_1 I \end{aligned}$$

This shows that A^{-1} can be constructed by applying the row operations which reduce A to I to the identity matrix. Thus we have proved Theorem 1.7.3.

EXAMPLE 5 Determine whether

$$A = \begin{pmatrix} 2 & 4 & 6 \\ 4 & 5 & 6 \\ 14 & 16 & 18 \end{pmatrix}$$

is invertible. If it is, calculate A^{-1}.

Solution Consider $(A|I)$ and row-reduce.

$$\left(\begin{array}{ccc|ccc} 2 & 4 & 6 & 1 & 0 & 0 \\ 4 & 5 & 6 & 0 & 1 & 0 \\ 14 & 16 & 18 & 0 & 0 & 1 \end{array}\right) \xrightarrow{\frac{1}{2}\text{R1}} \left(\begin{array}{ccc|ccc} 1 & 2 & 3 & \frac{1}{2} & 0 & 0 \\ 4 & 5 & 6 & 0 & 1 & 0 \\ 14 & 16 & 18 & 0 & 0 & 1 \end{array}\right)$$

$$\xrightarrow[-14\text{R1}+\text{R3}]{-4\text{R1}+\text{R2}} \left(\begin{array}{ccc|ccc} 1 & 2 & 3 & \frac{1}{2} & 0 & 0 \\ 0 & -3 & -6 & -2 & 1 & 0 \\ 0 & -12 & -24 & -7 & 0 & 1 \end{array}\right)$$

$$\xrightarrow[-\frac{1}{12}\text{R3}]{-\frac{1}{3}\text{R2}} \left(\begin{array}{ccc|ccc} 1 & 2 & 3 & \frac{1}{2} & 0 & 0 \\ 0 & 1 & 2 & \frac{2}{3} & -\frac{1}{3} & 0 \\ 0 & 1 & 2 & \frac{7}{12} & 0 & -\frac{1}{12} \end{array}\right)$$

$$\xrightarrow{-\text{R2}+\text{R3}} \left(\begin{array}{ccc|ccc} 1 & 2 & 3 & \frac{1}{2} & 0 & 0 \\ 0 & 1 & 2 & \frac{2}{3} & -\frac{1}{3} & 0 \\ 0 & 0 & 0 & -\frac{1}{12} & \frac{1}{3} & -\frac{1}{12} \end{array}\right)$$

Since the third row of the reduction of A is all zeros, $\det A = 0$ and so A^{-1} does not exist. Matrix A is singular.

EXAMPLE 6 The theory of statistics is used to estimate parameters in engineering problems.. In the case of autoregressive errors,[9] the $n \times n$ lower triangular matrix

$$D = \begin{pmatrix} 1 & 0 & 0 & 0 & \cdots & 0 \\ \rho & 1 & 0 & 0 & \cdots & 0 \\ \rho^2 & \rho & 1 & 0 & \cdots & 0 \\ \rho^3 & \rho^2 & \rho & 1 & \cdots & 0 \\ \vdots & \vdots & \vdots & \ddots & \ddots & \vdots \\ \rho^{n-1} & \rho^{n-2} & \rho^{n-3} & \rho^{n-4} & \cdots \rho & 1 \end{pmatrix}$$

[9] J. V. Beck and K. J. Arnold, *Parameter Estimation in Engineering and Science*, John Wiley & Sons, New York, 1977.

must be inverted (ρ is a real number related to errors in measurement of the process being studied). Find D^{-1}.

Solution To get an idea of what D^{-1} should be, we check the 3×3 case first:

$$\left(\begin{array}{ccc|ccc} 1 & 0 & 0 & 1 & 0 & 0 \\ \rho & 1 & 0 & 0 & 1 & 0 \\ \rho^2 & \rho & 1 & 0 & 0 & 1 \end{array}\right) \xrightarrow[-\rho^2 R1 + R3]{-\rho R1 + R2} \left(\begin{array}{ccc|ccc} 1 & 0 & 0 & 1 & 0 & 0 \\ 0 & 1 & 0 & -\rho & 1 & 0 \\ 0 & \rho & 1 & -\rho^2 & 0 & 1 \end{array}\right)$$

$$\xrightarrow{-\rho R2 + R3} \left(\begin{array}{ccc|ccc} 1 & 0 & 0 & 1 & 0 & 0 \\ 0 & 1 & 0 & -\rho & 1 & 0 \\ 0 & 0 & 1 & 0 & -\rho & 1 \end{array}\right)$$

Thus

$$D^{-1} = \begin{pmatrix} 1 & 0 & 0 \\ -\rho & 1 & 0 \\ 0 & -\rho & 1 \end{pmatrix}$$

If we work out the 4×4 case, we find

$$D^{-1} = \begin{pmatrix} 1 & 0 & 0 & 0 \\ -\rho & 1 & 0 & 0 \\ 0 & -\rho & 1 & 0 \\ 0 & 0 & -\rho & 1 \end{pmatrix}$$

A reasonable guess is that in the $n \times n$ case

$$D^{-1} = \begin{pmatrix} 1 & 0 & 0 & 0 & \cdots & 0 \\ -\rho & 1 & 0 & 0 & \cdots & 0 \\ 0 & -\rho & 1 & \ddots & & \vdots \\ \vdots & & \ddots & \ddots & \ddots & 0 \\ 0 & \cdots & 0 & \cdots & -\rho & 1 \end{pmatrix}$$

Direct multiplication shows $DD^{-1} = I$. Because the inverse of a matrix is unique, and $D^{-1}D = I$, the guess is correct.

The important results of this section allow us to state the basic result of the theory of linear equations.

THEOREM 1.7.5

Let A be an $n \times n$ matrix and B be an $n \times 1$ matrix. The following statements are equivalent.

(*a*) $AX = B$ has a unique solution.

(*b*) A is row-equivalent to I.

(*c*) A is invertible.

(*d*) $\det A \neq 0$.

Proof We will show that (a) implies (b), (b) implies (c), (c) implies (d), and (d) implies (a).

(a) implies (b): After row reduction of $(A|B)$ the reduced row echelon form of the augmented matrix, $(R|S)$, must represent an equivalent system. By equivalence, the reduced system has a unique solution. It follows that R must be I_n; thus A is row-equivalent to I.

(b) implies (c): If A is row-equivalent to I, $\det A$ is just some nonzero multiple of $\det I$. Therefore $\det A \neq 0$. By Theorem 1.7.2, A is invertible.

(c) implies (d): If A is invertible, then $AA^{-1} = I$ and $\det AA^{-1} = \det A \det A^{-1} = \det I = 1$. Therefore neither $\det A$ nor $\det A^{-1}$ is zero.

(d) implies (a): This follows from Cramer's rule.

We note that Theorem 1.7.5 implies a result which we will often use.

Let A be an $n \times n$ matrix. The homogeneous equation

$$AX = 0$$

has a nontrivial solution if and only if $\det A = 0$.

Finally, as promised in the last section, we give a proof of $\det AB = (\det A)(\det B)$. First we note that if E is an elementary matrix, then $\det EA = (\det E)(\det A)$. This is easily seen by considering three cases corresponding to which of the three elementary row operations E represents. Now, for the general case, suppose that the reduced row echelon forms of A and B are known. The reduced form will be I or

$$\begin{pmatrix} I_m & 0 \\ 0 & 0 \end{pmatrix}$$

Suppose for the first case that A has the reduced row echelon form of I. Then we have

$$\det (AB) = \det (E_1^{-1} \cdots E_k^{-1}B)$$

Because the E_i are elementary matrices, so are their inverses. Using our first remark in this paragraph, we have

$$\begin{aligned} \det AB &= \det E_1^{-1} \cdots \det E_k^{-1} \det B \\ &= \det (E_1^{-1} \cdots E_k^{-1}) \det B \\ &= \det A \det B \end{aligned}$$

In the second case, suppose A has the second possible reduced row echelon form. Then $\det A = 0$ and $(\det A)(\det B) = 0$. Also $\det (AB) = 0$ because AB has no inverse (if it did have an inverse, the inverse would have to be $B^{-1}A^{-1}$, but A^{-1} does not exist). Thus $\det (AB) = 0 = (\det A)(\det B)$.

PROBLEMS 1.7

1. For the following matrices A, determine whether A^{-1} exists. If A^{-1} exists, find it.

(a) $\begin{pmatrix} 1 & 2 \\ 3 & 4 \end{pmatrix}$ (b) $\begin{pmatrix} 2+i & 6 \\ 0 & -3 \end{pmatrix}$ (c) $\begin{pmatrix} 0 & 1 \\ 1 & 0 \end{pmatrix}$

(d) $\begin{pmatrix} 3 & -1 & 2 \\ -6 & 4 & 1 \\ -7 & -2 & 3 \end{pmatrix}$ (e) $\begin{pmatrix} 1 & -1 & 0 & 0 \\ 2 & 4 & 0 & 0 \\ 0 & 0 & 5 & 2 \\ 0 & 0 & -6 & -2 \end{pmatrix}$

2. Solve the following systems by using the inverse method.

(a) $\begin{aligned} x + y &= 7 \\ x - y &= 3 \end{aligned}$ (b) $\begin{aligned} x - 2y &= 6 \\ 3x + y &= 7 \end{aligned}$ (c) $\begin{aligned} x - 2y + z &= 6 \\ 3x + y - z &= 7 \\ -x + y + 7z &= 6 \end{aligned}$

3. If

$$\begin{pmatrix} 1 & 2 \\ -3 & 6 \end{pmatrix}$$

is the multiplicative inverse of A, what is the inverse of $3A$?

4. If $A_{n \times n}$ and $B_{n \times n}$ are invertible, is $A + B$ invertible?

5. Illustrate Theorem 1.7.4 with examples.

6. Let n be a positive integer greater than or equal to 2. Define

$$A^{-n} = (A^{-1})^n$$

when A is a nonsingular matrix. Let

$$A = \begin{pmatrix} 3 & 5 \\ 1 & 2 \end{pmatrix}$$

Calculate A^{-1}, A^{-2}, and A^{-3}.

7. Suppose A is invertible and B is not. Show that AB is not invertible. (*Hint*: Consider det AB.)

8. Solve

$$\begin{pmatrix} 3 & 5 \\ 1 & 2 \end{pmatrix}\begin{pmatrix} x \\ y \end{pmatrix} = \begin{pmatrix} a \\ b \end{pmatrix}$$

for the following cases:

(a) $\begin{pmatrix} a \\ b \end{pmatrix} = \begin{pmatrix} 0 \\ 0 \end{pmatrix}$ (b) $\begin{pmatrix} a \\ b \end{pmatrix} = \begin{pmatrix} 3 \\ 1 \end{pmatrix}$ (c) $\begin{pmatrix} a \\ b \end{pmatrix} = \begin{pmatrix} -6 \\ -7 \end{pmatrix}$

9. Let A be nonsingular and suppose that $AB = AC$. Show that $B = C$.

10. Suppose that A, B, and C are matrices and that C^{-1} exists. Show that if $C^{-1}AC = B$, then $\det A = \det B$.

11. Show for

$$A = \begin{pmatrix} \cos\theta & \sin\theta \\ -\sin\theta & \cos\theta \end{pmatrix}$$

that $A^{-1} = A^T$.

12. Show that if A is invertible and $A^{-1} = A^T$, then $\det A = \pm 1$. (*Hint*: Remember that $\det A^T = \det A$.)

13. (*a*) Show that if A is invertible and $A^{-1} = A^*$, then $|\det A| = +1$. (*Hint*: Remember, for complex numbers $z\bar{z} = |z|^2$.)

(*b*) Consider

$$A = \begin{pmatrix} 0 & i \\ -i & 0 \end{pmatrix}$$

Show that $A^*A = I$. Matrix A is a *Pauli spin matrix*. It is used to study electron spin in quantum mechanics.

14. If A is a symmetric nonsingular matrix, is A^{-1} symmetric? Answer the same question for skew-symmetric matrices.

15. Is it possible for a 3×3 matrix to be both nonsingular and skew-symmetric? Generalize the result.

16. Prove Theorem 1.7.4*d*, *e*, and *f*.

17. Show that every nilpotent matrix must be singular.

18. If A and AB are $n \times n$ and invertible, must B be invertible?

19. Let A and B be $n \times n$. If AB and BA are invertible, must both A and B be invertible?

20. If $A = B$ and A^{-1} exists, must $A^{-1} = B^{-1}$?

21. Let A and B be $n \times n$. If $A^{-1} = B^{-1}$, must $A = B$?

22. If A is upper triangular and nonsingular, is A^{-1} upper triangular?

23. Let $H_n = (h_{ij})_{n \times n}$, where $h_{ij} = 1/(i + j - 1)$. Here H_n is called a *Hilbert matrix*. In Prob. 25 of Sec. 1.6, the determinants of H_2, H_3, and H_4 were calculated. Are H_2, H_3, and H_4 invertible? If so, calculate the determinants of the inverses.

24. Let H_n be defined as in Prob. 23.

(*a*) Calculate H_3^{-1}.

(*b*) Use H_3^{-1} to solve

$$H_3 X = \begin{pmatrix} \frac{11}{6} \\ \frac{13}{12} \\ \frac{47}{60} \end{pmatrix}$$

(c) Calculate H_3^{-1}, keeping only three significant digits and rounding in each calculation. Keep only three significant digits in the entries of B. Solve the equation in (b) again.

1.8 *LU* DECOMPOSITION AND GAUSS-SEIDEL ITERATION, TWO MORE COMPUTATIONAL METHODS

Matrix algebra allows us to outline two important computational methods for solving systems of linear equations. First we discuss LU decomposition.

If the square matrix A can be written as LU, where L is lower triangular and U is an upper triangular matrix, and the system $AX = B$ has a unique solution, then the equations can be solved in two steps. Since $AX = LUX$, we can set UX equal to Y (Y is not known yet) and

1. Solve $LY = B$ for Y.

2. Solve $UX = Y$ for X.

At first glance, this may seem to be more work than solving the original equations. However, solving $LY = B$ and $UX = Y$ is easy since L and U are triangular matrices.

EXAMPLE 1 Given that

$$A = \begin{pmatrix} 1 & 2 & 1 \\ 2 & 2 & 3 \\ -1 & -3 & 0 \end{pmatrix} = \begin{pmatrix} 1 & 0 & 0 \\ 2 & 1 & 0 \\ -1 & \frac{1}{2} & 1 \end{pmatrix} \begin{pmatrix} 1 & 2 & 1 \\ 0 & -2 & 1 \\ 0 & 0 & \frac{1}{2} \end{pmatrix} = LU$$

use the two-step process described above to solve

$$AX = \begin{pmatrix} 0 \\ 3 \\ 2 \end{pmatrix} = B$$

Solution First we solve

$$LY = B \qquad \text{for } Y = \begin{pmatrix} y_1 \\ y_2 \\ y_3 \end{pmatrix}$$

The system in augmented matrix form is

$$\left(\begin{array}{ccc|c} 1 & 0 & 0 & 0 \\ 2 & 1 & 0 & 3 \\ -1 & \frac{1}{2} & 1 & 2 \end{array}\right)$$

To solve this, we use *forward substitution.* The first equation gives us immediately $y_1 = 0$. Substituting this into the second equation, we find $y_2 = 3$. Finally, putting $y_1 = 0$, $y_2 = 3$ into the third equation, we have $\frac{3}{2} + y_3 = 2$, or $y_3 = \frac{1}{2}$.

Step 2 involves solving

$$UX = Y = \begin{pmatrix} 0 \\ 3 \\ \frac{1}{2} \end{pmatrix}$$

This system is represented by

$$\left(\begin{array}{ccc|c} 1 & 2 & 1 & 0 \\ 0 & -2 & 1 & 3 \\ 0 & 0 & \frac{1}{2} & \frac{1}{2} \end{array}\right)$$

which we know is solved by *back substitution*: $y_3 = 1$ so $-2x_2 + 1 = 3$, and $x_2 = -1$. Finally $x_1 - 2(1) + 1 = 0$, so $x_1 = 1$ and the solution is

$$X = \begin{pmatrix} 1 \\ -1 \\ 1 \end{pmatrix}$$

Although the LU decomposition gives us a way to solve $AX = B$, we have not shown how to find L and U or whether they exist for a given matrix A. We will do this in a moment. First we state some situations in which the LU decomposition method is useful.

1. It is useful in solving $AX = B$ for several choices of B. Matrices L and U can be stored in the computer, and the two-step process

$$\text{Solve} \quad LY = B$$

$$\text{Solve} \quad UX = Y$$

involves only forward and back substitution, which is fast computationally. Thus $AX = B_1, AX = B_2, \ldots, AX = B_n$ can be solved quite quickly.

2. If A is tridiagonal, as often happens in applications to boundary-value problems for differential equations, the LU decomposition of A can be found easily and with very little error propagation. Therefore it is usually the method of choice for tridiagonal systems.

3. If A is *symmetric* and *positive definite* (A is defined to be positive definite if $X^TAX > 0$ for all $X \neq 0$), then a certain LU decomposition is very simple to obtain. It is

$$A = LL^T$$

and it requires less storage space than the general LU decomposition. The associated solution method is known as *Cholesky's method.* A necessary and sufficient condition for a symmetric matrix to be positive definite is given later in this section.

To find L and U in an LU decomposition of $A_{n\times n}$, we can try to use row reduction. To see this, we use elementary matrices. Upon obtaining an upper triangular form R for A by using m row operators of the form $aR_i + R_j$, where $i < j$ (this is *possible only when nonzero pivots are obtained after each row operation*), we have

$$E_m \cdots E_2 E_1 A = R$$

Now R is upper triangular and E_k is lower triangular, $k = 1, 2, \ldots, m$. Also E_k^{-1} is lower triangular, so

$$A = \underbrace{E_1^{-1}E_2^{-1}\cdots E_m^{-1}}_{\text{Lower triangular}}\ \underbrace{R}_{\text{Upper triangular}}$$

$$= \quad L \qquad U$$

Therefore,

> Given an $n \times n$ matrix A, A can be written as LU provided only nonzero pivots are obtained in reducing A to upper triangular form while using only row operations of the type $aR_i + R_j$.

EXAMPLE 2 Find an LU decomposition for the matrix from Example 1.

$$\begin{pmatrix} 1 & 2 & 1 \\ 2 & 2 & 3 \\ -1 & -3 & 0 \end{pmatrix} \xrightarrow[1R1 + R3]{-2R1 + R2} \begin{pmatrix} 1 & 2 & 1 \\ 0 & -2 & 1 \\ 0 & -1 & 1 \end{pmatrix} \xrightarrow{-\frac{1}{2}R2 + R3} \begin{pmatrix} 1 & 2 & 1 \\ 0 & -2 & 1 \\ 0 & 0 & \frac{1}{2} \end{pmatrix} = U$$

Calculating E_1, E_2, and E_3, we have

$$E_1^{-1}E_2^{-1}E_3^{-1} = \begin{pmatrix} 1 & 0 & 0 \\ +2 & 1 & 0 \\ -1 & +\frac{1}{2} & 1 \end{pmatrix} = L$$

Notice that the negatives of the multipliers in each row operation are elements of L. In fact, if we use $aR_j + R_k$, then $-a$ goes in the jk slot of L. We use this in the next example.

Checking our work, we have

$$LU = \begin{pmatrix} 1 & 0 & 0 \\ 2 & 1 & 0 \\ -1 & \frac{1}{2} & 1 \end{pmatrix}\begin{pmatrix} 1 & 2 & 1 \\ 0 & -2 & 1 \\ 0 & 0 & \frac{1}{2} \end{pmatrix} = \begin{pmatrix} 1 & 2 & 1 \\ 2 & 2 & 3 \\ -1 & -3 & 0 \end{pmatrix}$$

EXAMPLE 3 Find an LU decomposition for

$$A = \begin{pmatrix} 1 & -1 & 0 & 0 & 0 \\ -1 & 3 & -1 & 0 & 0 \\ 0 & -1 & 1 & -1 & 0 \\ 0 & 0 & -1 & 1 & -1 \\ 0 & 0 & 0 & -1 & 1 \end{pmatrix}$$

Solution Row-reducing gives

$$\begin{pmatrix} 1 & -1 & 0 & 0 & 0 \\ -1 & 3 & -1 & 0 & 0 \\ 0 & -1 & 1 & -1 & 0 \\ 0 & 0 & -1 & 1 & -1 \\ 0 & 0 & 0 & -1 & 1 \end{pmatrix} \xrightarrow[\substack{\mathbf{0}\,R1 + R3 \\ \mathbf{0}\,R1 + R4 \\ \mathbf{0}\,R1 + R5}]{\mathbf{+1}\,R1 + R2} \begin{pmatrix} 1 & -1 & 0 & 0 & 0 \\ 0 & 2 & -1 & 0 & 0 \\ 0 & -1 & 1 & -1 & 0 \\ 0 & 0 & -1 & 1 & -1 \\ 0 & 0 & 0 & -1 & 1 \end{pmatrix}$$

1st column of L below diagonal formed by negatives of these

$$\xrightarrow[\substack{\mathbf{0}\,R2 + R4 \\ \mathbf{0}\,R2 + R5}]{\mathbf{+\frac{1}{2}}\,R2 + R3} \begin{pmatrix} 1 & -1 & 0 & 0 & 0 \\ 0 & 2 & -1 & 0 & 0 \\ 0 & 0 & \frac{1}{2} & -1 & 0 \\ 0 & 0 & -1 & 1 & -1 \\ 0 & 0 & 0 & -1 & 1 \end{pmatrix}$$

2d column of L below diagonal formed by negatives of these

$$\xrightarrow[\mathbf{0}\,R3 + R5]{\mathbf{+2}\,R3 + R4} \begin{pmatrix} 1 & -1 & 0 & 0 & 0 \\ 0 & 2 & -1 & 0 & 0 \\ 0 & 0 & \frac{1}{2} & -1 & 0 \\ 0 & 0 & 0 & -1 & -1 \\ 0 & 0 & 0 & -1 & 1 \end{pmatrix}$$

3d column of L below diagonal formed by negatives of these

$$\xrightarrow{\mathbf{-1}\,R4 + R5} \begin{pmatrix} 1 & -1 & 0 & 0 & 0 \\ 0 & 2 & -1 & 0 & 0 \\ 0 & 0 & \frac{1}{2} & -1 & 0 \\ 0 & 0 & 0 & -1 & -1 \\ 0 & 0 & 0 & 0 & 2 \end{pmatrix} = U$$

4th column of L below diagonal formed by negatives of this

Therefore,

$$L = \begin{pmatrix} 1 & 0 & 0 & 0 & 0 \\ \mathbf{-1} & 1 & 0 & 0 & 0 \\ \mathbf{0} & \mathbf{-\frac{1}{2}} & 1 & 0 & 0 \\ \mathbf{0} & \mathbf{0} & \mathbf{-2} & 1 & 0 \\ \mathbf{0} & \mathbf{0} & \mathbf{0} & \mathbf{1} & 1 \end{pmatrix}$$

EXAMPLE 4 Solve Example 6 of Sec. 1.3 by using LU decomposition. Always round to four significant figures.

Solution We work as in Example 3:

$$\begin{pmatrix} 0.0001 & 1 \\ 1 & -1 \end{pmatrix} \xrightarrow{-10{,}000R1 + R2} \begin{pmatrix} 0.0001 & 1 \\ 0 & -10{,}000 \end{pmatrix} = U$$

$$L = \begin{pmatrix} 1 & 0 \\ 10{,}000 & 1 \end{pmatrix}$$

Therefore,

$$\begin{pmatrix} 0.0001 & 1 \\ 1 & -1 \end{pmatrix} \doteq \begin{pmatrix} 1 & 0 \\ 10{,}000 & 1 \end{pmatrix} \begin{pmatrix} 0.0001 & 1 \\ 0 & -10{,}000 \end{pmatrix}$$
$$= \begin{pmatrix} 0.0001 & 1 \\ 1 & 0 \end{pmatrix}$$

We note that roundoff error is already affecting the problem since LU is not exactly equal to A. Now we solve

$$LY = B = \begin{pmatrix} 1 \\ 0 \end{pmatrix}$$

and

$$UX = Y$$

For $LY = B$, we have

$$\left(\begin{array}{cc|c} 1 & 0 & 1 \\ 10{,}000 & 1 & 0 \end{array}\right) \rightarrow \begin{aligned} y_1 &= 1 \\ y_2 &= -10{,}000 \end{aligned}$$

and for $UX = Y$,

$$\left(\begin{array}{cc|c} 0.0001 & 1 & 1 \\ 0 & -10{,}000 & -10{,}000 \end{array}\right) \rightarrow x_2 = 1 \qquad x_1 = 0$$

which is the same answer as before.

Example 4 shows also that LU decomposition may need modification with a pivoting strategy; the interested reader should see *Computer Solution of Linear Algebraic Systems* by G. E. Forsythe and C. B. Moler (Prentice-Hall, Englewood Cliffs, N.J., 1967).

Roughly speaking, for solving one system $AX = B$ where A does not have many zero entries, gaussian elimination with partial pivoting is the method of choice. It is a "best buy" in the sense that you get good error control without spending too much computer time. If A is special (tridiagonal or positive definite and symmetric), then LU decomposition is faster than elimination and does not accumulate error as much. If we are solving $AX = B$ for several B's and A has an LU decomposition, then LU decomposition is the method of choice because it is faster to find L and U once and for all and then use the two-step process on each B.

Ill-Conditioned Matrices There is a class of matrices which are called *ill-conditioned.* If A is ill-conditioned, solving

$$AX = B$$

is difficult because small errors in rounded terms of B get increased many times. For example,

$$\begin{aligned} 9x + 8y &= 0.8 \\ 8x + 7y &= 0.7 \end{aligned} \tag{1.8.1}$$

has solution $x = 0$, $y = 0.1$. If we introduce an error of 0.01 in the right-hand sides and solve

$$\begin{aligned} 9x + 8y &= 0.81 \\ 8x + 7y &= 0.69 \end{aligned} \tag{1.8.2}$$

we find (using exact arithmetic) that

$$x = -0.15 \qquad y = 0.27$$

We see that the error of 0.01 in the right-hand side of (1.8.1) led to an error of 0.17 (17 times as great as the error in B) in the calculation of y. There are some special techniques for solving systems with ill-conditioned matrices. However, these techniques are beyond the scope of this text.

Checking the Computer's Answer To see whether a computed solution X_{comp} is a good solution to $AX = B$, we could calculate the *residual vector* $\mathbf{R}$ defined by

$$\mathbf{R} = AX_{\text{comp}} - B$$

and calculate its *magnitude* $|\mathbf{R}| = \sqrt{r_1^2 + r_2^2 + \cdots + r_n^2}$. If $|\mathbf{R}|$ is *small,* we might expect X_{comp} to be close to the actual solution. After all, if $\boldsymbol{R} \doteq 0$ (that is, if $|\boldsymbol{R}| \doteq 0$), then

$$AX_{\text{comp}} - B \doteq 0$$

and

$$AX_{\text{comp}} \doteq B$$

(Remember that $\doteq$ means "approximately equal to.")

Generally, calculating the residual and determining whether $|\boldsymbol{R}|$ is small (for example, $|\boldsymbol{R}| < 0.00001$) can be a good way to have the computer check its own answer. However, if A is ill-conditioned, $AX_{\text{comp}} - B$ (the residual) can be small even if X_{comp} is not close to the solution. For example, if the solution $x = -0.15$, $y = 0.27$ of (1.8.2) is proposed as the solution of (1.8.1), the residual is

$$\boldsymbol{R} = \begin{pmatrix} 9 & 8 \\ 8 & 7 \end{pmatrix} \begin{pmatrix} -0.15 \\ 0.27 \end{pmatrix} - \begin{pmatrix} 0.8 \\ 0.7 \end{pmatrix} = \begin{pmatrix} 0.01 \\ -0.01 \end{pmatrix}$$

Notice that $|\boldsymbol{R}| = \sqrt{(0.01)^2 + (0.01)^2} \doteq 0.014$. Thus the residual is small, but X_{comp} is not close to

$$X = \begin{pmatrix} 0 \\ 0.1 \end{pmatrix}$$

In fact,

$$|X_{\text{comp}} - X| = \left| \begin{pmatrix} -0.15 \\ 0.17 \end{pmatrix} \right| = 0.227$$

The preceding example and discussion support this advice: In solving $AX = B$, if A is ill-conditioned, treat computed answers with healthy suspicion. Detailed treatment of ill-conditioned systems can be found in numerical analysis texts or texts such as *Computational Methods of Linear Algebra*, by D. K. Fadeev and V. N. Fadeeva (Freeman, San Francisco, 1963).

Gauss-Seidel Iteration In some applications in physics and engineering, a system $AX = B$ must be solved in which A is *sparse*. A matrix is *sparse* if most of its entries are zeros. For example,

$$\begin{pmatrix} 0 & 0 & 1 & 0 \\ 0 & 0 & 0 & 1 \\ 1 & 0 & 0 & 0 \\ 0 & 0 & 1 & 0 \end{pmatrix}$$

is a sparse matrix. We do not quantify the word *most*, but certainly more than two-thirds of the entries of A should be zero for A to qualify as sparse.

Gaussian elimination is wasteful for systems with sparse A. Most of the row operations eliminate no variables, but the computer carries out the operations anyway. Therefore methods other than those of the last two sections have been developed for sparse systems. The Gauss-Seidel method is one of these.

The idea behind the Gauss-Seidel method is *iteration*—performing a sequence of similar calculations whose results get closer and closer to a desired number. For example, if we define

$$x_{n+1} = \frac{x_n}{2} + \frac{1}{x_n} \qquad n = 1, 2, 3, \ldots$$

and put $x_1 = 1$, we find

$$x_2 = \frac{x_1}{2} + \frac{1}{x_1} = \frac{1}{2} + \frac{1}{1} = 1.5$$

$$x_3 = \frac{x_2}{2} + \frac{1}{x_2} = \frac{1.5}{2} + \frac{1}{1.5} = 1.4166666\cdots$$

$$x_4 = \frac{x_3}{2} + \frac{1}{x_3} = 0.708333 + 0.7058823\cdots = 1.4142156\cdots$$

And we see that the sequence of calculated values seems to be homing in on $\sqrt{2}$. In fact, this is true (it is Newton's method from calculus), and the proof of convergence is found in most texts on numerical analysis. The spirit of the Gauss-Seidel method is similar.

In the Gauss-Seidel method, we proceed as follows: Let

$$A = \begin{pmatrix} a_{11} & \cdots & a_{1n} \\ \cdots & \cdots & \cdots \\ a_{n1} & \cdots & a_{nn} \end{pmatrix}$$

with $a_{11} \neq 0, \ldots, a_{nn} \neq 0$. If A is invertible, interchange of rows can always achieve $a_{kk} \neq 0$ (see the problems).

1. Write

$$A = \underbrace{\begin{pmatrix} a_{11} & 0 & \cdots & & 0 \\ a_{21} & a_{22} & 0 & \cdots & 0 \\ & & \ddots & \ddots & \vdots \\ \vdots & \vdots & & \ddots & 0 \\ a_{n1} & a_{n2} & \cdots & & a_{nn} \end{pmatrix}}_{N} - \underbrace{\begin{pmatrix} 0 & -a_{12} & -a_{13} & \cdots & -a_{1n} \\ 0 & 0 & -a_{23} & \cdots & \\ \vdots & & \ddots & \ddots & \vdots \\ 0 & \cdots & & 0 & -a_{n-1,n} \\ 0 & \cdots & & 0 & 0 \end{pmatrix}}_{P}$$

2. Write $AX = B$ as $(N - P)X = B$, and rewrite it as

$$NX = B + PX$$

3. Let X_1 be any $n \times 1$ column matrix, and define the iteration step by

$$NX_m = B + PX_{m-1} \qquad m = 2, 3, 4, \ldots \tag{1.8.3}$$

That is, solve (1.8.3) for X_m, once X_{m-1} is calculated. Since N is invertible, (1.8.3) will always be solvable. Notice here that we use the facts that $\det N = a_{11}a_{22}\cdots a_{nn} \neq 0$ and Theorem 1.7.5 to obtain the existence of N^{-1} and unique solvability of $NX = Y$. That is, *the theoretical results of Sec. 1.7 are quite useful to us now in developing a numerical method.*

4. Observe the behavior of the sequence $\{X_n\}$.

EXAMPLE 5 Use the Gauss-Seidel method to solve

$$AX = \begin{pmatrix} 2 & -1 & 0 \\ -1 & 2 & -1 \\ 0 & -1 & 2 \end{pmatrix} \begin{pmatrix} x_1 \\ x_2 \\ x_3 \end{pmatrix} = \begin{pmatrix} 1 \\ 0 \\ 1 \end{pmatrix} = B$$

Solution By step 1,

$$N = \begin{pmatrix} 2 & 0 & 0 \\ -1 & 2 & 0 \\ 0 & -1 & 2 \end{pmatrix} \quad \text{and} \quad P = \begin{pmatrix} 0 & 1 & 0 \\ 0 & 0 & 1 \\ 0 & 0 & 0 \end{pmatrix}$$

Let

$$X_1 = \begin{pmatrix} 1 \\ 2 \\ 3 \end{pmatrix}$$

and set up the iterative step (1.8.3) to obtain

$$NX_2 = B + PX_1 = \begin{pmatrix} 1 \\ 0 \\ 1 \end{pmatrix} + \begin{pmatrix} 0 & 1 & 0 \\ 0 & 0 & 1 \\ 0 & 0 & 0 \end{pmatrix} \begin{pmatrix} 1 \\ 2 \\ 3 \end{pmatrix} = \begin{pmatrix} 3 \\ 3 \\ 1 \end{pmatrix}$$

Solving the equation

$$\begin{pmatrix} 2 & 0 & 0 \\ -1 & 2 & 0 \\ 0 & -1 & 2 \end{pmatrix} X_2 = \begin{pmatrix} 3 \\ 3 \\ 1 \end{pmatrix}$$

by forward substitution gives

$$X_2 = \begin{pmatrix} \frac{3}{2} \\ \frac{9}{4} \\ \frac{13}{8} \end{pmatrix} = \begin{pmatrix} 1.5 \\ 2.25 \\ 1.625 \end{pmatrix}$$

Having X_2, we now solve for X_3:

$$NX_3 = B + PX_2$$

The solution is

$$X_3 = \begin{pmatrix} 1.625 \\ 1.625 \\ 1.3125 \end{pmatrix}$$

Continuing in this way, we find

$$X_4 = \begin{pmatrix} 1.3125 \\ 1.3125 \\ 1.15625 \end{pmatrix}$$

$$X_5 = \begin{pmatrix} 1.15625 \\ 1.15625 \\ 1.07813 \end{pmatrix}$$

$$X_6 = \begin{pmatrix} 1.07813 \\ 1.07813 \\ 1.03906 \end{pmatrix}$$

$$\vdots$$

$$X_{15} = \begin{pmatrix} 1.00015 \\ 1.00015 \\ 1.00008 \end{pmatrix}$$

Since the actual solution is

$$\begin{pmatrix} 1 \\ 1 \\ 1 \end{pmatrix}$$

we see that the sequence $X_1, X_2, X_3, \ldots$ is getting closer to the actual solution.

A basic result for the Gauss-Seidel method is as follows:

Let A be symmetric, positive definite, with all positive diagonal elements. With N and P chosen as above, the sequence $X_1, X_2, \ldots, X_n, \ldots$ defined by the iteration step (1.8.3) converges to the solution of $AX = B$, for *any* choice of X_1. (That is, as m increases, the numbers $|X_m - X|$ approach zero.)

The theorem says that if A is symmetric and positive definite, with $a_{kk} > 0$, $k = 1, 2, \ldots, n$, then the Gauss-Seidel sequence of approximations converges to the solution of $AX = B$ for *any* choice of X_1. However, the closer X_1 is to the solution, the faster the sequence $X_1, X_2, \ldots$ converges to the solution. For example, if in Example 1 we chose

$$X_1 = \begin{pmatrix} -300 \\ 2000 \\ -500 \end{pmatrix}$$

we would have to compute X_{23} to have an approximate solution comparable to the previous one. In fact, in this case we would have

$$X_{22} = \begin{pmatrix} 1.00023 \\ 1.00023 \\ 1.00017 \end{pmatrix} \qquad X_{23} = \begin{pmatrix} 1.00012 \\ 1.00012 \\ 1.00006 \end{pmatrix}$$

Since many of the sparse matrices in applications satisfy the hypotheses of this theorem, it is an important result. The proof requires methods which we have not developed.

A necessary and sufficient condition for a *symmetric* $n \times n$ matrix to be positive definite is

$$a_{11} > 0$$

$$\det\begin{pmatrix} a_{11} & a_{12} \\ a_{21} & a_{22} \end{pmatrix} > 0$$

$$\det\begin{pmatrix} a_{11} & a_{12} & a_{13} \\ a_{21} & a_{22} & a_{23} \\ a_{31} & a_{32} & a_{33} \end{pmatrix} > 0$$

$$\vdots$$

$$\det\begin{pmatrix} a_{11} & \cdots & a_{1n} \\ \cdots & \cdots & \cdots \\ a_{n1} & \cdots & a_{nn} \end{pmatrix} > 0$$

EXAMPLE 6 Show that

$$A = \begin{pmatrix} 2 & -1 & 0 \\ -1 & 2 & -1 \\ 0 & -1 & 2 \end{pmatrix}$$

is positive definite. Note that A is symmetric.

Solution Since $a_{11} = 2 > 0$ and

$$\det\begin{pmatrix} 2 & -1 \\ -1 & 2 \end{pmatrix} = 3 > 0 \quad \text{and} \quad \det\begin{pmatrix} 2 & -1 & 0 \\ -1 & 2 & -1 \\ 0 & -1 & 2 \end{pmatrix} = 4 > 0$$

matrix A is positive definite. Since A is symmetric and positive definite with positive diagonal entries, we can say that the Gauss-Seidel approximations in Example 5 are guaranteed to converge.

Other iterative methods exist which modify the Gauss-Seidel method to make the sequence $X_1, X_2, \ldots, X_n, \ldots$ approach the solution more quickly than Gauss-Seidel alone. One method which is particularly effective is the *successive overrelaxation* (SOR) *method.* The SOR method is explained in most texts on numerical analysis.

PROBLEMS 1.8

1. For the coefficient matrices from Probs. 5, 6, and 7 of Sec. 1.3, calculate LU decompositions.

2. Solve the systems in Probs. 5, 6, and 7 of Sec. 1.3 by using the LU decomposition.

3. It can be proved that if a matrix is strictly row-diagonally dominant, then the Gauss-Seidel method will converge. A matrix $A_{n \times n}$ is *strictly row-diagonally dominant* (SRD) if the absolute value of the diagonal element in every row is greater than the sum of the absolute values of the other elements in the same row. For example,

$$\begin{pmatrix} -3 & 1 & 0 \\ 5 & 10 & 2 \\ -1.99 & 2 & -4 \end{pmatrix}$$

is SRD since

$$\begin{aligned} |-3| &> |1| + |0| \\ |10| &> |5| + |2| \\ |-4| &> |-1.99| + |2| \end{aligned}$$

Determine which of the following matrices are SRD.

(a) $\begin{pmatrix} 2 & -1 & 0 \\ -1 & 2 & -1 \\ 0 & -1 & 2 \end{pmatrix}$ (b) Any diagonal matrix

(c) Any skew-symmetric matrix (d) $\begin{pmatrix} 3 & 1 & 0 & 1 \\ 1 & 4 & -2 & 0.5 \\ 2 & 2 & 6 & 1 \\ -5 & -6 & 1 & 15 \end{pmatrix}$

4. Is the sum of SRD matrices SRD?

5. Is the product of SRD matrices SRD?

6. Is the inverse of an SRD matrix SRD?

7. Show that

$$A = \begin{pmatrix} 1 & 1 & -1 \\ 2 & 3 & -5 \\ 3 & 5 & -9 \end{pmatrix}$$

has an LU decomposition. (Note: A is singular.)

8. Apply the Gauss-Seidel method to the following systems.

(a) $\begin{pmatrix} 4 & -1 & 0 & -1 \\ -1 & 4 & -1 & 0 \\ 0 & -1 & 4 & 0 \\ -1 & 0 & 0 & 4 \end{pmatrix} \begin{pmatrix} x_1 \\ x_2 \\ x_3 \\ x_4 \end{pmatrix} = \begin{pmatrix} 2 \\ 2 \\ 3 \\ 3 \end{pmatrix}$

Use $X_0 = 0$. (This is a small-scale version of problems arising in partial differential equations.)

(*b*)
$$\begin{pmatrix} 3 & -1 & 0 \\ -1 & 2 & -1 \\ 0 & -1 & 2 \end{pmatrix}\begin{pmatrix} x_1 \\ x_2 \\ x_3 \end{pmatrix} = \begin{pmatrix} 1 \\ 5 \\ -6 \end{pmatrix}$$

Matrix A is not quite SRD. However, A is positive definite with positive diagonal entries.

9. We know that the Gauss-Seidel method will work for $AX = B$ if A is SRD or if A is symmetric positive definite with positive diagonal entries.

(*a*) Give 2×2 and 3×3 examples of matrices which are SRD but are not positive definite with positive diagonal entries.

(*b*) Give 2×2 and 3×3 examples of matrices which are *not* SRD but are positive definite with positive diagonal entries.

10. Show that an invertible matrix $A_{n \times n}$ has a row-equivalent form B, with $b_{11}, b_{22}, \ldots, b_{nn}$ all nonzero, where B is obtained from A entirely by row interchanges. (*Hint*: If no row interchange can make $b_{kk} \neq 0$ for some k, what can you say about column k and det A?)

11. Give, by inspection, two different LU decompositions for I, thereby showing that the LU decomposition for a matrix is not unique unless further restrictions are put on the entries of L or U.

SUMMARY

The first and most basic problem of linear algebra is solving systems of linear equations. *Elimination of variables* is a method for solving linear equations which can be done by hand or implemented on a computer. When a system of equations is solved by using a computer, the phenomenon of *roundoff error* is encountered. The strategy of *partial pivoting* helps reduce the accumulation of roundoff error in the calculation of solutions by elimination of variables. Two other numerical methods utilize the *LU decomposition* and *Gauss-Seidel iteration.*

Matrices were used to store coefficients and right-hand sides of systems of linear equations so that elimination of variables could be carried out in an orderly way. *Gaussian elimination* led to *row echelon forms* of matrices, and Gauss-Jordan elimination to *reduced* row echelon forms. The echelon forms and the equivalent equations which they represented led us to various theorems about the solution of linear equations. However, the road to these theorems was paved by the concepts and results of *matrix algebra.* In particular, the ideas of *matrix multiplication, determinants,* and *matrix inverses* were extremely important. Moreover, *special matrices* such as *identity, zero, triangular,* and *symmetric* matrices were introduced.

The theory of linear equations developed in this way can be summarized as follows:

Let $A_{m \times n}$ and $B_{n \times 1}$ be given.

1. Let $n = m$; that is, A is square. Then $AX = B$ has a unique solution if and only if A^{-1} exists; $AX = B$ has a unique solution if and only if $\det A \neq 0$; and $AX = B$ has a unique solution if and only if A is row-equivalent to I.
2. Let $m \leq n$, $B = 0$. If $m = n$, then $AX = 0$ has a nontrivial solution if and only if $\det A = 0$. If $m < n$, then $AX = 0$ has a nontrivial solution.

The reader will see, as the text progresses, that the ability to solve both *homogeneous* and *nonhomogeneous* systems of linear equations is absolutely necessary to solve the more advanced problems of linear algebra. Matrices will play an important role as representors of certain functions (called *linear*) which arise in applied mathematics. A sampling of these developments is given in Chap. 2, in which we stay in the plane or space, delaying a journey into a more general setting until Chap. 3.

ADDITIONAL PROBLEMS

1. Given a system of linear equations which has a unique solution, can one always adjoin another equation so that the resulting system has no solution?
2. Given a system of linear equations which has a unique solution, can one always adjoin another equation so that the resulting system has an infinite number of solutions?
3. If a consistent system of linear equations $A_{n \times n}X_{n \times 1} = B_{n \times 1}$ has A and B with all real entries, must X have all real entries?
4. If the consistent system in Prob. 3 has A and B with all complex entries with nonzero imaginary parts, must X have entries with nonzero imaginary parts?
5. If A and B are invertible, is $A + iB$ invertible? What if A and B are required to be real?
6. A matrix $A_{n \times n}$ dominates $B_{n \times n}$ if $a_{ij} \geq b_{ij}$ for all i, j, and in this case we write $A \geq B$. If $A \geq B$, must it be true that $A^2 \geq B^2$? What if we require $B \geq 0$?
7. If $A \geq B$, is $\det A \geq \det B$?
8. A square matrix A is said to have a real square root if there exists a matrix B (real) such that $A = B^2$. For example, both I and $-I$ are square roots of I. Give an example of a real matrix which has no real square root. If a matrix has a real square root, what can you say about its determinant? If a matrix has all positive entries, must it have a real square root?
9. Let A be $n \times n$. Show that $(A + I_n)$ and $(A - I_n)$ commute.

10. *Ranking of competitors.* Suppose we have a number of competitors for a prize. We can use matrix A to represent relative pairwise strengths by putting $a_{ij} = 1$ and $a_{ji} = 0$ if competitor i is stronger than competitor j. We put $a_{ii} = 0$. Matrix A is called a *dominance matrix.* So if we have three competitors with 1 stronger than 2, 3 stronger than 1, and 2 stronger than 3, we have the dominance matrix

$$A = \begin{pmatrix} 0 & 1 & 0 \\ 0 & 0 & 1 \\ 1 & 0 & 0 \end{pmatrix}$$

In general, for n competitors if we form

$$S = A + A^2 + \cdots + A^{n-1}$$

then the sum of row i of S is called the *power* of competitor i. So for the competitors above, we have $S = A + A^2$. Show that the power of each individual is equal to 2 (no competitor will dominate). Suppose now that we have three competitors with 1 stronger than 2, 1 stronger than 3, and 3 stronger than 2. Compute the power of each competitor.

11. Four wrestlers are to compete for the world free-for-all championship. The dominance matrix is determined from past matches; 1 is stronger than either 2 or 3, 2 is stronger than 4, 3 is stronger than 2, and 4 is stronger than either 1 or 3. Compute the power of each competitor, and predict a winner.

12. For the circuit with resistor of R_1 ohms in Fig. P1.12a the matrix is

$$A = \begin{pmatrix} 1 & R_1 \\ 0 & 1 \end{pmatrix}$$

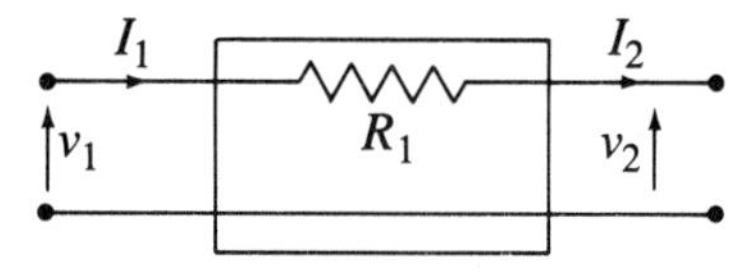

Figure P1.12*a*

For the circuit in Fig. P1.12b the matrix is

$$B = \begin{pmatrix} 1 & 0 \\ \dfrac{1}{R_2} & 1 \end{pmatrix}$$

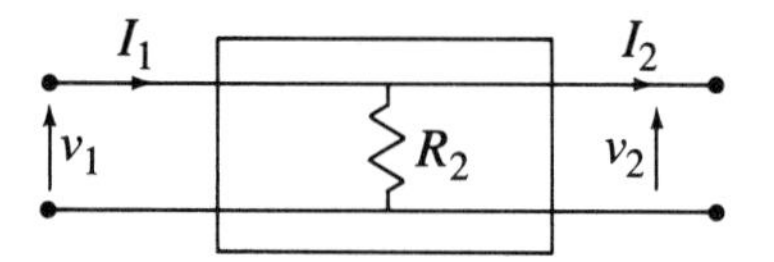

Figure P1.12*b*

If these circuits are put together in the order A then B from left to right, as in Fig. P1.12c, the representing matrix for the combined circuit is AB.

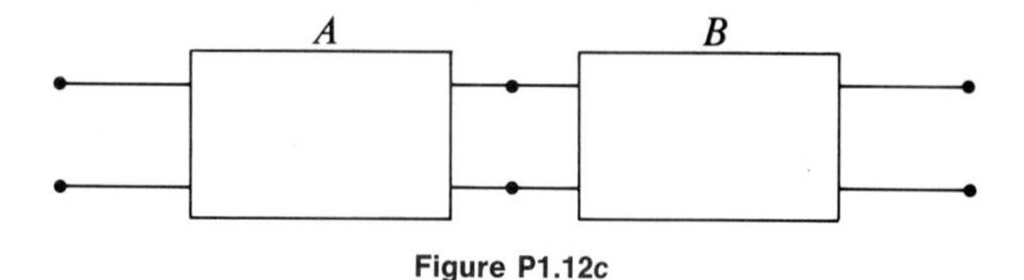

Figure P1.12c

Calculate AB. It is the matrix for the circuit shown in Fig. P1.12d.

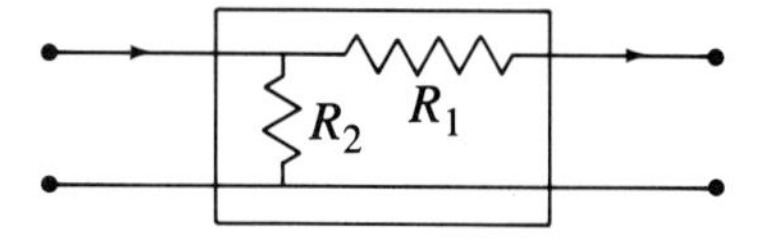

Figure P1.12d

13. Regarding Prob. 12, suppose the boxes are put together in the order B then A from left to right. The representing matrix for this circuit is BA. Is $AB = BA$? What does this say about the two combined circuits? Are they equivalent?

14. Associated with microwave transmission is a matrix called a *scattering matrix* S. It is of the form

$$S = (A - I)(A + I)^{-1}$$

The entries of A are determined by voltages and currents at a microwave junction, and A is symmetric. Show that S is symmetric.

15. If blackboxes with 2×2 representing matrices are connected in parallel, as in Fig. P1.15 then the matrix of the combined circuit is $A + B$. Calculate the matrix of the circuit, using A and B from Prob. 12. What does the commutativity of matrix addition say about the construction of parallel circuits?

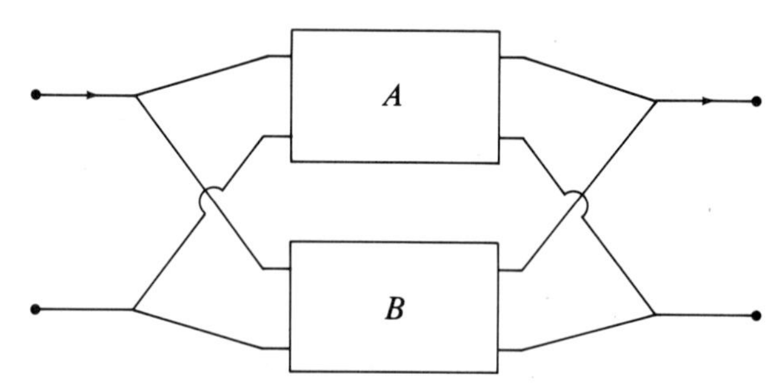

Figure P1.15

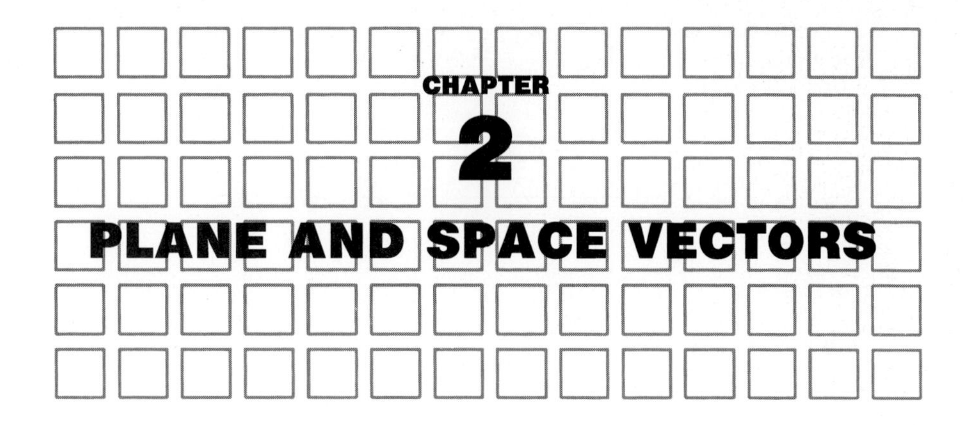

2.1 VECTORS

In chemical engineering one is interested in the concentrations of a chemical species at different levels of an absorption column (see Fig. 2.1.1). To record the concentrations at a certain time, we can write

$$\mathbf{C} = (c_1, c_2, \ldots, c_n)$$

where c_k is the concentration at the kth level. We can call $\mathbf{C}$ a *concentration vector*; it is just a $1 \times n$ matrix.

DEFINITION 2.1.1 An *n-vector* is a $1 \times n$ matrix.

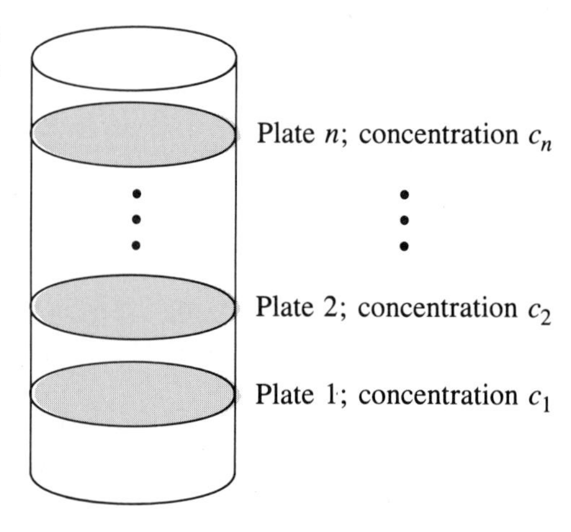

Figure 2.1.1
Absorption column

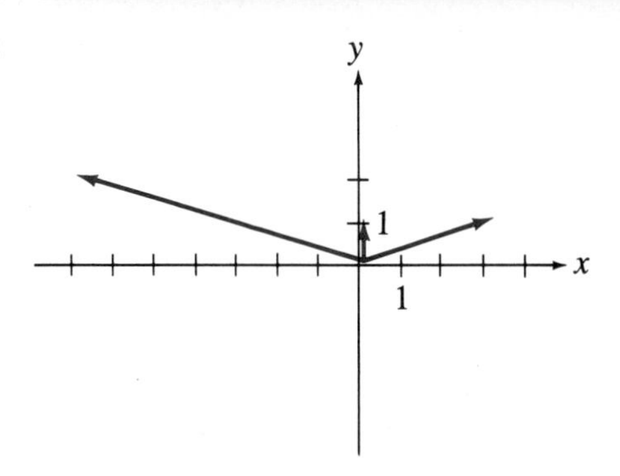

Figure 2.1.2
The vectors (0, 1), $(\pi, 1)$, $(-7, 2)$.

Before developing the algebra of n-vectors in Chap. 3, we study two-vectors and three-vectors because their structure can be visualized geometrically.

DEFINITION 2.1.2 A *plane vector* (or two-vector) is a 1×2 matrix with real entries. A *space vector* (or three-vector) is a 1×3 matrix with real entries. The set of all plane vectors is denoted $\mathbb{R}^2$; the set of all space vectors, $\mathbb{R}^3$.

The elements of the matrix are called the *components* of the vector. Two vectors are *equal* if they are equal as matrices.

EXAMPLE 1 The following are plane vectors: (0, 1), $(\pi, 1)$, $(-7, 2)$. Commas are placed between the components to emphasize the fact that plane vectors are associated with points in the standard xy plane (see Fig. 2.1.2). If we connect the point to the origin with an arrow (see Fig. 2.1.2), we call the arrow a *geometric representation* of the vector.

EXAMPLE 2 The following are space vectors: $(1, 1, -1)$, $(2, 0, 1)$, $(-1, 1, 1)$. We associate these vectors with points in the standard xyz coordinate system, as shown in Fig. 2.1.3.

A plane vector (a, b) is also associated with any arrow in the standard xy plane which begins at a point (x_0, y_0) and ends at $(x_0 + a, y_0 + b)$. In Fig. 2.1.4 we have drawn several arrows associated with the vector (1, 2).

In general, when we write a vector (a, b), we assume that the initial point of the associated vector is (0, 0) unless specified otherwise.

EXAMPLE 3 Find the vector associated with the arrow with initial point $(-2, 4)$ and terminal point $(-7, 3)$.

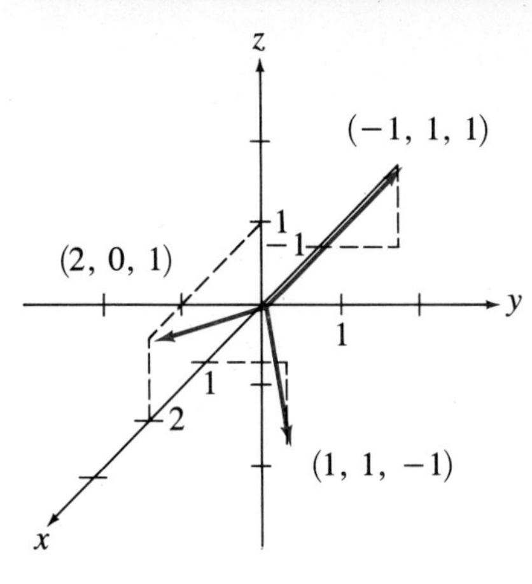

Figure 2.1.3
The vectors $(1, 1, -1)$, $(2, 0, 1)$, $(-1, 1, 1)$.

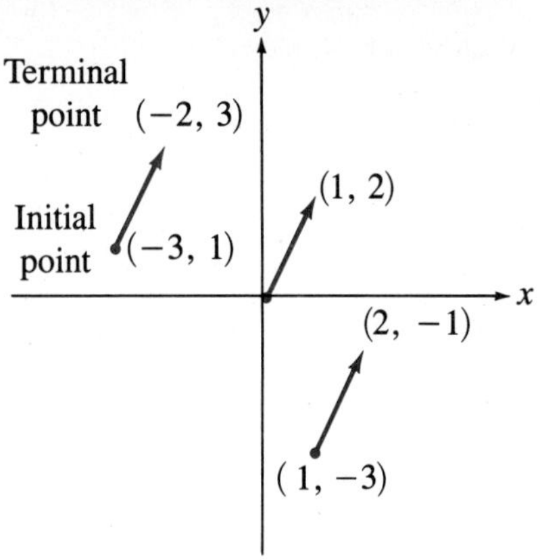

Figure 2.1.4
Copies of the same vector $(1, 2)$.

Solution Now $(-7, 3)$ must be equal to $(x_0 + a, y_0 + b)$. Since $(x_0, y_0) = (-2, 4)$, we have $(-7, 3) = (-2 + a, 4 + b)$. Thus $a = -5$, $b = -1$, so $(a, b) = (-5, -1)$. The easy way to solve this problem is to "subtract points": terminal point − initial point. In this way (just subtracting matrices)

$$(a, b) = (-7, 3) - (-2, 4) = (-5, -1)$$

See Fig. 2.1.5.

For space vectors we have similar methods and interpretations in the *xyz* coordinate system.

EXAMPLE 4 Sketch the arrow corresponding to the vector $(1, -1, 2)$ with initial point $(0, 0, 0)$ and with initial point $(1, 2, -1)$.

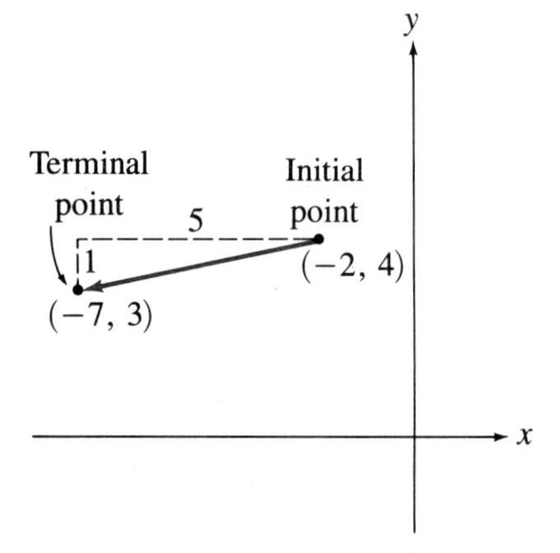

Figure 2.1.5
A version of the vector $(-5, -1)$.

Figure 2.1.6
Vectors in Example 4.

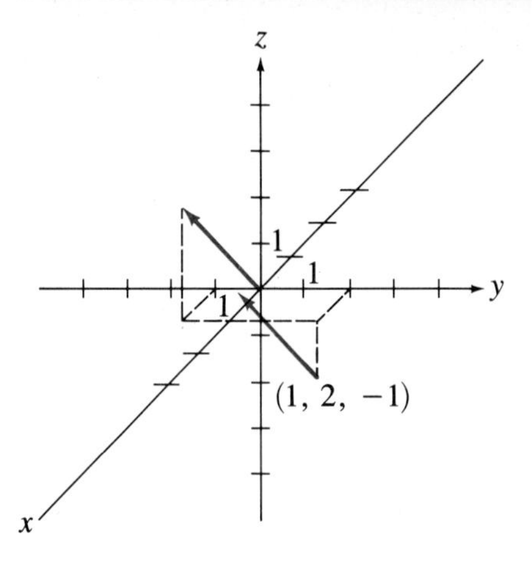

Solution We draw the standard x, y, and z axes, locate (0, 0, 0) and (1, −1, 2), and draw an arrow from (0, 0, 0) to (1, −1, 2). See Fig. 2.1.6. For the second case we locate the point (1, 2, −1) as the initial point. Let the terminal point be (a, b, c); then the vector (1, −1, 2) satisfies

$$(1, -1, 2) = (a, b, c) - (1, 2, -1)$$

This means that the terminal point is (2, 1, 1).

The length of the arrow representing a vector is defined to be the *length*, or *magnitude*, of the vector. Considering Fig. 2.1.4 again and using Pythagoras' theorem for right triangles, we find that the length of (1, 2) is $\sqrt{1^2 + 2^2} = \sqrt{5}$. So we have the following definition.

DEFINITION 2.1.3 The *length* (or *magnitude*) of the vector $\mathbf{A} = (a, b)$ is denoted $|A|$ or $|(a, b)|$ and is defined by

$$|A| = \sqrt{a^2 + b^2}$$

For three-vectors

$$|(a, b, c)| = \sqrt{a^2 + b^2 + c^2}$$

EXAMPLE 5 Calculate $|(0, 0)|$, $|(1, -1)|$, $|(3, 4)|$, $|(0, 1, 0)|$, and $|(\sqrt{11}, 3, 4)|$.

Solution

$$|(0, 0)| = \sqrt{0^2 + 0^2} = \sqrt{0} = 0$$
$$|(1, -1)| = \sqrt{1^2 + (-1)^2} = \sqrt{2}$$
$$|(3, 4)| = \sqrt{3^2 + 4^2} = \sqrt{25} = 5$$

$$|(0, 1, 0)| = \sqrt{0^2 + 1^2 + 0^2} = \sqrt{1} = 1$$
$$|(\sqrt{11}, 3, 4)| = \sqrt{(\sqrt{11})^2 + 3^2 + 4^2} = \sqrt{36} = 6$$

DEFINITION 2.1.4 The *sum* of vectors (a, b) and (c, d) is defined as

$$(a, b) + (c, d) = (a + c, b + d)$$

The *sum* of vectors (a, b, c) and (d, e, f) is defined as

$$(a, b, c) + (d, e, f) = (a + d, b + e, c + f)$$

EXAMPLE 6 Let $\mathbf{A} = (1, 0, -1)$, $\mathbf{B} = (3, -4, 2)$, $\mathbf{C} = (\pi, 0, 4)$, $\mathbf{D} = (0, 0, 0)$, and $\mathbf{E} = (-1, 0, 1)$. Calculate $\mathbf{A} + \mathbf{B}$, $\mathbf{B} + \mathbf{D}$, $\mathbf{A} + \mathbf{B} + \mathbf{C}$, and $\mathbf{A} + \mathbf{E}$.

Solution

$$\mathbf{A} + \mathbf{B} = (1 + 3, 0 + -4, -1 + 2) = (4, -4, 1)$$
$$\mathbf{B} + \mathbf{D} = (3 + 0, -4 + 0, 2 + 0) = (3, -4, 2) = \mathbf{B}$$
$$\mathbf{A} + \mathbf{B} + \mathbf{C} = (1 + 3 + \pi, 0 + (-4) + 0, -1 + 2 + 4) = (4 + \pi, -4, 5)$$
$$\mathbf{A} + \mathbf{E} = (1 + -1, 0 + 0, -1 + 1) = (0, 0, 0)$$

The definition of addition corresponds to the "addition" of forces represented by the vectors. In Fig. 2.1.7 we show the representation of two forces acting on a point at the origin; the sum of the two vectors is called the *resultant force*. The length of each vector corresponds to the magnitude of the force, and the direction to the direction of application of the force. The resultant vector is just the diagonal (having initial point in common with the two forces) of the parallelogram generated by the two forces.

Because vectors represent forces, some other concepts regarding forces trans-

Figure 2.1.7
Force diagram.

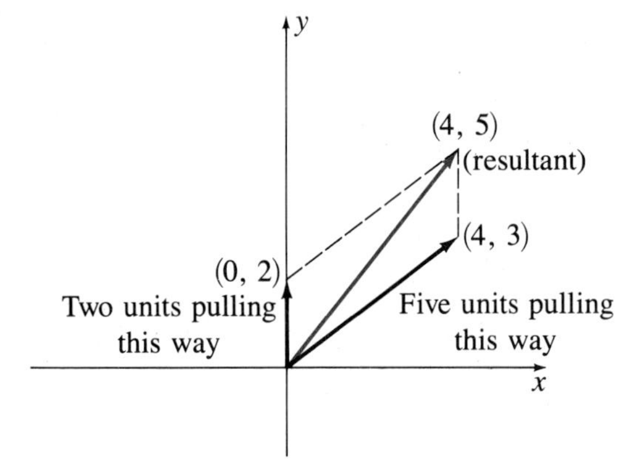

late to operations with vectors. For instance, if we pull "twice as hard" on an object, we think of doubling the force we are applying. If the original force is (a, b), the doubled force is $(2a, 2b)$. This leads to the definition of scalar multiple.

DEFINITION 2.1.5 Let r be a real number and (a, b) be a vector. The *scalar*[1] *multiple* $r(a, b)$ is defined as

$$r(a, b) = (ra, rb)$$

The reader should write a definition for a scalar multiple of (a, b, c). Note that this multiplication is the same as multiplication of a matrix by a scalar.

The idea of reversing a force or changing it to the opposite direction leads to a definition of the negative of a vector.

DEFINITION 2.1.6 The *negative* of a vector $\mathbf{V}$ is written $-\mathbf{V}$ and is defined as $(-1)\mathbf{V}$. The *difference* $\mathbf{A} - \mathbf{B}$ of two vectors $\mathbf{A}$ and $\mathbf{B}$ is defined by

$$\mathbf{A} - \mathbf{B} = \mathbf{A} + (-1)\mathbf{B}$$

EXAMPLE 7 Let $\mathbf{A} = (1, -1)$ and $\mathbf{B} = (2, 4)$. Calculate $4\mathbf{A} + \mathbf{B}$, $\mathbf{A} - \mathbf{B}$, and $\mathbf{A} - \mathbf{A}$.

Solution

$$4\mathbf{A} + \mathbf{B} = 4(1, -1) + (2, 4) = (4, -4) + (2, 4) = (6, 0)$$

$$\mathbf{A} - \mathbf{B} = \mathbf{A} + (-1)\mathbf{B} = (1, -1) + (-2, -4) = (-1, -5)$$

or more simply

$$= (1, -1) - (2, 4) = (1 - 2, -1 - 4) = (-1, -5)$$

Finally,

$$\mathbf{A} - \mathbf{A} = (1, -1) - (1, -1) = (1 - 1, -1 - (-1)) = (0, 0)$$

EXAMPLE 8 Represent geometrically the vector difference $\mathbf{A} - \mathbf{B}$.

Solution Now $\mathbf{A} - \mathbf{B} = \mathbf{A} + (-\mathbf{B})$. In Fig. 2.1.8 we show the resultant of $\mathbf{A}$ and $-\mathbf{B}$. Since the vector $\mathbf{A} - \mathbf{B}$ can be moved as long as we do not change direction or length, $\mathbf{A} - \mathbf{B}$ can be found geometrically by drawing an arrow from the end of $\mathbf{B}$ to the end of $\mathbf{A}$ ($\mathbf{A}$ and $\mathbf{B}$ have the same initial point). See Fig. 2.1.9.

Geometric interpretations of vector addition can be used to study the result

---

[1] In this chapter, *scalar* is restricted to mean "real number."

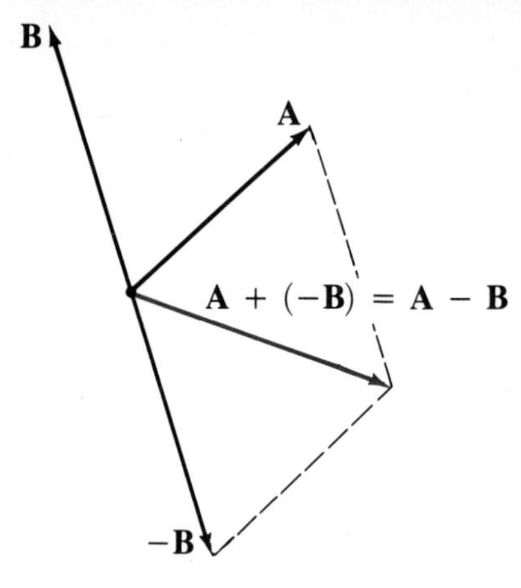

Figure 2.1.8
Geometric meaning of $A - B$.

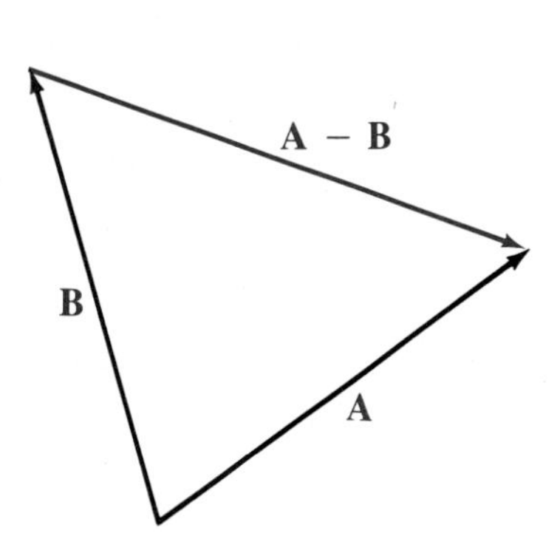

Figure 2.1.9
$A - B$ as third side of triangle.

of several forces acting on one body. These kinds of problems are considered in the study of mechanics, and we do not pursue them here.

So far, we have natural ways to add and subtract vectors as well as a natural way to multiply a vector by a scalar. Since vectors are just matrices, and the operations are the matrix operations all our properties from Chap. 1 hold for vectors. We repeat these below.

THEOREM 2.1.1 Let $\mathbf{U}$, $\mathbf{V}$, and $\mathbf{W}$ all be plane vectors or all space vectors. Let $\mathbf{0}$ be $(0, 0)$ or $(0, 0, 0)$, respectively, depending on the size of $\mathbf{U}$, $\mathbf{V}$, and $\mathbf{W}$. Let r and s be real numbers. Then we have

$$\begin{aligned}
\mathbf{U} + \mathbf{V} &= \mathbf{V} + \mathbf{U} \\
\mathbf{U} + (\mathbf{V} + \mathbf{W}) &= (\mathbf{U} + \mathbf{V}) + \mathbf{W} \\
\mathbf{U} + \mathbf{0} = \mathbf{0} + \mathbf{U} &= \mathbf{U} \\
\mathbf{U} + (-\mathbf{U}) &= \mathbf{0} \\
r(s\mathbf{U}) = (rs)\mathbf{U} &= s(r\mathbf{U}) \\
(r + s)\mathbf{U} &= r\mathbf{U} + s\mathbf{U} \\
r(\mathbf{U} + \mathbf{V}) &= r\mathbf{U} + r\mathbf{V} \\
0\mathbf{U} &= \mathbf{0} \\
1\mathbf{U} &= \mathbf{U}
\end{aligned}$$

The Vectors i, j, k A common way of representing three-vectors in engineering is to use the vectors

$$\begin{aligned}
\mathbf{i} &= (1, 0, 0) \\
\mathbf{j} &= (0, 1, 0) \\
\mathbf{k} &= (0, 0, 1)
\end{aligned}$$

In particular, using the results of matrix algebra, we have

$$(a, b, c) = a\mathbf{i} + b\mathbf{j} + c\mathbf{k}$$

For the two-vector case, **i** and **j** are defined as $\mathbf{i} = (1, 0)$ and $\mathbf{j} = (0, 1)$; thus $(a, b) = a\mathbf{i} + b\mathbf{j}$. In either case, the vectors **i**, **j**, and **k** are called *standard basis vectors* (the reason for this terminology is seen in Chap. 3), and the coefficients of **i**, **j**, and **k** are called the *standard coordinates* of the vector.

DEFINITION 2.1.7 Let $\mathbf{v} = a\mathbf{i} + b\mathbf{j} + c\mathbf{k}$ be a three-vector. The column matrix

$$\mathbf{v}_s = \begin{pmatrix} a \\ b \\ c \end{pmatrix}$$

is called the *standard coordinate matrix* of **v**. If $\mathbf{v} = a\mathbf{i} + b\mathbf{j}$ is a two-vector, its standard coordinate matrix is

$$\mathbf{v}_s = \begin{pmatrix} a \\ b \end{pmatrix}$$

EXAMPLE 9 Show that if **v** and **w** are the three-vectors $\mathbf{v} = v_1\mathbf{i} + v_2\mathbf{j} + v_3\mathbf{k}$ and $\mathbf{w} = w_1\mathbf{i} + w_2\mathbf{j} + w_3\mathbf{k}$, then the standard coordinate matrix of $\mathbf{v} + \mathbf{w}$ is the sum of the standard coordinate matrices of **v** and **w**. That is, show that

$$(\mathbf{v} + \mathbf{w})_s = \mathbf{v}_s + \mathbf{w}_s$$

Solution

$$\mathbf{v}_s = \begin{pmatrix} v_1 \\ v_2 \\ v_3 \end{pmatrix} \qquad \mathbf{w}_s = \begin{pmatrix} w_1 \\ w_2 \\ w_3 \end{pmatrix}$$

The vector $\mathbf{v} + \mathbf{w} = (v_1 + w_1)\mathbf{i} + (v_2 + w_2)\mathbf{j} + (v_3 + w_3)\mathbf{k}$, so that

$$(\mathbf{v} + \mathbf{w})_s = \begin{pmatrix} v_1 + w_1 \\ v_2 + w_2 \\ v_3 + w_3 \end{pmatrix} = \mathbf{v}_s + \mathbf{w}_s$$

EXAMPLE 10 Show that multiplication of the coordinate matrix of a two-vector **v** by

$$A = \begin{pmatrix} 3 & 0 \\ 0 & 3 \end{pmatrix}$$

results in the standard coordinate matrix of $3\mathbf{v}$.

Solution Let $\mathbf{v} = a\mathbf{i} + b\mathbf{j}$. Then

$$X = \begin{pmatrix} a \\ b \end{pmatrix}$$

is the standard coordinate matrix of $\mathbf{v}$. Now $3\mathbf{v} = 3a\mathbf{i} + 3b\mathbf{j}$ and

$$AX = \begin{pmatrix} 3a \\ 3b \end{pmatrix}$$

which is the standard coordinate matrix of $3\mathbf{v}$.

EXAMPLE 11 Let A be a 2×2 matrix and $\mathbf{v}$ a two-vector. Show that $A\mathbf{v}_s = (\mathbf{v}A^T)_s$.

Solution For any vector $\mathbf{w}$ with $\mathbf{w} = (a, b)$, we have

$$\mathbf{w}_s = \begin{pmatrix} a \\ b \end{pmatrix}$$

so that $\mathbf{w}_s = \mathbf{w}^T$. Therefore $(\mathbf{v}A^T)_s = (\mathbf{v}A^T)^T = A^{TT}\mathbf{v}^T = A\mathbf{v}^T = A\mathbf{v}_s$.

Introduction of standard coordinate matrices allows us to "stand (a, b, c) up" as

$$\begin{pmatrix} a \\ b \\ c \end{pmatrix}$$

The purpose is to allow multiplication of a vector on the left by a matrix. If A is 3×3 and $\mathbf{v} = (a, b, c)$, then $A\mathbf{v}$ makes no sense, but $A\mathbf{v}_s$ does. In Sec. 2.3 we will use this multiplication. Note that $\mathbf{v}_s \neq \mathbf{v}$ but $\mathbf{v}_s$ *represents* $\mathbf{v}$ in a natural way.

PROBLEMS 2.1

1. Sketch arrows with initial point (0, 0) corresponding to the following vectors:

   (*a*) (1, 2) (*b*) $(-1, -3)$ (*c*) $(-1, 2)$ (*d*) $(\pi, -1)$

2. Sketch the vectors in Prob. 1 so that their initial point is
   (*a*) (1, 2) (*b*) $(-3, -5)$

3. A pair of points is given. Find the vector beginning at the first point and ending at the second.

   (*a*) $(1, -1), (3, 5)$ (*b*) $(3, 2), (-3, -7)$
   (*c*) $(-1, -6), (2, -8)$ (*d*) $(1, 1), (7, -8)$
   (*e*) $(1, -1, 2), (0, -2, -6)$ (*f*) $(-3, 0, 1), (1, 1, 2)$

4. Sketch arrows corresponding to the following vectors: (*a*) $(1, -2, 0)$, (*b*) $(-2, -4, 3)$, and (*c*) (4, 4, 7).

5. Calculate the length of the vectors in Prob. 1.

6. Calculate the length of the vectors in Prob. 4.

7. What value should k have to make $|(k, 3k, 4k)| = 1$?

8. Let $\mathbf{A} = (1, -1, 2)$, $\mathbf{B} = (3, 0, 4)$, $\mathbf{C} = (4, 2)$, $\mathbf{D} = (-1, 1)$, $\mathbf{i} = (1, 0)$, and $\mathbf{j} = (0, 1)$. Calculate

(*a*) $\mathbf{A} + \mathbf{B}$ (*b*) $\mathbf{A} - \mathbf{B}$ (*c*) $2\mathbf{C} - 3\mathbf{D}$

(*d*) $4\mathbf{i} + 2\mathbf{j}$ (*e*) $|\mathbf{i}|$, $|\mathbf{j}|$, $|\mathbf{i} + \mathbf{j}|$

9. Let $\mathbf{A} = (1, 2)$ and $\mathbf{B} = (-3, 1)$. Calculate and sketch the given vector: (*a*) $\mathbf{A} + \mathbf{B}$, (*b*) $\mathbf{A} - \mathbf{B}$, (*c*) $\mathbf{B} - \mathbf{A}$, (*d*) $2\mathbf{A} + 3\mathbf{B}$.

10. Let $\mathbf{A} = (1, 2)$ and $\mathbf{B} = (-k^2, k^2 - k)$. Are there any real values of k which make $\mathbf{A} + \mathbf{B} = \mathbf{0}$?

11. Show that if $\mathbf{A} \neq \mathbf{0}$, then the vector $(1/|\mathbf{A}|)\,\mathbf{A}$ has length 1.

12. Show by example that $|\mathbf{A} + \mathbf{B}|$ is not necessarily equal to $|\mathbf{A}| + |\mathbf{B}|$. Using the geometric diagram for vector addition, reason that $|\mathbf{A} + \mathbf{B}| \leq |\mathbf{A}| + |\mathbf{B}|$.

13. Show that if $\mathbf{v}$ is a two- or three-vector and r is a real number, then $(r\mathbf{v})_s = r\mathbf{v}_s$.

14. Show that if $\mathbf{v}$ and $\mathbf{w}$ are three-vectors, then $(\mathbf{v} - \mathbf{w})_s = \mathbf{v}_s - \mathbf{w}_s$.

15. Show that if A is a 3×3 matrix and $\mathbf{v}$ is a three-vector, then $A\mathbf{v} = (\mathbf{v}A^T)_s$.

2.2 THE ANGLE BETWEEN VECTORS; PROJECTIONS

One of the most important problems in the analysis of vectors is the *angle problem:*

> Given two vectors $\mathbf{A}$ and $\mathbf{B}$, find the angle θ, $0 \leq \theta \leq \pi$, between $\mathbf{A}$ and $\mathbf{B}$. See Fig. 2.2.1.

If we can solve this problem, then we know whether $\mathbf{A}$ is parallel to $\mathbf{B}$ (θ is 0 or π) or $\mathbf{A}$ is perpendicular to $\mathbf{B}$ ($\theta = \pi/2$).

The solution to this problem for plane vectors can be found by using the law of cosines: For a triangle with sides of length a, b, and c as shown

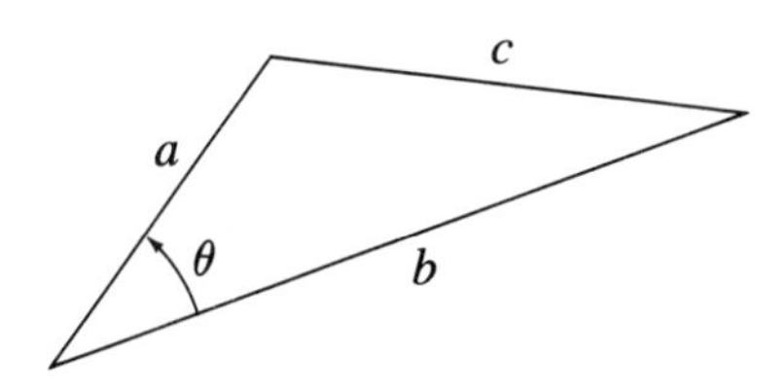

we have

$$c^2 = a^2 + b^2 - 2ab\cos\theta \tag{2.2.1}$$

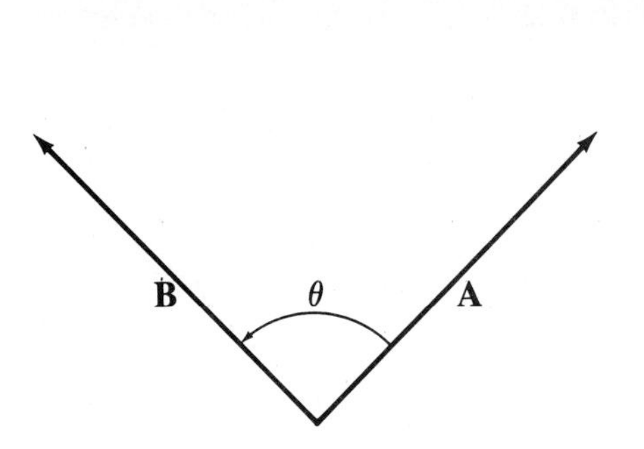

Figure 2.2.1
Vector Triangle

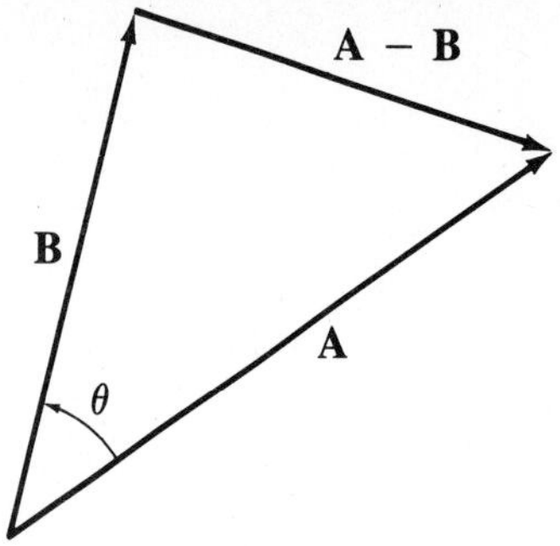

Figure 2.2.2
Vector Triangle

From Eq. (2.2.1) we find

$$\cos \theta = \frac{a^2 + b^2 - c^2}{2ab} \tag{2.2.2}$$

Therefore in any triangle with lengths of sides given, we can find the cosine of any of the angles. And this is just as good as finding θ—especially for the purposes of determining whether θ equals 0, $\pi/2$, or π.

To solve the angle problem, we consider the triangle formed by **A**, **B**, and **A** − **B** (see Fig. 2.2.2) and use the law of cosines. We have from (2.2.2)

$$\cos \theta = \frac{|\mathbf{A}|^2 + |\mathbf{B}|^2 - |\mathbf{A} - \mathbf{B}|^2}{2|\mathbf{A}|\,|\mathbf{B}|} \tag{2.2.3}$$

Suppose now that $\mathbf{A} = (a_1, a_2)$ and $\mathbf{B} = (b_1, b_2)$. Then $\mathbf{A} - \mathbf{B} = (a_1 - b_1, a_2 - b_2)$ and

$$\begin{aligned} |\mathbf{A}|^2 + |\mathbf{B}|^2 - |\mathbf{A} - \mathbf{B}|^2 &= a_1^{\,2} + a_2^{\,2} + b_1^{\,2} + b_2^{\,2} - [(a_1 - b_1)^2 + (a_2 - b_2)^2] \\ &= a_1^{\,2} + a_2^{\,2} + b_1^{\,2} + b_2^{\,2} \\ &\quad - (a_1^{\,2} - 2a_1b_1 + b_1^{\,2} + a_2^{\,2} - 2a_2b_2 + b_2^{\,2}) \\ &= 2(a_1b_1 + a_2b_2) \end{aligned}$$

This means that

$$\cos \theta = \frac{a_1b_1 + a_2b_2}{|\mathbf{A}|\,|\mathbf{B}|} \tag{2.2.4}$$

and the angle problem is solved for two-vectors.

EXAMPLE 1 Find the cosine of the angle between $\mathbf{A} = (3, 4)$ and $\mathbf{B} = (5, 12)$.

Solution Since $a_1 = 3$, $a_2 = 4$, $b_1 = 5$, and $b_2 = 12$, we have

$$\cos\theta = \frac{3\cdot 5 + 4\cdot 12}{\sqrt{3^2+4^2}\sqrt{5^2+12^2}} = \frac{63}{5\cdot 13} = \frac{63}{65}$$

EXAMPLE 2 Find the cosine of the angle between $\mathbf{A} = (1, 1)$ and $\mathbf{B} = (-2, 2)$. Sketch the vectors. Are they perpendicular?

Solution We have

$$\cos\theta = \frac{1(-2) + (+1)(2)}{\sqrt{1^2 + (-1)^2}\sqrt{(-2)^2 + 2^2}} = 0$$

The vectors are sketched in Fig. 2.2.3. Because $\cos\theta = 0$, $\theta = \pi/2$, so the vectors are perpendicular.

If we look at Eq. (2.2.4) again, we see that the numerator of the right-hand side, $a_1b_1 + a_2b_2$, is actually (in terms of matrix multiplication)

$$\mathbf{AB}^T = (a_1a_2)\begin{pmatrix} b_1 \\ b_2 \end{pmatrix} = a_1b_1 + a_2b_2$$

Therefore Eq. (2.2.4) can be written

$$\cos\theta = \frac{\mathbf{AB}^T}{|\mathbf{A}|\,|\mathbf{B}|} \qquad \text{solution to angle problem} \qquad (2.2.5)$$

It turns out that this formula also works for three-vectors.

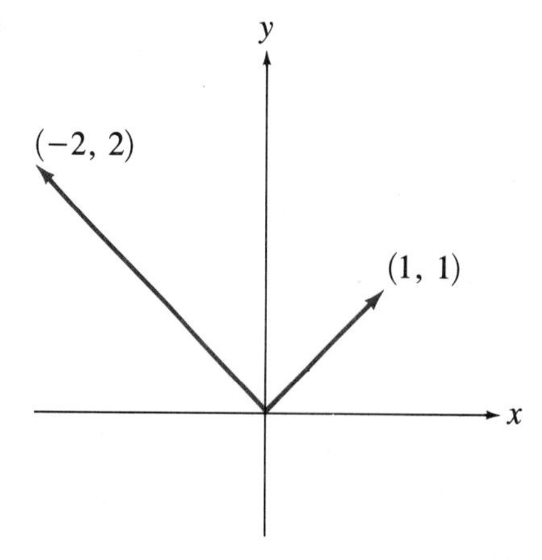

Figure 2.2.3
Vectors from Example 2

EXAMPLE 3 Find the cosine of the angle between $\mathbf{A} = (1, -1, 4)$ and $\mathbf{B} = (4, 0, -1)$.

Solution By 2.2.5

$$\cos\theta = \frac{\mathbf{A}\mathbf{B}^T}{|\mathbf{A}|\,|\mathbf{B}|} = \frac{4+0-4}{\sqrt{18}\sqrt{17}} = 0$$

Hence, these vectors are perpendicular.

Since the matrix product $\mathbf{A}\mathbf{B}^T$ is used to solve the angle problem for vectors, we give it a special name, the *dot product*.

DEFINITION 2.2.1 Given two plane or space vectors $\mathbf{A}$ and $\mathbf{B}$, the *dot product* $\mathbf{A}\cdot\mathbf{B}$ is defined by $\mathbf{A}\cdot\mathbf{B} = \mathbf{A}\mathbf{B}^T$. Because $\mathbf{A}\cdot\mathbf{B}$ is a real number, $(\mathbf{A}\cdot\mathbf{B})^T = \mathbf{A}\cdot\mathbf{B}$. Therefore $\mathbf{B}\cdot\mathbf{A} = \mathbf{B}\mathbf{A}^T = (\mathbf{A}\mathbf{B}^T)^T = (\mathbf{A}\cdot\mathbf{B})^T = \mathbf{A}\cdot\mathbf{B}$. This means that the dot product is commutative.

EXAMPLE 4 Let $\mathbf{A} = (-1, 4)$, $\mathbf{B} = (2, 3)$, $\mathbf{C} = (7, 0, 2)$, and $\mathbf{D} = (-1, 1, 1)$. Calculate, if possible, $\mathbf{A}\cdot\mathbf{B}$, $\mathbf{B}\cdot\mathbf{C}$, and $\mathbf{C}\cdot\mathbf{D}$.

Solution Now

$$\mathbf{A}\cdot\mathbf{B} = (-1 \quad 4)\begin{pmatrix}2\\3\end{pmatrix} = -2+12 = 10$$

and $\mathbf{B}\cdot\mathbf{C}$ cannot be calculated since $\mathbf{B}$ is 1×3 and $\mathbf{C}^T$ is 2×1. So $\mathbf{B}\mathbf{C}^T$ does not exist.

$$\mathbf{C}\cdot\mathbf{D} = \mathbf{C}\mathbf{D}^T = (7 \quad 0 \quad 2)\begin{pmatrix}-1\\1\\1\end{pmatrix} = -7+2 = -5$$

We can state several properties of the dot product now.

THEOREM 2.2.1 Let $\mathbf{A}$, $\mathbf{B}$, and $\mathbf{C}$ be vectors of the same order. Let r and s be real numbers.

(*a*) $\mathbf{A}\cdot\mathbf{B} = |\mathbf{A}|\,|\mathbf{B}|\cos\theta$, where θ is the angle between $\mathbf{A}$ and $\mathbf{B}$.

(*b*) $\mathbf{A}\cdot(\mathbf{B}+\mathbf{C}) = \mathbf{A}\cdot\mathbf{B} + \mathbf{A}\cdot\mathbf{C}$

(*c*) $(\mathbf{A}+\mathbf{B})\cdot\mathbf{C} = \mathbf{A}\cdot\mathbf{C} + \mathbf{B}\cdot\mathbf{C}$

(*d*) $r(\mathbf{A}\cdot\mathbf{B}) = (r\mathbf{A})\cdot\mathbf{B} = \mathbf{A}\cdot(r\mathbf{B})$

(*e*) $(r\mathbf{A})\cdot(s\mathbf{B}) = (rs)(\mathbf{A}\cdot\mathbf{B})$

(*f*) $\sqrt{\mathbf{A} \cdot \mathbf{A}} = |\mathbf{A}|$

(*g*) $\mathbf{A} \cdot \mathbf{A} = 0$ if and only if $\mathbf{A} = \mathbf{0}$ (the zero vector).

(*h*) Let $|\mathbf{A}| \neq 0$ and $|\mathbf{B}| \neq 0$. Then $\mathbf{A} \cdot \mathbf{B} = 0$ if and only if $\mathbf{A} \perp \mathbf{B}$.

(*i*) $|\mathbf{A} \cdot \mathbf{B}| \leq |\mathbf{A}|\,|\mathbf{B}|$ (Cauchy-Schwarz inequality)

(*j*) $|r\mathbf{A}| = |r|\,|\mathbf{A}|$

(*k*) $|\mathbf{A} + \mathbf{B}| \leq |\mathbf{A}| + |\mathbf{B}|$

Proof (*a*) This is just a restatement of Eq. (2.2.5).

(*b*) We have $\mathbf{A} \cdot (\mathbf{B} + \mathbf{C}) = \mathbf{A}(\mathbf{B} + \mathbf{C})^T = \mathbf{AB}^T + \mathbf{AC}^T$, where the last equality follows from the properties of the transpose and matrix multiplication. The last term is just $\mathbf{A} \cdot \mathbf{B} + \mathbf{A} \cdot \mathbf{C}$ by definition of dot product.

(*c*) (*d*), and (*e*) follow from matrix properties and are proved in a fashion similar to (*b*).

(*f*) Consider the case of three-vectors. Let $\mathbf{A} = (a_1, a_2, a_3)$. Then $\sqrt{\mathbf{A} \cdot \mathbf{A}} = \sqrt{{a_1}^2 + {a_2}^2 + {a_3}^2}$ which is just $|\mathbf{A}|$. For two-vectors the argument is the same.

(*g*) First we prove that if $\mathbf{A} = \mathbf{0}$, then $\mathbf{A} \cdot \mathbf{A} = 0$. If $\mathbf{A} = \mathbf{0} = (0, 0, 0)$, then $\mathbf{A} \cdot \mathbf{A} = 0^2 + 0^2 + 0^2 = 0$. Now we show that if $\mathbf{A} \cdot \mathbf{A} = 0$, then $\mathbf{A}$ must be the zero vector. Let $\mathbf{A} = (a_1, a_2, a_3)$, so that $\mathbf{A} \cdot \mathbf{A} = {a_1}^2 + {a_2}^2 + {a_3}^2$. Since ${a_1}^2$, ${a_2}^2$ and ${a_3}^2$ are all greater than or equal to zero, the only way the sum of these can be zero is if each term is zero. That is, we must have ${a_1}^2 = {a_2}^2 = {a_3}^2 = 0$. So $a_1 = a_2 = a_3 = 0$ and $\mathbf{A} = \mathbf{0}$.

(*h*) From (*a*), $\mathbf{A} \cdot \mathbf{B} = |\mathbf{A}|\,|\mathbf{B}| \cos \theta$. Since $|\mathbf{A}| \neq 0$ and $|\mathbf{B}| \neq 0$, the only way $\mathbf{A} \cdot \mathbf{B}$ can equal zero is if $\cos \theta = 0$. That is, $\theta = \pi/2$. Therefore, $\mathbf{A}$ is perpendicular to $\mathbf{B}$.

(*i*) From (*a*), $|\mathbf{A} \cdot \mathbf{B}| = |\mathbf{A}|\,|\mathbf{B}|\,|\cos \theta|$. Since $|\cos \theta| \leq 1$, $|\mathbf{A} \cdot \mathbf{B}| \leq |\mathbf{A}|\,|\mathbf{B}|$.

Proofs of (*j*) and (*k*) are left to the problems.

Projections are an important geometric idea which we can now discuss by using the dot product. Geometrically, the projection of a vector **B** on a vector **A** is shown in Fig. 2.2.4. Roughly speaking, the projection of **B** on **A** is the shadow which **B** casts on **A** due to light rays which hit **A**, the light rays being perpendicular to **A**. In mechanics, the projection is the component of the force **B** in the direction of **A**, and $\mathbf{B} \cdot \mathbf{A}$ is the work done by the force **B** in the direction of **A**.

We can determine the vector projection of **B** on **A** (written $\mathbf{B}_{\text{proj } \mathbf{A}}$) by finding its length; its direction is the same as **A**. Once we find $|\mathbf{B}_{\text{proj } \mathbf{A}}|$, we can multiply

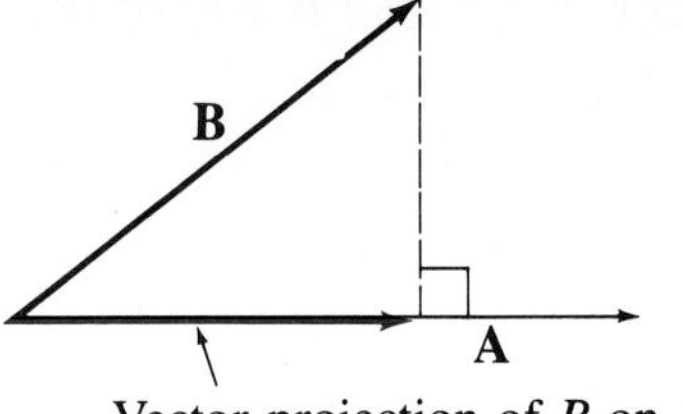

Figure 2.2.4
Vector projection.

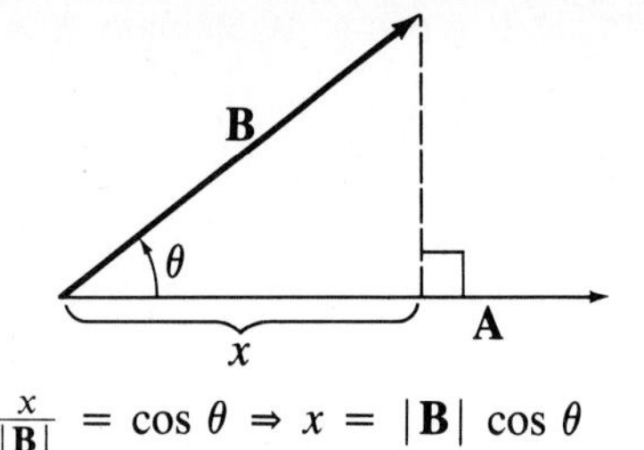

Figure 2.2.5
Length of projection.

this by $(1/|\mathbf{A}|)\mathbf{A}$ (the unit vector in the direction of **A**) to get $\mathbf{B}_{\text{proj }\mathbf{A}}$. We find $|\mathbf{B}_{\text{proj }\mathbf{A}}|$ by using trigonometry. From Fig. 2.2.5 we see that $|\mathbf{B}_{\text{proj }\mathbf{A}}| = |\mathbf{B}| \cos\theta$, where θ is the angle between **A** and **B**. Therefore, since

$$\frac{\mathbf{A}\cdot\mathbf{B}}{|\mathbf{A}|\,|\mathbf{B}|} = \cos\theta$$

we know that

$$|\mathbf{B}_{\text{proj }\mathbf{A}}| = \big||\mathbf{B}|\cos\theta\big| = |\mathbf{B}|\frac{\mathbf{A}\cdot\mathbf{B}}{|\mathbf{A}|\,|\mathbf{B}|} = \frac{\mathbf{A}\cdot\mathbf{B}}{|\mathbf{A}|}$$

and so

$$\mathbf{B}_{\text{proj }\mathbf{A}} = |\mathbf{B}_{\text{proj }\mathbf{A}}|\frac{1}{|\mathbf{A}|}\mathbf{A} = \frac{\mathbf{A}\cdot\mathbf{B}}{|\mathbf{A}|^2}\mathbf{A}$$

Sometimes $|\mathbf{B}_{\text{proj }\mathbf{A}}|$ is called the *scalar projection* of **B** on **A**.

EXAMPLE 5 Let $\mathbf{A} = (1, 0, 1)$ and $\mathbf{B} = (1, -1, 2)$. Find $\mathbf{B}_{\text{proj }\mathbf{A}}$.

Solution

$$\mathbf{B}_{\text{proj }\mathbf{A}} = \frac{\mathbf{A}\cdot\mathbf{B}}{|\mathbf{A}|^2}\mathbf{A} = \tfrac{3}{2}(1, 0, 1) = (\tfrac{3}{2}, 0, \tfrac{3}{2})$$

A vector product which is important in three-dimensional mechanics is the *cross product*. This product is specific to three-vectors and does not generalize in a natural way to higher dimensions. Because of this, our treatment of the cross product is brief.

DEFINITION 2.2.2 Let $\mathbf{A} = (a_1, a_2, a_3)$ and $\mathbf{B} = (b_1, b_2, b_3)$. Then the *cross product* of **A** and **B**, written $\mathbf{A} \times \mathbf{B}$, is a vector defined by

$$\mathbf{A} \times \mathbf{B} = (a_2b_3 - b_2a_3,\ -a_1b_3 + b_1a_3,\ a_1b_2 - a_2b_1)$$

EXAMPLE 6 Calculate $\mathbf{A} \times \mathbf{B}$ for $\mathbf{A} = (1, -1, 2)$ and $\mathbf{B} = (2, 2, 7)$.

Solution One way to do this is to form a symbolic determinant

$$\begin{matrix} \text{1st} & \text{2d} & \text{3d} & \text{component of } \mathbf{A} \times \mathbf{B} \end{matrix}$$

$$\det \begin{pmatrix} \mathbf{i} & \mathbf{j} & \mathbf{k} \\ 1 & -1 & 2 \\ 2 & 2 & 7 \end{pmatrix} \begin{matrix} \\ \leftarrow \mathbf{A} \text{ is row 2} \\ \leftarrow \mathbf{B} \text{ is row 3} \end{matrix}$$

and to expand by the first row:

$$\mathbf{A} \times \mathbf{B} = \mathbf{i} \det \begin{pmatrix} -1 & 2 \\ 2 & 7 \end{pmatrix} - \mathbf{j} \det \begin{pmatrix} 1 & 2 \\ 2 & 7 \end{pmatrix} + \mathbf{k} \det \begin{pmatrix} 1 & -1 \\ 2 & 2 \end{pmatrix} = (-11, -3, 4)$$

Notice that $\mathbf{A} \times \mathbf{B}$ is perpendicular to both $\mathbf{A}$ and $\mathbf{B}$ since

$$(\mathbf{A} \times \mathbf{B}) \cdot \mathbf{A} = (-11, -3, 4) \cdot (1, -1, 2) = -11 + 3 + 8 = 0$$
$$(\mathbf{A} \times \mathbf{B}) \cdot \mathbf{B} = (-11, -3, 4) \cdot (2, 2, 7) = -22 - 6 + 28 = 0$$

In fact, this is always true.

THEOREM 2.2.2 Let $\mathbf{A}$ and $\mathbf{B}$ be nonzero vectors. Then $(\mathbf{A} \times \mathbf{B}) \cdot \mathbf{A} = 0 = (\mathbf{A} \times \mathbf{B}) \cdot \mathbf{B}$.

Proof For $(\mathbf{A} \times \mathbf{B}) \cdot \mathbf{A}$ we have

$$\begin{aligned} (\mathbf{A} \times \mathbf{B}) \cdot \mathbf{A} &= (a_2 b_3 - b_2 a_3, -a_1 b_3 + b_1 a_3, a_1 b_2 - a_2 b_1) \cdot (a_1, a_2, a_3) \\ &= a_1 a_2 b_3 - a_1 b_2 a_3 - a_2 a_1 b_3 + a_2 b_1 a_3 \\ &\quad + a_3 a_1 b_2 - a_2 b_1 a_3 \\ &= 0 \end{aligned}$$

Geometrically $\mathbf{A} \times \mathbf{B}$ is the vector of length $|\mathbf{A}||\mathbf{B}| \sin \theta$, perpendicular to both $\mathbf{A}$ and $\mathbf{B}$ and pointing in the direction dictated by the right-hand rule: Using the right hand, curve your fingers from $\mathbf{A}$ to $\mathbf{B}$; $\mathbf{A} \times \mathbf{B}$ points in the direction of the thumb. (See Fig. 2.2.6.) This can be shown rigorously, but we omit the proof.

From the geometric meaning of $\mathbf{A} \times \mathbf{B}$ it is not hard to see that $\mathbf{B} \times \mathbf{A} =$

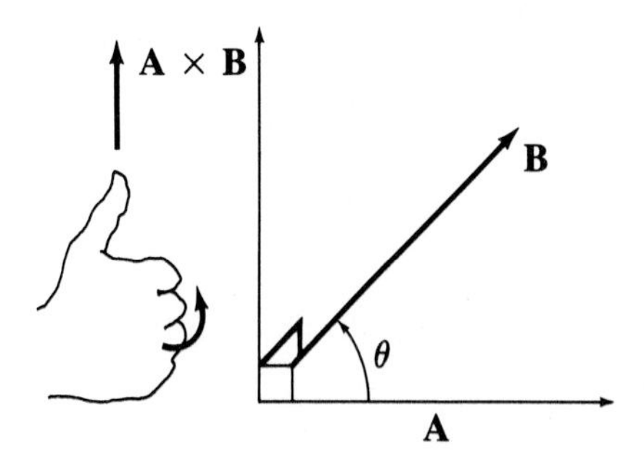

Figure 2.2.6 Cross product.

$-(\mathbf{A} \times \mathbf{B})$ and $\mathbf{A} \times \mathbf{B} = \mathbf{0}$ if $\sin\theta = 0$, that is, $\mathbf{A} \times \mathbf{B} = \mathbf{0}$ if $\mathbf{A}$ is parallel to $\mathbf{B}$ (the directions can be opposite).

PROBLEMS 2.2

1. Calculate $\mathbf{A} \cdot \mathbf{B}$.

(*a*) $\mathbf{A} = (1, 2), \mathbf{B} = (3, 7)$ (*b*) $\mathbf{A} = (-5, 2), \mathbf{B} = (3, 6)$

(*c*) $\mathbf{A} = (5, 4, 0), \mathbf{B} = (0, 1, 3)$ (*d*) $\mathbf{A} = (0, 1, -1), \mathbf{B} = (6, 6, 7)$

2. Find the angle between $\mathbf{A}$ and $\mathbf{B}$.

(*a*) $\mathbf{A} = (5, 0), \mathbf{B} = (2, 2)$

(*b*) $\mathbf{A} = (2, -2), \mathbf{B} = (-1, 1)$

(*c*) $\mathbf{A} = (3, 0, 4), \mathbf{B} = (-6, 0, -8)$

3. Find the cosine of the angle between the vectors in Prob. 1.

4. For the vectors in Prob. 1, verify the Cauchy-Schwarz inequality $|\mathbf{A} \cdot \mathbf{B}| \le |\mathbf{A}|\,|\mathbf{B}|$.

5. Find the projection of $\mathbf{B}$ on $\mathbf{A}$ for the vectors in Prob. 1.

6. Calculate $\mathbf{A} \times \mathbf{B}$ for the vectors in Prob. 1*c* and *d*.

7. Show that for the set of vectors $\{\mathbf{i}, \mathbf{j}, \mathbf{k}\}$ any vector from the set is perpendicular to all others in the set.

8. (*a*) Find a vector perpendicular to (1, 2).

(*b*) Find a vector of length 1 unit perpendicular to (1, 2).

(*c*) How many correct answers are there to part (*b*)?

9. Find the angle between the diagonal of a cube and one of its edges.

10. Let $\mathbf{A} \neq \mathbf{0}$, $\mathbf{B} \neq \mathbf{0}$, and $\mathbf{A} \neq \pm\mathbf{B}$. Show that the angle between $\mathbf{A} + \mathbf{B}$ and $\mathbf{B}$ is less than the angle between $\mathbf{A}$ and $\mathbf{B}$. *Hint*: The cosines of the angles are

$$\frac{\mathbf{A} \cdot \mathbf{B}}{|\mathbf{A}|\,|\mathbf{B}|} \quad \text{and} \quad \frac{(\mathbf{A} + \mathbf{B}) \cdot \mathbf{B}}{|\mathbf{A} + \mathbf{B}|\,|\mathbf{B}|}$$

since $0 \le \theta \le 180°$ if $\theta_1 < \theta_2$, $\cos\theta_1 > \cos\theta_2$. Show that

$$\frac{\mathbf{A} \cdot \mathbf{B}}{|\mathbf{A}|\,|\mathbf{B}|} < \frac{(\mathbf{A} + \mathbf{B}) \cdot \mathbf{B}}{|\mathbf{A} + \mathbf{B}|\,|\mathbf{B}|}$$

by using $|\mathbf{A} + \mathbf{B}| \le |\mathbf{A}| + |\mathbf{B}|$ and $\mathbf{A} \cdot \mathbf{B} \le |\mathbf{A}|\,|\mathbf{B}|$.

11. *Parallelogram identity.* Consider a parallelogram

Figure P2.2.11

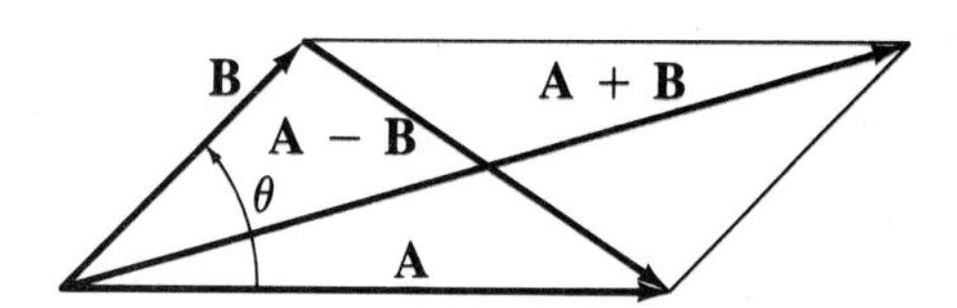

By the law of cosines

$$|\mathbf{A} - \mathbf{B}|^2 = |\mathbf{A}|^2 + |\mathbf{B}|^2 - 2|\mathbf{A}|\,|\mathbf{B}| \cos\theta$$
$$|\mathbf{A} + \mathbf{B}|^2 = |\mathbf{A}|^2 + |\mathbf{B}|^2 - 2|\mathbf{A}|\,|\mathbf{B}| \cos(180^\circ - \theta)$$

Use this to show that

(*a*) $\mathbf{A} \cdot \mathbf{B} = \frac{1}{4}|\mathbf{A} + \mathbf{B}|^2 - \frac{1}{4}|\mathbf{A} - \mathbf{B}|^2$

(*b*) $|\mathbf{A}|^2 + |\mathbf{B}|^2 = \frac{1}{2}(|\mathbf{A} + \mathbf{B}|^2 + |\mathbf{A} - \mathbf{B}|^2)$

In words, (*b*) says that the sum of the squares of the sides of a parallelogram is the average of the squares of the lengths of the diagonals.

12. Show that $\mathbf{B} \times \mathbf{A} = -(\mathbf{A} \times \mathbf{B})$, using Definition 2.2.2.

13. Show that $\mathbf{B} - \mathbf{B}_{\text{proj}\,\mathbf{A}}$ is perpendicular to $\mathbf{A}$.

2.3 MATRICES AS TRANSFORMERS OF SPACE

Analysis of functions is the lifeblood of applied mathematics. In calculus, functions with various domains and ranges were studied—a summary is given in Table 2.3.1. In linear algebra we analyze certain functions with domains and ranges which are like $\mathbb{R}^2$ and $\mathbb{R}^3$ in certain ways. The functions are called *linear transformations*, and the domains and ranges are called *vector spaces*. These are studied in detail in Chaps. 3, 4, and 5. In this section we present a "preview of coming attractions" in the concrete setting of $\mathbb{R}^2$ and $\mathbb{R}^3$.

Let a function f have domain either $\mathbb{R}^2$ or $\mathbb{R}^3$; this function is *linear* if, for every x and y in the domain of f and every real number r,

$$f(x + y) = f(x) + f(y)$$

Table 2.3.1 **CALCULUS**

FUNCTION NOTATION	DOMAIN	RANGE	ANALYSIS
$y = f(x)$	$\mathbb{R}$	$\mathbb{R}$	Derivative, tangent lines max-min, curve sketching, integrals, area between curves
$z = f(x, y)$	$\mathbb{R}^2$	$\mathbb{R}$	Partial derivatives, tangent planes, max-min, double or triple integrals, volumes
or			
$w = f(x, y, z)$	$\mathbb{R}^3$	$\mathbb{R}$	
$\vec{r}(t) = (f_1(t), f_2(t))$	$\mathbb{R}$	$\mathbb{R}^2$	Analysis of curves, vectors, curvature, line integrals
or			
$\vec{r}(t) = (f_1(t), f(t), f_3(t))$	$\mathbb{R}$	$\mathbb{R}^3$	

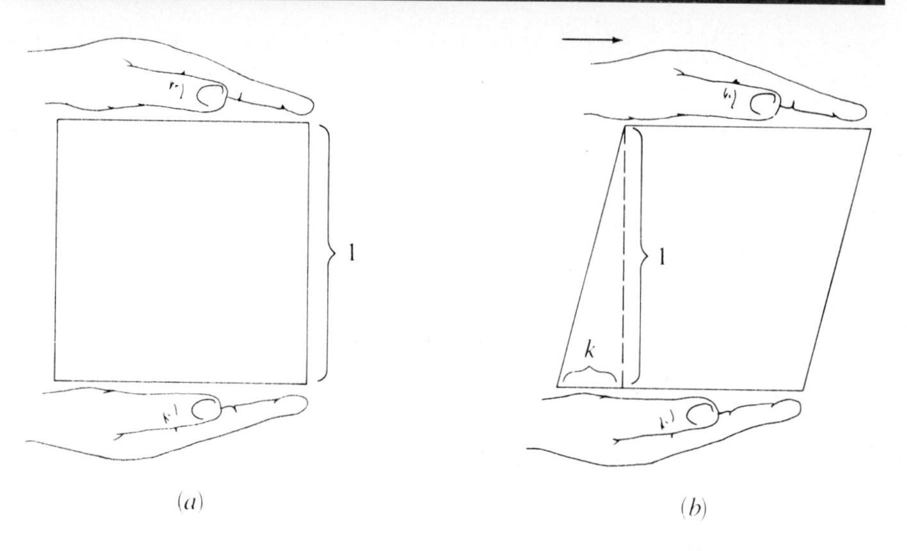

Figure 2.3.1
Shear. k is small.

and

$$f(rx) = rf(x)$$

Our first example of a linear function comes from the field of mechanics.

EXAMPLE 1 Imagine a side view of a cube of gelatin held between hands (see Fig. 2.3.1*a*). If the lower hand is held fixed and the upper hand is moved as in Fig. 2.3.1*b*, the cube deforms in a special way which leads to what is called *shear*. As an approximation, we assume in Fig. 2.3.1 that the height remains unchanged; this is reasonable if k is small. We notice in Fig. 2.3.2 that shear transforms $\mathbf{i}$ to $\mathbf{i}$ and $\mathbf{j}$ to $k\mathbf{i} + \mathbf{j}$ (we have placed the origin at the lower left corner of the face of the cube). In terms of standard coordinate matrices, shear transforms

$$\begin{pmatrix} 1 \\ 0 \end{pmatrix} \text{ to } \begin{pmatrix} 1 \\ 0 \end{pmatrix} \quad \text{and} \quad \begin{pmatrix} 0 \\ 1 \end{pmatrix} \text{ to } \begin{pmatrix} k \\ 1 \end{pmatrix}$$

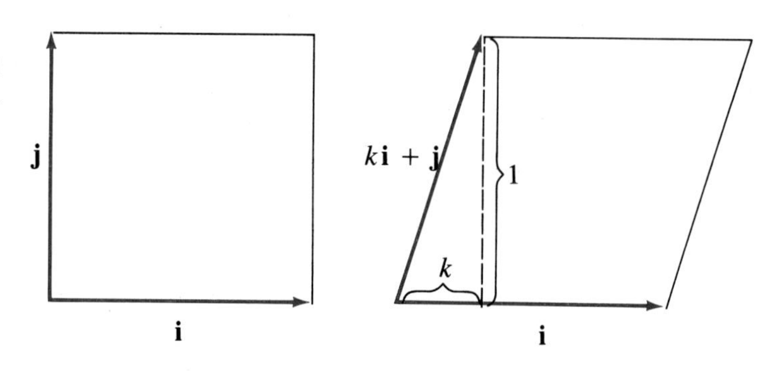

Figure 2.3.2
Vectors superimposed on shear.

The shear function $\mathscr{S}$ can be represented in standard coordinate matrices

$$\begin{pmatrix} a \\ b \end{pmatrix}$$

by

$$\mathscr{S}\left(\begin{pmatrix} a \\ b \end{pmatrix}\right) = \begin{pmatrix} 1 & k \\ 0 & 1 \end{pmatrix}\begin{pmatrix} a \\ b \end{pmatrix}$$

We note that under this definition

$$\mathscr{S}\left(\begin{pmatrix} 1 \\ 0 \end{pmatrix}\right) = \begin{pmatrix} 1 & k \\ 0 & 1 \end{pmatrix}\begin{pmatrix} 1 \\ 0 \end{pmatrix} = \begin{pmatrix} 1 \\ 0 \end{pmatrix}$$

and

$$\mathscr{S}\left(\begin{pmatrix} 0 \\ 1 \end{pmatrix}\right) = \begin{pmatrix} 1 & k \\ 0 & 1 \end{pmatrix}\begin{pmatrix} 0 \\ 1 \end{pmatrix} = \begin{pmatrix} k \\ 1 \end{pmatrix}$$

which matches the action shown in Fig. 2.3.2. As defined, the shear function is linear because by the laws of matrix algebra,

$$\begin{aligned} \mathscr{S}\left(\begin{pmatrix} a \\ b \end{pmatrix} + \begin{pmatrix} c \\ d \end{pmatrix}\right) &= \begin{pmatrix} 1 & k \\ 0 & 1 \end{pmatrix}\left(\begin{pmatrix} a \\ b \end{pmatrix} + \begin{pmatrix} c \\ d \end{pmatrix}\right) = \begin{pmatrix} 1 & k \\ 0 & 1 \end{pmatrix}\begin{pmatrix} a \\ b \end{pmatrix} + \begin{pmatrix} 1 & k \\ 0 & 1 \end{pmatrix}\begin{pmatrix} c \\ d \end{pmatrix} \\ &= \mathscr{S}\left(\begin{pmatrix} a \\ b \end{pmatrix}\right) + \mathscr{S}\left(\begin{pmatrix} c \\ d \end{pmatrix}\right) \end{aligned}$$

and

$$\mathscr{S}\left(r\begin{pmatrix} a \\ b \end{pmatrix}\right) = \begin{pmatrix} 1 & k \\ 0 & 1 \end{pmatrix}\left(r\begin{pmatrix} a \\ b \end{pmatrix}\right) = r\begin{pmatrix} 1 & k \\ 0 & 1 \end{pmatrix}\begin{pmatrix} a \\ b \end{pmatrix} = r\mathscr{S}\left(\begin{pmatrix} a \\ b \end{pmatrix}\right)$$

Shear as a linear function representable by a matrix is only one of several important applications. So it is important to study functions representable by matrices which multiply the standard coordinate matrices of vectors. We discuss several examples and to save words refer to standard coordinate matrices as just vectors.[2] In each example, since the function is defined by matrix multiplication, the linearity follows from the matrix algebra results:

$$A(X + Y) = AX + AY$$
$$A(rX) = rAX$$

---

[2] This is a slight abuse of terminology. A three-vector is 1×3, and its standard coordinate matrix is 3×1. However, it is easy to make the connection between

$$\begin{pmatrix} a \\ b \\ c \end{pmatrix}$$

and its vector (a, b, c).

EXAMPLE 2 (Production, engineering technology) Suppose a manufacturer makes two kinds of mattresses: soft and firm. The labor per mattress splits according to this chart:

	LABOR PER UNIT	
TYPE	SPRINGS	PADDING AND COVER
Soft	2 h	1 h
Firm	3 h	2 h

If x_1 soft mattresses and x_2 firm mattresses are produced, the hours of labor used for springs (S) and padding (P) are

$$S = 2x_1 + 3x_2$$
$$P = 1x_1 + 2x_2$$

These equations allow production planning when the amount of labor available for springs and padding is known. We can write this as

$$\begin{pmatrix} 2 & 3 \\ 1 & 2 \end{pmatrix}\begin{pmatrix} x_1 \\ x_2 \end{pmatrix} = \begin{pmatrix} S \\ P \end{pmatrix}$$

In this last form, the "labor matrix"

$$\begin{pmatrix} 2 & 3 \\ 1 & 2 \end{pmatrix}$$

represents a function which acts on a "production vector"

$$\begin{pmatrix} x_1 \\ x_2 \end{pmatrix}$$

and yields a "labor vector"

$$\begin{pmatrix} S \\ P \end{pmatrix}$$

EXAMPLE 3 Consider the linear function

$$f\left(\begin{pmatrix} x_1 \\ x_2 \end{pmatrix}\right) = \begin{pmatrix} 2 & 0 \\ 0 & 2 \end{pmatrix}\begin{pmatrix} x_1 \\ x_2 \end{pmatrix} = \begin{pmatrix} 2x_1 \\ 2x_2 \end{pmatrix} = 2\begin{pmatrix} x_1 \\ x_2 \end{pmatrix}$$

For this function, the action of f on

$$\begin{pmatrix} x_1 \\ x_2 \end{pmatrix}$$

is expressed by multiplication of

$$\begin{pmatrix} x_1 \\ x_2 \end{pmatrix} \quad \text{by the matrix} \begin{pmatrix} 2 & 0 \\ 0 & 2 \end{pmatrix}$$

We would like to be able to describe what f does to

$$\begin{pmatrix} x_1 \\ x_2 \end{pmatrix}$$

in geometric terms. This is, in fact, the main problem for this section.

Geometric analysis problem
Given a linear function f

$$f\left(\begin{pmatrix} x_1 \\ x_2 \end{pmatrix}\right) = \begin{pmatrix} a_{11} & a_{12} \\ a_{21} & a_{22} \end{pmatrix}\begin{pmatrix} x_1 \\ x_2 \end{pmatrix}$$

describe geometrically how f transforms vectors.

EXAMPLE 4 Solve the geometric analysis problem (GAP) for the function from Example 3:

$$f\left(\begin{pmatrix} x_1 \\ x_2 \end{pmatrix}\right) = \begin{pmatrix} 2 & 0 \\ 0 & 2 \end{pmatrix}\begin{pmatrix} x_1 \\ x_2 \end{pmatrix}$$

Solution In this case the problem is not as difficult as it will be later. In this case, by direct calculation

$$f\left(\begin{pmatrix} x_1 \\ x_2 \end{pmatrix}\right) = 2\begin{pmatrix} x_1 \\ x_2 \end{pmatrix}$$

So the action of f on a vector is to stretch the vector to twice its original length and leave the direction unchanged. (See Fig. 2.3.3.)

In Example 4 there is nothing special about the constant 2. If f is defined by

$$f\left(\begin{pmatrix} x_1 \\ x_2 \end{pmatrix}\right) = \begin{pmatrix} k & 0 \\ 0 & k \end{pmatrix}\begin{pmatrix} x_1 \\ x_2 \end{pmatrix}$$

where $k > 0$, then

$$f\left(\begin{pmatrix} x_1 \\ x_2 \end{pmatrix}\right) = k\begin{pmatrix} x_1 \\ x_2 \end{pmatrix}$$

and the action of f is to stretch ($k > 1$), shrink ($0 < k < 1$), or leave unchanged ($k = 1$) the original vector and, in any case, to leave the direction unchanged. Because of this we introduce this terminology:

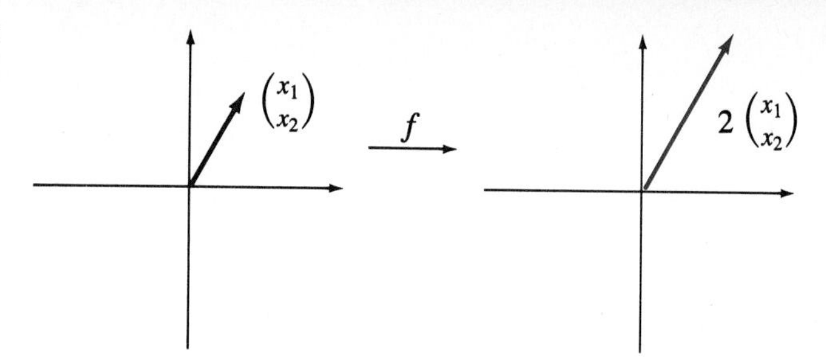

Figure 2.3.3 Action of f from Example 4.

If f is defined by a scalar matrix

$$\begin{pmatrix} k & 0 \\ 0 & k \end{pmatrix}$$

then it is called

A *dilation*, if $k > 1$
The *identity function*, if $k = 1$
A *contraction*, if $0 < k < 1$
The *zero function*, if $k = 0$

Now we consider a slightly more complicated function.

EXAMPLE 5 Solve the GAP for the function P defined by

$$P\left(\begin{pmatrix} x_1 \\ x_2 \end{pmatrix}\right) = \begin{pmatrix} 1 & 0 \\ 0 & 0 \end{pmatrix}\begin{pmatrix} x_1 \\ x_2 \end{pmatrix}$$

Solution As before, we calculate

$$P\left(\begin{pmatrix} x_1 \\ x_2 \end{pmatrix}\right) = \begin{pmatrix} 1 & 0 \\ 0 & 0 \end{pmatrix}\begin{pmatrix} x_1 \\ x_2 \end{pmatrix} = \begin{pmatrix} x_1 \\ 0 \end{pmatrix}$$

The effect of P is to wipe out the second component and preserve the first component. To see what this means geometrically, we consider some vectors in Fig. 2.3.4. We see that

$$P\left(\begin{pmatrix} x_1 \\ x_2 \end{pmatrix}\right)$$

is the *projection* of

$$\begin{pmatrix} x_1 \\ x_2 \end{pmatrix}$$

onto the x axis. Therefore,

Figure 2.3.4
Actions of P from Example 5.

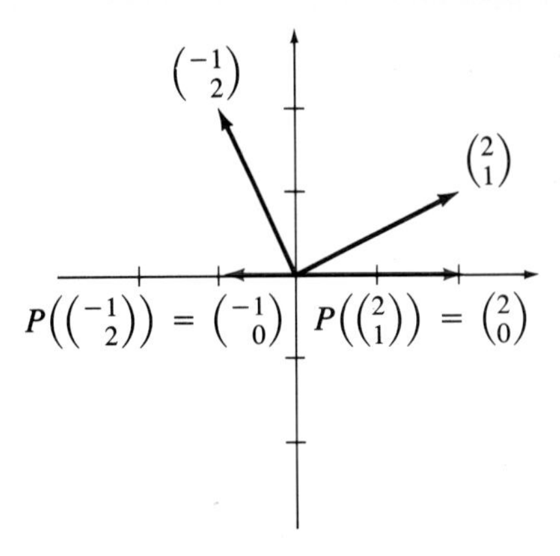

If

$$f\left(\begin{pmatrix}x_1\\x_2\end{pmatrix}\right)=\begin{pmatrix}1&0\\0&0\end{pmatrix}\begin{pmatrix}x_1\\x_2\end{pmatrix}$$

then f is a projection onto the x axis. Similarly if

$$f\left(\begin{pmatrix}x_1\\x_2\end{pmatrix}\right)=\begin{pmatrix}0&0\\0&1\end{pmatrix}\begin{pmatrix}x_1\\x_2\end{pmatrix}$$

then f is a projection onto the y axis.

Now we can analyze a slightly more complicated function. Consider

$$f\left(\begin{pmatrix}x_1\\x_2\end{pmatrix}\right)=\begin{pmatrix}2&0\\0&0\end{pmatrix}\begin{pmatrix}x_1\\x_2\end{pmatrix}$$

and calculate

$$f\left(\begin{pmatrix}x_1\\x_2\end{pmatrix}\right)=\begin{pmatrix}2x_1\\0\end{pmatrix}=\underbrace{2}_{\text{Dilation}}\underbrace{\begin{pmatrix}x_1\\0\end{pmatrix}}_{\text{Projection}}$$

In this case the original vector is first projected onto the x axis and then stretched to twice its length. That is, this function is geometrically a *projection* onto the x axis, followed by a *dilation* with constant 2. But the action could also be thought of as the *dilation first* and *projection second.*

In terms of the matrices defining the functions, this means that

$$\begin{pmatrix}2&0\\0&0\end{pmatrix}\begin{pmatrix}x_1\\ \end{pmatrix}=\underbrace{\begin{pmatrix}2&0\\0&2\end{pmatrix}}_{\substack{\text{Dilation}\\ \text{2d}}}\underbrace{\begin{pmatrix}1&0\\0&0\end{pmatrix}}_{\substack{\text{Projection}\\ \text{1st}}}\begin{pmatrix}x_1\\x_2\end{pmatrix}$$

but since we also have

$$\begin{pmatrix} 2 & 0 \\ 0 & 0 \end{pmatrix} = \begin{pmatrix} 1 & 0 \\ 0 & 0 \end{pmatrix}\begin{pmatrix} 2 & 0 \\ 0 & 2 \end{pmatrix}$$

we can write

$$\begin{pmatrix} 2 & 0 \\ 0 & 0 \end{pmatrix}\begin{pmatrix} x_1 \\ x_2 \end{pmatrix} = \underset{\substack{\uparrow \\ \text{Projection} \\ \text{2d}}}{\begin{pmatrix} 1 & 0 \\ 0 & 0 \end{pmatrix}}\left[\underset{\substack{\uparrow \\ \text{Dilation} \\ \text{1st}}}{\begin{pmatrix} 2 & 0 \\ 0 & 2 \end{pmatrix}}\begin{pmatrix} x_1 \\ x_2 \end{pmatrix}\right]$$

This last example illustrates the following general principle:

> To solve the geometric analysis problem, break the function f down into a sequence of simple functions such as dilations, contractions, projections, and others.

To use the principle, another simple linear function, rotation, is described.

EXAMPLE 6 For f defined by

$$f\left(\begin{pmatrix} x_1 \\ x_2 \end{pmatrix}\right) = \begin{pmatrix} 0 & -1 \\ 1 & 0 \end{pmatrix}\begin{pmatrix} x_1 \\ x_2 \end{pmatrix} \cdot$$

calculate

$$f\left(\begin{pmatrix} 1 \\ 1 \end{pmatrix}\right) \quad \text{and} \quad f\left(\begin{pmatrix} 0 \\ 1 \end{pmatrix}\right)$$

Solution We calculate

$$f\left(\begin{pmatrix} 1 \\ 1 \end{pmatrix}\right) = \begin{pmatrix} -1 \\ 1 \end{pmatrix}$$

and graph as in Fig. 2.3.5*a*. Also

$$f\begin{pmatrix} 0 \\ 1 \end{pmatrix} = \begin{pmatrix} -1 \\ 0 \end{pmatrix}$$

The graph is shown in Fig. 2.3.5*b*.

In Example 6 in each case the function gave a vector of the same length and rotated $\pi/2$ radians counterclockwise from the original vector. Of course, this does not prove that the function is a rotation of $\pi/2$ for *every* vector. However, we can show that the length of a vector is unchanged by this function and that

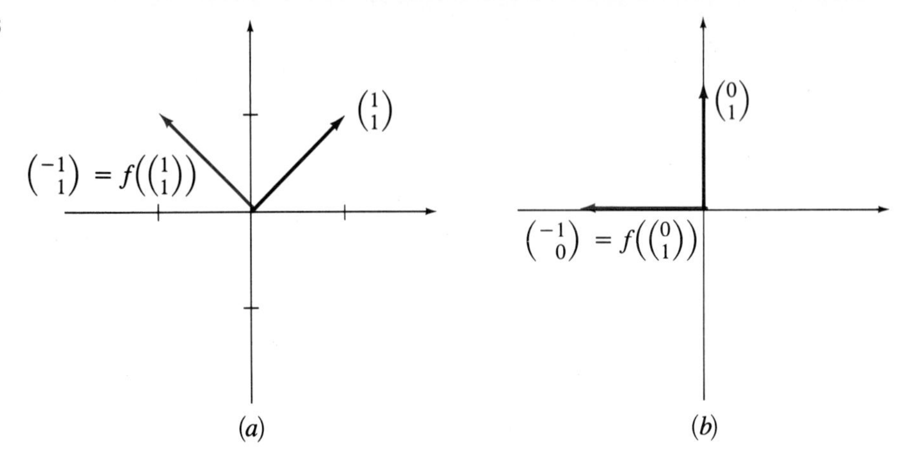

Figure 2.3.5 (*a*) Action of f from Example 6. (*b*) Action of f from Example 6.

$$f\left(\begin{pmatrix}x_1\\x_2\end{pmatrix}\right) \perp \begin{pmatrix}x_1\\x_2\end{pmatrix}$$

Since

$$f\left(\begin{pmatrix}x_1\\x_2\end{pmatrix}\right) = \begin{pmatrix}0 & -1\\1 & 0\end{pmatrix}\begin{pmatrix}x_1\\x_2\end{pmatrix} = \begin{pmatrix}-x_2\\x_1\end{pmatrix}$$

we have

$$\left|f\left(\begin{pmatrix}x_1\\x_2\end{pmatrix}\right)\right| = \left|\begin{matrix}-x_2\\x_1\end{matrix}\right| = \sqrt{(-x_2)^2 + {x_1}^2} = \sqrt{{x_2}^2 + {x_1}^2} = \left|\begin{pmatrix}x_1\\x_2\end{pmatrix}\right|$$

and

$$f\left(\begin{pmatrix}x_1\\x_2\end{pmatrix}\right) \cdot \begin{pmatrix}x_1\\x_2\end{pmatrix} = \begin{pmatrix}-x_2\\x_1\end{pmatrix} \cdot \begin{pmatrix}x_1\\x_2\end{pmatrix} = -x_2x_1 + x_1x_2 = 0$$

The last equation says only that

$$f\left(\begin{pmatrix}x_1\\x_2\end{pmatrix}\right) \quad \text{is perpendicular to} \quad \begin{pmatrix}x_1\\x_2\end{pmatrix}$$

The equation does not tell us that the right angle resulted from rotation. However, it is shown in trigonometry that a counterclockwise rotation of x, y coordinates through an angle of θ is defined by

$$x' = (\cos \theta)x - (\sin \theta)y$$
$$y' = (\sin \theta)x + (\cos \theta)y$$

which we can write in matrix form as

$$\begin{pmatrix}x'\\y'\end{pmatrix} = \begin{pmatrix}\cos \theta & -\sin \theta\\ \sin \theta & \cos \theta\end{pmatrix}\begin{pmatrix}x\\y\end{pmatrix}$$

If $\theta = \pi/2$, the matrix is

$$\begin{pmatrix} 0 & -1 \\ 1 & 0 \end{pmatrix}$$

which is exactly the matrix that defines the function we have been working with.

Therefore, the result from trigonometry tells us the following:

A function of the form

$$f\left(\begin{pmatrix} x_1 \\ x_2 \end{pmatrix}\right) = \begin{pmatrix} \cos\theta & -\sin\theta \\ \sin\theta & \cos\theta \end{pmatrix}\begin{pmatrix} x_1 \\ x_2 \end{pmatrix}$$

is a rotation counterclockwise through an angle of θ.

EXAMPLE 7 Identify the counterclockwise angle of rotation for the following functions.

(*a*) $f\left(\begin{pmatrix} x_1 \\ x_2 \end{pmatrix}\right) = \begin{pmatrix} 1 & 0 \\ 0 & 1 \end{pmatrix}\begin{pmatrix} x_1 \\ x_2 \end{pmatrix}$

(*b*) $f\left(\begin{pmatrix} x_1 \\ x_2 \end{pmatrix}\right) = \begin{pmatrix} -1 & 0 \\ 0 & -1 \end{pmatrix}\begin{pmatrix} x_1 \\ x_2 \end{pmatrix}$

(*c*) $f\left(\begin{pmatrix} x_1 \\ x_2 \end{pmatrix}\right) = \begin{pmatrix} 1/\sqrt{2} & -1/\sqrt{2} \\ 1/\sqrt{2} & 1/\sqrt{2} \end{pmatrix}\begin{pmatrix} x_1 \\ x_2 \end{pmatrix}$

(*d*) $f\left(\begin{pmatrix} x_1 \\ x_2 \end{pmatrix}\right) = \begin{pmatrix} -1/\sqrt{2} & -1/\sqrt{2} \\ 1/\sqrt{2} & -1/\sqrt{2} \end{pmatrix}\begin{pmatrix} x_1 \\ x_2 \end{pmatrix}$

Solution In general, we know that a rotation matrix appears as

$$\begin{pmatrix} a & -b \\ b & a \end{pmatrix} = \begin{pmatrix} \cos\theta & -\sin\theta \\ \sin\theta & \cos\theta \end{pmatrix}$$

To determine θ, we set $a = \cos\theta$ and $b = \sin\theta$ and solve for θ: $\theta = \cos^{-1} a$, $\theta = \sin^{-1} b$. This can be solved by using a table, a calculator, or (in simple cases) memory.

(*a*) We have $\cos\theta = 1$ and $\sin\theta = 0$, so $\theta = 0$. This makes sense because the function is the identity function.

(*b*) We have $\cos\theta = -1$ and $\sin\theta = 0$, so $\theta = \pi$. This makes sense because

$$f\left(\begin{pmatrix} x_1 \\ x_2 \end{pmatrix}\right) = \begin{pmatrix} -x_1 \\ -x_2 \end{pmatrix}$$

and

$$\begin{pmatrix} -x_1 \\ -x_2 \end{pmatrix}$$

is just the reversal of

$$\begin{pmatrix} x_1 \\ x_2 \end{pmatrix}$$

(*c*) In this case $\cos\theta = 1/\sqrt{2}$, and we find θ is $\pi/4$ or $7\pi/4$. Then using $\sin\theta = 1/\sqrt{2}$, we find that θ is $\pi/4$ or $3\pi/4$. Since $\pi/4$ is the common solution, that is the angle of rotation.

(*d*) Working as in (*c*), we have $\cos\theta = -1/\sqrt{2}$ so θ is $3\pi/4$ or $5\pi/4$. Also $\sin\theta = 1/\sqrt{2}$ which implies that θ is $\pi/4$ or $3\pi/4$. Therefore the angle of rotation is $3\pi/4$ (counterclockwise).

Now we can solve the geometric analysis problem for some more complicated functions. Consider

$$f\left(\begin{pmatrix} x_1 \\ x_2 \end{pmatrix}\right) = \begin{pmatrix} 0 & 0 \\ 2 & 0 \end{pmatrix}\begin{pmatrix} x_1 \\ x_2 \end{pmatrix} = \begin{pmatrix} 0 \\ 2x_1 \end{pmatrix}$$

This is an interesting function since f composed with itself is

$$f\left(f\left(\begin{pmatrix} x_1 \\ x_2 \end{pmatrix}\right)\right) = f\left(\begin{pmatrix} 0 \\ 2x_1 \end{pmatrix}\right) = \begin{pmatrix} 0 & 0 \\ 2 & 0 \end{pmatrix}\begin{pmatrix} 0 \\ 2x_1 \end{pmatrix} = \begin{pmatrix} 0 \\ 0 \end{pmatrix}$$

That is, if f is applied twice (or more), we get only the zero vector (note that the matrix is nilpotent). To analyze this function, we notice that x_1, the first component of

$$\begin{pmatrix} x_1 \\ x_2 \end{pmatrix}$$

is moved by f to the second component. Since this could be done by a rotation of $\pi/2$, we might expect this rotation to be involved. Also since x_1 is doubled by f, we might expect a dilation with constant 2 also to be involved. We further notice that f destroys x_2, so f probably contains a projection onto the x_1 axis. Now try to put this information together by representing the individual functions with matrices:

Projection onto x axis	$\begin{pmatrix} 1 & 0 \\ 0 & 0 \end{pmatrix}$
Dilation with constant 2	$\begin{pmatrix} 2 & 0 \\ 0 & 2 \end{pmatrix}$
Rotation of $\frac{\pi}{2}$	$\begin{pmatrix} 0 & -1 \\ 1 & 0 \end{pmatrix}$

There are several different orders in which these could be applied to

$$\begin{pmatrix} x_1 \\ x_2 \end{pmatrix}$$

For example,

$$\underset{\text{Projection}}{\begin{pmatrix} 1 & 0 \\ 0 & 0 \end{pmatrix}} \underset{\text{Dilation}}{\begin{pmatrix} 2 & 0 \\ 0 & 2 \end{pmatrix}} \underset{\text{Rotation}}{\begin{pmatrix} 0 & -1 \\ 1 & 0 \end{pmatrix}} \begin{pmatrix} x_1 \\ x_2 \end{pmatrix}$$

which gives

$$\begin{aligned} \begin{pmatrix} 1 & 0 \\ 0 & 0 \end{pmatrix} \begin{pmatrix} 2 & 0 \\ 0 & 2 \end{pmatrix} \begin{pmatrix} -x_2 \\ x_1 \end{pmatrix} &= \begin{pmatrix} 1 & 0 \\ 0 & 0 \end{pmatrix} \begin{pmatrix} -2x_2 \\ 2x_1 \end{pmatrix} \\ &= \begin{pmatrix} -2x_2 \\ 0 \end{pmatrix} \neq \begin{pmatrix} 0 \\ 2x_1 \end{pmatrix} = f\begin{pmatrix} x_1 \\ x_2 \end{pmatrix} \end{aligned}$$

So this order is not correct. However,

$$\begin{aligned} \underset{\substack{\text{Rotation} \\ \text{(3d)}}}{\begin{pmatrix} 0 & -1 \\ 1 & 0 \end{pmatrix}} \underset{\substack{\text{Projection} \\ \text{(2d)}}}{\begin{pmatrix} 1 & 0 \\ 0 & 0 \end{pmatrix}} \underset{\substack{\text{Dilation} \\ \text{(1st)}}}{\begin{pmatrix} 2 & 0 \\ 0 & 2 \end{pmatrix}} \begin{pmatrix} x_1 \\ x_2 \end{pmatrix} &= \begin{pmatrix} 0 & -1 \\ 1 & 0 \end{pmatrix} \begin{pmatrix} 1 & 0 \\ 0 & 0 \end{pmatrix} \begin{pmatrix} 2x_1 \\ 2x_1 \end{pmatrix} \\ &= \begin{pmatrix} 0 & -1 \\ 1 & 0 \end{pmatrix} \begin{pmatrix} 2x_1 \\ 0 \end{pmatrix} \\ &= \begin{pmatrix} 0 \\ 2x_1 \end{pmatrix} = f\begin{pmatrix} x_1 \\ x_2 \end{pmatrix} \end{aligned}$$

This means that f consists of first, a dilation with constant 2; second, a projection onto the x axis; and third, a rotation of $\pi/2$. See Fig. 2.3.6

Note that the matrix representing f satisfies

$$\begin{aligned} \begin{pmatrix} 0 & 0 \\ 2 & 0 \end{pmatrix} &= \begin{pmatrix} 0 & -1 \\ 1 & 0 \end{pmatrix} \begin{pmatrix} 1 & 0 \\ 0 & 0 \end{pmatrix} \begin{pmatrix} 2 & 0 \\ 0 & 2 \end{pmatrix} \\ &= \begin{pmatrix} 0 & -1 \\ 1 & 0 \end{pmatrix} \begin{pmatrix} 2 & 0 \\ 0 & 2 \end{pmatrix} \begin{pmatrix} 1 & 0 \\ 0 & 0 \end{pmatrix} \end{aligned}$$

So the dilation and projection can occur in reverse order. Since matrix multiplication is not commutative, only certain orders of the geometric operations will result in the action of f. In this last example, there are six possible ways to order the operations:

$$\begin{pmatrix} 0 & -1 \\ 1 & 0 \end{pmatrix} \begin{pmatrix} 1 & 0 \\ 0 & 0 \end{pmatrix} \begin{pmatrix} 2 & 0 \\ 0 & 2 \end{pmatrix} = \begin{pmatrix} 0 & 0 \\ 2 & 0 \end{pmatrix} \qquad \text{Yes}$$

$$\begin{pmatrix} 0 & -1 \\ 1 & 0 \end{pmatrix} \begin{pmatrix} 2 & 0 \\ 0 & 2 \end{pmatrix} \begin{pmatrix} 1 & 0 \\ 0 & 0 \end{pmatrix} = \begin{pmatrix} 0 & 0 \\ 2 & 0 \end{pmatrix} \qquad \text{Yes}$$

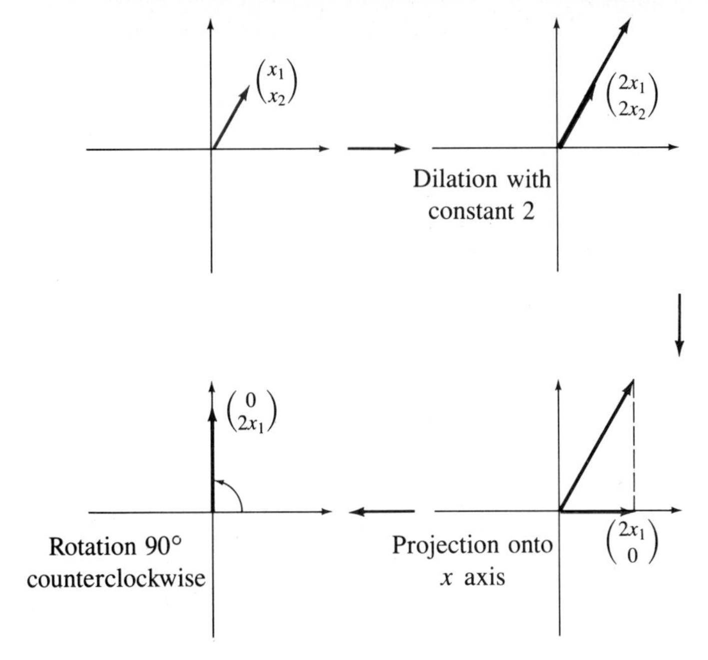

Figure 2.3.6 Geometrical analysis of

$$f\begin{bmatrix} x_1 \\ x_2 \end{bmatrix} = \begin{bmatrix} 0 & 0 \\ 2 & 0 \end{bmatrix}\begin{bmatrix} x_1 \\ x_2 \end{bmatrix} = \begin{bmatrix} 0 \\ 2x_1 \end{bmatrix}$$

$$\begin{pmatrix} 2 & 0 \\ 0 & 2 \end{pmatrix}\begin{pmatrix} 0 & -1 \\ 1 & 0 \end{pmatrix}\begin{pmatrix} 1 & 0 \\ 0 & 0 \end{pmatrix} = \begin{pmatrix} 2 & 0 \\ 0 & 0 \end{pmatrix} \qquad \text{No}$$

$$\begin{pmatrix} 1 & 0 \\ 0 & 0 \end{pmatrix}\begin{pmatrix} 0 & -1 \\ 1 & 0 \end{pmatrix}\begin{pmatrix} 2 & 0 \\ 0 & 2 \end{pmatrix} = \begin{pmatrix} 0 & -2 \\ 0 & 0 \end{pmatrix} \qquad \text{No}$$

$$\begin{pmatrix} 2 & 0 \\ 0 & 2 \end{pmatrix}\begin{pmatrix} 1 & 0 \\ 0 & 0 \end{pmatrix}\begin{pmatrix} 0 & -1 \\ 1 & 0 \end{pmatrix} = \begin{pmatrix} 0 & -2 \\ 0 & 0 \end{pmatrix} \qquad \text{No}$$

$$\begin{pmatrix} 1 & 0 \\ 0 & 0 \end{pmatrix}\begin{pmatrix} 2 & 0 \\ 0 & 2 \end{pmatrix}\begin{pmatrix} 0 & -1 \\ 0 & 0 \end{pmatrix} = \begin{pmatrix} 0 & -2 \\ 0 & 0 \end{pmatrix} \qquad \text{No}$$

We see from this chart that the rotation must occur last.

EXAMPLE 8 Analyze the linear function

$$f\left(\begin{pmatrix} x_1 \\ x_2 \end{pmatrix}\right) = \begin{pmatrix} 2 & 0 \\ 0 & 3 \end{pmatrix}\begin{pmatrix} x_1 \\ x_2 \end{pmatrix} = \begin{pmatrix} 2x_1 \\ 3x_2 \end{pmatrix}$$

Solution Note that

$$\begin{pmatrix} 2 & 0 \\ 0 & 3 \end{pmatrix} = \begin{pmatrix} 2 & 0 \\ 0 & 0 \end{pmatrix} + \begin{pmatrix} 0 & 0 \\ 0 & 3 \end{pmatrix}$$

so that

$$f\left(\begin{pmatrix} x_1 \\ x_2 \end{pmatrix}\right) = \begin{pmatrix} 2 & 0 \\ 0 & 0 \end{pmatrix}\begin{pmatrix} x_1 \\ x_2 \end{pmatrix} + \begin{pmatrix} 0 & 0 \\ 0 & 3 \end{pmatrix}\begin{pmatrix} x_1 \\ x_2 \end{pmatrix}$$

Since f can be split into the sum of the linear functions g and h, we know that f can be analyzed by analyzing g and h. We have already seen that g is a projection onto the x axis followed by a dilation with constant 2. The function h is similar: It is a projection onto the y axis followed by a dilation with constant 3. Therefore the action of f is the vector sum of (1) projection onto x followed by a dilation, constant 2, and (2) projection onto y followed by dilation, constant 3. The action of f is shown in Fig. 2.3.7.

EXAMPLE 9 (Computer graphics) A flat object can be projected on a computer terminal screen because the screen can be coordinatized. That is, a point on the screen can be set as the origin, and x and y axes can be declared and scaled. It may be desired to dilate, contract, or rotate the object on the screen. This can be done by multiplying the vectors representing the points on the object by the matrix representing the action, storing the new vectors, and finally projecting the new points (which are, of course, the terminal points of the vectors). One action which cannot be represented solely by multiplication by a matrix is translation:

$$T_{h,k}\left(\begin{pmatrix} x_1 \\ x_2 \end{pmatrix}\right) = \left(\begin{pmatrix} x_1 + h \\ x_2 + k \end{pmatrix}\right)$$

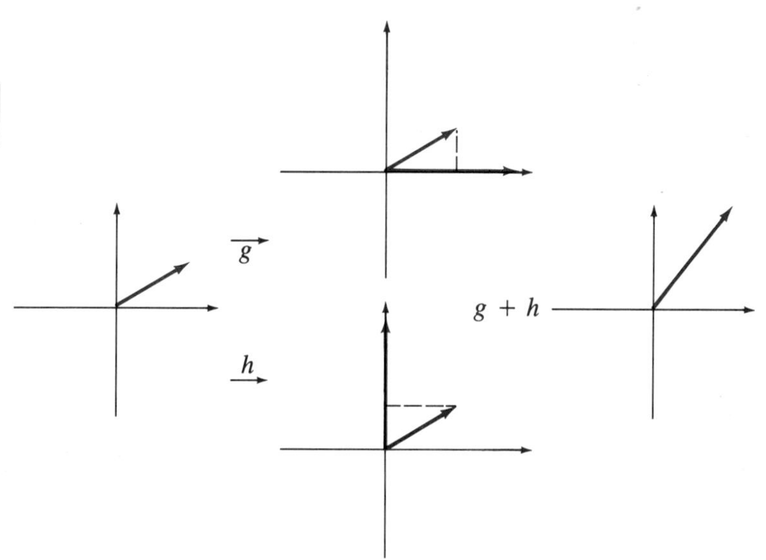

Figure 2.3.7 Action of $f\left[\begin{bmatrix} x_1 \\ x_2 \end{bmatrix}\right] = \begin{bmatrix} 2 & 0 \\ 0 & 3 \end{bmatrix}\begin{bmatrix} x_1 \\ x_2 \end{bmatrix}$

Table 2.3.2 **MATRICES M FOR FUNCTIONS $f\left(\begin{pmatrix} x_1 \\ x_2 \\ x_3 \end{pmatrix}\right) = M\begin{pmatrix} x_1 \\ x_2 \\ x_3 \end{pmatrix}$**

MATRICES	GEOMETRIC ACTION
$\begin{pmatrix} 1 & 0 & 0 \\ 0 & 0 & 0 \\ 0 & 0 & 0 \end{pmatrix} \begin{pmatrix} 0 & 0 & 0 \\ 0 & 1 & 0 \\ 0 & 0 & 0 \end{pmatrix} \begin{pmatrix} 0 & 0 & 0 \\ 0 & 0 & 0 \\ 0 & 0 & 1 \end{pmatrix}$	Projection onto x axis, y axis, z axis, respectively
$\begin{pmatrix} 1 & 0 & 0 \\ 0 & 1 & 0 \\ 0 & 0 & 0 \end{pmatrix} \begin{pmatrix} 1 & 0 & 0 \\ 0 & 0 & 0 \\ 0 & 0 & 1 \end{pmatrix} \begin{pmatrix} 0 & 0 & 0 \\ 0 & 1 & 0 \\ 0 & 0 & 1 \end{pmatrix}$	Projection onto xy plane, xz plane, yz plane, respectively
$\begin{pmatrix} \cos\theta & -\sin\theta & 0 \\ \sin\theta & \cos\theta & 0 \\ 0 & 0 & 1 \end{pmatrix}$	Rotation counterclockwise with z fixed
$\begin{pmatrix} k & 0 & 0 \\ 0 & k & 0 \\ 0 & 0 & k \end{pmatrix}$	$k > 1$ Dilation $k = 1$ Identity $0 < k < 1$ Contraction $k = 0$ Zero function

Translation is *non*linear, because

$$T_{h,k}\left(r\begin{pmatrix} x_1 \\ x_2 \end{pmatrix}\right) = T_{h,k}\left(\begin{pmatrix} rx_1 \\ rx_2 \end{pmatrix}\right) = \begin{pmatrix} rx_1 + h \\ rx_2 + k \end{pmatrix}$$

and

$$rT_{h,k}\left(\begin{pmatrix} x_1 \\ x_2 \end{pmatrix}\right) = \begin{pmatrix} rx + rh \\ rx + rk \end{pmatrix}$$

The examples so far have involved only linear functions with domain $\mathbb{R}^2$. Some simple examples of linear functions with domain $\mathbb{R}^3$ are given in Table 2.3.2.

In order to analyze more complicated linear functions generated by matrices or generated by matrices of larger size we need more sophisticated techniques which we will develop later in the text. This will be done by solving the *eigenvalue problem* in Chapter 5.

PROBLEMS 2.3

1. Solve the geometric analysis problem for the following functions.

(*a*) $f\left(\begin{pmatrix} x_1 \\ x_2 \end{pmatrix}\right) = \begin{pmatrix} 3 & 0 \\ 0 & 3 \end{pmatrix}\begin{pmatrix} x_1 \\ x_2 \end{pmatrix}$

(*b*) $f\left(\begin{pmatrix} x_1 \\ x_2 \end{pmatrix}\right) = \begin{pmatrix} \frac{1}{2} & 0 \\ 0 & \frac{1}{2} \end{pmatrix}\begin{pmatrix} x_1 \\ x_2 \end{pmatrix}$

(*c*) $f\left(\begin{pmatrix} x_1 \\ x_2 \end{pmatrix}\right) = \begin{pmatrix} -2 & 0 \\ 0 & -2 \end{pmatrix}\begin{pmatrix} x_1 \\ x_2 \end{pmatrix}$

2. Solve the geometric analysis problem for the following functions.

(*a*) $f\left(\begin{pmatrix} x_1 \\ x_2 \end{pmatrix}\right) = \begin{pmatrix} 0 & 0 \\ 0 & 1 \end{pmatrix}\begin{pmatrix} x_1 \\ x_2 \end{pmatrix}$ (*b*) $f\left(\begin{pmatrix} x_1 \\ x_2 \end{pmatrix}\right) = \begin{pmatrix} 0 & 0 \\ 0 & 3 \end{pmatrix}\begin{pmatrix} x_1 \\ x_2 \end{pmatrix}$

3. Solve the geometric analysis problem for the following functions.

(*a*) $f\left(\begin{pmatrix} x_1 \\ x_2 \end{pmatrix}\right) = \begin{pmatrix} 0 & 1 \\ 0 & 0 \end{pmatrix}\begin{pmatrix} x_1 \\ x_2 \end{pmatrix}$ (*b*) $f\left(\begin{pmatrix} x_1 \\ x_2 \end{pmatrix}\right) = \begin{pmatrix} 0 & 1 \\ -1 & 0 \end{pmatrix}\begin{pmatrix} x_1 \\ x_2 \end{pmatrix}$

4. Solve the geometric analysis problem for the following functions.

(*a*) $f\left(\begin{pmatrix} x_1 \\ x_2 \end{pmatrix}\right) = \begin{pmatrix} 3 & 0 \\ 0 & 5 \end{pmatrix}\begin{pmatrix} x_1 \\ x_2 \end{pmatrix}$ (*b*) $f\left(\begin{pmatrix} x_1 \\ x_2 \end{pmatrix}\right) = \begin{pmatrix} -1 & 0 \\ 0 & 2 \end{pmatrix}\begin{pmatrix} x_1 \\ x_2 \end{pmatrix}$

5. For linear functions of the form

$$f\left(\begin{pmatrix} x_1 \\ x_2 \end{pmatrix}\right) = \begin{pmatrix} a & b \\ c & d \end{pmatrix}\begin{pmatrix} x_1 \\ x_2 \end{pmatrix} = A\begin{pmatrix} x_1 \\ x_2 \end{pmatrix}$$

where A^{-1} exists, we say f is *invertible* and that

$$g\left(\begin{pmatrix} x_1 \\ x_2 \end{pmatrix}\right) = A^{-1}\begin{pmatrix} x_1 \\ x_2 \end{pmatrix}$$

is the inverse function of f. Show that (*a*) the inverse of dilation is contraction, (*b*) the inverse of rotation of angle θ is rotation of angle $-\theta$, and (*c*) projection onto the x axis is not invertible.

6. Show geometrically that

$$f\left(\begin{pmatrix} x_1 \\ x_2 \end{pmatrix}\right) = \begin{pmatrix} -1 & 0 \\ 0 & 1 \end{pmatrix}\begin{pmatrix} x_1 \\ x_2 \end{pmatrix}$$

represents a reflection about the y axis. What would

$$f\left(\begin{pmatrix} x_1 \\ x_2 \end{pmatrix}\right) = \begin{pmatrix} 1 & 0 \\ 0 & -1 \end{pmatrix}\begin{pmatrix} x_1 \\ x_2 \end{pmatrix}$$

represent?

7. Analyze

$$f\left(\begin{pmatrix} x_1 \\ x_2 \end{pmatrix}\right) = \begin{pmatrix} |x_1| \\ |x_2| \end{pmatrix}$$

even though it is not a linear function. (*Hint*: Consider cases such as $\{x_1 > 0, x_2 > 0\}$, $\{x_1 < 0, x_2 > 0\}$, and so on. Is f "piecewise linear"?)

8. Analyze

$$f\left(\begin{pmatrix} x_1 \\ x_2 \end{pmatrix}\right) = \begin{pmatrix} \sqrt{x_1^2 + x_2^2} \\ 0 \end{pmatrix}$$

even though it is not a linear function. (*Hint*: Consider what f does to circles of different radius.)

2.4 APPLICATIONS TO ANALYTIC GEOMETRY

We can use vectors to derive equations of lines and planes in three-space. Also vectors make the solution of some geometric problems fairly easy.

If we think of a desk top as representing a plane and stand a pencil (unsharpened) on it (see Fig. 2.4.1), the pencil points in a direction perpendicular to any line on the desk top. Such a vector is called a *normal vector*.

DEFINITION 2.4.1 A vector **N** which is perpendicular to all vectors in a plane P is called a *normal vector* for the plane. A normal vector for a plane is said to be perpendicular to the plane.

A plane can be specified by giving a normal to the plane and a point in the plane. This is similar to the case for lines in the plane for which two quantities specify a line: the slope and a point. In three-space, the normal vector serves the function of the slope in two-space. Note that two planes are parallel if their normals are parallel.

To find an equation for a plane P, let the point in P be (x_0, y_0, z_0) and the normal be $\mathbf{N} = (a, b, c)$. If (x, y, z) is any other point in P (see Fig. 2.4.2), then the vector $(x - x_0, y - y_0, z - z_0)$ lies in P and $\mathbf{N} \cdot (x - x_0, y - y_0, z - z_0) = 0$ since **N** is perpendicular to all vectors in P. Writing the dot product, we have

$$\begin{aligned} 0 = \mathbf{N} \cdot (x - x_0, y - y_0, z - z_0) &= (a, b, c) \cdot (x - x_0, y - y_0, z - z_0) \\ &= a(x - x_0) + b(y - y_0) + c(z - z_0) \end{aligned}$$

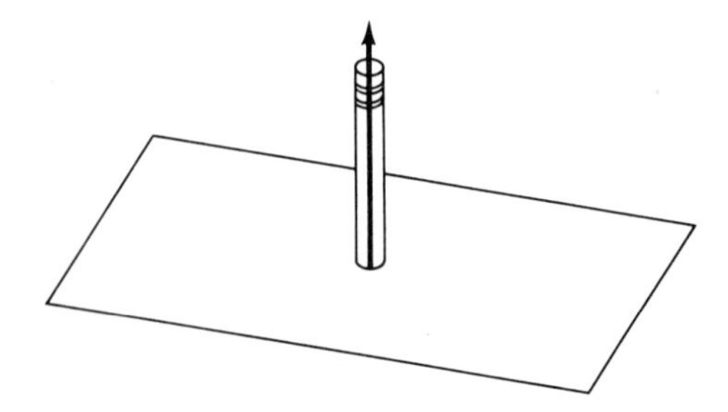

Figure 2.4.1
A pencil as normal vector.

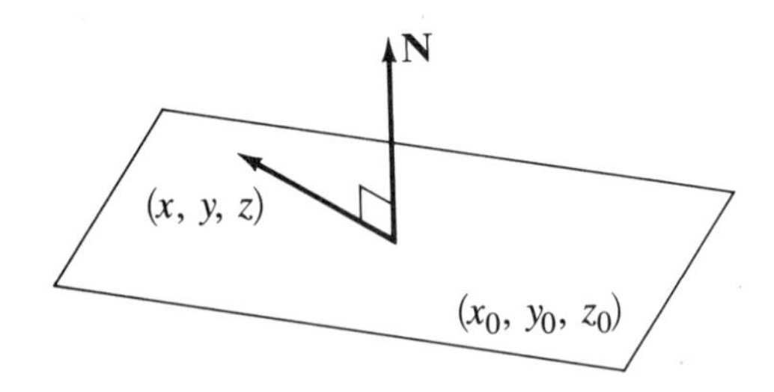

Figure 2.4.2
Normal vector.

Therefore,

> An equation of the plane containing (x_0, y_0, z_0) with normal (a, b, c) is
>
> $$a(x - x_0) + b(y - y_0) + c(z - z_0) = 0$$
>
> This form of the equation is called the *point-normal form* since the coordinates of the point and components of the normal appear explicitly. The equation
>
> $$(a, b, c) \cdot (x - x_0, y - y_0, z - z_0) = 0$$
>
> is called the *vector form* of equation for the plane.

EXAMPLE 1 Find the point-normal form of equation for the plane passing through $(1, -2, 4)$, having normal vector $(2, 3, -1)$.

Solution The vector form is

$$(2, 3, -1) \cdot (x - 1, y - (-2), z - 4) = 0$$

Writing out the dot product, we find

$$2(x - 1) + 3(y + 2) - 1(z - 4) = 0$$

EXAMPLE 2 A plane P has equation

$$2(x - 2) - 7(x - 3) + 4(x + 2) = 0$$

Find a normal vector to P and a point in P.

Solution Since the equation is in point-normal form, the desired information can be read off:

$$\mathbf{N} = (2, -7, 4)$$
$$\text{Point } (2, 3, -2)$$

The equation found in Example 1 could be simplified to

$$2x + 3y - z + 8 = 0$$

This equation fits the *general form*

$$ax + by + cz + d = 0$$

From the general form the normal vector can still be read. Because the equation is linear, a plane is called a *linear structure* in three-space.

EXAMPLE 3 A plane P has equation

$$2x - 3y + 4z + 12 = 0$$

Find a normal vector and two points in the plane.

Solution A normal is $(2, -3, 4)$. Any nonzero multiple of this vector is also a normal. To find points in the plane, we must find values for x, y, and z which satisfy the equation. To do this, we can give values to any two of the variables and solve for the third. If we let $x = 2$ and $y = 0$, we find $z = -4$, so $(2, 0, -4)$ is in the plane. If we let $y = 0$ and $z = 0$, we find $x = -6$ and obtain $(-6, 0, 0)$ as another point in the plane.

The cross product can sometimes be used to find a normal.

EXAMPLE 4 A plane P contains the points p_1: $(1, 1, 2)$, p_2: $(0, -3, 4)$ and p_3: $(2, 0, 3)$. Find a normal vector for P and an equation for P.

Solution By subtracting points we find as vectors in P: $(0, -3, 4) - (1, 1, 2) = (-1, -4, 2) = \mathbf{A}$ and $(2, 0, 3) - (1, 1, 2) = (1, -1, 1) = \mathbf{B}$. Now $\mathbf{A} \times \mathbf{B}$ is perpendicular to both $\mathbf{A}$ and $\mathbf{B}$ and hence is normal to P (see Fig. 2.4.3). Now

$$\mathbf{A} \times \mathbf{B} = (-2, 3, 5)$$

A vector equation is

$$(-2, 3, 5) \cdot (x - 1, y - 1, z - 2) = 0$$

which is equivalent to

$$-2(x - 1) + 3(y - 1) + 5(z - 2) = 2$$

or

$$-2x + 3y + 5z - 11 = 0$$

A straight line through the point (x_0, y_0, z) is the set of all points (x, y, z) such that the vector from (x_0, y_0, z_0) to (x, y, z) is a multiple of a given vector

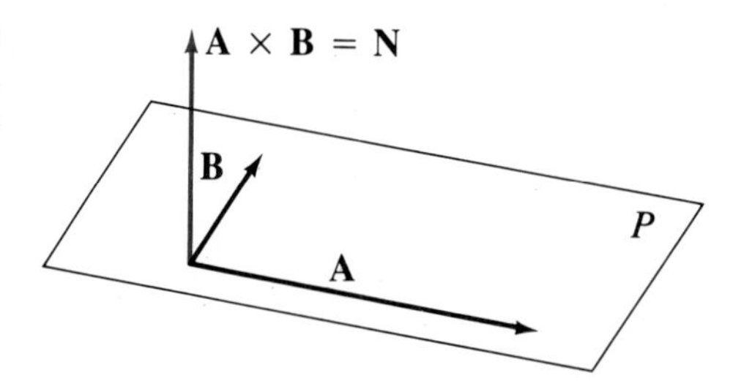

Figure 2.4.3 $A \times B$ as a normal vector.

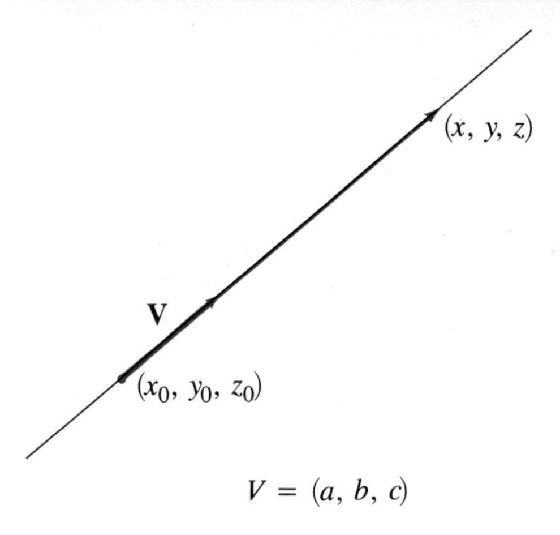

Figure 2.4.4
Line in space.

$\mathbf{V}$ (see Fig. 2.4.4). From Fig. 2.4.4 we see that

$$(x - x_0, y - y_0, z - z_0) = k\mathbf{V} = k(a, b, c)$$

where k is allowed to run through the real numbers, which means

$$x - x_0 = ka \qquad y - y_0 = kb \qquad z - z_0 = kc \qquad \text{(parametric equations)}$$

or

$$\frac{x - x_0}{a} = k = \frac{y - y_0}{b} = k = \frac{z - z_0}{c}$$

(unless a, b, or c is zero).

EXAMPLE 5 Find equations of the line passing through $(1, -1, 2)$ and $(7, 0, 5)$. Is $(-5, -2, -1)$ on the line? What about $(13, 1, 14)$?

Solution We have a point (in fact, we have two) and need a vector in the direction of the line. For this purpose a vector *in* the line will do. So we subtract the points to get

$$\mathbf{V} = (6, 1, 3)$$

Using $(1, -1, 2)$ as a point in the line, we see that

$$(x - 1, y + 1, z - 2) = k(6, 1, 3)$$

is a vector equation for the line. Parametric equations are

$$\begin{aligned} x - 1 &= 6k & \qquad x &= 6k + 1 \\ y + 1 &= k & \quad \text{or} \quad y &= k - 1 \qquad k \in \mathbb{R} \\ z - 2 &= 3k & \qquad z &= 3k + 2 \end{aligned}$$

To see whether $(-5, -2, -1)$ is on the line, we ask if there is a value for k (the parameter) which gives $x = -5$, $y = -2$, $z = -1$. Using the last equations, we try to solve

$$\begin{aligned} -5 &= 6k + 1 \\ -2 &= k - 1 \\ -1 &= 3k + 2 \end{aligned}$$

The first equation gives $k = -1$. This value for k also works in the second and third equations, so $(-5, -2, -1)$ is on the line.

Finally we consider the point $(13, 1, 14)$. We have this time

$$\begin{aligned} 13 &= 6k + 1 \\ 1 &= k - 1 \\ 14 &= 3k + 2 \end{aligned}$$

and the first equation implies that $k = 2$. This value for k satisfies the second equation but not the third. Thus $(13, 1, 14)$ is not on the line.

Lines can be generated as the intersection of planes. In general, two planes can intersect in one line, not intersect at all, or be the same plane.

EXAMPLE 6 Determine the line of intersection of $x + 2y - z = 3$ and $x + y - 3z = 5$.

Solution We solve the equations simultaneously:

$$\begin{aligned} x + 2y - z &= 3 \\ x + y - 3z &= 5 \end{aligned} \longrightarrow \begin{aligned} x + 2y - z &= 3 \\ -y - 2z &= 2 \end{aligned}$$

We have two equations in three unknowns. Let $z = t$, so that

$$\begin{aligned} y &= -2t - 2 \\ x &= 5t + 7 \end{aligned}$$

Thus, parametric equations for the line of intersection are

$$x = 5t + 7 \qquad y = -2t - 2 \qquad z = t \qquad t \in \mathbb{R}$$

EXAMPLE 7 Show that the planes $x - y + z = 3$ and $3x - 3y + 3z = 11$ do not intersect.

Solution Again we consider the equations simultaneously:

$$\begin{aligned} x - y + z &= 3 \\ 3x - 3y + 3z &= 11 \end{aligned} \longrightarrow \begin{aligned} x - y + z &= 3 \\ 0 &= 2 \end{aligned}$$

Since the reduced system is inconsistent, no points lie on both planes (they are parallel, since their normals are in the same direction) and so there is no line of intersection for these planes.

Note in Example 7 that if the second plane had had equation

$$3x - 3y + 3z = 9$$

then the planes would be not only parallel but also coincident, and there would be no unique line of intersection.

The use of vectors in the study of analytic geometry is a very effective method. Interested readers can find further information in texts on *vector geometry*.

PROBLEMS 2.4

1. Find vector and general forms of equations for the planes given. Sketch the planes.

(*a*) Point: $(1, -2, 1)$, normal: $(1, 2, 3)$

(*b*) Point: $(0, 0, 0)$, normal: $(3, -1, 0)$

(*c*) Point: $(-2, -3, 4)$, normal: $(1, 0, 0)$

(*d*) Point: $(1, 0, 0)$, normal: $(-1, -1, -1)$

2. For the following planes state a normal vector to the plane, a unit normal to the plane, and two points in the plane. Sketch the plane.

(*a*) $x - 3y + z = 7$ (*b*) $3x - 4y - 6z = 18$

(*c*) $-5x + 6y + z = 10$ (*d*) $x = 2$

3. Three points are given. Find vector forms and general forms of equation for the planes passing through the points. Sketch the planes.

(*a*) $(1, -1, 0)$ $(2, 1, 3)$ $(4, 6, 5)$

(*b*) $(-1, 0, 0)$ $(0, -1, 0)$ $(0, 0, -1)$

(*c*) $(2, 4, 6)$ $(2, -1, 3)$ $(2, 7, 13)$

4. Write equations for the line parallel to $\mathbf{V}$ and passing through the point P. Sketch the line.

(*a*) $\mathbf{V} = (1, -1, 3)$, P: $(0, 4, 7)$

(*b*) $\mathbf{V} = (1, 1, 0)$, P: $(-7, 1, 6)$

(*c*) $\mathbf{V} = (-1, -1, -2)$, P: $(0, 0, 0)$

5. Write equations for the lines passing through the given points. Sketch the lines.

(*a*) $(1, 0, 0)$ $(0, 1, 0)$ (*b*) $(0, 0, 0)$ $(1, 1, 1)$
(*c*) $(1, 1, 0)$ $(0, 0, 1)$ (*d*) $(1, 0, 1)$ $(0, 1, 1)$

6. Let l be the line passing through $(1, -1, 4)$ and $(2, 4, -2)$. Which of the following points lie on l?

(*a*) $(\frac{3}{2}, \frac{3}{2}, 1)$ (*b*) $(1, 5, -6)$ (*c*) $(3, 5, 7)$

(*d*) $(0, -6, 11)$ (*e*) $(0, -6, 10)$

7. Write equations for the (special) planes: the xy, xz, and yz planes.

8. Determine whether the following pairs of planes intersect. If they do and are not the same plane, give an equation for the line of intersection. Sketch the planes.

(*a*) $x - 3y + 4z = 11,\ 2x + 7y - z = 7$

(*b*) $x - 2y + z = 7,\ -5x + 10y - 5z = -30$

(*c*) $x = 2,\ x + y + z = 2$

9. Show that the line

$$\begin{aligned} x &= 2t - 6 \\ y &= t + 4 \qquad t \in \mathbb{R} \\ z &= 3t - 6 \end{aligned} \tag{1}$$

is parallel to

$$3x - 3y - z = 6 \tag{2}$$

by showing (*a*) that the normal to the plane is perpendicular to the line and (*b*) that points of form (1) do not satisfy Eq. (2) for any t.

10. Show that the line

$$\begin{aligned} x &= t - 3 \\ y &= 2t + 1 \\ z &= t + 6 \end{aligned} \tag{3}$$

intersects the plane

$$x - 2y + z = 6 \tag{4}$$

by substituting the expressions from (3) into (4) and solving for t. State the point of intersection. Sketch the line and plane.

SUMMARY We have reviewed and reinforced the reader's knowledge of elementary vector analysis; also we have introduced the idea of a *linear function generated by a matrix*. The *sum* and *difference* of vectors were given analytical definitions and geometric meaning. The *dot product* and corresponding idea of *perpendicularity* were developed.

In space and the plane, we have geometry to reinforce our analytic results about lines, planes, and linear functions. The first paragraph of this chapter points out that we must leave this comfortable situation now and make the leap into higher dimensions. Although we lose some geometric insight in our new setting and gain corresponding problems, the applications are rich and meaningful.

The point of departure in Chap. 3 lies in the generalization of our ideas of two and three vectors. This will give us new domains for linear functions of applied mathematics.

The idea of the geometric analysis of a linear function was introduced, and several examples were given. The idea of studying function geometrically is not new to the reader; after all, in high school algebra one learns to graph functions with domain and range in the real numbers. A graph is simply a geometric representation of a function. In this chapter we did not "graph" the linear functions in the examples. Instead we *described their action* in terms of definite geometric operations. As long as we stay in the plane or space, the ideas of *projection*, *rotation*, *dilation*, *contraction*, and *reflection* have a definite geometric, visual meaning for us. When we move to the higher dimensions mentioned above, we may still use the words *projection*, *dilation*, and so on, even though we cannot "see" what is happening (in five dimensions, for instance). Because of this it is important for us to now set up the machinery to study extended notions of projections, dilations, and so forth.

ADDITIONAL PROBLEMS

1. Consider the diagram of planar forces pulling on the origin in Fig. AP2.1. The forces are in equilibrium; that is, $\mathbf{A} + \mathbf{B} + \mathbf{C} = 0$. What is the vector $\boldsymbol{C}$?

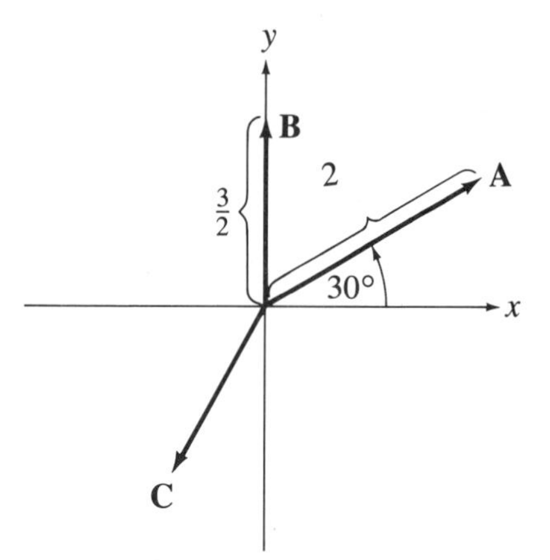

Figure AP2.1

2. If two planar forces are acting on a point in the plane, can one always impose a third force to create equilibrium? That is, can one always make the resultant the zero vector?

3. The principle of a teeter-totter is, from the picture in Fig. AP2.2, that the teeter-totter will balance if $w_1 x = w_2 y$. The principle also applies to the hanging weights in Fig. AP2.3: they will balance if $w_1 x = w_2 y$. For the more complicated system of hanging weights in Fig. AP2.4, the balancing equations, by the same principle, are

$$7w_1 = 6w_2 + 11w_3$$
$$2w_2 = 3w_3$$
$$7w_1 = 8(w_2 + w_3)$$

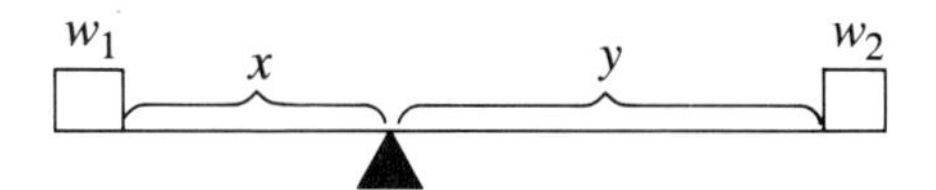

Figure AP2.2

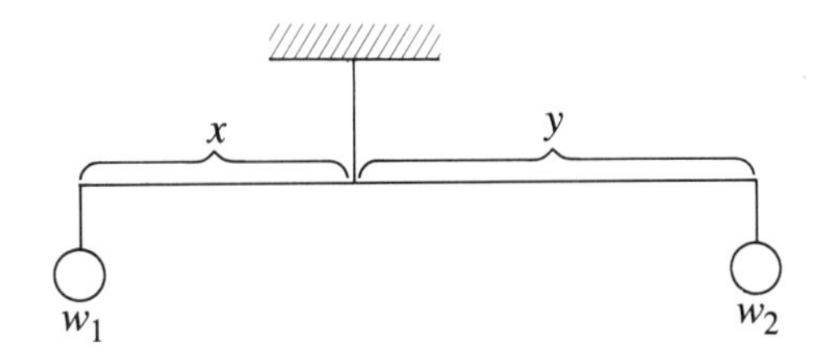

Figure AP2.3

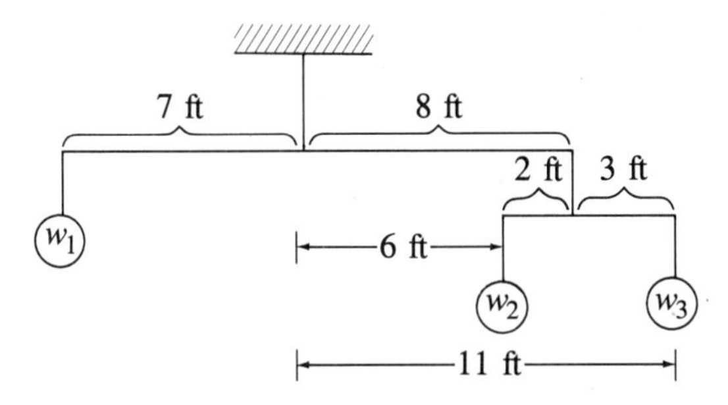

Figure AP2.4

Rewrite the equations as a homogeneous system. Solve the system. What can you say about the choice of weights to make the physical system balance?

4. Regarding the hanging weights in Prob. 3, suppose you were told that $w_1 =$ 5 lb. What would w_2 and w_3 be?

5. For the system of weights in Fig. AP2.5, the equations of balance are

$$w_3 = w_1 + w_2$$
$$2w_1 = kw_2$$
$$5w_3 = 7w_1 + (5 - k)w_2$$

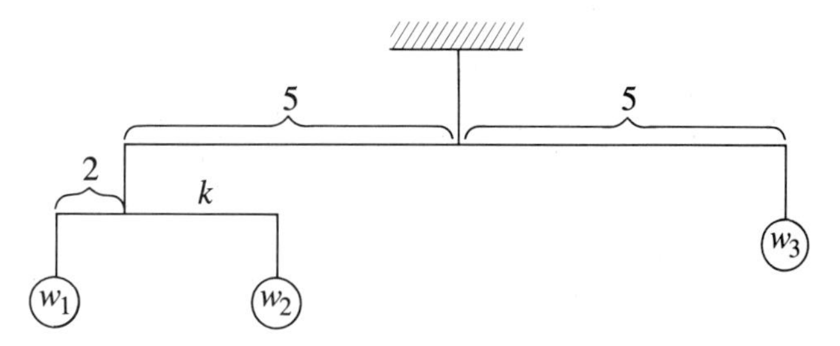

Figure AP2.5

What value of k guarantees a nontrivial solution and therefore a set of weights to balance the system?

6. Show that if each component planar force acting on a point is doubled, then the resultant is doubled. Use a matrix to represent the dilation.

7. Show that if one rotates by θ each component planar force acting on a point, then the resultant is rotated by that same angle.

8. Show that if you reflect about the x axis each component planar force acting on a point, then the resultant is reflected about the x axis.

9. Show that if you contract (multiply by k, $0 < k < 1$) each component planar force acting on a point, then the resultant force is similarly contracted.

10. We have discussed matrices as transformers of space. Matrices also act as transformers of outputs to inputs, as we saw in Chap. 1. The matrix

$$A = \begin{pmatrix} 1 & R \\ 0 & 1 \end{pmatrix}$$

Figure AP2.10

represents the box in Fig. AP2.10. Thus we have

$$\begin{pmatrix} V_1 \\ I_1 \end{pmatrix} = \begin{pmatrix} 1 & R \\ 0 & 1 \end{pmatrix} \begin{pmatrix} V_2 \\ I_2 \end{pmatrix}$$

If we calculate A^{-1}, we have a matrix which transforms the inputs to the outputs:

$$A^{-1}\begin{pmatrix} V_1 \\ I_1 \end{pmatrix} = (A^{-1}A)\begin{pmatrix} V_2 \\ I_2 \end{pmatrix} = \begin{pmatrix} V_2 \\ I_2 \end{pmatrix}$$

Calculate A^{-1}. Note that A^{-1} has some negative entries and is in the same upper triangular form as A. Whether we can construct a blackbox represented by A^{-1} depends on what we can put in the box.

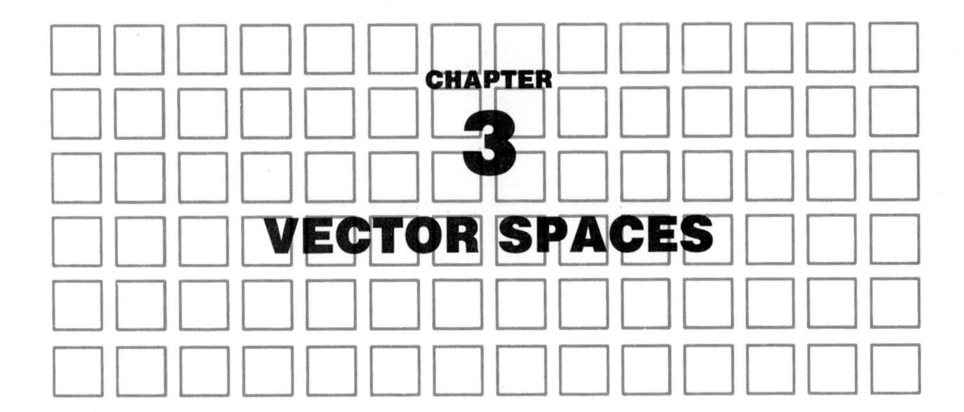

3.1 EUCLIDEAN SPACE E^n

Real euclidean space is a generalization of two-space and three-space. Generalization is needed since in some applications more than three variables may be needed to describe a situation. Recall the example of an absorption column at the beginning of Chap. 2: The number of plates in the column dictates the number of components in the concentration vector. Or consider the problem of estimating the thermal conductivity (a real number associated with a given material) of a heat-conducting body. Solution may require that temperatures be measured at several points at several times. Arrangement of these data in a vector requires much more than three components.

DEFINITION 3.1.1

REAL EUCLIDEAN n-SPACE, DENOTED E^n, CONSISTS OF	PARALLEL SITUATION FOR THREE-SPACE
1. The set of all ordered n-tuples $(x_1, x_2, \ldots, x_n)$, where $x_1, x_2, \ldots, x_n$ are all real numbers. An n-tuple is called a *vector*, and the numbers $x_1, \ldots, x_n$ are called *components* of the vector.	The set of ordered triples (x_1, x_2, x_3).
2. An operation of *vector addition* + defined by $(x_1, x_2, \ldots, x_n) + (y_1, y_2, \ldots, y_n) = (x_1 + y_1, x_2 + y_2, \ldots, x_n + y_n)$	Addition here was given by $(x_1, x_2, x_3) + (y_1, y_2, y_3) = (x_1 + y_1, x_2 + y_2, x_3 + y_3)$

REAL EUCLIDEAN n-SPACE, DENOTED E^n, CONSISTS OF	PARALLEL SITUATION FOR THREE-SPACE
3. An operation of *scalar multiplication* defined by $r(x_1, \ldots, x_n) = (rx_1, \ldots, rx_n)$ where r is any real number.	Multiplication here was given by $r(x_1, x_2, x_3) = (rx_1, rx_2, rx_3)$

DEFINITION 3.1.2 Two vectors $(x_1, \ldots, x_n)$ and $(y_1, \ldots, y_n)$ in E^n are called *equal* if $x_1 = y_1$, $x_2 = y_2, \ldots, x_n = y_n$.

EXAMPLE 1 In E^5 let $\mathbf{v} = (-1, 0, 2, 3, -6)$, $\mathbf{w} = (1, 0, -2, -3, 6)$, $\mathbf{u} = (3, 7, -6, 2, -1)$, and $\mathbf{z} = (0, 0, 0, 0, 0)$. Calculate (*a*) $\mathbf{u} + \mathbf{v}$, (*b*) $\mathbf{v} + \mathbf{u}$, (*c*) $\mathbf{u} + \mathbf{z}$, (*d*) $3\mathbf{u}$, (*e*) $-7\mathbf{v}$, (*f*) $3\mathbf{u} + 7\mathbf{v}$, (*g*) $\mathbf{v} + \mathbf{w}$, (*h*) $0\mathbf{u}$, and (*i*) $2\mathbf{z}$.

Solution

(*a*)
$$\begin{aligned}(3, 7, -6, 2, -1) &+ (-1, 0, 2, 3, -6)\\ &= (3 + (-1), 7 + 0, -6 + 2, 2 + 3, -1 + (-6))\\ &= (2, 7, -4, 5, -7)\end{aligned}$$

(*b*)
$$\begin{aligned}(-1, 0, 2, 3, -6) &+ (3, 7, -6, 2, -1)\\ &= (-1 + 3, 0 + 7, 2 + (-6), 3 + 2, -6 + (-1))\\ &= (2, 7, -4, 5, -7)\end{aligned}$$

Note that $\mathbf{u} + \mathbf{v} = \mathbf{v} + \mathbf{u}$.

(*c*)
$$\begin{aligned}\mathbf{u} + \mathbf{z} &= (3, 7, -6, 2, -1) + (0, 0, 0, 0, 0)\\ &= (3 + 0, 7 + 0, -6 + 0, 2 + 0, -1 + 0)\\ &= (3, 7, -6, 2, -1)\end{aligned}$$

(*d*)
$$\begin{aligned}3\mathbf{u} = 3(3, 7, -6, 2, -1) &= (3 \cdot 3, 3 \cdot 7, 3 \cdot (-6), 3 \cdot 2, 3 \cdot (-1))\\ &= (9, 21, -18, 6, -3)\end{aligned}$$

(*e*)
$$\begin{aligned}-7\mathbf{v} = -7(-1, 0, 2, 3, -6) &= (-7 \cdot (-1), -7 \cdot 0, -7 \cdot 2, -7 \cdot 3, -7 \cdot (-6))\\ &= (7, 0, -14, -21, 42)\end{aligned}$$

(*f*)
$$\begin{aligned}3\mathbf{u} + 7\mathbf{v} &= (9, 21, -18, 6, -3) + (-7, 0, 14, 21, -42)\\ &= (2, 21, -4, 27, -45)\end{aligned}$$

(*g*)
$$\begin{aligned}\mathbf{v} + \mathbf{w} &= (-1, 0, 2, 3, -6) + (1, 0, -2, -3, 6)\\ &= (0, 0, 0, 0, 0)\end{aligned}$$

(*h*)
$$\begin{aligned}0\mathbf{u} &= 0(3, 7, -6, 2, -1) = (0 \cdot 3, 0 \cdot 7, 0 \cdot (-6), 0 \cdot 2, 0 \cdot (-1))\\ &= (0, 0, 0, 0, 0)\end{aligned}$$

Notice that $0(x_1, x_2, x_3, x_4, x_5) = (0, 0, 0, 0, 0)$ regardless of the values for x_1, x_2, x_3, x_4, and x_5.

(*i*)
$$\begin{aligned}2\mathbf{z} = 2(0, 0, 0, 0, 0) &= (2 \cdot 0, 2 \cdot 0, 2 \cdot 0, 2 \cdot 0, 2 \cdot 0)\\ &= (0, 0, 0, 0, 0) = \mathbf{z}\end{aligned}$$

Note that $r\mathbf{z} = \mathbf{z}$ for any real number r.

Example 1b leads us to believe that the commutative property for addition of vectors in three-space carries over to E^n. Also the vector with all zero components may have a special role (Example 1c) as an additive identity. In fact, this intuition is correct. Let us define some special vectors first.

DEFINITION 3.1.3 The *zero vector* in E^n is denoted by $\boldsymbol{\theta}$ and is defined by $\boldsymbol{\theta} = (0, 0, \ldots, 0)$.

DEFINITION 3.1.4 Given a vector $\mathbf{x} = (x_1, \ldots, x_n)$ in E^n, the *negative of* $\mathbf{x}$ in E^n is denoted $-\mathbf{x}$ and is defined by

$$-\mathbf{x} = (-x_1, -x_2, \ldots, -x_n)$$

EXAMPLE 2 Let $\mathbf{x} = (1, 2, -3, 4)$. Find $-\mathbf{x}$. Calculate $\mathbf{x} + (-\mathbf{x})$.

Solution By definition $-\mathbf{x} = (-1, -2, -(-3), -4) = (-1, -2, 3, -4)$. Now $\mathbf{x} + (-\mathbf{x}) = (1, 2, -3, 4) + (-1, -2, 3, -4) = (0, 0, 0, 0) = \boldsymbol{\theta}$.

Example 2 illustrates the fact that for any $\mathbf{x}$ in E^n, $\mathbf{x} + (-\mathbf{x}) = \boldsymbol{\theta}$. This is part ($d$) of the next theorem.

THEOREM 3.1.1 Let $\mathbf{x} = (x_1, x_2, \ldots, x_n)$, $\mathbf{y} = (y_1, y_2, \ldots, y_n)$, and $\mathbf{z} = (z_1, z_2, \ldots, z_n)$ be vectors in E^n, and let r and s be real numbers. Then

(a) $\mathbf{x} + \mathbf{y} = \mathbf{y} + \mathbf{x}$ — Commutative law

(b) $(\mathbf{x} + \mathbf{y}) + \mathbf{z} = \mathbf{x} + (\mathbf{y} + \mathbf{z})$ — Associative law

(c) $\mathbf{x} + \boldsymbol{\theta} = \mathbf{x}$ — Additive identity

(d) $\mathbf{x} + (-\mathbf{x}) = \boldsymbol{\theta}$ — Additive inverse

(e) $(rs)\mathbf{x} = r(s\mathbf{x})$ — Associative law

(f) $(r + s)\mathbf{x} = r\mathbf{x} + s\mathbf{x}$ — Distributive laws

(g) $r(\mathbf{x} + \mathbf{y}) = r\mathbf{x} + r\mathbf{y}$

(h) $1\mathbf{x} = \mathbf{x}$ — Multiplicative identity

Proof (a) $\mathbf{x} + \mathbf{y} = (x_1 + y_1, x_2 + y_2, \ldots, x_n + y_n)$. The additions in each component are additions in the set of real numbers. Since addition is commutative for real numbers, we have

$$\mathbf{x} + \mathbf{y} = (x_1 + y_1, \ldots, x_n + y_n) = (y_1 + x_1, \ldots, y_n + x_n) = \mathbf{y} + \mathbf{x}$$

(d) Since $-\mathbf{x} = (-x_1, -x_2, \ldots, -x_n)$, we know that

$$\begin{aligned}\mathbf{x} + (-\mathbf{x}) &= (x_1, x_2, \ldots, x_n) + (-x_1, -x_2, \ldots, -x_n)\\ &= (x_1 + (-x_1), x_2 + (-x_2), \ldots, x_n + (-x_n))\\ &= (0, 0, \ldots, 0) = \boldsymbol{\theta}\end{aligned}$$

(g) $$\begin{aligned}r(\mathbf{x} + \mathbf{y}) &= r(x_1 + y_1, x_2 + y_2, \ldots, x_n + y_n)\\ &= (r(x_1 + y_1), r(x_2 + y_2), \ldots, r(x_n + y_n))\end{aligned}$$

Now since in each component we can apply the distributive law for real numbers, the last vector is

$$r(\mathbf{x} + \mathbf{y}) = (rx_1 + ry_1, rx_2 + ry_2, \ldots, rx_n + ry_n)$$

However,

$$\begin{aligned}r\mathbf{x} + r\mathbf{y} &= r(x_1, x_2, \ldots, x_n) + r(y_1, y_2, \ldots, y_n)\\ &= (rx_1, rx_2, \ldots, rx_n) + (ry_1, ry_2, \ldots, ry_n)\\ &= (rx_1 + ry_1, rx_2 + ry_2, \ldots, rx_n + ry_n)\end{aligned}$$

Therefore $r(\mathbf{x} + \mathbf{y}) = r\mathbf{x} + r\mathbf{y}$.

Proofs of the other parts are left to the problems.

EXAMPLE 3 Illustrate parts (b), (e) and (f) of Theorem 3.1.1 by means of examples from E^3.

Solution For (b), let $\mathbf{x} = (1, -1, 2)$, $\mathbf{y} = (0, 3, 6)$, and $\mathbf{z} = (-5, 0, 2)$. Then

$$\begin{aligned}(\mathbf{x} + \mathbf{y}) + \mathbf{z} &= ((1, -1, 2) + (0, 3, 6)) + (-5, 0, 2)\\ &= (1, 2, 8) + (-5, 0, 2)\\ &= (-4, 2, 10)\end{aligned}$$

and

$$\begin{aligned}\mathbf{x} + (\mathbf{y} + \mathbf{z}) &= (1, -1, 2) + ((0, 3, 6) + (-5, 0, 2))\\ &= (1, -1, 2) + (-5, 3, 8)\\ &= (-4, 2, 10)\end{aligned}$$

For (e) let $r = 3$, $s = -5$, and $\mathbf{x} = (4, -2, 6)$. Then

$$(rs)\mathbf{x} = [3(-5)](4, -2, 6) = -15(4, -2, 6) = (-60, 30, -90)$$
$$r(s\mathbf{x}) = 3[-5(4, -2, 6)] = 3(-20, 10, -30) = (-60, 30, -90)$$

For (f) use the same values as in (e). Then

$$\begin{aligned}(r + s)\mathbf{x} &= -2(4, -2, 6) = (-8, 4, -12)\\ r\mathbf{x} + s\mathbf{x} &= 3(4, -2, 6) + (-5)(4, -2, 6) = (12, -6, 18) + (-20, 10, -30)\\ &= (-8, 4, -12)\end{aligned}$$

Theorem 3.1.1 shows that E^n has the same generic properties as the set of two- and three-vectors had. In the next section we will see how the important properties of E^n outlined in Theorem 3.1.1 can be used to define the concept of a vector space. We postpone the discussion of generalizing the dot product until Sec. 3.6.

PROBLEMS 3.1

1. From the following list of vectors pick all possible pairs of vectors which are equal.

$$\mathbf{A} = (1, -1, 0) \qquad \mathbf{B} = (a^2 - b^2, 1, 2) \qquad \mathbf{C} = (\sin \pi/2, \cos \pi, \sin \pi)$$
$$\mathbf{D} = ((a-b)^2, 1, 2) \qquad \mathbf{E} = ((a-b)(a+b), (-1)^2, 2)$$

2. Let $\mathbf{x}$ and $\mathbf{y}$ be vectors in E^n. Define $\mathbf{x} - \mathbf{y}$ as $\mathbf{x} + (-\mathbf{y})$. Setting $\mathbf{x} = (x_1, x_2, \ldots, x_n)$ and $\mathbf{y} = (y_1, y_2, \ldots, y_n)$, write out the vector $\mathbf{x} - \mathbf{y}$.

3. Show by example that $(\mathbf{x} - \mathbf{y}) - \mathbf{z} \neq \mathbf{x} - (\mathbf{y} - \mathbf{z})$ in E^n. That is, the associative law does not hold for vector subtraction. (*Note:* To show that a statement does not hold in E^n, a particular value of n can be chosen. So in this problem the example can come from E^1, E^2, or E^3.)

4. Now E^1 is just the set of real numbers. In E^1 multiplication by a scalar is quite simple. What is it?

5. In E^3 consider the vectors $\mathbf{e}_1 = (1, 0, 0)$, $\mathbf{e}_2 = (0, 1, 0)$, and $\mathbf{e}_3 = (0, 0, 1)$. A sum of the form $a\mathbf{e}_1 + b\mathbf{e}_2 + c\mathbf{e}_3$, where a, b, and c are real numbers, is called a *linear combination* of $\mathbf{e}_1$, $\mathbf{e}_2$, and $\mathbf{e}_3$. Write the vector $(3, -2, 1)$ as a linear combination of $\mathbf{e}_1$, $\mathbf{e}_2$, and $\mathbf{e}_3$. (These vectors are also called $\mathbf{i}$, $\mathbf{j}$, and $\mathbf{k}$.)

6. In E^n consider the vectors

$$\mathbf{e}_1 = (1, 0, 0, \ldots, 0), \mathbf{e}_2 = (0, 1, 0, \ldots, 0), \ldots,$$
$$\mathbf{e}_k = (0, 0, \ldots, 0, \underset{\substack{\uparrow \\ k\text{th component}}}{1}, 0, \ldots, 0), \ldots, \mathbf{e}_n = (0, 0, \ldots, 1)$$

A sum of the form $a_1\mathbf{e}_1 + a_2\mathbf{e}_2 + \cdots + a_n\mathbf{e}_n$ is called a *linear combination* of $\mathbf{e}_1, \mathbf{e}_2, \ldots, \mathbf{e}_n$. Show that any vector $\mathbf{x} = (x_1, x_2, x_3, \ldots, x_n)$ can be written as a linear combination of $\mathbf{e}_1, \mathbf{e}_2, \ldots, \mathbf{e}_n$.

7. Let $\mathbf{x} = (1, -2, 3)$ and $\mathbf{y} = (3, 1, 0)$. Calculate $\mathbf{x} + \mathbf{y}$. Place the three vectors $\mathbf{x}$, $\mathbf{y}$, and $\mathbf{x} + \mathbf{y}$ in a matrix as rows of the matrix (remove the commas). Calculate the determinant of the matrix.

8. Work Prob. 7 with $\mathbf{x}$ and $\mathbf{y}$ as given, but replace $\mathbf{x} + \mathbf{y}$ with $3\mathbf{x} - 2\mathbf{y}$.

9. Work Prob. 7 with $\mathbf{x}$ and $\mathbf{y}$ as given, but replace $\mathbf{x} + \mathbf{y}$ with $a\mathbf{x} + b\mathbf{y}$, where a and b are nonzero real numbers.

10. Find constants c_1 and c_2 which satisfy

$$(3, 5) = c_1(1, -2) + c_2(2, -3)$$

11. Show that there do not exist constants c_1, c_2, c_3 which satisfy

$$(1, -1, 4) = c_1(2, 3, 5) + c_2(-1, 0, 6) + c_3(1, 3, 11)$$

12. Prove part (*b*) of Theorem 3.1.1. Also illustrate with an example.

13. Prove part (*c*) of Theorem 3.1.1, and illustrate with an example.

14. Prove part (*e*) of Theorem 3.1.1. Illustrate with an example.

15. Prove part (*f*) of Theorem 3.1.1. Also illustrate with an example.

16. Prove part (*h*) of Theorem 3.1.1, and illustrate with an example.

17. State the commutative, associative, and distributive laws in your own words.

3.2 VECTOR SPACES

The structure of euclidean space (as outlined in Theorem 3.1.1) is important for other "spaces" used in applied mathematics. For example, consider the mass-spring arrangement in Fig. 3.2.1*a*. If the body of mass m is displaced from rest position and released or pushed, oscillatory motion results. If this is done in the absence of external forces and friction is ignored, the displacement $y(t)$ of the body from rest is given by

$$y(t) = C_1 \cos \omega t + C_2 \sin \omega t$$

where $\omega = \sqrt{k/m}$ with k being the spring constant. Constants C_1 and C_2 are determined by how far the body is started from rest and how fast and in what direction it is pushed initially. The set of functions

$$V = \{f \mid f(t) = C_1 \cos \omega t + C_2 \sin \omega t,\ C_1 \in \mathbb{R},\ C_2 \in \mathbb{R}\}$$

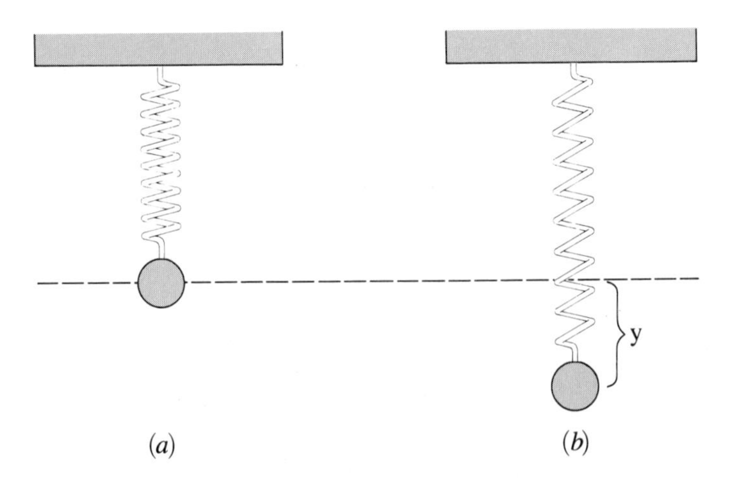

Figure 3.2.1 (*a*) Mass attached to a spring (in rest position). *b*) Mass in non-rest position

along with the usual definition (from calculus) of addition of functions and multiplication of a function by a constant turns out to have the properties of E^n outlined in Sec. 3.1.

Because there are many important structures with the properties of euclidean spaces, we put them all under the umbrella concept of vector space.

DEFINITION 3.2.1 A *real vector space* consists of the following.

(*a*) A set V of objects. These objects are called *vectors* even though they may be functions or matrices in a specific case.

(*b*) An operation denoted by $+$ which associates with each pair of vectors $\mathbf{v}$, $\mathbf{w}$ in V a vector $\mathbf{v} + \mathbf{w}$ in V, called the *sum* of $\mathbf{v}$ and $\mathbf{w}$.

(*c*) An operation called *scalar multiplication* which associates with each real number r and vector $\mathbf{v}$ in V a vector $r\mathbf{v}$ in V that is called the *product* of r and $\mathbf{v}$.

The operations must be defined in such a way that

$$\left.\begin{array}{l}\textbf{1. } \text{Addition is commutative: } \mathbf{v} + \mathbf{w} = \mathbf{w} + \mathbf{v}. \\ \textbf{2. } \text{Addition is associative: } \mathbf{v} + (\mathbf{w} + \mathbf{u}) = (\mathbf{y} + \mathbf{w}) + \mathbf{u}.\end{array}\right\} \mathbf{v}, \mathbf{w}, \mathbf{u} \text{ in } V$$

3. There exists a zero vector $\boldsymbol{\theta}$ in V such that $\mathbf{u} + \boldsymbol{\theta} = \mathbf{u}$ for all vectors $\mathbf{u}$ in V. Vector $\boldsymbol{\theta}$ is called an *additive identity*.

4. For each vector $\mathbf{v}$ in V there exists an additive inverse $-\mathbf{v}$ in V such that $\mathbf{v} + (-\mathbf{v}) = \boldsymbol{\theta}$.

$$\left.\begin{array}{l}\textbf{5. } (rs)\mathbf{v} = r(s\mathbf{v}) \\ \textbf{6. } (r + s)\mathbf{v} = r\mathbf{v} + s\mathbf{v} \\ \textbf{7. } r(\mathbf{v} + \mathbf{w}) = r\mathbf{v} + r\mathbf{w}\end{array}\right\} r, s \in \mathbb{R}$$

8. $1\mathbf{v} = \mathbf{v}$ for every $\mathbf{v}$ in V.

It is important that a real vector space consist of the set of vectors and the two operations with certain properties. The same set of vectors with different operations may not satisfy the required properties.

Note that the properties which we *derived* for E^n have become the *defining properties*, or *axioms*, for a vector space. The reason that this works well is that E^n, although a specific example, possesses the important qualities for generalization. This happens often in applied mathematics: A specific problem leads to a specific solution—yet the solution actually solves many more problems when it is seen in a larger context.

To show that an object is a vector space, we must show that *closure for both operations* holds [parts (*b*) and (*c*) of the definition] and that properties 1 through 8 hold. Altogether these 10 properties are called the *axioms for a real vector space.* To show that an object is not a vector space, we need only show that 1 of the 10 axioms fails to hold.

EXAMPLE 1 Is the set of ordered pairs $V = \{(x_1, x_2) | x_1 \in \mathbb{R}, x_2 \in \mathbb{R}\}$ a vector space?

Solution Until operations of vector addition and scalar multiplication are specified, we cannot test for vector space structure. Therefore, we do not have a vector space.

EXAMPLE 2 Now E^n is a vector space, because the definitions of the operations imply closure and Theorem 3.1.1 shows that properties 1 to 8 hold.

At this point, we recall that in E^n only the vector $(0, 0, \ldots, 0)$ had the property of being an additive identity. Also, given $(x_1, \ldots, x_n)$ in E^n, only $(-x_1, \ldots, -x_n)$ is an additive inverse. In general, this uniqueness holds in any vector space.

THEOREM 3.2.1 Let V be a vector space.

(*a*) There exists only one additive identity in V.
(*b*) Given $\mathbf{x}$ in V, there is only additive inverse of $\mathbf{x}$ in V.

Proof (*a*) Let $\boldsymbol{\theta}_1$ and $\boldsymbol{\theta}_2$ be additive identities for V. We will show that they are equal. By the additive-identity axiom

$$\boldsymbol{\theta}_1 + \boldsymbol{\theta}_2 = \boldsymbol{\theta}_1$$

However,

$$\boldsymbol{\theta}_1 + \boldsymbol{\theta}_2 \underset{\uparrow}{=} \boldsymbol{\theta}_2 + \boldsymbol{\theta}_1 \underset{\uparrow}{=} \boldsymbol{\theta}_2$$

Commutativity — Additive-identity axiom

Therefore $\boldsymbol{\theta}_1 = \boldsymbol{\theta}_1 + \boldsymbol{\theta}_2 = \boldsymbol{\theta}_2$.

(*b*) Let $\mathbf{x}$ be in V. Let $\mathbf{w}$ and $\mathbf{u}$ be additive inverses. We will show that $\mathbf{w} = \mathbf{u}$. We have

$$\mathbf{x} + \mathbf{w} = \boldsymbol{\theta} = \mathbf{x} + \mathbf{u}$$

By commutativity, $\mathbf{w} + \mathbf{x} = \mathbf{u} + \mathbf{x}$, and upon adding $\mathbf{w}$ to both sides of this last equation, we have

$$(\mathbf{w} + \mathbf{x}) + \mathbf{w} = (\mathbf{u} + \mathbf{x}) + \mathbf{w}$$

Finally, by associativity

$$\mathbf{w} + (\mathbf{x} + \mathbf{w}) = \mathbf{u} + (\mathbf{x} + \mathbf{w})$$
$$\mathbf{w} + \boldsymbol{\theta} = \mathbf{u} + \boldsymbol{\theta}$$
$$\mathbf{w} = \mathbf{u}$$

EXAMPLE 3 Let $V = \mathscr{M}_{mn} = \{m \times n$ matrices with real entries$\}$, let vector addition be the addition of matrices, and let scalar multiplication be the multiplication of matrices by scalars. Is V with these operations a vector space?

Solution From our work with matrices we know that closure, commutativity, associativity, and distributivity hold. The additive identity is the $m \times n$ matrix with all entries zero. The additive inverse of $A_{m \times n}$ is $-A_{m \times n}$. Clearly $1A_{m \times n} = A_{m \times n}$. Therefore this is a vector space.

EXAMPLE 4 Let $\mathscr{P}_2 = \{$polynomials $f(x) = a_2x^2 + a_1x + a_0$, where $a_0, a_1, a_2 \in \mathbb{R}\}$, and define addition and scalar multiplication as follows:

(*a*) Addition. Let $f(x) = a_2x^2 + a_1x + a_0$ and $g(x) = b_2x^2 + b_1x + b_0$. Then define

$$(f + g)(x) = (a_2 + b_2)x^2 + (a_1 + b_1)x + (a_0 + b_0)$$

(*b*) Scalar multiplication. Let $r \in \mathbb{R}$, $f(x) = a_2x^2 + a_1x + a_0$. Define

$$(rf)(x) = (ra_2)x^2 + (ra_1)x + (ra_0)$$

Show that $\mathscr{P}_2$ is a vector space.

Solution Closure follows easily from the definitions since the right-hand sides of the equations in (*a*) and (*b*) are polynomials of degree ≤ 2. For commutativity

$$\begin{aligned}(f + g)(x) &= (a_2 + b_2)x^2 + (a_1 + b_1)x + (a_0 + b_0)\\ &= (b_2 + a_2)x^2 + (b_1 + a_1)x + (b_0 + a_0)\\ &= (g + f)(x)\end{aligned}$$

Associativity follows similarly. The $\boldsymbol{\theta}$ is $0x^2 + 0x + 0$, which is the function $f(x) \equiv 0$. The additive inverse $-f$ is simply $-a_2x^2 - a_1x - a_0$. The other axioms are easily verified.

EXAMPLE 5 Let $\mathscr{P}_n = \{$polynomials $f(x) = a_nx^n + a_{n-1}x^{n-1} + \cdots + a_2x^2 + a_1x + a_0\}$ with operations defined by [let $g(x) = b_nx^n + \cdots + b_1x + b_0$ and $r \in \mathbb{R}$]

$$(f + g)(x) = (a_n + b_n)x^n + \cdots + (a_1 + b_1)x + (a_0 + b_0)$$
$$(rf)(x) = (ra_n)x^n + \cdots + (ra_1)x + (ra_0)$$

So $\mathscr{P}_n$ is a real vector space.

As we see more examples of vector spaces, we will be led to theorems about their structure. Theorems are formed by considering examples. For instance, in calculus after showing

$$\frac{d}{dx}x^2 = 2x \qquad \frac{d}{dx}x^3 = 3x^2$$

by using the definition of derivative, we can guess that

$$\frac{d}{dx}x^4 = 4x^3$$

and so on until we formulate the theorem:

$$\frac{d}{dx}x^n = nx^{n-1} \qquad n \geq 1, n \text{ an integer}$$

In proving theorems we can use axioms, previous theorems, and facts from earlier mathematics courses.

EXAMPLE 6 In E^n, $\mathscr{M}_{mn}$, and $\mathscr{P}_n$, what is the result of multiplying a vector by the scalar $r = 0$? State a possible theorem.

Solution In E^n,

$$0(x_1, \ldots, x_n) = (0x_1, \ldots, 0x_n) = (0, 0, \ldots, 0) = \boldsymbol{\theta}$$

In $\mathscr{M}_{mn}$,

$$0\begin{pmatrix} a_{11} & \cdots & a_{1n} \\ \cdots & \cdots & \cdots \\ a_{m1} & \cdots & a_{mn} \end{pmatrix} = \begin{pmatrix} 0 & \cdots & 0 \\ \cdots & \cdots & \cdots \\ 0 & \cdots & 0 \end{pmatrix}$$

And in $\mathscr{P}_n$,

$$0(a_n x^n + \cdots + a_1 x + a_0) = 0x^n + \cdots + 0x + 0 = \boldsymbol{\theta}$$

A possible theorem is: If V is a vector space and $\mathbf{x} \in V$, then $0\mathbf{x} = \boldsymbol{\theta}$.

The guess in the solution to Example 6 is actually correct.

THEOREM 3.2.2 If V is a real vector space and $\mathbf{x} \in V$, then $0\mathbf{x} = \boldsymbol{\theta}$.

Proof

$$\begin{aligned} 0\mathbf{x} &= (0 + 0)\mathbf{x} = 0\mathbf{x} + 0\mathbf{x} \\ \boldsymbol{\theta} &= 0\mathbf{x} + [-(0\mathbf{x})] = (0\mathbf{x} + 0\mathbf{x}) + [-(0\mathbf{x})] \\ &= 0\mathbf{x} + \{0\mathbf{x} + [-(0\mathbf{x})]\} \\ &= 0\mathbf{x} + \boldsymbol{\theta} \\ &= 0\mathbf{x} \end{aligned}$$

Therefore, $\boldsymbol{\theta} = 0\mathbf{x}$.

EXAMPLE 7 This is an important example because it shows that vector addition need not be related to ordinary addition and the zero vector $\boldsymbol{\theta}$ need not involve the real number 0. Let $V = \{x \mid x \in \mathbb{R}, x > 0\}$. Define addition and scalar multiplication as follows:

Addition. For $x \in V$, $y \in V$, define

$$\underbrace{x \oplus y}_{\text{Vector sum}} = \underbrace{xy}_{\text{Ordinary multiplication}}$$

Multiplication. For $r \in \mathbb{R}$, $x \in V$, define

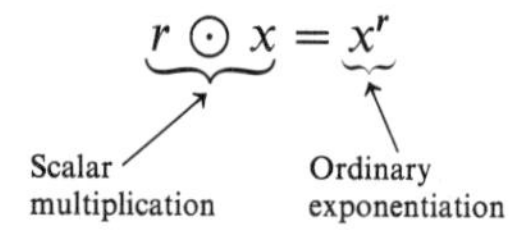

Show that V with these operations is a vector space. We have used $\oplus$ and $\odot$ to denote the vector space operations, to distinguish them from the ordinary operations in the solution.

Solution Before showing that V with these operations is a vector space, we look at some specific vector sums and scalar multiples. First, note that the vectors are just positive real numbers. So, for addition we have

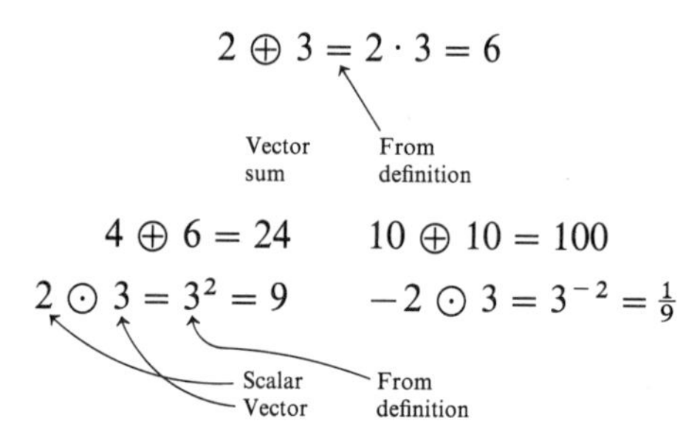

Now to show that we have a vector space, we must show that all the properties of the definition are satisfied.

Closure for addition. Let $x \in V$, $y \in V$. Then $x \oplus y = xy$. Since x and y are positive real numbers and the product of positive real numbers is a positive real number, $xy \in V$. Therefore, $x \oplus y \in V$, and we have closure for addition.

Closure for scalar multiplication. Let $r \in \mathbb{R}$ and $x \in V$. Then $r \odot x = x^r$. Since x is a positive real number, any real power of it is also a positive real number. Therefore $r \odot x \in V$, and we have closure for addition.

1. *Commutativity for addition.* Let $x \in V$, $y \in V$. Then

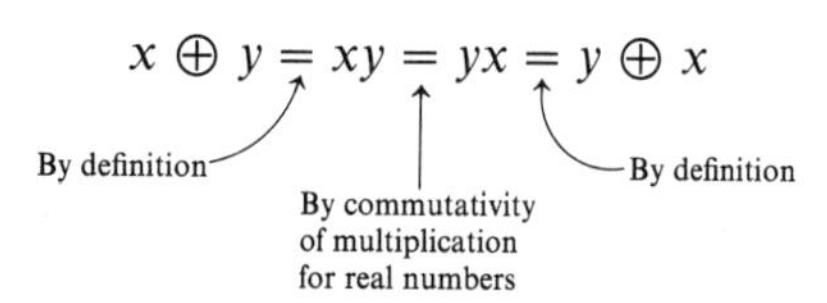

Therefore $x \oplus y = y \oplus x$.

2. *Associativity for addition.* Let $x \in V$, $y \in V$, $z \in V$. Then

$$(x \oplus y) \oplus z = (xy) \oplus z = (xy)z = x(yz) = x \oplus (yz) = x \oplus (y \oplus z)$$

By definition — Associativity for multiplication of real numbers — By definition

3. *Additive identity.* To determine $\boldsymbol{\theta}$, we ask for the equation

$$x \oplus \boldsymbol{\theta} = x$$

to hold. The left-hand side is $x\boldsymbol{\theta}$ by definition of $\oplus$. Therefore, $\boldsymbol{\theta} = 1$. That is, the positive real number 1 is the only additive identity or "zero" for this space. That is,

$$x \oplus 1 = x$$

4. *Additive inverse.* For any $x \in V$, $-x$ must satisfy $x \oplus (-x) = \boldsymbol{\theta}$. Now we must be very careful: $-x$ is not "minus x"; it is the symbol for additive inverse, and $\boldsymbol{\theta}$ is not the real number zero, it is 1. So the equation is

$$x \oplus (-x) = \boldsymbol{\theta}$$
$$x \cdot (-x) = 1$$

By definition — From 3 above

Therefore the unique solution for $-x$ is $(-x) = 1/x$. Since $x > 0$, we know that $1/x > 0$, so $-x \in V$.

5. *Associative law for scalar multiplication.* Let $r \in \mathbb{R}$, $s \in \mathbb{R}$, $x \in V$. Then

$$(rs) \odot x = x^{rs} \qquad \text{and} \qquad r \odot (s \odot x) = r \odot (x^s) = (x^s)^r$$

By laws of exponents, $(x^s)^r = x^{sr} = x^{rs}$. Therefore, $(rs) \odot x = r \odot (s \odot x)$.

6, 7. *Distributive laws.* Let $r \in \mathbb{R}$, $s \in \mathbb{R}$, $x \in V$, $y \in V$. Then

$$(r + s) \odot x = x^{r+s} = x^r x^s = x^r \oplus x^s = (r \odot x) \oplus (s \odot x)$$

By definition of scalar multiplication — Properties of exponents — Definition of addition — By definition of scalar multiplication

$$r \odot (x \oplus y) = r \odot (xy) = (xy)^r = x^r y^r = x^r \oplus y^r = (r \odot x) \oplus (r \odot y)$$

8. Let $x \in V$. Then $1 \odot x = x^1 = x$.

Therefore, V with the operations as defined is a vector space.

EXAMPLE 8 Let $V = \{x | x \in \mathbb{R},\ x > 0\}$, and define addition and scalar multiplication as follows:

1. *Addition.* Let $x \in V$, $y \in V$. Define

$$x + y = x + y$$

Vector addition — Ordinary addition of real numbers

2. *Scalar multiplication.* Let $r \in \mathbb{R}$, $x \in V$. Define

$$rx = r \cdot x$$

Ordinary multiplication

Show that V is not a vector space.

Solution To show that the structure is not a vector space, all we have to show is that at least one of the axioms fails to hold true. In this case closure for scalar multiplication fails to hold since if $r < 0$, $rx = r \cdot x < 0$ and $rx \notin V$. Again we see that the truth of the axioms depends on the set V *and* the operations.

EXAMPLE 9 Regarding the mass spring apparatus in Fig. 3.2.1 and the related discussion,

$$V = \{f \mid f(t) = c_1 \cos wt + c_2 \sin wt, t \in \mathbb{R}\}$$

with the usual definition of addition of functions and multiplication of functions is a vector space.

Solution *Closure.* Let $f(t) = c_1 \cos wt + c_2 \sin wt$ and $g(t) = c_3 \cos wt + c_4 \sin wt$. We have

$$(f + g)(t) = (c_1 + c_3)(\cos wt) + (c_2 + c_4)(\sin wt)$$
$$(rf)(t) = rc_1 \cos wt + rc_2 \sin wt$$

which are in V. Commutative and associative properties hold. The zero vector is $0 \cos wt + 0 \sin wt$, which is the identically zero function. The additive inverse of f is

$$-c_1 \cos wt - c_2 \sin wt$$

The other properties are left to the reader.

Complex Vector Spaces If in the definition of real vector space we replace real numbers by complex numbers, we have the notion of a complex vector space. The complex vector spaces we use most often in this text are $\mathbb{C}^n$ and $\mathscr{C}_{mn}$, which are defined below.

DEFINITION 3.2.2 Vector space $\mathbb{C}^n$ is the complex vector space consisting of n-tuples $(z_1, z_2, \ldots, z_n)$ of complex numbers with the operations

$$(z_1, \ldots, z_n) + (w_1, \ldots, w_n) = (z_1 + w_1, \ldots, z_n + w_n)$$
$$c(z_1, \ldots, z_n) = (cz_1, \ldots, cz_n)$$

where $(w_1, \ldots, w_n)$ is a complex n-tuple and c is any complex number.

DEFINITION 3.2.3 $\mathscr{C}_{mn}$ is the complex vector space of $m \times n$ matrices with complex number entries along with the standard matrix operations of addition and scalar multiplication.

The zero vectors for $\mathbb{C}^n$ and $\mathscr{C}_{mn}$ are, respectively, the same as the zero vectors for E^n and $\mathscr{M}_{mn}$, as can be verified directly.

In the remainder of the text, the term *vector space* will mean the *real* vector space unless we are working specifically with a complex vector space. When you are working with complex vector spaces, it is important to remember that the vectors can be constructed by using complex numbers and that the scalars for scalar multiplication can be any complex number.

EXAMPLE 10 Let $V = \{\text{hermitian } n \times n \text{ matrices}\}$, and give V the usual matrix operations. Is V a vector space?

Solution Since there is no restriction on the entries of the matrices, we are checking to see whether V with these operations is a complex vector space. Recall that the main-diagonal entries of a hermitian matrix are real numbers. Thus, in general, if A is hermitian, the scalar multiple iA is not hermitian. Therefore V is not closed under scalar multiplication and is not a vector space.

PROBLEMS 3.2

In Probs. 1 to 20, a set V and operations of scalar multiplication and vector addition are given. Determine whether V is a vector space. If V fails to be a vector space, state an axiom which fails to hold. In a given problem, if the objects in V can be constructed by using complex numbers, the problem is to determine whether V is a complex vector space.

1. Let $V = \{\text{ordered triples } (x_1, x_2, 0);\ x_1, x_2 \in \mathbb{R}\}$, the operations as in E^3.
2. Let $V = \{\text{ordered triples } (x_1, x_2, 1);\ x_1, x_2 \in \mathbb{R}\}$, the operations as in E^3.
3. Let $V = \{\text{ordered pairs } (x_1, x_2);\ x_1, x_2 \in \mathbb{R}\}$ with the operations $(x_1, x_2) + (y_1, y_2) = (x_1 + y_1, y_2 + x_2)$ and $r(x_1, x_2) = (rx_1, rx_2)$.
4. Let $V = \{\text{ordered pairs } (x_1, x_2);\ x_1, x_2 \in \mathbb{R}\}$ with the operations $(x_1, x_2) + (y_1, y_2) = (x_1 + y_1, x_2 + y_2)$ and $r(x_1, x_2) = (rx_1, x_2)$.
5. Let $V = \mathbb{R}$, and let the operations be the standard addition and multiplication in $\mathbb{R}$.
6. Let $V = \{n \times n \text{ matrices with positive entries}\}$ with the usual matrix operations. (Entries must be real to be compared to zero.)
7. Let $V = \{n \times n \text{ real symmetric matrices}\}$ with the usual matrix operations.

8. Let $V = \{n \times n$ real skew-symmetric matrices$\}$ with the usual matrix operations.

9. Let $V = \{n \times n$ upper triangular matrices$\}$ with the usual matrix operations.

10. Let $V = \{n \times n$ diagonal matrices$\}$ with the usual matrix operations.

11. Let $V = \{$functions defined for all x in $\mathbb{R}$ with $f(0) = 0\}$ with operations $(f + g)(x) = f(x) + g(x)$ and $(rf)(x) = r[f(x)]$.

12. Let $V = \{$functions defined for all x with $f(0) = 1\}$ with operations as in Prob. 11.

13. Let $V = \{n \times n$ nonsingular matrices$\}$ with the usual matrix operations.

14. Let $V = \{n \times n$ singular matrices$\}$ with the usual matrix operations.

15. Let $V = \{n \times n$ nilpotent matrices$\}$ with the usual matrix operations.

16. Let $V = \{n \times n$ idempotent matrices$\}$ with the usual matrix operations.

17. Let $V = \{n \times n$ matrices A with $A^2 = I\}$ with the usual matrix operations.

18. Let $V = \{$real-valued functions defined on $\mathbb{R}$ with $f(x) > 0$, for all $x\}$ with the usual operations.

19. Let V be as in Prob. 18, but give V the operations $(f + g)(x) = f(g)g(x)$ and $(rf)(x) = [f(x)]^r$, for r a real number.

20. Let $V = \{n \times n$ matrices with sum of main diagonal entries equal to zero$\}$ with the usual matrix operations.

21. Let C be a fixed $n \times n$ matrix. Let $V = \{A_{n \times n}$ such that $AC = 0\}$. Given the usual matrix operations, is V a vector space?

22. Prove that in a real vector space V, $r\boldsymbol{\theta} = \boldsymbol{\theta}$ for all $r \in \mathbb{R}$. (*Hint:* Mimic the proof of Theorem 3.2.2.)

23. Prove that in a real vector space V, $(-1)\mathbf{x} = -\mathbf{x}$ for all $x \in V$. (*Hint:* Mimic the proof of Theorem 3.2.2.)

24. Prove that in a real vector space V, if $\mathbf{x} \in V$, $r \in \mathbb{R}$, and $r\mathbf{x} = \boldsymbol{\theta}$ then $\mathbf{x} = \boldsymbol{\theta}$ or $r = 0$.

25. Let $V = \{Z\}$ (a set consisting of one element) and define

$$Z + Z = Z$$
$$rZ = Z \qquad \text{For all } r \in \mathbb{R}$$

Is V a real vector space?

26. Work Prob. 25 with the second operation being $cZ = Z$ for all $c \in \mathbb{C}$. Is V a complex vector space? (Check the vector space axioms.)

27. In the list of properties of vector space operations, could 3 and 4 be reversed in order? Explain.

28. Let $V = \{(x_1, x_2) | x_1 + x_2 = 1, 0 \leq x_1 \leq 1, 0 \leq x_2 \leq 1\}$ with the operations $(x_1, x_2) + (y_1, y_2) = (\frac{1}{2}(x_1 + y_1), \frac{1}{2}(x_2 + y_2))$ and $r(x_1, x_2) = (x_1, x_2)$. Are the closure axioms satisfied? Is V a real vector space?

3.3 THE SUBSPACE PROBLEM

Certain applications involve the use of subsets of vector spaces which are vector spaces also. We will see an example from coding theory at the end of the section. Another example is the set $\mathscr{S}_n$ of all $n \times n$ symmetric matrices (used in least squares problems; see App. III) with the usual matrix operations which is a subset of $\mathscr{M}_{nn}$. $\mathscr{S}_n$ is a vector space (Prob. 7 of Sec. 3.2). In this case we say that $\mathscr{S}_n$ is a *subspace* of $\mathscr{M}_{nn}$.

DEFINITION 3.3.1 Let V be a real or complex vector space, and let W be a subset of V with W inheriting the operations of V. We say that W is a *subspace* of V if W with the inherited operations is a vector space.

Given a subset of a real or complex vector space, we often need to know whether the subset is a subspace. Thus we have the following.

Subspace problem.
Given a subset W of a vector space V, with W having the same operations as V, determine whether W is a subspace of V.

EXAMPLE 1 Let $V = E^3$, and let $W = \{\mathbf{x} \in E^3 | \mathbf{x} = (x_1, x_2, 0)\}$. Solve the subspace problem for W and V.

Solution By its definition W is a subset of V; we must determine whether W with the operations inherited from E^3 is a vector space.

Closure. Let $r \in \mathbb{R}$, $\mathbf{x} \in W$, $\mathbf{y} \in W$. We check to see whether $\mathbf{x} + \mathbf{y} \in W$ and $r\mathbf{x} \in W$. We have

$$\mathbf{x} + \mathbf{y} = (x_1, x_2, 0) + (y_1, y_2, 0) = (x_1 + y_1, x_2 + y_2, 0)$$

Since the third component is 0, $\mathbf{x} + \mathbf{y} \in W$. For the scalar multiplication

$$r\mathbf{x} = (rx_1, rx_2, r0) = (rx_1, rx_2, 0)$$

and since the third component is 0, $r\mathbf{x} \in W$.

1. *Commutativity of addition.* Let $\mathbf{x} \in W$, $\mathbf{y} \in W$. Then

$$\mathbf{x} + \mathbf{y} = (x_1 + y_1, x_2 + y_2, 0) = (y_1 + x_1, y_2 + x_2, 0) = \mathbf{y} + \mathbf{x}$$

2. *Associativity of addition.* This is similar to 1; see the problems.

3. *Existence of zero.* We let $\mathbf{x} \in W$ and see whether there is any vector in W which acts as an additive identity. Because $(0, 0, 0) \in W$ (the last component is zero) and $\mathbf{x} + (0, 0, 0) = \mathbf{x}$, $\boldsymbol{\theta} = (0, 0, 0)$ acts as zero for W.

4. *Additive inverse.* Let $\mathbf{x} = (x_1, x_2, 0)$. Since $(-x_1, -x_2, 0)$ is also in W and

$$(x_1, x_2, 0) + (-x_1, -x_2, 0) = (0, 0, 0) = \boldsymbol{\theta}$$

we know that $-\mathbf{x} = (-x_1, -x_2, 0)$ is in W.

5–8. These are straightforward and left to the problems.

A close examination of Example 1 shows that most axioms actually are true by "inheritance." For example, in showing $\mathbf{x} + \mathbf{y} = \mathbf{y} + \mathbf{x}$, the fact that $\mathbf{x}$ and $\mathbf{y}$ were in W was not important; since they were in V and commutativity holds for all $\mathbf{x}$ and $\mathbf{y}$ in V, commutativity holds in *any* subset of V. Thus it is the properties of closure and the existence of zero and additive inverse in W that must be checked. However, we can show that closure leads to the existence of zero and an additive inverse in W.

THEOREM 3.3.1 Let V be a real vector space and W a nonempty subset of V. Then W is a subspace of V if and only if (1) $\mathbf{x}$, $\mathbf{y}$ in W implies $\mathbf{x} + \mathbf{y}$ is in W and (2) r in $\mathbb{R}$ and $\mathbf{x}$ in W implies $r\mathbf{x}$ is in W.

This theorem contains the solution to the subspace problem. It is emphasized that the theorem holds for complex vector spaces with $\mathbb{R}$ replaced by $\mathbb{C}$. Before proving this theorem, we give some examples.

EXAMPLE 2 Let $V = E^3$ and $W = \{\mathbf{x} \mid \mathbf{x} = a(1, 0, 2) + b(1, -1, 3)\}$, where a and b can be any real numbers. Is W a subspace of V?

Solution Let $\mathbf{x} \in W$, $\mathbf{y} \in W$, and consider $\mathbf{x} + \mathbf{y}$. We have

$$\begin{aligned}\mathbf{x} + \mathbf{y} &= (a(1, 0, 2) + b(1, -1, 3)) + (c(1, 0, 2) + d(1, -1, 3)) \\ &= a(1, 0, 2) + b(1, -1, 3) + c(1, 0, 2) + d(1, -1, 3) \\ &= (a + c)(1, 0, 2) + (b + d)(1, -1, 3)\end{aligned}$$

Hence $\mathbf{x} + \mathbf{y}$ is in the form of an element of W; that is, $\mathbf{x} + \mathbf{y} \in W$. Consider now $r\mathbf{x} = r(a(1, 0, 2) + b(1, -1, 3)) = ra(1, 0, 2) + rb(1, -1, 3)$. Thus $r\mathbf{x} \in W$, and W is a subspace by Theorem 3.3.1.

EXAMPLE 3 Let $V = E^3$ and $W = \{\mathbf{x} \mid \mathbf{x} = (x_1, x_2, 1)\}$, and consider $\mathbf{x} + \mathbf{y}$. Now

$$\mathbf{x} + \mathbf{y} = (x_1, x_2, 1) + (y_1, y_2, 1) = (x_1 + y_1, x_2 + y_2, 2)$$

The last component of $\mathbf{x} + \mathbf{y}$ is not equal to 1. Therefore $\mathbf{x} + \mathbf{y} \notin W$, and so W is *not* a subspace of V.

EXAMPLE 4 Let $V = \mathscr{C}_{nn}$ and let $W = \{n \times n \text{ hermitian matrices}\}$. Is W a subspace of V?

Solution From the solution of Example 10 of Sec. 3.2, We know that if A is hermitian, then iA need not be hermitian. For example,

$$A = \begin{pmatrix} 1 & i \\ -i & 2 \end{pmatrix}$$

is hermitian, but

$$iA = \begin{pmatrix} i & -1 \\ 1 & 2i \end{pmatrix}$$

is not hermitian. Therefore, W is not closed under scalar multiplication and is not a subspace of $\mathscr{C}_{nn}$.

Proof of Theorem 3.3.1 This is an "if and only if" theorem: p holds if and only if q holds. To prove it, we must prove two things:

1. If p holds, then q holds.

2. If q holds, then p holds.

Part 1. If W is a subspace, then $\mathbf{x} + \mathbf{y} \in W$ and $r\mathbf{x} \in W$ for all $r \in \mathbb{R}$, $\mathbf{x} \in W$, $\mathbf{y} \in W$.

If W is a subspace of V, then W is a vector space and all axioms hold. In particular, closure holds. Therefore $r\mathbf{x} \in W$ by closure for scalar multiplication, and $\mathbf{x} + \mathbf{y} \in W$ by closure for vector addition.

Part 2. If $\mathbf{x} + \mathbf{y} \in W$ and $r\mathbf{x} \in W$ for all $r \in \mathbb{R}$, $\mathbf{x} \in W$, $\mathbf{y} \in W$, then W is a subspace of V.

We must check all the vector space axioms for W, keeping in mind the hypothesis that $\mathbf{x} + \mathbf{y} \in W$ and $r\mathbf{x} \in W$ for all $r \in \mathbb{R}$, $\mathbf{x} \in W$, $\mathbf{y} \in W$. As we have mentioned, all axioms except closure, zero, and additive inverse follow by heredity. The closure is just restatement of our hypothesis for this part of the theorem. Now let $\mathbf{x} \in W$ and $r = -1$. Then by hypothesis $(-1)\mathbf{x} \in W$, so $(-1)\mathbf{x} + \mathbf{x} \in W$. But

$$\begin{aligned} (-1)\mathbf{x} + \mathbf{x} &= (-1)\mathbf{x} + 1\mathbf{x} \\ &= (-1 + 1)\mathbf{x} && \text{(Distributive law)} \\ &= 0\mathbf{x} \\ &= \boldsymbol{\theta} && \text{(Theorem 3.2.2)} \end{aligned}$$

Therefore $\boldsymbol{\theta} \in W$. (We note that Theorem 3.2.2 holds for all of V, so we can apply it to W.)

To show additive inverses are in W, let $\mathbf{x} \in W$ and $r = -1$. We have $(-1)\mathbf{x} \in W$ and

$$\mathbf{x} + (-1)\mathbf{x} = [1 + (-1)]\mathbf{x} = 0\mathbf{x} = \boldsymbol{\theta} \in W$$

Therefore, additive inverses are in W.

EXAMPLE 5 Let V be any vector space. Then V itself is a subspace.

EXAMPLE 6 Let V be any vector space, and $W = \{\boldsymbol{\theta}\}$. Then W is a subspace because $r\boldsymbol{\theta} = \boldsymbol{\theta}$, $\boldsymbol{\theta} + \boldsymbol{\theta} = \boldsymbol{\theta}$, for r real or complex.

Because any vector space V has V and $\{\boldsymbol{\theta}\}$ as subspaces, these are called the *trivial subspaces* of V. All other subspaces of V are called *proper subspaces*, or nontrivial subspaces, of V.

EXAMPLE 7 *For the real case*: Let $\mathbf{x} = a_1\mathbf{u} + a_2\mathbf{z}$, and $\mathbf{y} = b_1\mathbf{u} + b_2\mathbf{z}$, so that $\mathbf{x}$ and $\mathbf{y}$ are in in V. Let $W = \{\mathbf{x} \mid \mathbf{x} = c_1\mathbf{u} + c_2\mathbf{z}$, where c_1 and c_2 can be any real (complex) numbers$\}$. Show that W is a subspace of V.

Solution *For the real case*: Let $\mathbf{x} = a_1\mathbf{u} + a_2\mathbf{z}$, and $\mathbf{y} = b_1\mathbf{u} + b_2\mathbf{z}$, so that $\mathbf{x}$ and $\mathbf{y}$ are in W. Consider $\mathbf{x} + \mathbf{y}$:

$$\mathbf{x} + \mathbf{y} = \underbrace{(a_1 + b_1)}_{\text{Real numbers}}\mathbf{u} + \underbrace{(a_2 + b_2)}_{\text{Real numbers}}\mathbf{z} \in W$$

Closure for scalar multiplication is similar. By Theorem 3.3.1, W is a subspace of V.

EXAMPLE 8 Let $V = \{x \in \mathbb{C}^2 \mid x$ has purely imaginary components$\}$. Now V is not a subspace of $\mathbb{C}^2$ because $i(i, i) = (-1, -1)$ which is not in V.

EXAMPLE 9 Let $V = \mathscr{M}_{22}$ and $W = \{$invertible 2×2 matrices$\}$. Determine whether W is a subspace of $\mathscr{M}_{22}$.

Solution 1 If W were to be a subspace, $\boldsymbol{\theta}$ would have to be in W. But

$$\boldsymbol{\theta} = \begin{pmatrix} 0 & 0 \\ 0 & 0 \end{pmatrix}$$

is not invertible and cannot be in W. So W is not a subspace.

Solution 2 Let

$$\mathbf{x} = \begin{pmatrix} a & b \\ c & d \end{pmatrix} \quad \text{and} \quad \mathbf{y} = \begin{pmatrix} e & f \\ g & h \end{pmatrix}$$

be invertible, and consider $\mathbf{x} + \mathbf{y}$. Now

$$\mathbf{x} + \mathbf{y} = \begin{pmatrix} a+e & b+f \\ c+g & d+h \end{pmatrix}$$

which is invertible if and only if its determinant is nonzero, that is, if and only if

$$(a+e)(d+h) - (c+g)(b+f) \neq 0$$
$$ad + de + ah + eh - bc - gb - cf - gf \neq 0$$
$$\underbrace{(ad - bc)}_{\neq 0} + (de + ah - gb - cf) + \underbrace{(ef - gh)}_{\neq 0} \neq 0$$

It is not clear that the sum on the left-hand side should be nonzero. This solution shows that testing $\mathbf{x} + \mathbf{y}$ may be tedious. The first solution is preferable in this case.

EXAMPLE 10 Let $A \in \mathscr{M}_{nn}$. Show that the set $W\ (\subseteq \mathscr{M}_{n1})$ of all solutions to $A_{n\times n}X_{n\times 1} = 0_{n\times 1}$ is a subspace of $\mathscr{M}_{n1}$.

Solution Let $U_{n\times 1}$ and $V_{n\times 1}$ be solutions of the homogeneous equation $A_{n\times n}X_{n\times 1} = 0_{n\times 1}$. To check the closures, let $r \in \mathbb{R}$ and calculate

$$A(U+V) = AU + AV = 0 + 0 = 0$$
$$A(rU) = rAU = r0 = 0$$

Therefore $U + V$ and rU are solutions in W; W is a subspace.

EXAMPLE 11 Let W be the subset of $\mathscr{C}_{nn}$ defined by $W = \{Z \mid \bar{Z} = Z\}$. Is W a subspace of $\mathscr{C}_{nn}$? (Note that W is not $\mathscr{M}_{nn}$ because our "scalars" for scalar multiplication come from $\mathbb{C}$ now.)

Solution First we check for closure of addition. Let Y and Z be in W. We have

$$\overline{Y+Z} \underset{\substack{\uparrow \\ \text{Previous property}}}{=} \bar{Y} + \bar{Z} \underset{\substack{\uparrow \\ Y,Z \in W}}{=} Y + Z$$

so W is closed under addition. Now check multiplication by letting $c \in \mathbb{C}$ and $Z \in W$; we must see whether $\overline{cZ} = cZ$. However $\overline{cZ} = \bar{c}\bar{Z} = \bar{c}Z$. Because $\bar{c} \neq c$ unless c is real, closure of multiplication fails and W is not a subspace.

Table 3.3.1 **ADDITION AND MULTIPLICATION FOR $F = \{0, 1\}$**

+	0	1
0	0	1
1	1	0

·	0	1
0	0	0
1	0	1

The idea of subspace has important applications. In fact, subspaces are used to define certain concepts in coding theory.

EXAMPLE 12 (Linear codes) This example is of an unusual vector space for which subspaces have applications in coding theory. First, consider $F = \{0, 1\}$ with the operations of multiplication and addition defined as in $\mathbb{R}$ except that we define $1 + 1 = 0$ (see Table 3.3.1). The set F along with these operations is called a *commutative field*. This just means that it enjoys the properties of the real number system. For the vector space let F^n be the set of n-tuples of elements of F with operations

$$(a_1, \ldots, a_n) + (b_1, \ldots, b_n) = (a_1 + b_1, \ldots, a_n + b_n)$$
$$r(a_1, \ldots, a_n) = (ra_1, \ldots, ra_n)$$

Note that a_k, b_k, and r can be only 0 or 1. F^n with these operations is a vector space. How is it used in coding theory? Well, in this case we have what is called a *binary channel* because the components of the vectors in F^n can be chosen from a set of only *two* symbols. A subset of F^n is called a *linear code* if and only if it is a *subspace* of F^n. For example, F^3 contains eight possible vectors. The subset $V = \{(0, 0, 0), (1, 0, 0)\}$ is a linear code because we have closure: we can just check all possible sums and scalar products. Note that $(1, 0, 0) + (1, 0, 0) = (0, 0, 0)$.

PROBLEMS 3.3 In Probs. 1 to 14, a vector space V and subset W are given. Determine whether W is a subspace.

1. $V = \mathscr{M}_{mn}$, $W = \{x \in \mathscr{M}_{mn}$ with nonnegative entries$\}$
2. $V = \mathscr{M}_{nn}$, $W = \{$symmetric $n \times n$ matrices$\}$
3. (Define the trace of an $n \times n$ matrix A as $\operatorname{tr} A = a_{11} + a_{22} + \cdots + a_{nn}$.) Let $V = \mathscr{M}_{nn}$ and $W = \{A \in \mathscr{M}_{nn}$ with $\operatorname{tr} A = 0\}$
4. $V = E^3$, $W = \{(x_1, x_2\ x_3) | ax_1 + bx_2 + cx_3 = 0$, where a, b, c are fixed numbers$\}$
5. $V = \mathscr{M}_{22}$, $W = \{$noninvertible matrices$\}$
6. $V = E^2$, $W = \{(x_1, x_2) | {x_1}^2 + {x_2}^2 = 1\}$

7. $V = \mathscr{M}_{nn}$, $W = \{A \in \mathscr{M}_{nn} | A = -A^T\}$

8. $V = E^3$, $W = \{\mathbf{x} \in E^3 | \mathbf{x} \text{ is perpendicular to } (a, b, c)\}$

9. $V = \mathscr{C}_{nn}$, $W = \{n \times n \text{ matrices with real entries}\}$

10. $V = \mathbb{C}^n$, $W = \{n\text{-tuples with real entries}\}$

11. $V = \mathscr{P}_n$, $W = \{f \text{ in } \mathscr{P}_n | f(0) = 0\}$

12. $V = \mathscr{P}_n$, $W = \{f \text{ in } \mathscr{P}_n | f(0) \neq 0\}$

13. $V = \mathscr{P}_n$, $W = \{f \text{ in } \mathscr{P}_n | f(1) = 0\}$

14. $V = \mathscr{P}_n$, $W = \{f \text{ in } \mathscr{P}_n | f(1) \neq 0\}$

15. Consider the system of equations $AX = 0$, where A is in $\mathscr{C}_{nn}$. Show that the set of all solutions of $AX = 0$ is a vector space under the usual operations.

16. If $B_{n \times 1} \neq 0$ in the system of equations $AX = B$, where A is $n \times n$ show that the set of solutions cannot be a vector space, given the standard matrix operations.

17. Let A and B be square matrices. Show that $\operatorname{tr}(A + B) = \operatorname{tr} A + \operatorname{tr} B$, $\operatorname{tr}(rA) = r(\operatorname{tr} A)$, $\operatorname{tr}(AB) = \operatorname{tr}(BA)$. (See Prob. 3 for the definition of tr A.)

18. Using the terminology of Example 12, show that

$$V = \{(0, 0, 0, 0), (1, 1, 0, 1), (1, 0, 0, 1), (0, 1, 0, 0)\}$$

is a linear code.

19. Let $V = E^2$, which can be associated with the plane. Show that $W = \{(x_1, x_2) | x_2 = mx_1\}$ is a subspace of V. Show that $U = \{(x_1\ x_2) | x_2 = mx_1 + b, b \neq 0\}$ is not a subspace of V. This shows that if a straight line is to represent a subspace of the plane, the line must pass through the origin.

20. Let $V = E^3$. Show that $W = \{(x_1, x_2, x_3) | ax_1 + bx_2 + cx_3 = d\}$ is a subspace of V if and only if $d = 0$. This shows that if a plane is to represent a subspace of three-space, it must pass through the origin.

21. Let $V = E^3$. Show that $W = \{(x_1, x_2, x_3) | x_1 = at + k_1, x_2 = bt + k_2, x_3 = ct + k_3, t \text{ real}\}$ is a proper subspace of V if and only if there exists T such that $aT + k_1 = bT + k_2 = cT + k_3 = 0$. This shows that a line represents a subspace of three-space if and only if the line passes through the origin.

22. Fill in the details for the solution of Example 1.

3.4 LINEARLY INDEPENDENT SETS OF VECTORS

The equation $x + 2y - z = 0$ has general solution $(x, y, z) = (s - 2r, r, s)$, where r and s are any numbers. Any solution can be written in terms of the vectors

(1, 0, 1) and (−2, 1, 0) from E^3 as

$$(s - 2r, r, s) = s(1, 0, 1) + r(-2, 1, 0)$$

That is, an infinite number of solutions can be constructed in terms of just two vectors, and analysis of the solutions can be performed by considering just these two vectors. To use similar methods of analysis in vector spaces, we will need the concepts of span and linear independence of sets of vectors. Both concepts involve linear combinations of vectors.

DEFINITION 3.4.1 Let $\mathbf{v}_1, \mathbf{v}_2, \ldots, \mathbf{v}_n$ be vectors in a vector space V. A *linear combination* of the vectors $\mathbf{v}_1, \mathbf{v}_2, \ldots, \mathbf{v}_n$ is any sum of the form

$$c_1\mathbf{v}_1 + c_2\mathbf{v}_2 + \cdots + c_n\mathbf{v}_n$$

where the numbers $c_1, c_2, \ldots, c_n$ are called the *coefficients* of the linear combination.

EXAMPLE 1 Write five linear combinations of the vectors (1, −1), (1, 2), and (3, 0) in E^2.

Solution Five possibilities are

$$0(1, -1) + 0(1, 2) + 0(3, 0) = (0, 0)$$
$$3(1, -1) - 1(1, 2) + 7(3, 0) = (23, -5)$$
$$1(1, -1) + 0(1, 2) + 0(3, 0) = (1, -1)$$
$$2(1, -1) + 1(1, 2) + 5(3, 0) = (18, 0)$$
$$2(1, -1) + 1(1, 2) - 1(3, 0) = (0, 0)$$

Note that the first and last linear combinations yield the same vector (0, 0), even though the coefficients are not the same. The last four linear combinations are called nontrivial because in each at least one coefficient is nonzero.

EXAMPLE 2 Write (7, −2, 2) in E^3 as a linear combination of (1, −1, 0), (0, 1, 1), and (2, 0, 1).

Solution We want to find c_1, c_2, c_3 so that

$$(7, -2, 2) = c_1(1, -1, 0) + c_2(0, 1, 1) + c_3(2, 0, 1)$$

or

$$(7, -2, 2) = (c_1 + 2c_3, -c_1 + c_2, c_2 + c_3)$$

which yields equations

$$\begin{aligned} 7 &= c_1 + 2c_3 \\ -2 &= -c_1 + c_2 \\ 2 &= c_2 + c_3 \end{aligned}$$

The solution is $c_1 = 1$, $c_2 = -1$, $c_3 = 3$, so

$$(7, -2, 2) = (1, -1, 0) - (0, 1, 1) + 3(2, 0, 1)$$

EXAMPLE 3 Can $(3, -1, 4)$ be written as a linear combination of $(1, -1, 0)$, $(0, 1, 1)$, and $(3, -5, -2)$?

Solution We check to see whether the equation

$$(3, -1, 4) = c_1(1, -1, 0) + c_2(0, 1, 1) + c_3(3, -5, -2)$$

has a solution. This is equivalent to

$$\begin{aligned} 3 &= c_1 \qquad + 3c_3 \\ -1 &= -c_1 + c_2 - 5c_3 \\ 4 &= \qquad c_2 - 2c_3 \end{aligned}$$

In reduced form this is

$$\left(\begin{array}{ccc|c} 1 & 0 & 3 & 3 \\ 0 & 1 & -2 & 2 \\ 0 & 0 & 0 & 2 \end{array}\right)$$

and there is no solution. Hence $(3, -1, 4)$ cannot be written as a linear combination of the given vectors.

DEFINITION 3.4.2 Let $\{\mathbf{v}_1, \mathbf{v}_2, \ldots, \mathbf{v}_n\}$ be a set of vectors in the vector space V. The *span* of $\{\mathbf{v}_1, \mathbf{v}_2, \ldots, \mathbf{v}_n\}$ is the set of all possible linear combinations of $\mathbf{v}_1, \mathbf{v}_2, \ldots, \mathbf{v}_n$. The notation is span $\{\mathbf{v}_1, \mathbf{v}_2, \ldots, \mathbf{v}_n\}$.

The span of a set of vectors from V is actually a subspace of V.

THEOREM 3.4.1 If $\mathscr{S} = \{\mathbf{v}_1, \mathbf{v}_2, \ldots, \mathbf{v}_n\} \subset V$ with V a vector space, then span $\mathscr{S}$ is a subspace of V.

Proof Let $\mathbf{x}$ and $\mathbf{y}$ be in span $\mathscr{S}$. Then $\mathbf{x}$ and $\mathbf{y}$ are linear combinations:

$$\begin{aligned} \mathbf{x} &= a_1\mathbf{v}_1 + \cdots + a_n\mathbf{v}_n \\ \mathbf{y} &= b_1\mathbf{v}_1 + \cdots + b_n\mathbf{v}_n \end{aligned}$$

Now

$$\mathbf{x} + \mathbf{y} = (a_1 + b_1)\mathbf{v}_1 + \cdots + (a_n + b_n)\mathbf{v}_n$$

and

$$r\mathbf{x} = (ra_1)\mathbf{v}_1 + \cdots + (ra_n)\mathbf{v}_n \qquad r \text{ a scalar}^1$$

which are also in span $\mathscr{S}$. Therefore, span $\mathscr{S}$ is a subspace of V. It is called the *subspace of V spanned by $\mathscr{S}$*.

EXAMPLE 4 Example 2 could have been worded as follows: Show that $(7, -2, 2)$ is in span $\{(1, -1, 0), (0, 1, 1), (2, 0, 1)\}$. The solution would be exactly the same.

EXAMPLE 5 Example 3 could have been worded as follows: Is $(3, -1, 4)$ in span $\{(1, -1, 0), (0, 1, 1), (3, -5, -2)\}$? The solution would be exactly the same. The answer is that $(3, -1, 4)$ is not in span $\{(1, -1, 0), (0, 1, 1), (3, -5, -2)\}$.

EXAMPLE 6 We have already seen that the solutions of $x + 2y - z = 0$ can be written as

$$s(1, 0, 1) + r(-2, 1, 0)$$

for all complex s and r. Another way to say this is that all solutions form the subspace span $\{(1, 0, 1), (-2, 1, 0)\}$.

In some instances span $\mathscr{S}$ may be all of V.

EXAMPLE 7 Let $S = \{1, x, x^2\}$, which is a set of vectors in $\mathscr{P}_2$. Then span $S = \{a_0 1 + a_1 x + a_2 x^2$, where a_0, a_1, a_2 can be any real numbers$\}$. Thus span $S = \mathscr{P}_2$.

EXAMPLE 8 Let $S = \{(1, 0, 0, 0), (0, 1, 0, 0), (0, 0, 1, 0), (0, 0, 0, 1)\} \subseteq E^4$. Span S is all E^4, since span S is all vectors of the form

$$x_1(1, 0, 0, 0) + x_2(0, 1, 0, 0) + x_3(0, 0, 1, 0) + x_4(0, 0, 0, 1) = (x_1, x_2, x_3, x_4)$$

where x_1, x_2, x_3, and x_4 can be any real numbers.

EXAMPLE 9 Show that span $\{(1, 0, 1), (-1, 2, 3), (0, 1, -1)\}$ is all of E^3.

Solution We must show that any vector (a, b, c) in E^3 can be written as a linear combination of the three given vectors. That is, we must show that there are constants c_1, c_2, c_3 so that

---

[1] The statement of the theorem is for any vector space. In the real case the scalars would come from $\mathbb{R}$; in the complex case, from $\mathbb{C}$.

$$(a, b, c) = c_1 (1, 0, 1) + c_2(-1, 2, 3) + c_3(0, 1, -1)$$

regardless of what real values a, b, and c take. The last equation is equivalent to

$$\begin{aligned} a &= c_1 - c_2 \\ b &= \quad\; 2c_2 + c_3 \\ c &= c_1 + 3c_2 - c_3 \end{aligned}$$

which has solutions

$$c_1 = \frac{5a + b + c}{6} \qquad c_2 = \frac{b + c - a}{6} \qquad c_3 = \frac{a + 2b - c}{3}$$

Therefore $(a, b, c) \in \text{span } \{(1, 0, 1), (-1, 2, 3), (0, 1, -1)\}$, and the span is all of E^3.

EXAMPLE 10 Determine whether span $\{(1, -1, 0), (0, 1, 1), (3, -5, -2)\}$ is all of E^3.

Solution Let (a, b, c) be an arbitrary vector in E^3. We want to know whether it is possible to write

$$(a, b, c) = c_1(1, -1, 0) + c_2(0, 1, 1) + c_3(3, -5, -2)$$

The last equation is equivalent to

$$\left(\begin{array}{rrr|c} 1 & 0 & 3 & a \\ -1 & 1 & -5 & b \\ 0 & 1 & -2 & c \end{array}\right)$$

which reduces to

$$\left(\begin{array}{rrr|c} 1 & 0 & 3 & a \\ 0 & 1 & -2 & a + b \\ 0 & 0 & 0 & c - a - b \end{array}\right)$$

Therefore a solution exists only if $c - a - b = 0$; but this places a restriction on (a, b, c), and so the very first equation cannot be solved for an arbitrary vector (a, b, c). Therefore the span of the given vectors is not all E^3.

The question in Example 10 could have been asked in a slightly different way.

EXAMPLE 11 Describe span S, where $S = \{(1, -1, 0), (0, 1, 1), (3, -5, -2)\}$.

Solution Suppose (a, b, c) is in span S. Then the equation

$$(a, b, c) = c_1(1, -1, 0) + c_2(0, 1, 1) + c_3(3, -5, -2)$$

must be solvable. Working as in Example 10, we conclude that $c - a - b = 0$.

Thus span $S = \{(a, b, c) | c = a + b\}$. That is, the span of S is all vectors whose third component is the sum of the first two components. So, for example, $(1, 3, 5) \notin$ span S and $(1, 3, 4) \in$ span S.

The vector space E^2 is spanned by $S = \{(1, 0), (0, 1)\}$. It is also spanned by a larger set $S' = \{(1, 0), (0, 1), (1, 1)\}$. As we will see later, E^2 and functions on E^2 can be analyzed by using spanning sets; hence for economy's sake, we want to be able to find the smallest possible spanning sets for vector spaces. To do this, the idea of *linear independence* is required.

DEFINITION 3.4.3 A set $\{\mathbf{v}_1, \ldots, \mathbf{v}_n\}$ of vectors in a vector space V is called *linearly independent* if the only solution to the equation

$$c_1\mathbf{v}_1 + c_2\mathbf{v}_2 + \cdots + c_n\mathbf{v}_n = \boldsymbol{\theta}$$

is $c_1 = c_2 = \cdots = c_n = 0$. If the set is not linearly independent, it is called *linearly dependent*.

To determine whether a set $S = \{\mathbf{v}_1, \mathbf{v}_2, \ldots, \mathbf{v}_n\}$ is linearly independent or linearly dependent, we need to find out about the solution of

$$c_1\mathbf{v}_1 + c_2\mathbf{v}_2 + \cdots + c_n\mathbf{v}_n = \boldsymbol{\theta}$$

If we find (by actually solving the resulting system or by any other technique) that only the trivial solution $c_1 = c_2 = \cdots = c_n = 0$ exists, then S is linearly independent. However, if one or more of the c_k's is nonzero, then the set S is linearly dependent.

EXAMPLE 12 Determine whether $S = \{(1, 0), (0, 1)\}$ is linearly independent.

Solution Consider

$$c_1(1, 0) + c_2(0, 1) = \boldsymbol{\theta} = (0, 0)$$

This equation is equivalent to

$$(c_1, c_2) = (0, 0)$$

which has only $c_1 = 0$, $c_2 = 0$ as a solution. Therefore, S is linearly independent.

EXAMPLE 13 Is $S = \{(1, 0), (0, 1), (1, -1)\}$ linearly independent?

Solution Consider

$$c_1(1, 0) + c_2(0, 1) + c_3(1, -1) = \boldsymbol{\theta} = (0, 0)$$

which is equivalent to

$$c_1 + c_3 = 0$$
$$c_2 - c_3 = 0$$

This system has solution $c_3 = k$, $c_1 = -k$, $c_2 = k$; if $k \neq 0$, then we have a nontrivial solution, and so S is not linearly independent—it is linearly dependent.

EXAMPLE 14 Determine whether $S = \{1 + x, x + x^2, 1 + x^2\}$ is linearly independent in $\mathscr{P}_2$.

Solution Consider

$$c_1(1 + x) + c_2(x + x^2) + c_3(1 + x^2) = \boldsymbol{\theta} = 0 + 0x + 0x^2$$

By collecting terms on the left-hand side, this equation can be rewritten

$$(c_1 + c_3) + (c_1 + c_2)x + (c_2 + c_3)x^2 = 0 + 0x + 0x^2 = \boldsymbol{\theta}$$

From algebra we know that a polynomial is identically zero only when all the coefficients are zero. So we have

$$\begin{aligned} c_1 \qquad + c_3 &= 0 \\ c_1 + c_2 \qquad &= 0 \\ c_2 + c_3 &= 0 \end{aligned}$$

which has only the trivial solution. Therefore, S is linearly independent.

Figure 3.4.1 Vectors in $S = \{(1, 0), (0, 1), (1, -1)\}$ as linear combinations of the others.

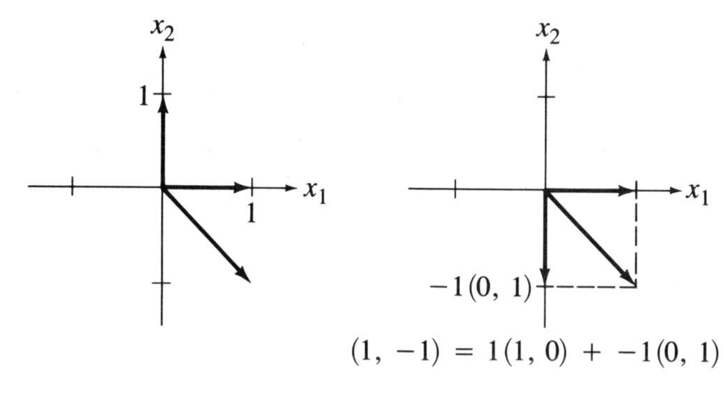

$(1, -1) = 1(1, 0) + -1(0, 1)$

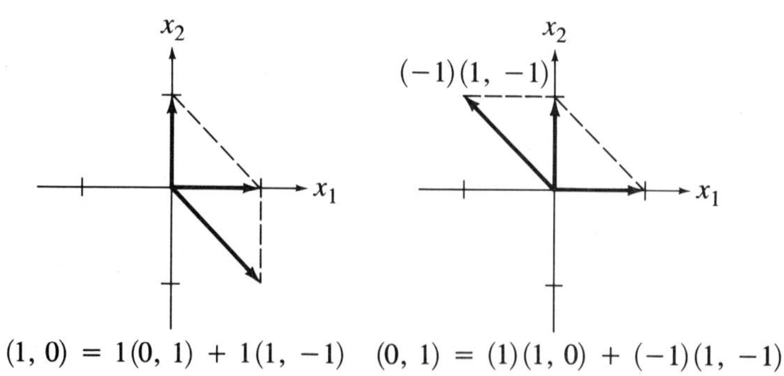

$(1, 0) = 1(0, 1) + 1(1, -1)$ $(0, 1) = (1)(1, 0) + (-1)(1, -1)$

EXAMPLE 15 The set $S = \{\boldsymbol{\theta}\}$ is linearly dependent in any real or complex vector space because $c_1\boldsymbol{\theta} = \boldsymbol{\theta}$ has nontrivial solution $c_1 = 1$.

Linear dependence of a set of two or more vectors means that at least one of the vectors in the set can be written as a linear combination of the others. Recall Example 13 and the set $S = \{(1, 0), (0, 1), (1, -1)\}$. In Fig. 3.4.1 we have shown geometrically the dependence of the vectors in S. A general statement of this situation is as follows:

THEOREM 3.4.2 Let $S = \{\mathbf{v}_1, \mathbf{v}_2, \ldots, \mathbf{v}_n\}$ be a set of at least two vectors ($n \geq 2$) in a vector space V. Then S is linearly dependent if and only if one of the vectors in S can be written as a linear combination of the rest.

Proof ($\Rightarrow$) If S is linearly dependent, then there are constants $c_1, c_2, \ldots, c_n$, some of which are nonzero, such that

$$c_1\mathbf{v}_1 + c_2\mathbf{v}_2 + \cdots + c_n\mathbf{v}_n = \boldsymbol{\theta}$$

Suppose c_k ($1 \leq k \leq n$) is a nonzero coefficient in the linear combination. Then

$$c_k\mathbf{v}_k = -c_1\mathbf{v}_1 - c_2\mathbf{v}_2 - \cdots - c_{k-1}\mathbf{v}_{k-1} - c_{k+1}\mathbf{v}_{k+1} - \cdots - c_n\mathbf{v}_n$$

and since $c_k \neq 0$,

$$\mathbf{v}_k = -\frac{c_1}{c_k}\mathbf{v}_1 - \frac{c_2}{c_k}\mathbf{v}_2 - \cdots - \frac{c_{k-1}}{c_k}\mathbf{v}_{k-1} - \frac{c_{k+1}}{c_k}\mathbf{v}_{k+1} - \cdots - \frac{c_n}{c_k}\mathbf{v}_n$$

Therefore, $\mathbf{v}_k$ is a linear combination of the other vectors in S.

($\Leftarrow$) Suppose $\mathbf{v}_k = d_1\mathbf{v}_1 + d_2\mathbf{v}_2 + \cdots + d_{k-1}\mathbf{v}_{k-1} + d_{k+1}\mathbf{v}_{k+1} + \cdots + d_n\mathbf{v}_n$. Then, adding $(-1)\mathbf{v}_k$ to both sides, we have

$$\boldsymbol{\theta} = d_1\mathbf{v}_1 + \cdots + (-1)\mathbf{v}_k + \cdots + d_n\mathbf{v}_n$$

Because the coefficient of $\mathbf{v}_k$ is nonzero, the set S is linearly dependent.

EXAMPLE 16 Show that

$$S = \left\{\begin{pmatrix} 1 & 1 \\ 0 & 3 \end{pmatrix}, \begin{pmatrix} -1 & 0 \\ 2 & 1 \end{pmatrix}, \begin{pmatrix} -1 & 2 \\ 6 & 9 \end{pmatrix}\right\}$$

is linearly dependent in $\mathcal{M}_{22}$. Write one of the vectors as a linear combination of the others.

Solution Consider

$$c_1\begin{pmatrix} 1 & 1 \\ 0 & 3 \end{pmatrix} + c_2\begin{pmatrix} -1 & 0 \\ 2 & 1 \end{pmatrix} + c_3\begin{pmatrix} -1 & 2 \\ 6 & 9 \end{pmatrix} = \boldsymbol{\theta} = \begin{pmatrix} 0 & 0 \\ 0 & 0 \end{pmatrix}$$

This equation is equivalent to

$$\begin{aligned} c_1 - c_2 - c_3 &= 0 \\ c_1 \quad\quad + 2c_3 &= 0 \\ 2c_2 + 6c_3 &= 0 \\ 3c_1 + c_2 + 9c_3 &= 0 \end{aligned}$$

which reduces to

$$\left(\begin{array}{ccc|c} 1 & -1 & -1 & 0 \\ 0 & 1 & 3 & 0 \\ 0 & 0 & 0 & 0 \\ 0 & 0 & 0 & 0 \end{array}\right)$$

Therefore $c_1 = -2k$, $c_2 = -3k$, $c_3 = k$ is a solution, where k is arbitrary. Thus the set S is linearly dependent. Choosing $k = 1$, we have

$$-2\begin{pmatrix} 1 & 1 \\ 0 & 3 \end{pmatrix} - 3\begin{pmatrix} -1 & 0 \\ 2 & 1 \end{pmatrix} + \begin{pmatrix} -1 & 2 \\ 6 & 9 \end{pmatrix} = \begin{pmatrix} 0 & 0 \\ 0 & 0 \end{pmatrix}$$

and we can write

$$\begin{pmatrix} 1 & 1 \\ 0 & 3 \end{pmatrix} = -\frac{3}{2}\begin{pmatrix} -1 & 0 \\ 2 & 1 \end{pmatrix} + \frac{1}{2}\begin{pmatrix} -1 & 2 \\ 6 & 9 \end{pmatrix}$$

Of course, we could also write

$$\begin{pmatrix} -1 & 2 \\ 6 & 9 \end{pmatrix} = 2\begin{pmatrix} 1 & 1 \\ 0 & 3 \end{pmatrix} + 3\begin{pmatrix} -1 & 0 \\ 2 & 1 \end{pmatrix}$$

or

$$\begin{pmatrix} -1 & 0 \\ 2 & 1 \end{pmatrix} = -\frac{2}{3}\begin{pmatrix} 1 & 1 \\ 0 & 3 \end{pmatrix} + \frac{1}{3}\begin{pmatrix} -1 & 2 \\ 6 & 9 \end{pmatrix}$$

Some Geometry of Spanning Sets in E^2 and E^3 The span of a single nonzero vector is a line containing the origin. Span $\{\mathbf{v}\}$ is all multiples of $\mathbf{v}$, which is all position vectors in the same direction as $\mathbf{v}$ (see Fig. 3.4.2). The terminal points of these vectors form the line with vector equation

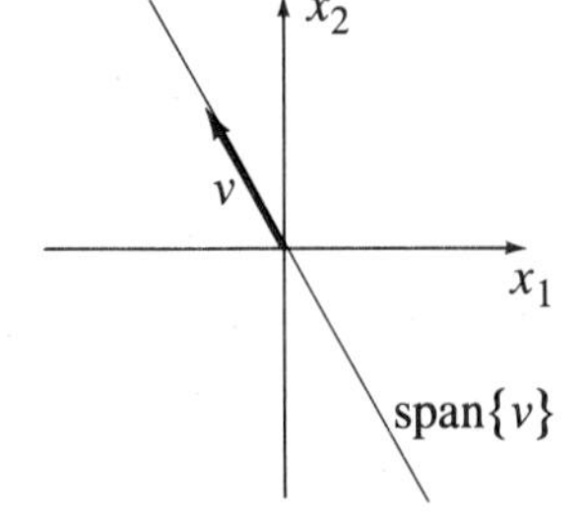

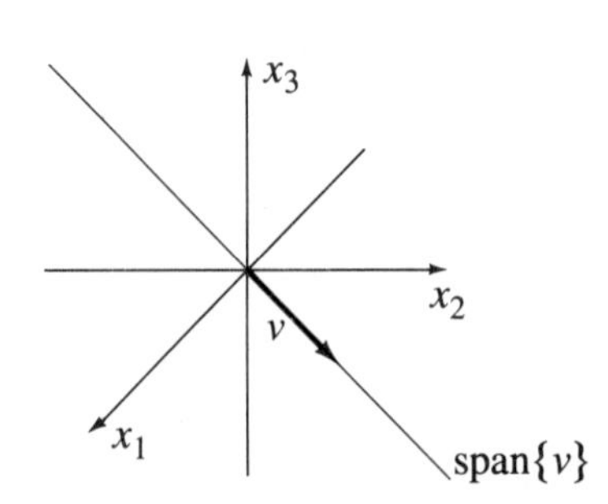

Figure 3.4.2 Span $\{v\}$ in E^2 and E^3.

$$\mathbf{r} = t\mathbf{v} + \boldsymbol{\theta}$$

The span of two independent vectors is a plane containing the origin. To see this in E^3, let $\mathbf{v}$ and $\mathbf{w}$ be given by (a, b, c) and (d, e, f), respectively. The plane containing $\mathbf{v}$ and $\mathbf{w}$ has normal vector $\mathbf{v} \times \mathbf{w}$ and vector equation

$$(\mathbf{v} \times \mathbf{w}) \cdot (x - 0, y - 0, z - 0) = 0$$

If we calculate $\mathbf{v} \times \mathbf{w}$ and write the vector equation, we find

$$(bf - ce)x + (cd - af)y + (ae - bd)z = 0$$

where (x, y, z) is a vector in the plane. However, if (x, y, z) is to be a linear combination of $\mathbf{v}$ and $\mathbf{w}$, we must have

$$\begin{aligned} x &= c_1 a + c_2 d \\ y &= c_1 b + c_2 e \\ z &= c_1 c + c_2 f \end{aligned}$$

This system reduces to $(a \neq 0)$

$$\left(\begin{array}{cc|c} 1 & d/a & x/a \\ 0 & 1 & (ay - bx)(ax - cd) \\ 0 & 0 & (az - cx)(ae - bd) - (ay - bx)(af - cd) \end{array}\right)$$

which has a solution if and only if

$$(az - cx)(ae - bd) - (ay - bx)(af - cd) = 0$$

which holds if and only if the vector equation holds. So the span of two independent vectors is the plane containing the vectors. See Fig. 3.4.3.

Figure 3.4.3
Span $\{v_1, v_2\}$ in E^3.

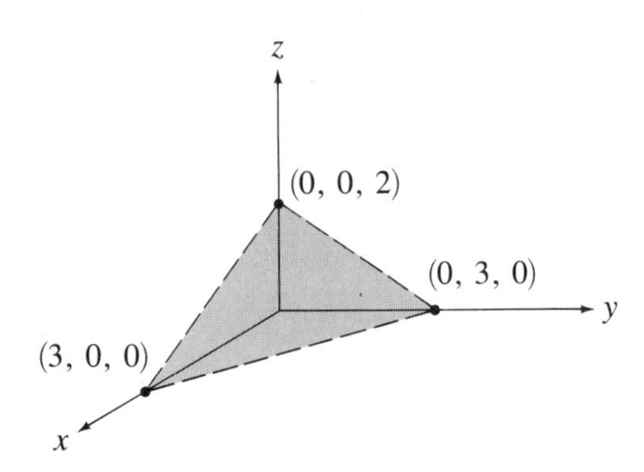

The span of three nonzero vectors in E^3 can be a line, a plane, or all of E^3, depending on the degree of dependence of the three vectors. If all three are multiples of each other, we have only a line. If two of the vectors $\mathbf{v}_1$ and $\mathbf{v}_2$ are independent but the entire set is linearly dependent, then $\mathbf{v}_3$ is a linear combination of $\mathbf{v}_1$ and $\mathbf{v}_2$ and $\mathbf{v}_3$ lies in the plane defined by $\mathbf{v}_1$ and $\mathbf{v}_2$. That is, the vectors are *coplanar*. Lay three pencils on a tabletop with erasers joined for a graphic example of coplanar vectors. If $\{\mathbf{v}_1, \mathbf{v}_2, \mathbf{v}_3\}$ is linearly independent, then the span is all E^3. This can be verified directly in individual cases; to show it in general requires methods of the next section.

Linear combinations in complex vector spaces have important applications, as the next examples illustrate.

EXAMPLE 17 The set of Pauli spin matrices, used in the study of electron spin in quantum chemistry, is

$$S = \left\{ \begin{pmatrix} 0 & 1 \\ 1 & 0 \end{pmatrix}, \begin{pmatrix} 0 & -i \\ i & 0 \end{pmatrix}, \begin{pmatrix} 1 & 0 \\ 0 & -1 \end{pmatrix} \right\}$$

Show that S is linearly independent in $\mathscr{C}_{22}$. Discuss the importance of the independence.

Solution We consider the equation

$$c_1 \begin{pmatrix} 0 & 1 \\ 1 & 0 \end{pmatrix} + c_2 \begin{pmatrix} 0 & -i \\ i & 0 \end{pmatrix} + c_3 \begin{pmatrix} 1 & 0 \\ 0 & -1 \end{pmatrix} = \boldsymbol{\theta} = \begin{pmatrix} 0 & 0 \\ 0 & 0 \end{pmatrix}$$

because the zero for $\mathscr{C}_{22}$ is

$$\begin{pmatrix} 0 & 0 \\ 0 & 0 \end{pmatrix}$$

just as for $\mathscr{M}_{22}$. The matrix equation gives us

$$\begin{aligned} c_3 &= 0 \\ c_1 - ic_2 \qquad &= 0 \\ c_1 + ic_2 \qquad &= 0 \\ c_3 &= 0 \end{aligned}$$

which reduces to

$$\left(\begin{array}{ccc|c} 1 & -i & 0 & 0 \\ 1 & i & 0 & 0 \\ 0 & 0 & 1 & 0 \end{array}\right)$$

And

$$\det \begin{pmatrix} 1 & -i & 0 \\ 1 & i & 0 \\ 0 & 0 & 1 \end{pmatrix} = 2i \neq 0$$

so we have only the trivial solution, and S is a linearly independent set. The importance of the independence is that none of the matrices can be written in terms of the others; so the study of electron spin by Pauli matrices cannot, in general, be conducted with a proper subset of S.

EXAMPLE 18 Let

$$\mathcal{S} = \left\{ \begin{pmatrix} 0 & 1 \\ 1 & 0 \end{pmatrix}, \begin{pmatrix} 0 & -i \\ i & 0 \end{pmatrix}, \begin{pmatrix} 1 & 0 \\ 0 & -1 \end{pmatrix}, \begin{pmatrix} 1 & 0 \\ 0 & 1 \end{pmatrix} \right\}$$

Show that $\mathcal{S}$ spans all $\mathcal{C}_{22}$. Show that $\mathcal{S}$ is linearly independent. Note that $\mathcal{S}$ is the set of Pauli spin matrices with I_2 adjoined.

Solution Consider the equation

$$\begin{pmatrix} z_{11} & z_{12} \\ z_{21} & z_{22} \end{pmatrix} = c_1 \begin{pmatrix} 0 & 1 \\ 1 & 0 \end{pmatrix} + c_2 \begin{pmatrix} 0 & -i \\ i & 0 \end{pmatrix} + c_3 \begin{pmatrix} 1 & 0 \\ 0 & -1 \end{pmatrix} + c_4 \begin{pmatrix} 1 & 0 \\ 0 & 1 \end{pmatrix}$$

which is equivalent to

$$\begin{aligned} z_{11} &= \qquad\qquad\quad c_3 + c_4 \\ z_{12} &= c_1 - ic_2 \\ z_{21} &= c_1 + ic_2 \\ z_{22} &= \qquad\qquad -c_3 + c_4 \end{aligned}$$

which has solution $c_1 = (z_{12} + z_{21})/2$, $c_2 = (z_{21} - z_{12})/(2i)$, $c_3 = (z_{11} - z_{22})/2$, and $c_4 = (z_{11} + z_{22})/2$. Therefore $\mathcal{S}$ spans $\mathcal{C}_{22}$. For the independence, note that if $z_{11} = z_{12} = z_{21} = z_{22} = 0$, we have only the trivial solution.

PROBLEMS 3.4 In Probs. 1 to 9, a set S of vectors in a vector space is given. (*a*) Describe span S (see Example 11); (*b*) determine whether S is linearly independent; and (*c*) if S is linearly dependent, write one vector as a linear combination of the others.

1. $S = \{(1, 2, 0), (0, 3, 2)\}$ in E^3

2. $S = \{1, x^2\}$ in $\mathcal{P}_2$

3. $S = \{(1, i, 0), (0, 1, i), (i, i-1, -1)\}$ in $\mathbb{C}^3$

4. $S = \left\{ \begin{pmatrix} 1 & 0 \\ 0 & 0 \end{pmatrix}, \begin{pmatrix} 0 & 0 \\ 0 & 1 \end{pmatrix}, \begin{pmatrix} 0 & 1 \\ 1 & 0 \end{pmatrix} \right\}$ in $\mathcal{M}_{22}$

5. $S = \{(1, 1), (2, 6), (7, 12)\}$ in E^2

6. $S = \{(1, 0, 2), (3, 2, -7), (-1, -2, 11)\}$ in E^3

7. $S = \{(1, -1, 2), (0, 0, 0)\}$ in E^3

8. $S = \{x + 1, x^2 - 2, x - 1, 3\}$ in $\mathcal{P}_2$

9. $S = \left\{ \begin{pmatrix} 1 & 0 \\ 0 & -1 \end{pmatrix} \begin{pmatrix} 0 & 1 \\ 0 & 0 \end{pmatrix} \begin{pmatrix} 0 & 0 \\ 1 & 0 \end{pmatrix} \right\}$ in $\mathscr{M}_{22}$

10. Which of the following sets of vectors span E^3?

(*a*) $\{(1, -1, 1), (2, 0, 3), (3, -1, 4)\}$

(*b*) $\{(0, 0, 0), (1, 0, 0), (1, 1, 0)\}$

(*c*) $\{(1, 0, 0), (1, 1, 0), (1, 1, 1)\}$

(*d*) $\{(1, -1, 1), (1, 0, 1), (3, -1, 2)\}$

11. Show that a set of two vectors from E^3 cannot span E^3.

12. Write any three distinct vectors from E^2. Show that they are linearly dependent.

13. Show that, in general, any three vectors from E^2 must be linearly dependent.

14. Show that any set S of vectors which contains $\boldsymbol{\theta}$ is linearly dependent.

15. A polynomial is called *even* if its terms are constants and constants times *even* powers of x. Show that span $\{1, x^2\}$ is all the even polynomials in $\mathscr{P}_3$.

16. A polynomial is called *odd* if its terms are only constants times *odd* powers of x. Show that span $\{x, x^3\}$ is all the odd polynomials in $\mathscr{P}_3$.

17. The set

$$S = \left\{ \begin{pmatrix} 0 & 1 \\ 1 & 0 \end{pmatrix}, \begin{pmatrix} 1 & 0 \\ 0 & -1 \end{pmatrix} \right\}$$

is the set of real Pauli spin matrices used in the study of electron spin. Show that span S is the set of all 2×2 symmetric matrices with trace zero.

18. Let S be a linearly independent set of vectors from a vector space V. Show that any subset of S is linearly independent.

19. Let S be a linearly dependent set of vectors from a vector space V. Show that any set T with $S \subseteq T$ is a linearly dependent set.

20. Matrices A and B are said to *anticommute* if $AB = -BA$. Show that any pair of the Pauli spin matrices (see Example 17) anticommute.

21. The *commutator* of two square matrices of the same size is defined to be $AB - BA$ and is denoted $[A, B]$. Let

$$A = \begin{pmatrix} 2 & 1 \\ 1 & -1 \end{pmatrix} \quad \text{and} \quad B = \begin{pmatrix} 1 & 5 \\ 4 & 1 \end{pmatrix}$$

Calculate $[A, I]$, $[A, B]$, $[B, A]$, and $[A, A]$.

22. For any square matrix A, what are $[A, I]$, $[A, A]$, $[A, O]$, and $[A, A^2]$? If A is invertible, what is $[A, A^{-1}]$?

23. Show that A and B commute for multiplication if and only if the commutator of A and B is 0.

24. Show, for $n \times n$ matrices A, B, and C and any scalar r, that

(*a*) $[A, B + C] = [A, B] + [A, C]$

(*b*) $[rA, B] = r[A, B] = [A, rB]$

(*c*) $[A, B] = -[B, A]$

(*d*) $[A, [B, C]] + [B, [C, A]] + [C, [A, B]] = 0$

(*e*) $[A, B]^T = [B^T, A^T]$

(*f*) $\text{tr}\,[A, B] = 0$

3.5 BASES OF VECTOR SPACES; THE BASIS PROBLEM

The set of vectors $S = \{(1, 1), (1, -1)\}$ spans E^2. That is, any vector in E^2 is a linear combination of (1, 1) and (1, −1). The set of vectors $T = \{(1, 1), (1, -1), (1, 0)\}$ also spans E^2. Sets S and T differ in that S is linearly independent while T is linearly dependent. This makes a difference in writing a vector as a linear combination of vectors in the set. For example, writing (2, 4) in terms of the vectors in S, we have for the only possibility

$$(2, 4) = 3(1, 1) - 1(1, -1)$$

However, in terms of vectors from T, we have several possibilities:

$$(2, 4) = 3(1, 1) - 1(1, -1) + 0(1, 0)$$
$$(2, 4) = 0(1, 1) - 4(1, -1) + 6(1, 0)$$
$$(2, 4) = 4(1, 1) + 0(1, -1) - 2(1, 0)$$

Or, in general,

$$(2, 4) = (k + 4)(1, 1) + k(1, -1) + (-2 - 2k)(1, 0)$$

The point is: If a set S of vectors spans V and S is linearly dependent, then representation of a vector $\mathbf{x}$ in terms of vectors in S is not unique. If we want uniqueness, the spanning set must also be linearly independent. Such a set is called a *basis* for V. Bases are used in coding theory, as we see later in this section.

DEFINITION 3.5.1 A vector space V is said to be *finitely generated* if there exists a finite set of vectors $S = \{\mathbf{v}_1, \mathbf{v}_2, \ldots, \mathbf{v}_n\}$ in V such that span $S = V$. If the set S is also linearly independent, then S is called a *basis* for V.

This definition says the following:

A finite set is a basis for V if it (1) spans V and (2) is linearly independent.

EXAMPLE 1 $S = \{(1, 2), (3, -1)\}$ is a basis for E^2.

Solution We must show that the set is linearly independent and spans E^2. That is, we must show that

$$c_1(1, 2) + c_2(3, -1) = (a, b)$$

has a solution for any (a, b) and that

$$c_1(1, 2) + c_2(3, -1) = (0, 0)$$

has only the solution $c_1 = c_2 = 0$. These equations, respectively, in augmented form are

$$\left(\begin{array}{cc|c} 1 & 3 & a \\ 2 & -1 & b \end{array}\right) \quad \text{and} \quad \left(\begin{array}{cc|c} 1 & 3 & 0 \\ 2 & -1 & 0 \end{array}\right)$$

Instead of solving both sets of equations separately, we solve both at once by working with the doubly augmented matrix

$$\left(\begin{array}{cc|c|c} 1 & 3 & a & 0 \\ 2 & -1 & b & 0 \end{array}\right)$$

Doing this, we find that this matrix reduces to

Check linear independence with this part

$$\left(\begin{array}{cc|c|c} 1 & 3 & a & 0 \\ 0 & 1 & \dfrac{-b + 2a}{7} & 0 \end{array}\right)$$

Check span with this part

We find for the span that $c_2 = (2a - b)/7$, $c_1 = (a + 3b)/7$; for linear independence we find that $c_1 = c_2 = 0$. Since S is linearly independent and spans E^2, it is a basis for E^2.

In Example 1, the coefficients in the linear combination of basis elements were unique for any given vector (a, b). This is true in general.

THEOREM 3.5.1 Let $S = \{\mathbf{v}_1, \ldots, \mathbf{v}_n\}$ be a basis for a vector space V. Let $\mathbf{v}$ be in V. The coefficients in the representation

$$\mathbf{v} = c_1\mathbf{v}_1 + \cdots + c_n\mathbf{v}_n$$

are unique.

Proof Suppose we have two representations

$$\mathbf{v} = a_1\mathbf{v}_1 + \cdots + a_n\mathbf{v}_n$$
$$\mathbf{v} = b_1\mathbf{v}_1 + \cdots + b_n\mathbf{v}_n$$

for $\mathbf{v}$; we will show that the coefficients are actually equal. To do this, form $\mathbf{v} + (-\mathbf{v})$, which equals $\boldsymbol{\theta}$, and combine terms to obtain

$$\boldsymbol{\theta} = (a_1 - b_1)\mathbf{v}_1 + \cdots + (a_n - b_n)\mathbf{v}_n$$

Since S is a basis, it is a linearly independent set. Thus, the coefficients in the last linear combination must all be zero. That is, $a_1 = b_1, \ldots, a_n = b_n$, and the original linear combinations are the same.

EXAMPLE 2 $S = \{(i, 1 + i), (2, 1 - i)\}$ is a basis for $\mathbb{C}^2$.

Solution Proceeding as in Example 1, we form the doubly augmented matrix and row-reduce:

$$\left(\begin{array}{cc|c|c} i & 2 & a & 0 \\ 1+i & 1-i & b & 0 \end{array}\right) \xrightarrow{-iR1} \left(\begin{array}{cc|c|c} 1 & -2i & -ai & 0 \\ 1+i & 1-i & b & 0 \end{array}\right)$$
$$\xrightarrow{-(1+i)R1 + R2} \left(\begin{array}{cc|c|c} 1 & -2i & -ai & 0 \\ 0 & -1+i & b+ai-a & 0 \end{array}\right)$$

This system has a unique solution. Therefore, S is linearly independent and spans $\mathbb{C}^2$; it is a basis for $\mathbb{C}^2$.

EXAMPLE 3 Show that the set $S = \{(1, 2), (3, -1), (1, 0)\}$ is not a basis for E^2.

Solution The set S is linearly dependent because, for example,

$$(1, 2) + 2(3, -1) - 7(1, 0) = (0, 0)$$

So S cannot be a basis for E^2.

EXAMPLE 4 The zero vector space has no basis, because any subset contains the zero vector and must be linearly dependent.

Example 4 shows that a vector space may fail to have a basis. We need to decide whether a given vector space has a basis or not. Spanning sets of vectors help us answer the question.

THEOREM 3.5.2 If $S = \{\mathbf{v}_1, \ldots, \mathbf{v}_m\}$ is a set of nonzero vectors which spans a subspace W of a vector space V, then some subset of S is a basis for W. (*Note*: This means V itself, being a trivial subspace, has a basis if it is spanned by S.)

Proof If S is a linearly independent set, then by definition S is a basis for W. If S is linearly dependent, then one of the vectors can be written as a linear combination of the others. Suppose $\mathbf{v}_m$ is such a vector (if not, shift the vectors in S around and relabel so that this is true). We claim that $S' = \{\mathbf{v}_1, \ldots, \mathbf{v}_{m-1}\}$ still spans W. To see this, let $\mathbf{x}$ be in W with

$$\mathbf{x} = c_1\mathbf{v}_1 + \cdots + c_{m-1}\mathbf{v}_{m-1} + c_m\mathbf{v}_m$$

Now $\mathbf{v}_m = d_1\mathbf{v}_1 + \cdots + d_{m-1}\mathbf{v}_{m-1}$, so we can substitute this expression into the former linear combination to obtain

$$\mathbf{x} = (c_1 + c_m d_1)\mathbf{v}_1 + \cdots + (c_{m-1} + c_m d_{m-1})\mathbf{v}_{m-1}$$

Thus S' spans W. If S' is linearly independent, S' is a basis for W. If S' is linearly dependent, one of the vectors in S' is a linear combination of the others. Now we argue as before. In this way we must arrive eventually at a linearly independent set which spans W. (If we reduce to a set with a single vector, that set is linearly independent because S was a set of nonzero vectors.) The resulting set is a basis for W.

Thus we have the following fundamental result:

> Any finitely generated vector space, generated by a set of nonzero vectors, has a basis.

EXAMPLE 5 Let V be the set of all polynomials with the usual operations. The vector space V is not finitely generated. In fact, if we take any finite subset S of V, then there will be a term of maximum degree, say x^p, in the set. The polynomial x^{p+1} is not in span S, and S cannot span V.

We are particularly interested in bases of finitely generated vector spaces. Example 3 illustrated the fact that any set of three or more vectors from E^2 cannot be a basis for E^2. After all, these vectors would be coplanar and form a linearly dependent set. A set of only one vector cannot be a basis for E^2 because it spans only a line through the origin. Thus it appears that any basis for E^2 must contain exactly two vectors. This follows from Theorem 3.5.3.

THEOREM 3.5.3 If $\mathscr{S} = \{\mathbf{v}_1, \mathbf{v}_2, \ldots, \mathbf{v}_n\}$ is a basis for V, then (*a*) any set of $n + 1$ (or more) vectors is linearly dependent and therefore is not a basis for V and (*b*) any set of $n - 1$ (or less) vectors fails to span V and therefore is not a basis for V.

This theorem means that the number of vectors in a basis is unique. If we find a basis $\mathscr{S}$ for V and $\mathscr{S}$ has eight vectors in it, then *every* basis has eight vectors in it. Because of this we can define the *dimension* of a vector space V to be the number of vectors in a basis for V. If a basis $\mathscr{S}$ has n vectors in it, the dimension of V (dim V) is n, we write $\dim V = n$, and we say V is *finite-dimensional.* More particularly, V is called an *n-dimensional vector space* when a basis for V has n vectors in it. Example 1 shows that $\dim E^2 = 2$. The dimension of the zero vector space is defined to be zero.

Proof (*a*) Let $T = \{\mathbf{w}_1, \mathbf{w}_2, \ldots, \mathbf{w}_n, \mathbf{w}_{n+1}\}$. That is, let T contain exactly $n + 1$ vectors. We will show that T cannot be a basis by showing that T is linearly dependent. To do this, we consider

$$c_1\mathbf{w}_1 + c_2\mathbf{w}_2 + \cdots + c_n\mathbf{w}_n + c_{n+1}\mathbf{w}_{n+1} = \boldsymbol{\theta} \tag{3.5.1}$$

Now each $\mathbf{w}_k$ can be written as

$$\mathbf{w}_k = a_{1k}\mathbf{v}_1 + a_{2k}\mathbf{v}_2 + \cdots + a_{nk}\mathbf{v}_n \qquad k = 1, 2, \ldots, n+1$$

since $\mathscr{S}$ spans V. Substituting this into the Eq. (3.5.1), we have

$$\begin{aligned}\sum_{k=1}^{n+1} c_k\mathbf{w}_k &= \sum_{k=1}^{n+1} c_k \sum_{j=1}^{n} a_{jk}\mathbf{v}_j \\ &= \sum_{j=1}^{n}\left(\sum_{k=1}^{n+1} a_{jk}c_k\right)\mathbf{v}_j = \boldsymbol{\theta}\end{aligned}$$

Since $\mathscr{S}$ is linearly independent,

$$\sum_{k=1}^{n+1} a_{jk}c_k = 0 \qquad \text{for all } j = 1, \ldots, n$$

That is,

$$\begin{aligned} a_{11}c_1 + a_{12}c_2 + \cdots + a_{1,n+1}c_{n+1} &= 0 \\ a_{21}c_1 + a_{22}c_2 + \cdots + a_{2,n+1}c_{n+1} &= 0 \\ \cdots\cdots\cdots\cdots\cdots\cdots\cdots\cdots\cdots\cdots \\ a_{n1}c_1 + a_{n2}c_2 + \cdots + a_{n,n+1}c_{n+1} &= 0 \end{aligned}$$

This last homogeneous system has fewer equations than unknowns; so there exists a nontrivial solution for $c_1, c_2, \ldots, c_{n+1}$. This means that T is linearly dependent. Now suppose T contains *more than* $n + 1$ vectors. Let $\tilde{T}$ be a subset of $n + 1$ vectors. It must be linearly dependent (as was just shown). Since $\tilde{T} \subset T$, set T contains a linearly dependent subset and must itself be linearly dependent (see Prob. 19 of Sec. 3.4).

(*b*) Suppose T contains $n - 1$ vectors and spans V. Then by Theorem 3.5.2 T must contain a basis $\mathscr{T}$ for V. If $\mathscr{T}$ contains r vectors, we must have $r \leq n - 1$. Since $\mathscr{T}$ is a basis and $\mathscr{S}$ has $r + 1$ or more vectors in it, it follows from part (*a*) that $\mathscr{S}$ is linearly dependent. This contradicts the fact that $\mathscr{S}$ is a basis.

We now know that finitely generated vector spaces which are generated by a set of nonzero vectors have bases and are finite-dimensional. Clearly, finite-dimensional vector spaces are finitely generated by a basis. Now we give several examples.

EXAMPLE 6 (Standard basis) Show that E^n has the (standard) basis

$\mathscr{E} = \{\mathbf{e}_1, \mathbf{e}_2, \ldots, \mathbf{e}_n\}$, where

$$\mathbf{e}_j = (0, 0, \ldots, 0, \underset{j\text{th component}}{1}, 0, \ldots, 0)$$

Solution For the span consider any $\mathbf{x} = (x_1, x_2, \ldots, x_n) \in E^n$ and note that

$$\begin{aligned}\mathbf{x} &= x_1(1, 0, \ldots, 0) + x_2(0, 1, 0, \ldots, 0) + \cdots + x_n(0, \ldots, 0, 1)\\ &= x_1\mathbf{e}_1 + x_2\mathbf{e}_2 + \cdots + x_n\mathbf{e}_n\end{aligned}$$

For linear independence consider

$$c_1\mathbf{e}_1 + c_2\mathbf{e}_2 + \cdots + c_n\mathbf{e}_n = \boldsymbol{\theta} = (0, 0, \ldots, 0)$$

so that $c_1 = c_2 = \cdots = c_n = 0$. Therefore, $\mathscr{E}$ is a basis for E^n, and $\dim E^n = n$. This is reasonable since we associate (see Fig. 3.5.1)

E^1	Line	One-dimensional object
E^2	Plane	Two-dimensional object
E^3	Space	Three-dimensional object

Comparing Examples 1 and 4, we see that E^2 has more than one possible basis. In general, a vector space (nonzero) has an infinite number of bases. However, the number of elements in any basis is always the same: Remember, this number is the dimension of the space.

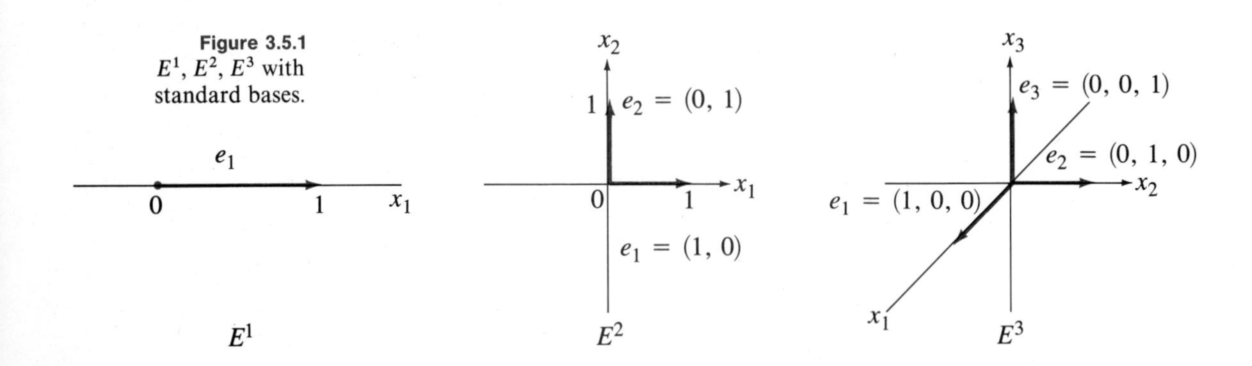

Figure 3.5.1 E^1, E^2, E^3 with standard bases.

EXAMPLE 7 Show that $\mathscr{M}_{23}$ has dimension 6.

Solution A basis is (in fact, this is the standard basis)

$$S = \left\{ \begin{pmatrix} 1 & 0 & 0 \\ 0 & 0 & 0 \end{pmatrix}, \begin{pmatrix} 0 & 1 & 0 \\ 0 & 0 & 0 \end{pmatrix}, \begin{pmatrix} 0 & 0 & 1 \\ 0 & 0 & 0 \end{pmatrix}, \begin{pmatrix} 0 & 0 & 0 \\ 1 & 0 & 0 \end{pmatrix}, \begin{pmatrix} 0 & 0 & 0 \\ 0 & 1 & 0 \end{pmatrix}, \begin{pmatrix} 0 & 0 & 0 \\ 0 & 0 & 1 \end{pmatrix} \right\}$$

so $\dim \mathscr{M}_{23} =$ number of vectors in $S = 6$.

EXAMPLE 8 The vector space $\mathscr{P}_n$ has dimension $n + 1$.

Solution A basis is

$$S = \{1, x, \ldots, x^n\}$$

To see this, we check first for linear independence. The equation

$$a_1 1 + a_2 x + a_3 x^2 + \cdots + a_{n+1} x^{n+1} = 0$$

holds only if the polynomial on the left is zero for all real x. From algebra this occurs only if all the coefficients are zero, that is, only if $a_1 = a_2 = \cdots = a_{n+1} = 0$. Therefore, S is linearly independent. That S spans $\mathscr{P}_n$ follows from the fact that any polynomial in $\mathscr{P}_n$ is of the form

$$a_0 + a_1 x + \cdots + a_n x^n$$

EXAMPLE 9

$$\mathscr{S} = \left\{ \begin{pmatrix} 0 & 1 \\ 1 & 0 \end{pmatrix}, \begin{pmatrix} 0 & -i \\ i & 0 \end{pmatrix}, \begin{pmatrix} 1 & 0 \\ 0 & -1 \end{pmatrix}, \begin{pmatrix} 1 & 0 \\ 0 & 1 \end{pmatrix} \right\}$$

is a basis for $\mathscr{C}_{22}$. Example 18 of Sec. 3.4 showed that $\mathscr{S}$ is linearly independent and spans $\mathscr{C}_{22}$.

Now that we know what a basis of a vector space is, we can state *the second fundamental problem of linear algebra.*

Basis Problem.

Let V be a vector space. The basis problem may take one of the following forms.

Problem 1. Construct a basis for V, by choosing vectors from V.

Problem 2. Given a set S of vectors in V, construct a basis for V by enlarging S, or deleting some (but not all) vectors from S, or both.

Before we try to solve this problem, we might ask, Is a solution even possible? Theorem 3.5.2, which tells us to "throw out dependent vectors from a spanning set" to get a basis, helps us here.

Problem 1. If we can pick a set of vectors from V which spans V, then by throwing out dependent vectors we will arrive at a basis for V.

Problem 2. If the given set S spans V, we proceed as in problem 1. If not, we enlarge S by putting in more vectors until a spanning set is achieved. Then we proceed as in problem 1.

EXAMPLE 10 (Problem 2 of the basis problem) Let $S = \{(1, 0, 3), (2, 1, 4)\}$. Find a basis T for E^3 containing S.

Solution 1 Since E^3 has dimension 3, we know that T must contain exactly three vectors. Set S is already linearly independent, so we have to add only one more vector to S. However, we must be careful. The new vector we join to the set S must not make the set T linearly dependent. So the new vector must not be in the span of the vectors already in S.

Now span $\{(1, 0, 3), (2, 1, 4)\} = \{\mathbf{x} | \mathbf{x} = a(1, 0, 3) + b(2, 1, 4)\} = \{\mathbf{x} | \mathbf{x} = (a + 2b, b, 3a + 4b)\}$. We must make sure that the new vector is not of the form $(a + 2b, b, 3a + 4b)$. To do this, we suppose our new vector is (x_1, x_2, x_3), and we force the equation

$$(a + 2b, b, 3a + 4b) = (x_1, x_2, x_3)$$

to have no solution for a and b. Matching components yields the equations

$$\left(\begin{array}{cc|c} 1 & 2 & x_1 \\ 0 & 1 & x_2 \\ 3 & 4 & x_3 \end{array}\right)$$

which reduces to

$$\left(\begin{array}{cc|c} 1 & 2 & x_1 \\ 0 & 1 & x_2 \\ 0 & 0 & x_3 - 3x_1 + 2x_2 \end{array}\right)$$

So if we choose x_1, x_2, x_3 with $x_3 - 3x_1 + 2x_2 \neq 0$, we have a vector not in span $\{(1, 0, 3), (2, 1, 4)\}$. Therefore, $\mathbf{x} = (0, 1, 0)$ works and

$$T = \{(1, 0, 3), (2, 1, 4), (0, 1, 0)\}$$

is a basis for E^3.

Solution 2 If the third vector (x_1, x_2, x_3) were to make $\{(1, 0, 3), (2, 1, 4), (x_1, x_2, x_3)\}$ linearly dependent, then

$$\det\begin{pmatrix} 1 & 0 & 3 \\ 2 & 1 & 4 \\ x_1 & x_2 & x_3 \end{pmatrix} = 0$$

since one of the rows would be a linear combination of the others. So we require

$$\det\begin{pmatrix} 1 & 0 & 3 \\ 2 & 1 & 4 \\ x_1 & x_2 & x_3 \end{pmatrix} \neq 0$$

Calculating the determinant, we have

$$x_3 + 2x_2 - 3x_1 \neq 0$$

which is the same condition obtained in solution 1. Possible third vectors are (1, 0, 0), (0, 1, 0), (0, 0, 1), (0, 1, 1), or (1, 1, 0); actually there are an infinite number of choices.

Solution 3 (Trial and error) We just try standard basis vectors until one works. Try first (1, 0, 0) and check for linear dependence:

$$c_1(1, 0, 3) + c_2(2, 1, 4) + c_2(1, 0, 0) = (0, 0, 0)$$

$$\Rightarrow \left(\begin{array}{ccc|c} 1 & 2 & 1 & 0 \\ 0 & 1 & 0 & 0 \\ 3 & 4 & 0 & 0 \end{array}\right) \rightarrow \left(\begin{array}{ccc|c} 1 & 2 & 1 & 0 \\ 0 & 1 & 0 & 0 \\ 0 & 0 & 1 & 0 \end{array}\right)$$

Therefore, the vectors are linearly independent and form a basis $T = \{(1, 0, 3), (2, 1, 4), (1, 0, 0)\}$.

In Example 10 we have implicitly used the following theorem.

THEOREM 3.5.4 Let $\dim V = n$, and let $S = \{\mathbf{v}_1, \ldots, \mathbf{v}_n\}$ be a subset of V. The following are equivalent:

1. Set S is a basis for V,
2. Set S is linearly independent,
3. Set S spans V.

Proof $(1 \Rightarrow 2)$ This follows from the definition of basis.
$(2 \Rightarrow 3)$ Suppose S is linearly independent and S does *not* span V. Then there is a vector $\mathbf{v}_{n+1} \in V$ which is not in span S. That is,

$$T = \{\mathbf{v}_1, \mathbf{v}_2, \ldots, \mathbf{v}_n, \mathbf{v}_{n+1}\}$$

is a linearly independent set from V. But then $\dim V \geq n + 1$, which contradicts the hypothesis that $\dim V = n$.

($3 \Rightarrow 1$) Suppose S spans V and is not a basis for V. Then it must be linearly dependent. By Theorem 3.5.2 there is a subset of S which is a basis of V. However, this subset must have less than n vectors in it, which implies that $\dim V < n$, a contradiction.

Note that had Theorem 3.5.4 been available prior to Examples 1 and 2, the solutions would have required half the work. Showing linear independence would have sufficed.

EXAMPLE 11 (Problem 1 of the basis problem) Find a basis for the solution space of

$$\begin{aligned} x_1 + x_2 - x_3 + 2x_4 &= 0 \\ x_2 + x_3 - x_4 &= 0 \\ 3x_1 + 4x_2 - 2x_3 + 5x_4 &= 0 \end{aligned}$$

Solution The equations in augmented matrix form are

$$\left(\begin{array}{cccc|c} 1 & 1 & -1 & 2 & 0 \\ 0 & 1 & 1 & -1 & 0 \\ 3 & 4 & -2 & 5 & 0 \end{array}\right) \xrightarrow{\text{Row reduction}} \left(\begin{array}{cccc|c} 1 & 1 & -1 & 2 & 0 \\ 0 & 1 & 1 & -1 & 0 \\ 0 & 0 & 0 & 0 & 0 \end{array}\right)$$

and we can let $x_4 = k$, $x_3 = j$ to find the solutions

$$\begin{aligned} (x_1, x_2, x_3, x_4) &= (2j - 3k, k - j, k, j) \\ &= j(2, -1, 0, 1) + k(-3, 1, 1, 0) \end{aligned}$$

Since $S = \{(2, -1, 0, 1), (-3, 1, 1, 0)\}$ spans the solution space and is linearly independent, S is a basis for the solution space. Therefore the dimension of the solution space is 2.

Examples 10 and 11 show that solution of the basis problem is usually not found by formula or rote methods. It requires thought, versatility, and the ability to use almost all the preceding material.

EXAMPLE 12 The set

$$\text{span } S = \text{span } \{(1, -1, 2), (0, 5, -8), (3, 2, -2), (8, 2, 0)\}$$

is a vector space. Find a basis for it.

Solution We delete vectors which are linear combinations of the others. To see the dependencies, we consider

$$c_1(1, -1, 2) + c_2(0, 5, -8) + c_3(3, 2, -2) + c_4(8, 2, 0) = (0, 0, 0)$$

which is equivalent to

$$\left(\begin{array}{rrrr|r} 1 & 0 & 3 & 8 & 0 \\ -1 & 5 & 2 & 2 & 0 \\ 2 & -8 & -2 & 0 & 0 \end{array}\right) \xrightarrow{\text{Row reduction}} \left(\begin{array}{rrrr|r} 1 & 0 & 3 & 8 & 0 \\ 0 & 1 & 1 & 2 & 0 \\ 0 & 0 & 0 & 0 & 0 \end{array}\right) \tag{3.5.2}$$

Solutions are $c_4 = k$, $c_3 = j$, $c_2 = -j - 2k$, $c_1 = -3j - 8k$. Choosing $k = 1$, $j = 0$, we have

$$8(1, -1, 2) + 2(0, 5, -8) = (8, 2, 0)$$

Choosing $k = 0$ and $j = 1$, we have

$$3(1, -1, 2) + (0, 5, -8) = (3, 2, -2)$$

Thus $(8, 2, 0)$ and $(3, 2, -2)$ depend on $(1, -1, 2)$ and $(0, 5, -8)$, and

$$T = \{(1, -1, 2), (0, 5, -8)\}$$

is a basis for span S. The dimension of span S is 2.

In Example 12, the augmented matrix on the left-hand side of Eq. (3.5.2) has columns consisting of the vectors from S. Also the number of rows (two) in the row-reduced form is equal to the dimension of span S.

In general, for problems in E^n of the type in Example 12, we have a helpful theorem. Before stating it, we note that for an $m \times n$ matrix A, if we consider the rows as vectors from E^n then the span of those vectors is called the *row space of A*.

THEOREM 3.5.5 If $S = \{\mathbf{v}_1, \ldots, \mathbf{v}_m\}$ is a set of vectors from E^n and A is the matrix formed by putting $\mathbf{v}_1$ in row 1, $\mathbf{v}_2$ in row 2, and so on, and if B is the reduced row echelon form of A, then the nonzero rows of B form a basis for the row space of A. That is, the nonzero rows of B form a basis for span S.

Proof Let the matrix be

$$A_{m \times n} = \begin{pmatrix} \mathbf{v}_1 \\ \mathbf{v}_2 \\ \vdots \\ \mathbf{v}_m \end{pmatrix}$$

By the definition of row operations, if a row of zeros is obtained, that row was equal to a linear combination of other vectors in the set. The remaining rows are therefore all linear combinations of the independent vectors from the original set. Thus the span of the nonzero rows is equal to span S. Thus the nonzero rows, being independent, form a basis for span S and dim (span S) = number of nonzero rows.

EXAMPLE 13 Work Example 12 by using Theorem 3.5.5.

Solution Form A and row-reduce.

$$A = \begin{pmatrix} 1 & -1 & 2 \\ 0 & 5 & -8 \\ 3 & 2 & -2 \\ 8 & 2 & 0 \end{pmatrix} \longrightarrow \begin{pmatrix} 1 & -1 & 2 \\ 3 & 2 & -2 \\ 8 & 2 & 0 \\ 0 & 5 & -8 \end{pmatrix} \longrightarrow \begin{pmatrix} 1 & -1 & 2 \\ 0 & 5 & -8 \\ 0 & 10 & -16 \\ 0 & 5 & -8 \end{pmatrix}$$

$$\longrightarrow \begin{pmatrix} 1 & -1 & 2 \\ 0 & 1 & -\frac{8}{5} \\ 0 & 0 & 0 \\ 0 & 0 & 0 \end{pmatrix}$$

By Theorem 3.5.5, $T = \{(1, -1, 2), (0, 1, -\frac{8}{5})\}$ is a basis for span S, and $\dim(\text{span } S) = 2$.

Theorem 3.5.5 can be stated in terms of the *rank* of a matrix A.

DEFINITION 3.5.2 Let A be an $m \times n$ matrix. The *row rank* of a matrix is the number of nonzero rows in the reduced row echelon form of A. The *column rank* of a matrix is the number of nonzero rows in the reduced row echelon form of A^T. The row and column ranks of the zero matrix are defined to be zero.

EXAMPLE 14 Calculate the row and column ranks of

$$\begin{pmatrix} 1 & -1 & 2 \\ 0 & 5 & -8 \\ 3 & 2 & -2 \\ 8 & 2 & 0 \end{pmatrix}$$

Solution In Example 13 we found the row rank to be 2. For the column rank we can do column operations or form A^T, do row operations, and transpose. We will use column operations.

$$\begin{pmatrix} 1 & -1 & 2 \\ 0 & 5 & -8 \\ 3 & 2 & -2 \\ 8 & 2 & 0 \end{pmatrix} \xrightarrow[-2C1 + C3]{C1 + C2} \begin{pmatrix} 1 & 0 & 0 \\ 0 & 5 & -8 \\ 3 & 5 & -8 \\ 8 & 10 & -16 \end{pmatrix} \xrightarrow{\frac{8}{5}C2 + C3} \begin{pmatrix} 1 & 0 & 0 \\ 0 & 5 & 0 \\ 3 & 5 & 0 \\ 8 & 10 & 0 \end{pmatrix}$$

$$\xrightarrow{\frac{1}{5}C2} \begin{pmatrix} 1 & 0 & 0 \\ 0 & 1 & 0 \\ 3 & 1 & 0 \\ 8 & 2 & 0 \end{pmatrix}$$

The column rank of A is 2 also.

The column rank and row rank of A in Example 14 were equal. This is always true.

THEOREM 3.5.6 For any matrix A,

$$\text{row rank } A = \text{column rank } A$$

Proof Let $\mathbf{v}_1, \ldots, \mathbf{v}_m$ be the rows of $A_{m \times n}$. After reduction a basis $\{\mathbf{w}_1, \mathbf{w}_2, \ldots, \mathbf{w}_k\}$, $k \le m$, is found for span $\{\mathbf{v}_1, \ldots, \mathbf{v}_m\}$. Thus

$$\begin{aligned} \mathbf{v}_1 &= a_{11}\mathbf{w}_1 + a_{12}\mathbf{w}_2 + \cdots + a_{1k}\mathbf{w}_k \\ \mathbf{v}_2 &= a_{21}\mathbf{w}_1 + a_{22}\mathbf{w}_2 + \cdots + a_{2k}\mathbf{w}_k \\ &\cdots\cdots\cdots\cdots\cdots\cdots \\ \mathbf{v}_m &= a_{m1}\mathbf{w}_1 + a_{m2}\mathbf{w}_2 + \cdots + a_{mk}\mathbf{w}_k \end{aligned} \tag{3.5.3}$$

Now writing

$$\mathbf{v}_j = (b_{j1}, b_{j2}, \ldots, b_{jn}) \qquad 1 \le j \le n$$

and

$$\mathbf{w}_i = (c_{i1}, c_{i2}, \ldots, c_{in}) \qquad 1 \le i \le n$$

we find after substitution into Eq. (3.5.3) that

$$\underbrace{\begin{pmatrix} b_{1j} \\ b_{2j} \\ \vdots \\ b_{mj} \end{pmatrix}}_{\substack{\text{This is } j\text{th} \\ \text{column of } A}} = c_{1j}\begin{pmatrix} a_{11} \\ a_{21} \\ \vdots \\ a_{m1} \end{pmatrix} + \cdots + c_{kj}\begin{pmatrix} a_{1k} \\ a_{2k} \\ \vdots \\ a_{mk} \end{pmatrix}$$

k of these

This means that the transpose of each column is a linear combination of k vectors; therefore the column rank of A is less than or equal to k. That is,

$$\text{column rank } A \le \text{row rank } A$$

In the same way, we find row rank $A \le$ column rank A. Therefore the ranks are equal.

As a result of this theorem, we define the *rank of a matrix A* as just the row rank of A.

Now results regarding linear equations and matrices can be concisely stated.

1. $A_{m\times n}X_{n\times 1} = B_{m\times 1}$ has a solution if and only if rank A = rank $(A|B)$
2. Matrix $A_{n\times n}$ is invertible if and only if rank $A = n$. Also det $A_{n\times n} \neq 0$ if and only if rank $A = n$.
3. $A_{m\times n}X_{n\times 1} = 0_{n\times 1}$ has a nontrivial solution if and only if rank $A < n$.

Now let us return to coding theory to see another way that bases are used.

EXAMPLE 15 As defined in Sec. 3.3, a binary channel linear code V is a subspace of F^n. Code V, being finitely generated by nonzero vectors, has a basis. When the elements of a basis of V are arranged in rows of a matrix, the matrix is called the *generator matrix* of V. The span of the rows, therefore, is the entire code; the matrix furnishes a compact storage mechanism for a code. For example, the matrix

$$M = \begin{pmatrix} 1 & 0 & 1 \\ 0 & 1 & 1 \end{pmatrix}$$

generates the code in this table:

VECTOR IN CODE	GENERATION
(0, 0, 0)	0(1, 0, 1) + 0(0, 1, 1)
(0, 1, 1)	0(1, 0, 1) + 1(0, 1, 1)
(1, 0, 1)	1(1, 0, 1) + 0(0, 1, 1)
(1, 1, 0)	1(1, 0, 1) + 1(0, 1, 1)

A 30×20 matrix (30 vectors of 20 elements each) generates a code of more than 10^9 code vectors. After all, each element of the code is a sum of 30 basis vectors, and there are 2 choices for each of the 30 coefficients. Thus there are $2^{30} = 1{,}073{,}741{,}824 > 10^9$ code vectors in V.

Example 15 raises an important question for decoding: Is each code vector a unique linear combination of the rows of the generator matrix? The answer is yes and was proved in Theorem 3.5.1.

Another question generated by the coding example is: If a code C in F^n is "smaller" than another code D in F^n, in the sense that C is a proper subspace of D, then is a basis for C "smaller" than a basis for D? This seems reasonable, and the answer is yes, as shown in Theorem 3.5.7.

THEOREM 3.5.7 If V is a finite-dimensional vector space and W is a subspace of V, then $\dim W \le \dim V$.

Proof If $W = \{\boldsymbol{\theta}\}$, then the result is true. Suppose $W \neq \{\boldsymbol{\theta}\}$. We first show that W is finitely generated. Since $W \neq \{\boldsymbol{\theta}\}$, there is a nonzero vector in $\mathbf{v}_1$ in W. Now either span $\{\mathbf{v}_1\} = W$ or not. If so, W is finitely generated by $\{\mathbf{v}_1\}$. If not, there is a vector $\mathbf{v}_2$ not in span $\{\mathbf{v}_1\}$ which is still in W. The set $\{\mathbf{v}_1, \mathbf{v}_2\}$ is linearly independent; otherwise, $\mathbf{v}_2$ would be in span $\{\mathbf{v}_1\}$. Now if span $\{\mathbf{v}_1, \mathbf{v}_2\} = W$, then we are done; if not, we continue to add vectors and check linear independence. This process must stop at some point because if V has dimension n, then any set of $n + 1$ vectors from V (and therefore from W) must be linearly dependent. Say that the process stops after k steps. Then we have $W =$ span $\{\mathbf{v}_1, \mathbf{v}_2, \ldots, \mathbf{v}_k\}$ and $k \le n$.

PROBLEMS 3.5

1. Which of the following sets are bases for E^2?

(*a*) $\{(1, -3), (-17, 54)\}$ (*b*) $\{(1, 0), (0, 0)\}$ (*c*) $\{(1, 2)\}$

(*d*) $\{(1, -1), (2, -2)\}$ (*e*) $\{(1, 0), (0, 1), (1, -1)\}$

2. Which of the following sets are bases for E^3?

(*a*) $S = \{(1, -1, 2), (2, 3, 5), (-3, 0, 2)\}$

(*b*) $S = \{(1, 2, 4), (-6, 2, 3), (-4, 6, 11)\}$

(*c*) $S = \{(1, -1, 2), (0, 3, 6)\}$

(*d*) $S = \{(1, 0, 1), (0, 1, 1), (1, 1, 0), (1, 0, 0)\}$

In Probs. 3 to 7, a set S and a vector space V are given. Find a basis for V containing S.

3. $V = E^3$, $S = \{(1, -1, 1), (0, 1, -1)\}$

4. $V = E^4$, $S = \{(1, 0, 2, 2), (1, 1, 0, 0)\}$

5. $V = \mathscr{M}_{22}$, $S = \left\{\begin{pmatrix} 1 & 1 \\ 0 & 0 \end{pmatrix}, \begin{pmatrix} 0 & 0 \\ 1 & 1 \end{pmatrix}, \begin{pmatrix} 1 & 0 \\ 0 & 1 \end{pmatrix}\right\}$

6. $V = \mathscr{P}_2$, $S = \{1 - x + x^2, x - x^2\}$

7. $V = \mathbb{C}^3$, $S = \{(i, 1, 0), (0, i, 1)\}$

In Probs. 8 to 12, a set S and a vector space V are given. Find a basis for V by deleting vectors from S.

8. $V = E^2$, $S = \{(1, -1), (1, 2), (3, 4)\}$

9. $V =$ span $\{(1, 2, 0), (1, 5, 6), (2, 13, 18)\} =$ span S

10. $V = E^3$, $S = \{(1, -1, 2), (4, -3, 7), (2, 0, 5), (1, 2, 6)\}$

11. $V = \mathscr{P}_2$, $S = \{1 + x, x + x^2, 1 + x^2, x - x^2\}$

12. $V = \mathbb{C}^3$, $S = \{(1, 0, 0), (i, 1, 0), (1, i, 0), (0, 0, 1)\}$

In Probs. 13 to 18, a vector space V is given. Find a basis and state dim V.

13. $V = E^5$

14. $V = \mathscr{P}_4$

15. $V = \mathscr{M}_{24}$

16. $V = \text{span}\ \{(1, -1, 2), (5, 5, 5), (0, 6, 3)\}$

17. $V = \mathscr{C}_{32}$

18. $V =$ solution space of

$$\begin{aligned} x_1 - x_2 + x_3 &= 0 \\ x_1 + x_2 - x_3 &= 0 \\ x_1 + 3x_2 - 3x_3 &= 0 \end{aligned}$$

19. Show that dim $\mathscr{M}_{nn} = n^2$. What is the dimension of the subspace of symmetric matrices?

20. Show that dim $\mathscr{P}_n = n + 1$. What is the dimension of the subspace of even polynomials (polynomials with even exponents)?

21. What is dim $\mathscr{M}_{mn}$? What is dim $\mathscr{C}_{mn}$?

22. Show that

dim $\mathscr{P}_n$ = dim (subspace of even polynomials) + dim (subspace of odd polynomials)

23. Calculate the rank of the given matrices.

(a) $A = \begin{pmatrix} 1 & 2 & 3 \\ 4 & 5 & 6 \\ 7 & 8 & 9 \end{pmatrix}$ (b) $A = \begin{pmatrix} 1 & -1 & 2 & 3 \\ 0 & 5 & 6 & 2 \\ -1 & 2 & 4 & 3 \\ 1 & 2 & -1 & 2 \end{pmatrix}$

(c) $A = \begin{pmatrix} 1 & 2 & 3 & 4 \\ 5 & 6 & 7 & 8 \\ 9 & 10 & 11 & 12 \\ 13 & 14 & 15 & 16 \end{pmatrix}$

24. Show that the standard basis for E^n is also a basis for $\mathbb{C}^n$.

25. Show that dim $\mathbb{C}^n = n$.

26. Show that

$$M = \begin{pmatrix} 1 & 1 & 0 & 1 \\ 1 & 0 & 0 & 1 \\ 1 & 0 & 1 & 0 \end{pmatrix}$$

generates the linear code

$$V = \{(0, 0, 0, 0), (1, 1, 0, 1), (1, 0, 0, 1), (1, 0, 1, 0), (0, 1, 0, 0), (0, 1, 1, 1), (0, 0, 1, 1), (1, 1, 1, 0)\}$$

Show that

$$N = \begin{pmatrix} 1 & 1 & 0 & 1 \\ 1 & 0 & 0 & 1 \end{pmatrix}$$

generates a subcode W of V. What are dim W and dim V?

27. Compare rank A and rank $(A|B)$ to determine whether the following systems have solutions.

(*a*)
$$\begin{aligned} x - y + z &= 2 \\ 2x + y - z &= 3 \\ x + 2y + 4z &= 7 \end{aligned}$$

(*b*)
$$\begin{aligned} x + y + 2z &= 3 \\ -x - 3y + 4z &= 2 \\ -x - 5y + 10z &= 7 \end{aligned}$$

(*c*)
$$\begin{aligned} x + y + 2z &= 3 \\ -x - 3y + 4z &= 2 \\ -x - 5y + 10z &= 11 \end{aligned}$$

28. Find bases for the row spaces of the matrices in Prob. 23.

3.6 PERPENDICULARITY IN VECTOR SPACES

E^3 has a standard basis of $\mathscr{E} = \{(1, 0, 0), (0, 1, 0), (0, 0, 1)\}$ which can be pictured as in Fig. 3.5.1. These vectors have length 1 and are perpendicular to each other. Such a basis is called *orthonormal*: *ortho* for orthogonal (perpendicular) and *normal* for normalized (length 1). To discuss orthonormal bases for other vector spaces, we must extend the idea of dot product to what is called an *inner product*.

DEFINITION 3.6.1 Let V be a real or complex vector space. An *inner product* on V is a function which associates with each pair $\mathbf{x}$, $\mathbf{y}$ of vectors a number $\langle \mathbf{x}, \mathbf{y} \rangle$ and satisfies the following properties. (On the right-hand side we list the situation in E^2. The overbar denotes complex conjugate.)

Let r be a scalar, $\mathbf{x}, \mathbf{y}, \mathbf{z} \in V$	In E^2, $\langle \mathbf{x}, \mathbf{y} \rangle = \mathbf{x} \cdot \mathbf{y}$ $\quad \mathbf{x} = (x_1, x_2)$ $\quad \mathbf{y} = (y_1, y_2)$
1. $\langle \mathbf{x}, \mathbf{y} \rangle = \overline{\langle \mathbf{y}, \mathbf{x} \rangle}$	**1.** $\langle \mathbf{x}, \mathbf{y} \rangle = \mathbf{x} \cdot \mathbf{y} = x_1y_1 + x_2y_2 = y_1x_1 + y_2x_2 = \mathbf{y} \cdot \mathbf{x} = \overline{\langle \mathbf{y}, \mathbf{x} \rangle}$
2. $\langle \mathbf{x} + \mathbf{z}, \mathbf{y} \rangle = \langle \mathbf{x}, \mathbf{y} \rangle + \langle \mathbf{z}, \mathbf{y} \rangle$	**2.** $\langle \mathbf{x} + \mathbf{z}, \mathbf{y} \rangle = (\mathbf{x} + \mathbf{z}) \cdot \mathbf{y} = (x_1 + z_1)y_1 + (x_2 + z_2)y_2 = x_1y_1 + x_2y_2 + z_1y_1 + z_2y_2 = \langle \mathbf{x}, \mathbf{y} \rangle + \langle \mathbf{z}, \mathbf{y} \rangle$
3. $\langle r\mathbf{x}, \mathbf{y}) = r\langle \mathbf{x}, \mathbf{y} \rangle$	**3.** $\langle r\mathbf{x}, \mathbf{y} \rangle = (r\mathbf{x}) \cdot \mathbf{y} = (rx_1, rx_2) \cdot (y_1, y_2) = rx_1y_1 + rx_2y_2 = r(x_1y_1 + x_2y_2) = r\langle \mathbf{x}, \mathbf{y} \rangle$
4. $\langle \mathbf{x}, \mathbf{x} \rangle > 0$ unless $\mathbf{x} = \boldsymbol{\theta}$	**4.** $\langle \mathbf{x}, \mathbf{x} \rangle = \mathbf{x} \cdot \mathbf{x} = x_1{}^2 + x_2{}^2$ which is a sum of nonnegative terms. The sum can be zero if and only if $x_1 = 0$ and $x_2 = 0$, that is, when $\mathbf{x} = 0$.

EXAMPLE 1 In E^n, if $\mathbf{x} = (x_1, \ldots, x_n)$ and $\mathbf{y} = (y_1, \ldots, y_n)$, then an *inner product* can be defined as

$$\langle \mathbf{x}, \mathbf{y} \rangle \equiv x_1y_1 + x_2y_2 + \cdots + x_ny_n$$

This inner product reduces to the ordinary dot product in E^2 and E^3. It is called the *standard inner product for E^n*.

EXAMPLE 2 In $\mathbb{C}^n$, if $\mathbf{x} = (x_1, \ldots, x_n)$ and $\mathbf{y} = (y_1, \ldots, y_n)$, then an *inner product* can be defined as

$$\langle \mathbf{x}, \mathbf{y} \rangle \equiv x_1\bar{y}_1 + x_2\bar{y}_2 + \cdots + x_n\bar{y}_n$$

This is the *standard inner product for $\mathbb{C}^n$*. Note that $\langle c\mathbf{x}, \mathbf{y} \rangle = c\langle \mathbf{x}, \mathbf{y} \rangle$, but $\langle \mathbf{x}, c\mathbf{y} \rangle = x_1\overline{cy}_1 + x_2\overline{cy}_2 + \cdots + x_n\overline{cy}_n = \bar{c}\langle \mathbf{x}, \mathbf{y} \rangle$.

EXAMPLE 3 In $\mathscr{M}_{23}$ if

$$A = \begin{pmatrix} a_{11} & a_{12} & a_{13} \\ a_{21} & a_{22} & a_{23} \end{pmatrix} \quad \text{and} \quad B = \begin{pmatrix} b_{11} & b_{12} & b_{13} \\ b_{21} & b_{22} & b_{23} \end{pmatrix}$$

then we can define an inner product as

$$\langle A, B \rangle = a_{11}b_{11} + a_{12}b_{12} + a_{13}b_{13} + a_{21}b_{21} + a_{22}b_{22} + a_{23}b_{23}$$

EXAMPLE 4 In E^2, let $\mathbf{x} = (x_1, x_2)$, $\mathbf{y} = (y_1, y_2)$, and define $\langle \mathbf{x}, \mathbf{y} \rangle = 2x_1y_1 + 2x_2y_2$. $\langle \mathbf{x}, \mathbf{y} \rangle$ is an inner product.

Example 4 shows that there may be more than one inner product defined on a vector space; E^2 also has the standard inner product.

In E^2 and E^3, we have

$$|x| = \sqrt{\mathbf{x} \cdot \mathbf{x}} = \sqrt{\langle \mathbf{x}, \mathbf{x} \rangle}$$

so that it is natural to define the length of a vector $\mathbf{x}$ in any vector space with an inner product as $\sqrt{\langle \mathbf{x}, \mathbf{x} \rangle}$. Vector spaces with an inner product are also called *inner product spaces.*

DEFINITION 3.6.2 Let V be a vector space with inner product $\langle \cdot, \cdot \rangle$. If $\mathbf{x}$, $\mathbf{y}$ are in V,

1. The *norm* (length) of $\mathbf{x}$ is denoted by $\|\mathbf{x}\|$ and is defined as

$$\|\mathbf{x}\| = \sqrt{\langle \mathbf{x}, \mathbf{x} \rangle}$$

2. The *distance* between $\mathbf{x}$ and $\mathbf{y}$ is defined as $\|\mathbf{x} - \mathbf{y}\|$.
3. Nonzero $\mathbf{x}$ and $\mathbf{y}$ are defined to be *orthogonal* if $\langle \mathbf{x}, \mathbf{y} \rangle = 0$.

Note that $\|x\|$ may be different for different inner products, as the next example shows.

EXAMPLE 5 In E^2 let $\mathbf{x} = (1, 1)$ and $\mathbf{y} = (2, 3)$. Calculate $\|\mathbf{x}\|$ and $\|\mathbf{x} - \mathbf{y}\|$ for norms generated by the standard inner product and the inner product in Example 4.

Solution

Standard inner product	Inner product from Example 4
$\langle (1, 1), (2, 3) \rangle = 5$	$\langle (1, 1), (2, 3) \rangle = 10$
$\|\mathbf{x}\| = \sqrt{\langle \mathbf{x}, \mathbf{x} \rangle} = \sqrt{2}$	$\|\mathbf{x}\| = \sqrt{\langle \mathbf{x}, \mathbf{x} \rangle} = \sqrt{2 + 2} = 2$
$\|\mathbf{x} - \mathbf{y}\| = \|(-1, -2)\| = \sqrt{5}$	$\|\mathbf{x} - \mathbf{y}\| = \|(-1, -2)\| = \sqrt{10}$

Several properties of inner products are generalizations of properties of the dot product in E^2 and E^3. Some of these are stated in Theorem 3.6.1.

THEOREM 3.6.1 Let V be a vector space with inner product $\langle \cdot, \cdot \rangle$. The following properties hold for all $\mathbf{x}$ and $\mathbf{y}$ in V.

(*a*) $\langle \mathbf{x}, \boldsymbol{\theta} \rangle = 0$

(*b*) $\|\mathbf{x}\| \geq 0$, $\|\mathbf{x}\| = 0$ if and only if $\mathbf{x} = \boldsymbol{\theta}$

(*c*) $\langle \mathbf{x}, \mathbf{y} + \mathbf{z} \rangle = \langle \mathbf{x}, \mathbf{y} \rangle + \langle \mathbf{x}, \mathbf{z} \rangle$ and $\langle \mathbf{x}, r\mathbf{y} \rangle = \bar{r} \langle \mathbf{x}, \mathbf{y} \rangle$

(*d*) $\|k\mathbf{x}\| = |k| \, \|\mathbf{x}\|$

(*e*) $|\langle \mathbf{x}, \mathbf{y} \rangle| \leq \|\mathbf{x}\| \, \|\mathbf{y}\|$ (Cauchy-Schwarz inequality)

(*f*) $\|\mathbf{x} + \mathbf{y}\| \leq \|\mathbf{x}\| + \|\mathbf{y}\|$ (triangle inequality)

Proof (*a*) We have

$$\langle \mathbf{x}, \boldsymbol{\theta} \rangle = \langle \mathbf{x}, \boldsymbol{\theta} + \boldsymbol{\theta} \rangle = \langle \mathbf{x}, \boldsymbol{\theta} \rangle + \langle \mathbf{x}, \boldsymbol{\theta} \rangle$$

The only solution to the equation $q = q + q$ is $q = 0$. Thus $\langle \mathbf{x}, \boldsymbol{\theta} \rangle = 0$.

(*b*), (*c*) See the problems.

(*d*) $\|k\mathbf{x}\| = \sqrt{\langle k\mathbf{x}, k\mathbf{x} \rangle} = \sqrt{k\bar{k}\langle \mathbf{x}, \mathbf{x} \rangle} = \sqrt{|k|^2}\sqrt{\langle \mathbf{x}, \mathbf{x} \rangle} = \sqrt{|k|^2}\|\mathbf{x}\|$.

Since the radical signifies the positive square root, $\sqrt{k^2} = |k|$.

(*e*) See the problems.

(*f*)
$$\begin{aligned} \|\mathbf{x} + \mathbf{y}\|^2 &= \langle \mathbf{x} + \mathbf{y}, \mathbf{x} + \mathbf{y} \rangle = \langle \mathbf{x}, \mathbf{x} + \mathbf{y} \rangle + \langle \mathbf{y}, \mathbf{x} + \mathbf{y} \rangle \\ &= \langle \mathbf{x}, \mathbf{x} \rangle + \langle \mathbf{x}, \mathbf{y} \rangle + \overline{\langle \mathbf{x}, \mathbf{y} \rangle} + \langle \mathbf{y}, \mathbf{y} \rangle \\ &\le \langle \mathbf{x}, \mathbf{x} \rangle + 2\|\mathbf{x}\|\,\|\mathbf{y}\| + \langle \mathbf{y}, \mathbf{y} \rangle \qquad \text{by } (e) \\ &= \|\mathbf{x}\|^2 + 2\|\mathbf{x}\|\,\|\mathbf{y}\| + \|\mathbf{y}\|^2 \\ &= (\|\mathbf{x}\| + \|\mathbf{y}\|)^2 \end{aligned}$$

Taking the square root of both sides gives

$$\|\mathbf{x} + \mathbf{y}\| \le \|\mathbf{x}\| + \|\mathbf{y}\|$$

If $\mathbf{x}$ and $\mathbf{y}$ are vectors in a real vector space, we can define the angle between $\mathbf{x}$ and $\mathbf{y}$. Notice that the Cauchy-Schwarz inequality allows us to define

$$\cos\theta = \frac{\langle \mathbf{x}, \mathbf{y} \rangle}{\|\mathbf{x}\|\,\|\mathbf{y}\|}$$

because $|\langle \mathbf{x}, \mathbf{y} \rangle| \le \|\mathbf{x}\|\,\|\mathbf{y}\|$ implies that

$$-1 \le \frac{\langle \mathbf{x}, \mathbf{y} \rangle}{\|\mathbf{x}\|\,\|\mathbf{y}\|} \le 1$$

Now that orthogonality and norm have been defined, we can discuss bases of orthonormal vectors. A set $\mathcal{O} = \{\mathbf{v}_1, \ldots, \mathbf{v}_n\}$ of vectors is an *orthonormal set* in a vector space V with inner product $\langle \cdot, \cdot \rangle$ if

$$\|\mathbf{v}_k\| = 1 \qquad k = 1, 2, \ldots, n$$

and

$$\langle \mathbf{v}_k, \mathbf{v}_j \rangle = 0 \qquad k \ne j$$

That is,

$$\langle \mathbf{v}_k, \mathbf{v}_j \rangle = \delta_{kj}$$

One of the most important results of linear algebra is that *orthonormal sets are always linearly independent.* Consider

$$c_1\mathbf{v}_1 + c_2\mathbf{v}_2 + \cdots + c_k\mathbf{v}_k + \cdots + c_n\mathbf{v}_n = \boldsymbol{\theta}$$

Even though we do not know what V is or what the $\mathbf{v}_k$'s look like, we can solve for the coefficients. We use the important technique of taking the inner product of both sides with a particular vector, in this case $\mathbf{v}_k$. We have

$$c_1\langle \mathbf{v}_1, \mathbf{v}_k\rangle + c_2\langle \mathbf{v}_2, \mathbf{v}_k\rangle + \cdots + c_k\langle \mathbf{v}_k, \mathbf{v}_k\rangle + \cdots + c_n\langle \mathbf{v}_k, \mathbf{v}_n\rangle = \langle \boldsymbol{\theta}, \mathbf{v}_k\rangle = 0 \tag{3.6.1}$$

and since $\langle \mathbf{v}_k, \mathbf{v}_j\rangle = 0$, for $j \neq k$, Eq. (3.6.1) becomes

$$c_k = 0$$

Since this works for any $k = 1, 2, \ldots, n$, $c_1 = c_2 = \cdots = c_n = 0$, and the set $\mathcal{O}$ is linearly independent.

The fact that orthonormal sets are always linearly independent makes them nice to work with when we are finding bases for vector spaces with inner products. All we have to worry about is spanning V. Thus in E^n the standard basis is an *orthonormal basis.* If a basis is made up of orthogonal vectors but not all are of norm 1, then the basis is an *orthogonal* basis.

EXAMPLE 6 Let $V = \mathscr{P}_2$, $p(x) = a_0 + a_1x + a_2x^2$, and $q(x) = b_0 + b_1x + b_2x^2$. Define $\langle p, q\rangle = a_0b_0 + a_1b_1 + a_2b_2$. Show that $\langle \cdot, \cdot\rangle$ is an inner product. Give an example of two orthogonal vectors from $\mathscr{P}_2$. Find an orthonormal basis for $\mathscr{P}_2$.

Solution Verifying properties 1 to 4 of the definition of inner product (Definition 3.6.1) is straightforward. The vectors

$$p(x) = 1 + x^2 \qquad \text{and} \qquad q(x) = x$$

are orthogonal. The vector $r(x) = 1 - x^2$ is orthogonal to both p and q. The set

$$S = \{1 + x^2, x, 1 - x^2\}$$

is orthogonal and linearly independent. Since $\dim \mathscr{P}_2 = 3$, set S must be a basis. However, S is only an orthogonal basis. To make it an orthonormal basis, the vectors must normalized. We put

$$u_1 = \frac{p}{\|p\|} = \frac{1 + x^2}{\sqrt{\langle 1 + x^2, 1 + x^2\rangle}} = \frac{1 + x^2}{\sqrt{2}} = \frac{1}{\sqrt{2}} + \frac{1}{\sqrt{2}}x^2$$

$$u_2 = q = x \qquad (q \text{ is already of norm } 1)$$

$$u_3 = \frac{r}{\|r\|} = \frac{1 - x^2}{\sqrt{\langle 1 - x^2, 1 - x^2\rangle}} = \frac{1}{\sqrt{2}} - \frac{1}{\sqrt{2}}x^2$$

Now

$$\mathcal{O} = \{u_1, u_2, u_3\} = \left\{\frac{1}{\sqrt{2}} + \frac{1}{\sqrt{2}}x^2, x, \frac{1}{\sqrt{2}} - \frac{1}{\sqrt{2}}x^2\right\}$$

is an orthonormal basis for $\mathscr{P}_2$. $\mathcal{O}$ is not the only orthonormal basis for $\mathscr{P}_2$

with this inner product; the standard basis $\mathscr{S} = \{1, x, x^2\}$ is also an orthonormal basis.

PROBLEMS 3.6

1. Which of the following are inner products on E^2? Let $\mathbf{x} = (x_1, x_2)$ and $\mathbf{y} = (y_1, y_2)$.

(a) $\langle \mathbf{x}, \mathbf{y} \rangle = 4x_1y_1 + x_2y_2$ (b) $\langle \mathbf{x}, \mathbf{y} \rangle = x_1 + y_1$

(c) $\langle \mathbf{x}, \mathbf{y} \rangle = x_1y_1 - x_2y_2$ (d) $\langle \mathbf{x}, \mathbf{y} \rangle = x_1{}^2y_1 + x_2y_2$

(e) $\langle \mathbf{x}, \mathbf{y} \rangle = x_1y_2 + x_2y_1$ (f) $\langle \mathbf{x}, \mathbf{y} \rangle = x_1x_2y_1y_2$

(g) $\langle \mathbf{x}, \mathbf{y} \rangle = x_1 + x_2 + y_1 + y_2$ (h) $\langle \mathbf{x}, \mathbf{y} \rangle = |x_1y_1 + x_2y_2|$

2. Using the inner product from Prob. 1(*a*) on E^2, show that $\mathcal{O} = \{(\frac{1}{2}, 0), (0, 1)\}$ is an orthonormal basis for E^2 with that inner product.

3. In E^2 let $\mathbf{x} = (x_1, x_2)$ and $\mathbf{y} = (y_1, y_2)$. Show that the function

$$\langle \mathbf{x}, \mathbf{y} \rangle = ax_1y_1 + bx_2y_2 \qquad \text{where } a > 0, b > 0$$

is an inner product. What happens if either $a < 0$ or $b < 0$?

4. *Difficulty norm.* Consider a way of "norming" a hike up a mountain. Difficulty indices are as follows.

CATEGORY	ANGLE OF CLIMB	DIFFICULTY INDEX
1	0°–10°	1
2	10°–25°	2
3	25°–45°	4

A difficulty norm for a trip of x_1 mi in category 1, x_2 mi in category 2, and x_3 mi in category 3 is

$$\|(x_1, x_2, x_3)\| = \sqrt{x_1{}^2 + 2x_2{}^2 + 4x_3{}^2}$$

which is generated by the inner product

$$\langle (x_1, x_2, x_3), (y_1, y_2, y_3) \rangle = x_1y_1 + 2x_2y_2 + 4x_3y_3$$

Which of the following hikes is most difficult (has the largest difficulty norm)?

	MILES IN		
TRIP	CATEGORY 1	CATEGORY 2	CATEGORY 3
T_1	3	2	1
T_2	0	6	0
T_3	5	0	1

Note that all hikes are 6 mi long.

5. Let V be a vector space with inner product $\langle \mathbf{x}, \mathbf{y} \rangle$. Let a new function $\langle\ ,\ \rangle_k$ be defined on V by $\langle \mathbf{x}, \mathbf{y} \rangle_k = k\langle \mathbf{x}, \mathbf{y} \rangle$, where $k > 0$. Show that $\langle\ ,\ \rangle_k$ is also an inner product. Define $\|\mathbf{x}\|_k = \sqrt{\langle \mathbf{x}, \mathbf{x} \rangle_k}$. Compare $\|\mathbf{x}\|_k$ with $\|\mathbf{x}\|_1$ (the original norm) for $k > 1$ and $k < 1$.

6. Let

$$A = \begin{pmatrix} 1 & 1 \\ 0 & 1 \end{pmatrix} \quad \text{and} \quad B = \begin{pmatrix} 0 & 0 \\ 1 & 0 \end{pmatrix}$$

in $\mathscr{M}_{22}$ with standard inner product. Calculate $\langle A, B \rangle$, $\|A\|$, $\|B\|$, $\|A - B\|$, and $\cos\theta$, where θ is the angle between A and B.

7. Consider E^3 with the standard inner product.

(*a*) Find a vector orthogonal to both $(3, -1, 0)$ and $(1, 2, 3)$.

(*b*) Find vectors x and y orthogonal to $(1, -1, 2)$ *and* independent of each other.

8. Let V be a vector space with inner product $\langle\ ,\ \rangle$. Let $\mathbf{x} \perp \mathbf{y}$ mean that $\langle \mathbf{x}, \mathbf{y} \rangle = 0$. Show that

(*a*) If $\mathbf{x} \perp \mathbf{y}$ and $\mathbf{x} \perp \mathbf{z}$, then $\mathbf{x} \perp (a\mathbf{y} + b\mathbf{z})$ for all a and b.

(*b*) If $\mathbf{x} \perp \mathbf{v}_1$, $\mathbf{x} \perp \mathbf{v}_2, \ldots, \mathbf{x} \perp \mathbf{v}_n$, then $\mathbf{x}$ is orthogonal to any vector in span $\{\mathbf{v}_1, \mathbf{v}_2, \ldots, \mathbf{v}_n\}$.

9. Furnish details for verifying the inner product in Example 1.

10. Furnish details for verifying the inner product in Example 2.

11. Prove part (*b*) and (*c*) of Theorem 3.6.1.

12. Prove part (*e*) of Theorem 3.6.1 for real vector spaces by following these steps.

(*a*) Show that the inequality is true for $\mathbf{x} = \boldsymbol{\theta}$.

(*b*) Assume $\mathbf{x} \neq \boldsymbol{\theta}$ and use

$$0 \leq \langle r\mathbf{x} + \mathbf{y}, r\mathbf{x} + \mathbf{y} \rangle \qquad r \text{ real}$$

to find

$$0 \leq \underbrace{\|x\|^2}_{\text{Call } A} r^2 + \underbrace{2\langle \mathbf{x}, \mathbf{y} \rangle}_{\text{Call } B} r + \underbrace{\|y\|^2}_{\text{Call } C}$$

(*c*) Recall that a quadratic

$$Ar^2 + Br + C \tag{1}$$

which is greater than or equal to 0 has either no or one real root. Calculate the discriminant (from the quadratic formula) for Eq. (1). It must be less than or equal to zero.

(*d*) Use $B^2 - 4AC \leq 0$ to obtain the Cauchy-Schwarz inequality.

13. Show that if $\mathbf{x} \perp \mathbf{y}$, then $\|\mathbf{x} + \mathbf{y}\|^2 = \|\mathbf{x}\|^2 + \|\mathbf{y}\|^2$.

14. Show that if $S = \{\mathbf{v}_1, \mathbf{v}_2, \ldots, \mathbf{v}_n\}$ is an orthogonal set, then

$$\|\mathbf{v}_1 + \mathbf{v}_2 + \cdots + \mathbf{v}_n\|^2 = \|\mathbf{v}_1\|^2 + \|\mathbf{v}_2\|^2 + \cdots + \|\mathbf{v}_n\|^2$$

15. In $\mathscr{M}_{n1}$, $\langle X, Y \rangle \equiv X^T Y$ is an inner product. Calculate $X^T Y$ for

$$X = \begin{pmatrix} 1 \\ 2 \\ -1 \end{pmatrix} \qquad Y = \begin{pmatrix} -2 \\ 4 \\ -3 \end{pmatrix}$$

16. In $\mathscr{C}_{n1}$, $\langle X, Y \rangle \equiv X^* Y$ is an inner product (recall $X^* = \bar{X}^T$). Calculate $X^* Y$ for

$$X = \begin{pmatrix} 1 \\ i \end{pmatrix} \qquad Y = \begin{pmatrix} 1 - i \\ 2 \end{pmatrix}.$$

Calculate $U^* V$ for

$$U = \begin{pmatrix} 1 - i \\ 1 \end{pmatrix} \qquad V = \begin{pmatrix} 1 - i \\ 2 \end{pmatrix}$$

17. Let A and B be matrices in $\mathscr{M}_{mn}$. and define $\langle A, B \rangle = \operatorname{tr} A^T B$.

(*a*) What choice of m and n makes $\langle A, B \rangle$ equivalent to the inner product in Example 3?

(*b*) Show that $\langle A, B \rangle$ as defined is an inner product on $\mathscr{M}_{mn}$.

3.7 THE BASIS PROBLEM REVISITED

We consider in this section the following form of the basis problem:

> Given a basis $S = \{\mathbf{v}_1, \ldots, \mathbf{v}_n\}$ of a vector space V with inner product $\langle\ ,\ \rangle$ find an orthonormal basis $\mathscr{O} = \{\mathbf{u}_1, \mathbf{u}_2, \ldots, \mathbf{u}_n\}$ of V.

One reason for wanting an orthonormal basis for V is that for any vector $\mathbf{x} \in V$ it is easy to calculate the coefficients of $\mathbf{x}$ in the linear combination

$$\mathbf{x} = c_1\mathbf{u}_1 + c_2\mathbf{u}_2 + \cdots + c_n\mathbf{u}_n$$

In fact,

$$\begin{aligned} \langle \mathbf{x}, \mathbf{u}_k \rangle &= c_1\langle \mathbf{u}_1, \mathbf{u}_k \rangle + \cdots + c_k\langle \mathbf{u}_k, \mathbf{u}_k \rangle + \cdots + c_n\langle \mathbf{u}_n, \mathbf{u}_k \rangle \\ &= c_k \end{aligned}$$

EXAMPLE 1 The set $\mathscr{O} = \{\mathbf{u}_1, \mathbf{u}_2, \mathbf{u}_3\} = \{(1/\sqrt{2}, 0, 1/\sqrt{2}), (0, 1, 0), (1/\sqrt{2}, 0, -1/\sqrt{2})\}$ is an orthonormal basis for E^3. Write $(2, -1, 4)$ as a linear combination of the basis elements.

Solution We have

$$\mathbf{x} = \langle \mathbf{x}, \mathbf{u}_1 \rangle \mathbf{u}_1 + \langle \mathbf{x}, \mathbf{u}_2 \rangle \mathbf{u}_2 + \langle \mathbf{x}, \mathbf{u}_3 \rangle \mathbf{u}_3$$
$$= \frac{6}{\sqrt{2}} \mathbf{u}_1 + (-1)\mathbf{u}_2 + \frac{-2}{\sqrt{2}} \mathbf{u}_3$$

A solution of the basis problem stated above involves what is known as the *Gram-Schmidt procedure.*

To get an idea of this procedure, consider the problem in E^3 and let $S = \{\mathbf{v}_1, \mathbf{v}_2, \mathbf{v}_3\}$, as pictured in Fig. 3.7.1. What we can do is to first obtain an orthogonal basis $T = \{\mathbf{w}_1, \mathbf{w}_2, \mathbf{w}_3\}$ and then to normalize all the vectors to obtain $\mathcal{O}$. For the first vector $\mathbf{w}_1$ in T, choose $\mathbf{w}_1 = \mathbf{v}_1$. Now $\mathbf{w}_2$ must be perpendicular to $\mathbf{w}_1$. That is, it must lie in the plane P_1 shown in Fig. 3.7.2. We see that there are an infinite number of choices for $\mathbf{w}_2$. The idea in the Gram-Schmidt process is to choose $\mathbf{w}_2$ as a linear combination of $\mathbf{w}_1$ and $\mathbf{v}_2$. Since span $\{\mathbf{w}_1, \mathbf{v}_2\}$ is the plane containing $\mathbf{w}_1$ and $\mathbf{v}_2$, we know that $\mathbf{w}_2$ must lie in the intersection of the plane P_1 and span $\{\mathbf{w}_1, \mathbf{v}_2\}$ (see Fig. 3.7.3). Once $\mathbf{w}_2$ is

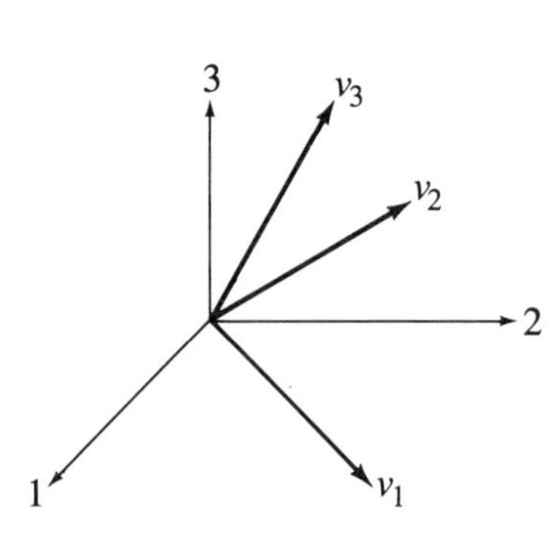

Figure 3.7.1
Picture of $S = \{v_1, v_2, v_3\}$ in E^3.

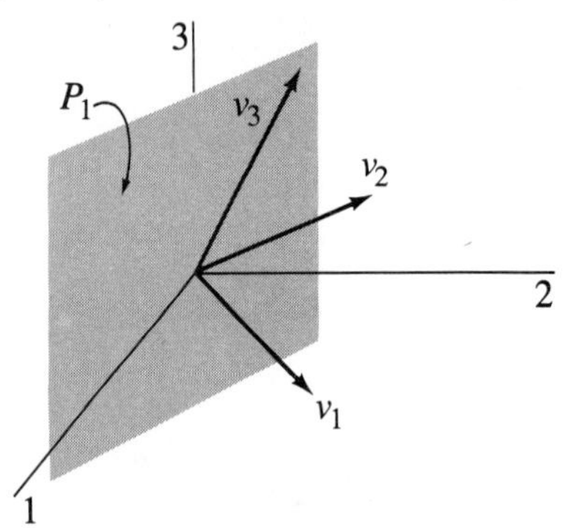

Figure 3.7.2
Plane where w_2 must lie. The plane of all vectors perpendicular to v_1.

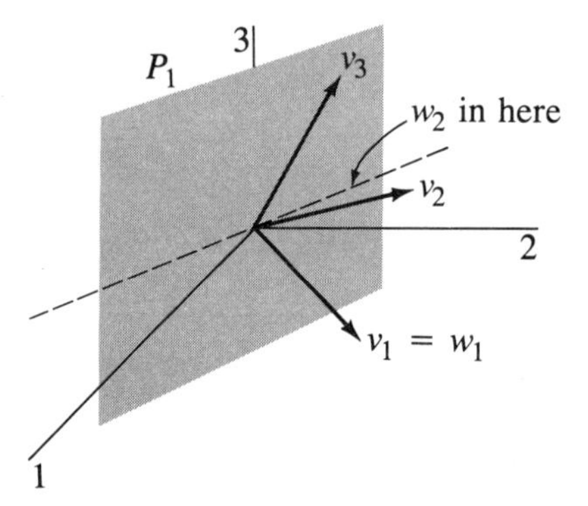

Figure 3.7.3
Intersection of P_1 and span $\{v_1, v_2\}$, where w_2 must lie in Gram-Schmidt process.

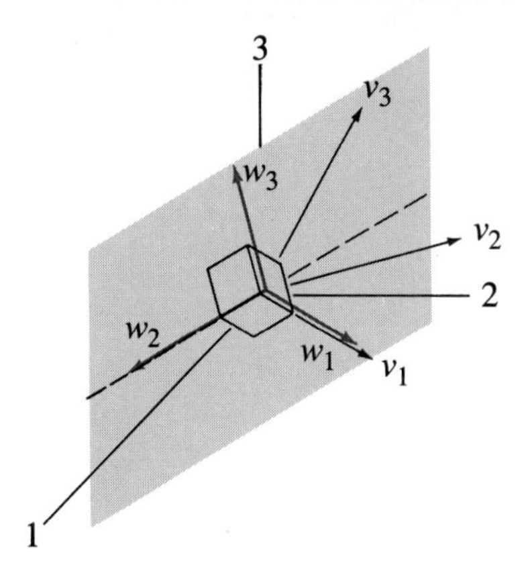

Figure 3.7.4 Orthogonal basis constructed from $\{v_1, v_2, v_3\}$ by the Gram-Schmidt process.

chosen, $\mathbf{w}_3$ must be found perpendicular to $\mathbf{w}_1$ and $\mathbf{w}_2$. In the Gram-Schmidt process, $\mathbf{w}_3$ is required to lie in span $\{\mathbf{w}_1, \mathbf{w}_2, \mathbf{v}_3\}$ and to be perpendicular to $\mathbf{w}_1$ and $\mathbf{w}_2$ (see Fig. 3.7.4). Finally then

$$\mathcal{O} = \left\{\frac{\mathbf{w}_1}{\|\mathbf{w}_1\|}, \frac{\mathbf{w}_2}{\|\mathbf{w}_2\|}, \frac{\mathbf{w}_3}{\|\mathbf{w}_3\|}\right\}$$

Let us actually carry out this process in E^3 before stating the general result.

EXAMPLE 2 Using the procedure just outlined, construct an orthogonal basis from $S = \{(1, 1, 0), (1, 3, 1), (2, 2, 3)\} = \{\mathbf{v}_1, \mathbf{v}_2, \mathbf{v}_3\}$ in E^3 with the standard inner product.

Solution We choose $\mathbf{w}_1 = (1, 1, 0)$ and require of $\mathbf{w}_2$ that

$$\langle \mathbf{w}_2, \mathbf{w}_1 \rangle = 0$$

$$\mathbf{w}_2 = c_1\mathbf{w}_1 + c_2\mathbf{v}_2 \tag{3.7.1}$$

Now take the inner product of both sides of Eq. (3.7.1) with $\mathbf{w}_1$ to find

$$0 = \langle \mathbf{w}_2, \mathbf{w}_1 \rangle = c_1\langle \mathbf{w}_1, \mathbf{w}_1 \rangle + c_2\langle \mathbf{v}_2, \mathbf{w}_1 \rangle$$

which is one homogeneous equation in two unknowns. The solutions are $c_1 = -c_2\langle \mathbf{v}_2, \mathbf{w}_1 \rangle / \|\mathbf{w}_1\|^2$. Choosing $c_2 = 1$, we have

$$\mathbf{w}_2 = \frac{-\langle \mathbf{v}_2, \mathbf{w}_1 \rangle}{\|\mathbf{w}_1\|^2}\mathbf{w}_1 + \mathbf{v}_2 = \mathbf{v}_2 - \boxed{\frac{\langle \mathbf{v}_2, \mathbf{w}_1 \rangle}{\|\mathbf{w}_1\|^2}\mathbf{w}_1} \tag{3.7.2}$$

Projection of v_2 on w_1

$$= (1, 3, 1) - \tfrac{4}{2}(1, 1, 0) = (-1, 1, 1)$$

Now having $\mathbf{w}_1 = (1, 1, 0)$ and $\mathbf{w}_2 = (-1, 1, 1)$, we require of $\mathbf{w}_3$ that

$$\langle \mathbf{w}_1, \mathbf{w}_3 \rangle = 0$$
$$\langle \mathbf{w}_2, \mathbf{w}_3 \rangle = 0$$
$$\mathbf{w}_3 = c_1\mathbf{w}_1 + c_2\mathbf{w}_2 + c_3\mathbf{v}_3 \tag{3.7.3}$$

Now take inner products of both sides of Eqs. (3.7.3) with $\mathbf{w}_1$ and $\mathbf{w}_2$ to find equations for c_1, c_2, c_3:

$$0 = \langle \mathbf{w}_1, \mathbf{w}_3 \rangle = c_1\langle \mathbf{w}_1, \mathbf{w}_1 \rangle + c_2\langle \mathbf{w}_2, \mathbf{w}_1 \rangle + c_3\langle \mathbf{w}_1, \mathbf{v}_3 \rangle$$
$$0 = \langle \mathbf{w}_2, \mathbf{w}_3 \rangle = c_1\langle \mathbf{w}_2, \mathbf{w}_1 \rangle + c_2\langle \mathbf{w}_2, \mathbf{w}_2 \rangle + c_3\langle \mathbf{w}_2, \mathbf{v}_3 \rangle$$

Since $\langle \mathbf{w}_1, \mathbf{w}_2 \rangle = 0$, these last equations reduce to

$$0 = c_1\|\mathbf{w}_1\|^2 \qquad\qquad + c_3\langle \mathbf{w}_1, \mathbf{v}_3 \rangle$$
$$0 = \qquad\qquad c_2\|\mathbf{w}_2\|^2 + c_3\langle \mathbf{w}_2, \mathbf{v}_3 \rangle$$

Putting $c_3 = 1$, we have

$$c_1 = \frac{-\langle \mathbf{w}_1, \mathbf{v}_3 \rangle}{\|\mathbf{w}_1\|^2} \qquad c_2 = \frac{-\langle \mathbf{w}_2, \mathbf{v}_3 \rangle}{\|\mathbf{w}_2\|^2}$$

Projection of v_3 on w_2 — Projection of v_3 on w_1

$$\mathbf{w}_3 = \mathbf{v}_3 - \frac{\langle \mathbf{w}_2, \mathbf{v}_3 \rangle}{\|\mathbf{w}_2\|^2}\mathbf{w}_2 - \frac{\langle \mathbf{w}_1, \mathbf{v}_3 \rangle}{\|\mathbf{w}_1\|^2}\mathbf{w}_1 \tag{3.7.4}$$

$$= (2, 2, 3) - \tfrac{3}{3}(-1, 1, 1) - \tfrac{4}{2}(1, 1, 0)$$
$$= (1, -1, 2)$$

Thus an orthogonal basis is

$$T = \{(1, 1, 0), (-1, 1, 1), (1, -1, 2)\}$$

By normalizing each vector in T, we find an orthonormal basis

$$\left\{\left(\frac{1}{\sqrt{2}}, \frac{1}{\sqrt{2}}, 0\right), \left(-\frac{1}{\sqrt{3}}, \frac{1}{\sqrt{3}}, \frac{1}{\sqrt{3}}\right), \left(\frac{1}{\sqrt{6}}, -\frac{1}{\sqrt{6}}, \frac{1}{\sqrt{6}}\right)\right\}$$

Graphs of S and $\mathcal{O}$ are shown in Fig. 3.7.5.

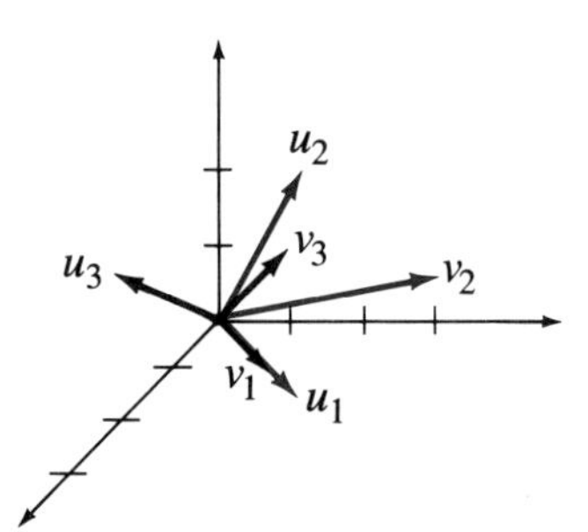

Figure 3.7.5 Graphs of S (blue vectors) and $\mathcal{O}$ (black vectors.)

The method of Example 2 actually furnishes a method of proof for a general result.

THEOREM 3.7.1 (Gram-Schmidt procedure) Let $S = \{\mathbf{v}_1, \mathbf{v}_2, \ldots, \mathbf{v}_n\}$ be a basis for a vector space V with inner product $\langle\ ,\ \rangle$. The set $T = \{\mathbf{w}_1, \ldots, \mathbf{w}_n\}$, where

$$\begin{aligned}
\mathbf{w}_1 &= \mathbf{v}_1 \\
\mathbf{w}_2 &= \mathbf{v}_2 - \frac{\langle \mathbf{v}_2, \mathbf{w}_1 \rangle}{\|\mathbf{w}_1\|^2}\mathbf{w}_1 \\
\mathbf{w}_3 &= \mathbf{v}_3 - \frac{\langle \mathbf{v}_3, \mathbf{w}_2 \rangle}{\|\mathbf{w}_2\|^2}\mathbf{w}_2 - \frac{\langle \mathbf{v}_3, \mathbf{w}_1 \rangle}{\|\mathbf{w}_1\|^2}\mathbf{w}_1 \\
\mathbf{w}_n &= \mathbf{v}_n - \frac{\langle \mathbf{v}_n, \mathbf{w}_{n-1} \rangle}{\|\mathbf{w}_{n-1}\|^2}\mathbf{w}_{n-1} - \cdots - \frac{\langle \mathbf{v}_n, \mathbf{w}_2 \rangle}{\|\mathbf{w}_2\|^2}\mathbf{w}_2 - \frac{\langle \mathbf{v}_n, \mathbf{w}_1 \rangle}{\|\mathbf{w}_1\|^2}\mathbf{w}_1
\end{aligned} \tag{3.7.5}$$

is an orthogonal basis for V. The set

$$\mathcal{O} = \left\{ \frac{\mathbf{w}_1}{\|\mathbf{w}_1\|}, \frac{\mathbf{w}_2}{\|\mathbf{w}_2\|}, \ldots, \frac{\mathbf{w}_n}{\|\mathbf{w}_n\|} \right\}$$

is an orthonormal basis for V.

Equations (3.7.5) are a generalization of Eqs. (3.7.2) and (3.7.4) in Example 2.

EXAMPLE 3 Find an orthonormal basis for $V = \text{span}\,\{(1, -1, 0, 2), (3, 2, 1, 2), (-2, 1, 0, 1)\}$ in E^4 with the standard inner product.

Solution Let $\mathbf{v}_1 = (1, -1, 0, 2)$, $\mathbf{v}_2 = (3, 2, 1, 2)$, and $\mathbf{v}_3 = (-2, 1, 0, 1)$. Using the formulas from Theorem 3.7.1, we have

$$\begin{aligned}
\mathbf{w}_1 &= (1, -1, 0, 2) \\
\mathbf{w}_2 &= (3, 2, 1, 2) - \tfrac{5}{6}(1, -1, 0, 2) = (\tfrac{13}{6}, \tfrac{17}{6}, 1, \tfrac{1}{3}) = \tfrac{1}{6}(13, 17, 6, 2) \\
\mathbf{w}_3 &= (-2, 1, 0, 1) - \frac{-\frac{7}{6}}{\frac{498}{36}}\left(\frac{13}{6}, \frac{17}{6}, 1, \frac{1}{3}\right) - \frac{-1}{6}(1, -1, 0, 2) \\
&= \tfrac{1}{498}(-822, 534, 42, 678)
\end{aligned}$$

An orthonormal basis is

$$\left\{ \frac{\mathbf{w}_1}{\|\mathbf{w}_1\|}, \frac{\mathbf{w}_2}{\|\mathbf{w}_2\|}, \frac{\mathbf{w}_3}{\|\mathbf{w}_3\|} \right\}$$

$$= \left\{ \frac{1}{\sqrt{6}}(1, -1, 0, 2), \frac{1}{\sqrt{498}}(13, 17, 6, 2), \frac{1}{\sqrt{355{,}572}}(-411, 267, 21, 339) \right\}$$

Now that we know what an orthonormal basis is, we can describe an important class of matrices which we will use heavily later on.

DEFINITION 3.7.1 A square matrix A is called *orthogonal* if $A^T = A^{-1}$, that is, if $AA^T = A^TA = I$.

EXAMPLE 4 Show that

$$A = \begin{pmatrix} \frac{3}{5} & \frac{4}{5} \\ -\frac{4}{5} & \frac{3}{5} \end{pmatrix}$$

is orthogonal.

Solution

$$A^TA = \begin{pmatrix} \frac{3}{5} & -\frac{4}{5} \\ \frac{4}{5} & \frac{3}{5} \end{pmatrix}\begin{pmatrix} \frac{3}{5} & \frac{4}{5} \\ -\frac{4}{5} & \frac{3}{5} \end{pmatrix} = \begin{pmatrix} 1 & 0 \\ 0 & 1 \end{pmatrix}$$

Therefore $A^T = A^{-1}$ and A is orthogonal.

The rows of A in Example 4 are, as vectors, $\mathbf{v}_1 = (\frac{3}{5}, \frac{4}{5})$ and $\mathbf{v}_2 = (-\frac{4}{5}, \frac{3}{5})$. Notice that in the standard inner product on E^2, $\mathbf{v}_1 \perp \mathbf{v}_2$, $\|\mathbf{v}_1\| = \|\mathbf{v}_2\| = 1$. That is, $\{\mathbf{v}_1, \mathbf{v}_2\}$ is an orthonormal basis for E^2. In fact, the same can be said for the columns of A. This observation carries over to all orthogonal matrices, as the next theorems show.

THEOREM 3.7.2 A matrix $A_{n\times n}$ is orthogonal if and only if the rows of A form an orthonormal basis for E^n with the standard inner product.

Proof ($\Leftarrow$) Suppose the rows of A form an orthonormal basis $\{\mathbf{v}_1, \mathbf{v}_2, \ldots, \mathbf{v}_n\}$. Then

$$AA^T = \begin{pmatrix} - & \mathbf{v}_1 & - \\ - & \mathbf{v}_2 & - \\ & \vdots & \\ - & \mathbf{v}_n & - \end{pmatrix}\begin{pmatrix} | & | & & | \\ \mathbf{v}_1 & \mathbf{v}_2 & \cdots & \mathbf{v}_n \\ | & | & & | \end{pmatrix}$$

Given the definition of matrix multiplication, the ij entry of the product is $\langle \mathbf{v}_i, \mathbf{v}_j \rangle$, which is 1 if $i = j$ and 0 if $i \neq j$. Therefore $AA^T = I$.

($\Rightarrow$) Suppose A is orthogonal. Then $AA^T = I$. Then c_{ij}, the ij entry of AA^T, is 1 if $i = j$ and 0 if $i \neq j$. By the definitions of matrix multiplication and the transpose of a matrix, we see that $\langle \text{row } i, \text{row } j \rangle$ is 1 if $i = j$ and is 0 if $i \neq j$. Therefore the rows of $A_{n\times n}$ form an orthonormal basis for E^n.

THEOREM 3.7.3 Matrix A is orthogonal if and only if A^T is orthogonal.

Proof If A is orthogonal, then $AA^T = A^TA = I$. Thus

$$A^T(A^T)^T = A^TA = I$$

and

$$(A^T)^TA^T = AA^T = I$$

Therefore A^T is orthogonal. The argument is reversible (see the problems).

THEOREM 3.7.4 A matrix $A_{n\times n}$ is orthogonal if and only if the columns of A form an orthonormal basis for E^n with the standard inner product.

Proof Matrix A is orthogonal if and only if A^T is orthogonal. The columns of A are the rows of A^T, which form an orthonormal basis for E^n.

EXAMPLE 5 The identity matrix is an orthogonal matrix.

EXAMPLE 6 Give an example of a 3×3 orthogonal matrix (not I).

Solution There are an infinite number of possibilities. By using the standard basis for E^3 we can obtain

$$\begin{pmatrix}1&0&0\\0&0&1\\0&1&0\end{pmatrix}, \quad \begin{pmatrix}0&1&0\\1&0&0\\0&0&1\end{pmatrix}, \quad \begin{pmatrix}0&0&1\\1&0&0\\0&1&0\end{pmatrix}, \quad \begin{pmatrix}0&0&1\\0&1&0\\1&0&0\end{pmatrix}, \quad \begin{pmatrix}0&1&0\\0&0&1\\1&0&0\end{pmatrix}$$

Another possibility is

$$A = \begin{pmatrix}1/\sqrt{2} & 1/\sqrt{2} & 0\\ 1/\sqrt{2} & -1/\sqrt{2} & 0\\ 0 & 0 & 1\end{pmatrix}$$

Notice that $A^T = A$, so that $A^2 = I$. Such matrices are called *involutory.*

In the complex case, a square matrix A is called *unitary* if $A^*A = I$. For example,

$$A = \begin{pmatrix}0 & -i\\ i & 0\end{pmatrix}$$

is unitary. We have virtually identical theorems for unitary matrices, and their proofs are virtually the same.

THEOREM 3.7.5 A square matrix A is unitary if and only if its rows and columns are mutually orthogonal with respect to the standard inner product in $\mathbb{C}^n$.

EXAMPLE 7 The matrix

$$\begin{pmatrix}i/\sqrt{2} & 1/\sqrt{2}\\ i/\sqrt{2} & -1/\sqrt{2}\end{pmatrix}$$

is unitary.

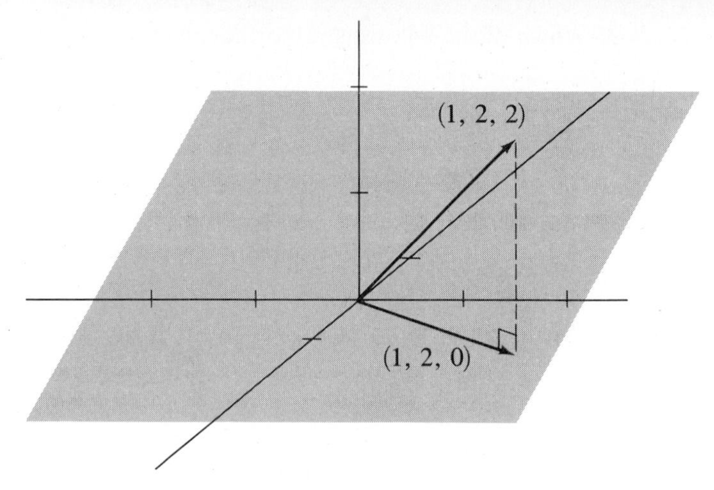

Figure 3.7.6 Orthogonal projection of (1, 2, 2) onto $W = E^2$ (the base plane).

Orthogonal Projections If W is a subspace of E^n with an orthonormal basis $\{\mathbf{v}_1, \mathbf{v}_2, \ldots, \mathbf{v}_m\}$, then the *orthogonal projection of* $\mathbf{v}$ *onto* W is defined as

$$\text{proj}_W \mathbf{v} = \langle \mathbf{v} \cdot \mathbf{v}_1 \rangle \mathbf{v}_1 + \cdots + \langle \mathbf{v} \cdot \mathbf{v}_m \rangle \mathbf{v}_m$$

where the dot product is the standard dot product on E^n. For example, if $\mathbf{v} = (1, 2, 2)$ and $W = \{(\mathbf{x}_1, \mathbf{x}_2, \mathbf{x}_3) | \mathbf{x}_3 = 0\}$ with basis $S = \{(1, 0, 0), (0, 1, 0)\}$, then $\text{proj}_W (1, 2, 2) = (1, 2, 0)$ (see Fig. 3.7.6). From Fig. 3.7.6 it is not too hard to see that $\text{proj}_W \mathbf{v}$ is the vector in W "closest" to $\mathbf{v}$, in the sense that the distance from the "end of $\mathbf{v}$" to the "end of $\text{proj}_W \mathbf{v}$" is the smallest compared to all other vectors from W. That is,

$$\|\mathbf{v} - \text{proj}_W \mathbf{v}\| \leq \|\mathbf{w} - \mathbf{v}\|$$

for all vectors $\mathbf{w}$ in W. Although this is true in general, we do not prove it here. We note that if nonstandard inner products are used, the norms generated need not represent distance in the usual way.

PROBLEMS 3.7

1. Verify that the following are orthogonal bases of E^3. Convert them to orthonormal bases.

   (*a*) $\{(1, 1, 0), (0, 0, 1), (-1, 1, 0)\}$

   (*b*) $\{(1, 1, 1), (0, 1, -1), (1, -\frac{1}{2}, -\frac{1}{2})\}$

2. Use the Gram-Schmidt procedure to construct orthonormal bases for the following subspaces of E^4.

   (*a*) span $\{(1, 0, 0, 0), (1, 0, 1, 0), (1, 0, 1, 1)\}$

   (*b*) span $\{(1, 0, -1, 0), (1, 0, 1, 1), (1, -1, 0, 0)\}$

3. Which of the following are orthogonal matrices?

(*a*) $\begin{pmatrix} 1 & 1 \\ 1 & -1 \end{pmatrix}$ (*b*) $\begin{pmatrix} 0 & 0 \\ 0 & 0 \end{pmatrix}$

(c) $\begin{pmatrix} \frac{5}{13} & -\frac{12}{13} & 0 \\ \frac{12}{13} & \frac{5}{13} & 0 \\ 0 & 0 & 1 \end{pmatrix}$ (*d*) $\begin{pmatrix} 1/\sqrt{2} & 1/\sqrt{2} & 0 & 0 \\ -1/\sqrt{2} & 1/\sqrt{2} & 0 & 0 \\ 0 & 0 & 0 & 1 \\ 0 & 0 & -1 & 0 \end{pmatrix}$

4. Which of the following are unitary matrices?

(*a*) $\begin{pmatrix} 1 & i \\ i & 1 \end{pmatrix}$ (*b*) $\begin{pmatrix} 1 & 1+i \\ 1-i & 2 \end{pmatrix}$

(c) $\begin{pmatrix} i & 0 \\ 0 & 1 \end{pmatrix}$ (*d*) $\begin{pmatrix} 1 & i & 3+i \\ -i & -1 & 2-2i \\ 3-i & 2+2i & 2 \end{pmatrix}$

5. Let $\mathcal{O} = \{\mathbf{v}_1, \mathbf{v}_2, \ldots, \mathbf{v}_n\}$ be an orthonormal basis for V. Let $\mathbf{x} \in V$. Show that if $\mathbf{x} = c_1\mathbf{v}_1 + c_2\mathbf{v}_2 + \cdots + c_n\mathbf{v}_n$, then $\|\mathbf{x}\|^2 = c_1{}^2 + c_2{}^2 + \cdots + c_n{}^2$.

6. Let $S = \{\mathbf{v}_1, \mathbf{v}_2, \ldots, \mathbf{v}_k\}$ be a linearly independent set of vectors in a vector space V of dimension n, $n > k$, with inner product. Show that if $\mathbf{x}$ is orthogonal to all the vectors in S, then $T = \{\mathbf{v}_1, \mathbf{v}_2, \ldots, \mathbf{v}_k, \mathbf{x}\}$ is also linearly independent ($\mathbf{x} \neq \boldsymbol{\theta}$).

7. Show that

$$\mathcal{O} = \left\{ \begin{pmatrix} 1 & 0 \\ 0 & 0 \end{pmatrix}, \begin{pmatrix} 0 & 0 \\ 0 & 1 \end{pmatrix}, \begin{pmatrix} 0 & 1/\sqrt{2} \\ 1/\sqrt{2} & 0 \end{pmatrix}, \begin{pmatrix} 0 & -1/\sqrt{2} \\ 1/\sqrt{2} & 0 \end{pmatrix} \right\}$$

is an orthonormal basis for $\mathcal{M}_{22}$, with the inner product

$$\left\langle \begin{pmatrix} a & b \\ c & d \end{pmatrix}, \begin{pmatrix} e & f \\ g & h \end{pmatrix} \right\rangle = ae + bf + cg + dh$$

Write

$$\begin{pmatrix} 2 & 1 \\ -3 & 2 \end{pmatrix}$$

as a linear combination of the basis vectors.

8. Use the Gram-Schmidt procedure to transform the given bases to orthonormal bases in the spaces given. (Use the standard inner product.)

(*a*) $\{(1, 1), (1, 2)\}$ in E^2

(*b*) $\{(1, 0), (3, 7)\}$ in E^2

(c) $\{(1, 1, -1), (0, 1, -1), (1, 1, 0)\}$ in E^3

(*d*) $\{(1, 2, 3), (4, 5, 6), (1, 1, 0)\}$ in E^3

9. Reverse the argument in Theorem 3.7.3 to complete the proof.

10. Use the Gram-Schmidt procedure to find an orthonormal basis for span $\{(1, 3, 1), (1, 2, 1)\}$ in E^3 with the standard inner product. Do the same in E^3 with the inner product $\langle \mathbf{x}, \mathbf{y} \rangle = 3x_1y_1 + 2x_2y_2 + x_3y_3$.

11. Find an orthonormal basis for E^3 which includes $\mathbf{v} = (1/\sqrt{2}, 1/\sqrt{2}, 0)$ (do this by inspection).

12. Find an orthogonal basis for E^3 which includes $\mathbf{v}_1 = (1, 1, 1)$. First find any basis which includes $\mathbf{y}_1$ as the first vector; then use the Gram-Schmidt procedure, but do not normalize $\mathbf{v}_1$.

13. (*a*) Let A be a unitary matrix. Show that $|\det A| = 1$.
(*b*) Show by example that for a unitary matrix A, $\det A$ need not be 1 or -1. (*Note*: Remember that A can have complex entries.)
(*c*) Let A be an orthogonal matrix. Show that $\det A = \pm 1$. In general, a matrix A with $\det A = 1$ is called *unimodular*.

14. What conditions on a and b guarantee that

$$A = \begin{pmatrix} 0 & 1 & 0 & b \\ a & 0 & b & 0 \\ 0 & b & 0 & a \\ b & 0 & a & 0 \end{pmatrix}$$

is orthogonal? Is there a condition which will make A unitary?

15. Is the square of an orthogonal matrix orthogonal? Is the square of a unitary matrix also unitary?

16. If A is orthogonal, is A^n orthogonal $(n \geq 2)$?

17. (*a*) Show that if U is orthogonal and $B = U^TAU$, then $\det B = \det A$.
(*b*) Show that if U is unitary and $B = U^*AU$, then $\det B = \det A$.

18. If A and B are orthogonal what about $A + B$? What about $-A$?

19. Let A be involutory. Show that if n is odd, then $A^n = A$.

20. If A is unitary, is $-A$ unitary?

21. If A is idempotent, must A be symmetric?

22. Show that if $S = \{\mathbf{v}_1, \ldots, \mathbf{v}_n\}$ is an orthogonal basis for V, then for any $\mathbf{x}$ in V,

$$\mathbf{x} = \frac{\langle \mathbf{x}, \mathbf{v}_1 \rangle}{\|\mathbf{v}_1\|^2}\mathbf{v}_1 + \frac{\langle \mathbf{x}, \mathbf{v}_2 \rangle}{\|\mathbf{v}_2\|^2}\mathbf{v}_2 + \cdots + \frac{\langle \mathbf{x}, \mathbf{v}_n \rangle}{\|\mathbf{v}_n\|^2}\mathbf{v}_n$$

23. Develop a proof of Theorem 3.7.5 by mimicking the proofs of Theorems 3.7.2, 3.7.3, and 3.7.4.

24. Show that the rotation matrix

$$\begin{pmatrix} \cos\alpha & -\sin\alpha \\ \sin\alpha & \cos\alpha \end{pmatrix}$$

is orthogonal. Is the matrix unitary?

25. Let a and b be real numbers. Under what conditions is

$$\begin{pmatrix} 0 & a & 0 & ib \\ a & 0 & ib & 0 \\ 0 & ib & 0 & a \\ ib & 0 & a & 0 \end{pmatrix}$$

unitary? (This is an example of a *scattering matrix*.)

26. The set $S = \{(1/\sqrt{2}, -1/\sqrt{2}, 0), (0, 1, 0)\}$ does not span E^3. Find the orthogonal projections of the following vectors on span S. Sketch span S and the vector. (*a*) $(1, 0, 0)$ (*b*) $(1, -1, 0)$ (*c*) $(1, 1, 2)$

3.8 CHANGING BASES IN VECTOR SPACES

Ask several people directions to a certain location, and the instructions may sound different yet be equivalent. In Fig. 3.8.1 we see a map of a small town with an origin 0, where we ask directions to the corner of McDaniel and Walker streets, marked by an X. Two possible sets of directions are

1. Go 8 blocks east and 6 blocks north.

2. Go 6 blocks north and 8 blocks east.

Both sets of directions are correct. What is the difference? Each set is related to a different basis of E^2! In Fig. 3.8.2 we have drawn two bases S_1 and S_2

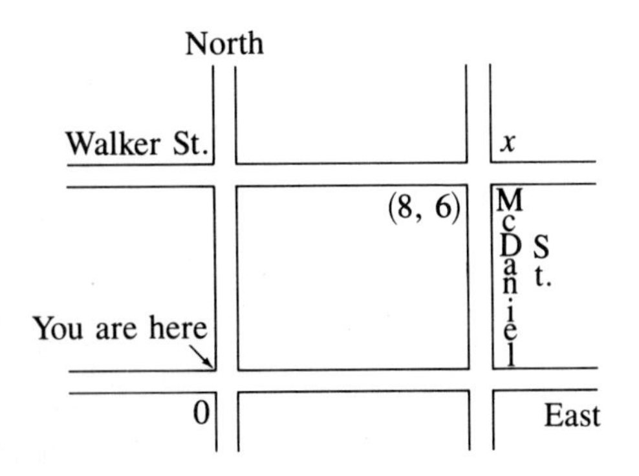

Figure 3.8.1
A small town. 1 block = $\frac{1}{4}$ mile.

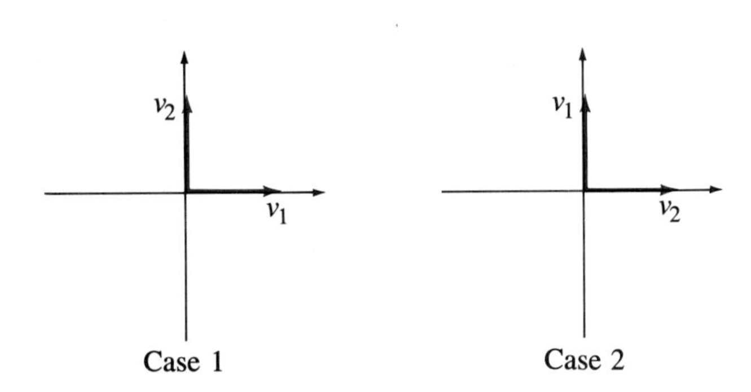

Figure 3.8.2
Different bases for directions.

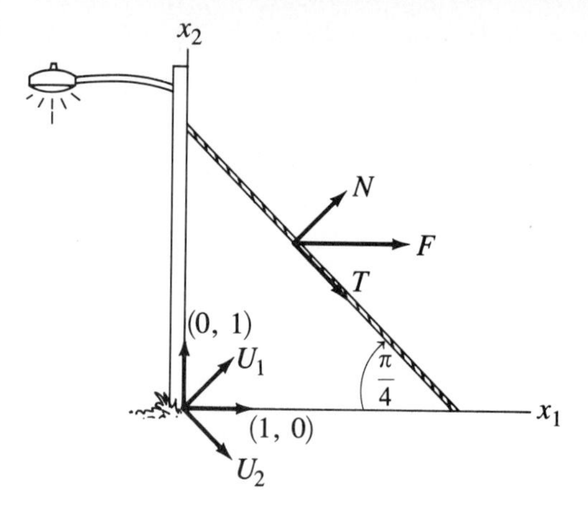

Figure 3.8.3
Forces on a guy wire.

which correspond to cases 1 and 2 respectively, above. We have in units of blocks

$$S_1 = \{(1, 0), (0, 1)\} \qquad S_2 = \{(0, 1), (1, 0)\}$$

The ordering of the basis vectors is switched in S_1 and S_2. The corner is at (8, 6), yet in terms of the different bases (8, 6) has different coefficients, in terms of order.

$$(8, 6) = 8(1, 0) + 6(0, 1)$$
$$(8, 6) = 6(0, 1) + 8(1, 0)$$

Note that the coefficients in the linear combinations above are actually the "directions" in terms of the given basis. Comparing S_1 and S_2, we see that *order* is important in a basis when we are using the basis to describe vectors. We also note that the "direction" vectors

$$\begin{pmatrix} 8 \\ 6 \end{pmatrix} \quad \text{and} \quad \begin{pmatrix} 6 \\ 8 \end{pmatrix}$$

can be related by means of matrices:

$$\begin{pmatrix} 8 \\ 6 \end{pmatrix} = \begin{pmatrix} 0 & 1 \\ 1 & 0 \end{pmatrix} \begin{pmatrix} 6 \\ 8 \end{pmatrix}$$

In this section we analyze the *problem of describing a given vector in terms of different bases.* This type of problem is of interest in mechanics. Consider a guy wire, as shown in Fig. 3.8.3. The vector $\mathbf{F}$ represents a force acting horizontally on the wire. In terms of the standard basis (with x_1 and x_2 axes as shown), $\mathbf{F} = 2(1, 0) + 0(0, 1)$. Of interest are the normal and tangential forces, $\mathbf{T}$ and $\mathbf{N}$, on the wire. In terms of the basis $\{\mathbf{U}_1, \mathbf{U}_2\} = \{(1/\sqrt{2}, -1/\sqrt{2}), (1/\sqrt{2}, 1/\sqrt{2})\}$, however, $\mathbf{F} = \sqrt{2}\mathbf{U}_1 + \sqrt{2}\mathbf{U}_2$. That is, the tangential and normal components are both $\sqrt{2}$.

DEFINITION 3.8.1 An *ordered basis* is a basis $S = \{\mathbf{v}_1, \mathbf{v}_2, \ldots, \mathbf{v}_n\}$ of vectors along with an ordering[2] of the vectors in the basis.

DEFINITION 3.8.2 Let $S = \{\mathbf{v}_1, \mathbf{v}_2, \ldots, \mathbf{v}_n\}$ be an ordered basis for a vector space V, and let $\mathbf{x} \in V$. If

$$\mathbf{x} = c_1\mathbf{v}_1 + c_2\mathbf{v}_2 + \cdots + c_n\mathbf{v}_n \tag{3.8.1}$$

then the coefficients $c_1, c_2, \ldots, c_n$ are called the *coordinates of* $\mathbf{x}$ *with respect to* S. The matrix

$$(\mathbf{x})_S = \begin{pmatrix} c_1 \\ c_2 \\ \vdots \\ c_n \end{pmatrix}$$

is called the *coordinate matrix of* x *with respect to* S.

A coordinate matrix of x for a given ordered basis is unique, because the linear combination in Eq. (3.8.1) is unique (by Theorem 3.5.1).

EXAMPLE 1 Let $\mathbf{x} = (2, 3)$ in E^2. Find the coordinate matrix for $\mathbf{x}$ with respect to (*a*) $S =$ standard basis, (*b*) $S = \{(1, 1), (1, -1)\}$, (*c*) $S = \{(1, 0), (1, 1)\}$ and (*d*) $S = \{(1, 1), (1, 0)\}$.

Solution (*a*) Since $(2, 3) = 2(1, 0) + 3(0, 1)$,

$$((2, 3))_S = \begin{pmatrix} 2 \\ 3 \end{pmatrix}$$

(*b*) Since $\{(1, 1), (1, -1)\}$ is an orthogonal basis,

$$(2, 3) = \frac{\langle (2, 3), (1, 1) \rangle}{\|(1, 1)\|^2}(1, 1) + \frac{\langle (2, 3), (1, -1) \rangle}{\|(1, -1)\|^2}(1, -1)$$

(see Prob. 22 of Sec. 3.7). Thus

$$((2, 3))_S = \begin{pmatrix} \frac{5}{2} \\ -\frac{1}{2} \end{pmatrix}$$

(*c*) We solve $(2, 3) = c_1(1, 0) + c_2(1, 1)$ to find $c_1 = -1$ and $c_2 = 3$, so

$$((2, 3))_S = \begin{pmatrix} -1 \\ 3 \end{pmatrix}$$

(*d*) In this case $(2, 3) = c_1(1, 1) + c_2(1, 0)$. Solving, we find $c_1 = 3$ and $c_2 = -1$.

---

[2] An ordering of a set S of n objects is a function from the set $\{1, 2, \ldots, n\}$ onto S. It is simply a rule for telling which object of S is to be called the first, the second, and so on.

Therefore

$$((2, 3))_S = \begin{pmatrix} 3 \\ -1 \end{pmatrix}$$

Notice in (*c*) and (*d*) that the reversal of the order of the basis elements resulted in a reversal of the elements of the coordinate matrix.

EXAMPLE 2 $S = \{(1, 0, 1), (0, 1, 1), (1, 2, 4)\}$ is a basis for E^3. Suppose

$$(\mathbf{x})_S = \begin{pmatrix} -1 \\ 2 \\ 4 \end{pmatrix}$$

Find $\mathbf{x}$.

Solution In terms of the basis S,

$$\begin{pmatrix} -1 \\ 2 \\ 4 \end{pmatrix} \to -1(1, 0, 1) + 2(0, 1, 1) + 4(1, 2, 4)$$
$$= (3, 10, 17)$$

The arrow simply replaces the word "represents."

From Example 1 we see that if we change the basis for a vector space V, then the coordinate matrix for a vector changes. This poses the problem:

> Given a vector space V with basis S, we can calculate $(\mathbf{x})_S$. If we give V a new basis T, can we calculate $(\mathbf{x})_T$ by using $(\mathbf{x})_S$? That is, is there a simple relationship between $(\mathbf{x})_S$ and $(\mathbf{x})_T$?

Let us consider the problem in E^3. Let $S = \{\mathbf{v}_1, \mathbf{v}_2, \mathbf{v}_3\}$ and $T = \{\mathbf{w}_1, \mathbf{w}_2, \mathbf{w}_3\}$ be two bases for E^3. Suppose $\mathbf{x} \in E^3$ and

$$(\mathbf{x})_S = \begin{pmatrix} c_1 \\ c_2 \\ c_3 \end{pmatrix}$$

This is,

$$\mathbf{x} = c_1\mathbf{v}_1 + c_2\mathbf{v}_2 + c_3\mathbf{v}_3 \tag{3.8.2}$$

Now we want to "convert the $\mathbf{v}$'s to $\mathbf{w}$'s." Since T is a basis, there are coefficients $a_{11}, a_{12}, \ldots, a_{33}$ such that

$$\begin{aligned} \mathbf{v}_1 &= a_{11}\mathbf{w}_1 + a_{21}\mathbf{w}_2 + a_{31}\mathbf{w}_3 \\ \mathbf{v}_2 &= a_{12}\mathbf{w}_1 + a_{22}\mathbf{w}_2 + a_{32}\mathbf{w}_3 \\ \mathbf{v}_3 &= a_{13}\mathbf{w}_1 + a_{23}\mathbf{w}_2 + a_{33}\mathbf{w}_3 \end{aligned} \tag{3.8.3}$$

Substituting these into (3.8.2), we find

$$\mathbf{x} = (a_{11}c_1 + a_{12}c_2 + a_{13}c_3)\mathbf{w}_1 + (a_{21}c_1 + a_{22}c_2 + a_{23}c_3)\mathbf{w}_2 + (a_{31}c_1 + a_{32}c_2 + a_{33}c_3)\mathbf{w}_3$$

and so

$$(\mathbf{x})_T = \begin{pmatrix} a_{11}c_1 + a_{12}c_2 + a_{13}c_3 \\ a_{21}c_1 + a_{22}c_2 + a_{23}c_3 \\ a_{31}c_1 + a_{32}c_2 + a_{33}c_3 \end{pmatrix} = \begin{pmatrix} a_{11} & a_{12} & a_{13} \\ a_{21} & a_{22} & a_{23} \\ a_{31} & a_{32} & a_{33} \end{pmatrix} \begin{pmatrix} c_1 \\ c_2 \\ c_3 \end{pmatrix}$$

By Eq. (3.8.3) $= ((\mathbf{v}_1)_T(\mathbf{v}_2)_T(\mathbf{v}_3)_T)(\mathbf{x})_S$

Therefore,

$$(\mathbf{x})_T = ((\mathbf{v}_1)_T(\mathbf{v}_2)_T(\mathbf{v}_3)_T)(\mathbf{x})_S$$

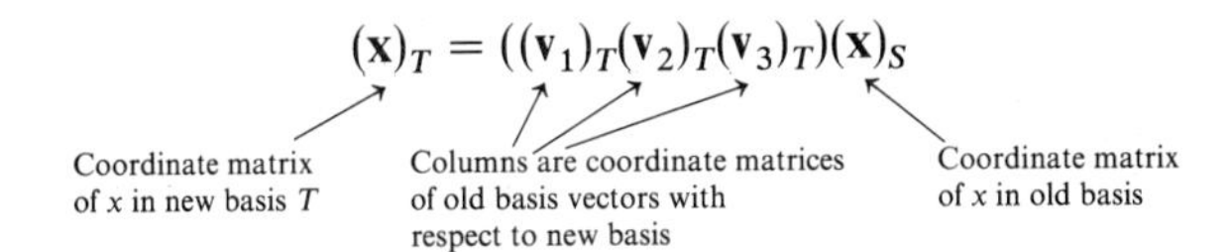

In an n-dimensional vector space V, if $S = \{\mathbf{v}_1, \ldots, \mathbf{v}_n\}$ is the old basis and T is the new basis, then

$$(\mathbf{x})_T = ((\mathbf{v}_1)_T(\mathbf{v}_2)_T \cdots (\mathbf{v}_n)_T)(\mathbf{x})_S \tag{3.8.4}$$

This can be derived in the same way as in the E^3 case. The matrix in Eq. (3.8.4) is called the *transition matrix from S to T*, and it is customarily denoted P. Thus we have $(\mathbf{x})_T = P(\mathbf{x})_S$. Occasionally we write $P_{T \leftarrow S}$ when it is necessary to emphasize the fact that P is the transition matrix *from S to T*.

These are two important facts about transition matrices:

1. Matrix P is invertible, and P^{-1} is the transition matrix from T to S.

2. If S and T are orthonormal bases, then P is an orthogonal matrix.

The proofs of these facts are outlined in the problems.

EXAMPLE 3 Consider E^2 and bases $S = \{(1, 1), (1, -1)\}$ and $T = \{(1, 0), (1, 1)\}$.

(*a*) Find the transition matrix from basis S to basis T.

(*b*) Find the transition matrix from basis T to basis S.

(*c*) Show that the matrices found in (*a*) and (*b*) are inverses of each other.

(*d*) Find $((3, -5))_T$.

(*e*) Find $((3, -5))_T$ by using the transition matrix from (*a*).

Solution (*a*)

$$P = ((\mathbf{v}_1)_T(\mathbf{v}_2)_T) = (((1, 1))_T((1, -1))_T).$$

Since

$$\mathbf{v}_1 = (1, 1) = 0(1, 0) + 1(1, 1)$$

and

$$\mathbf{v}_2 = (1, -1) = 2(1, 0) - 1(1, 1)$$

we have

$$P_{T \leftarrow S} = \begin{pmatrix} 0 & 2 \\ 1 & -1 \end{pmatrix}$$

(*b*) Let $Q = ((\mathbf{w}_1)_S(\mathbf{w}_2)_S) = (((1, 0))_S((1, 1))_S)$. Since

$$\mathbf{w}_1 = (1, 0) = \tfrac{1}{2}(1, 1) + \tfrac{1}{2}(1, -1)$$

and

$$\mathbf{w}_2 = (1, 1) = 1(1, 1) + 0(1, -1)$$

we have

$$Q_{S \leftarrow T} = \begin{pmatrix} \frac{1}{2} & 1 \\ \frac{1}{2} & 0 \end{pmatrix}$$

(*c*) Since

$$PQ = \begin{pmatrix} 0 & 2 \\ 1 & -1 \end{pmatrix}\begin{pmatrix} \frac{1}{2} & 1 \\ \frac{1}{2} & 0 \end{pmatrix} = \begin{pmatrix} 1 & 0 \\ 0 & 1 \end{pmatrix}$$

we know that $P^{-1} = Q$ and $Q^{-1} = P$.

(*d*) Now $(3, -5) = -1(1, 1) + 4(1, -1)$, so

$$((3, -5))_S = \begin{pmatrix} -1 \\ 4 \end{pmatrix}$$

(*e*) Now

$$((3, -5))_T = P((3, -5))_S = \begin{pmatrix} 0 & 2 \\ 1 & -1 \end{pmatrix}\begin{pmatrix} -1 \\ 4 \end{pmatrix} = \begin{pmatrix} 8 \\ -5 \end{pmatrix}$$

We can check our answer by computing

$$8\mathbf{w}_1 - 5\mathbf{w}_2 = 8(1, 0) - 5(1, 1) = (3, -5)$$

$$\begin{pmatrix} 8 \\ -5 \end{pmatrix}$$

EXAMPLE 4 Consider E^3 with orthonormal bases (in the standard inner product)

$$S = \{(1/\sqrt{2}, -1/\sqrt{2}, 0), (1/\sqrt{2}, 1/\sqrt{2}, 0)(0, 0, 1)\}$$
$$T = \{(1/\sqrt{2}, 0, 1/\sqrt{2}), (0, 1, 0), (-1/\sqrt{2}, 0, 1/\sqrt{2})\}$$

Show that the transition matrix P from S to T is orthogonal. Write the transition matrix Q from T to S.

Solution Using inner products

$$(v_1)_T = \begin{pmatrix} \langle \mathbf{v}_1, \mathbf{w}_1 \rangle \\ \langle \mathbf{v}_1, \mathbf{w}_2 \rangle \\ \langle \mathbf{v}_1, \mathbf{w}_3 \rangle \end{pmatrix} = \begin{pmatrix} \frac{1}{2} \\ -1/\sqrt{2} \\ -\frac{1}{2} \end{pmatrix}$$

$$(v_2)_T = \begin{pmatrix} \langle \mathbf{v}_2, \mathbf{w}_1 \rangle \\ \langle \mathbf{v}_2, \mathbf{w}_2 \rangle \\ \langle \mathbf{v}_2, \mathbf{w}_3 \rangle \end{pmatrix} = \begin{pmatrix} \frac{1}{2} \\ 1/\sqrt{2} \\ -\frac{1}{2} \end{pmatrix}$$

$$(v_3)_T = \begin{pmatrix} \langle \mathbf{v}_3, \mathbf{w}_1 \rangle \\ \langle \mathbf{v}_3, \mathbf{w}_2 \rangle \\ \langle \mathbf{v}_3, \mathbf{w}_3 \rangle \end{pmatrix} = \begin{pmatrix} 1/\sqrt{2} \\ 0 \\ 1/\sqrt{2} \end{pmatrix}$$

we know that

$$P = \begin{pmatrix} \frac{1}{2} & \frac{1}{2} & 1/\sqrt{2} \\ -1/\sqrt{2} & 1/\sqrt{2} & 0 \\ -\frac{1}{2} & -\frac{1}{2} & 1/\sqrt{2} \end{pmatrix}$$

Matrix P is orthogonal since its columns are orthogonal and have norm 1. Since P^{-1} is the transition matrix from T to S, we know that $Q = P^{-1}$. However, P is orthogonal so $P^{-1} = P^T$ and

$$Q = P^T = \begin{pmatrix} \frac{1}{2} & -1/\sqrt{2} & -\frac{1}{2} \\ \frac{1}{2} & 1/\sqrt{2} & -\frac{1}{2} \\ 1/\sqrt{2} & 0 & 1/\sqrt{2} \end{pmatrix}$$

EXAMPLE 5 In Example 4, if the first and second vectors in S are interchanged, how is P changed?

Solution Examining the solution to Example 4, we see that the roles of $\mathbf{v}_1$ and $\mathbf{v}_2$ are reversed. So the transition matrix is

$$P = \begin{pmatrix} \frac{1}{2} & \frac{1}{2} & 1/\sqrt{2} \\ 1/\sqrt{2} & -1/\sqrt{2} & 0 \\ -\frac{1}{2} & -\frac{1}{2} & 1/\sqrt{2} \end{pmatrix}$$

That is, the first and second columns are interchanged.

EXAMPLE 6 In Example 4, if the first and second vectors in T are interchanged, how is P changed?

Solution Examining the solution of Example 4, we see that since the roles of $\mathbf{w}_1$ and $\mathbf{w}_2$ are reversed, the first two entries in each coordinate matrix $(\mathbf{v}_1)_T$, $(\mathbf{v}_2)_T$, and $(\mathbf{v}_3)_T$ are reversed. Therefore the transition matrix is

$$\begin{pmatrix} -1/\sqrt{2} & 1/\sqrt{2} & 0 \\ \frac{1}{2} & \frac{1}{2} & 1/\sqrt{2} \\ -\frac{1}{2} & -\frac{1}{2} & 1/\sqrt{2} \end{pmatrix}$$

That is, the first and second *rows* are interchanged.

Examples 5 and 6 illustrate the general principle that if *vectors in basis S are interchanged*, then *corresponding columns in $P_{T \leftarrow S}$ are interchanged.* If *vectors in basis T are interchanged*, then *corresponding rows in $P_{T \leftarrow S}$ are interchanged.*

EXAMPLE 7 Consider the basis $T = \{(-1/\sqrt{2}, 1/\sqrt{2})(-1/\sqrt{2}, -1/\sqrt{2})\}$ in E^2 obtained from the basis $S = \{(0, 1), (-1, 0)\}$ by rotating the vectors in S by $\pi/4$ radians (Fig. 3.8.4). Find the transition matrix from S to T.

Solution Since the bases are orthonormal, we can calculate the coordinate matrices by using inner products:

$$(\mathbf{v}_1)_T = ((0, 1))_T = \begin{pmatrix} 1/\sqrt{2} \\ -1/\sqrt{2} \end{pmatrix}$$

$$(\mathbf{v}_2)_T = ((-1, 0))_T = \begin{pmatrix} 1/\sqrt{2} \\ 1/\sqrt{2} \end{pmatrix}$$

Therefore

$$P = \begin{pmatrix} 1/\sqrt{2} & 1/\sqrt{2} \\ -1/\sqrt{2} & 1/\sqrt{2} \end{pmatrix}$$

Figure 3.8.4

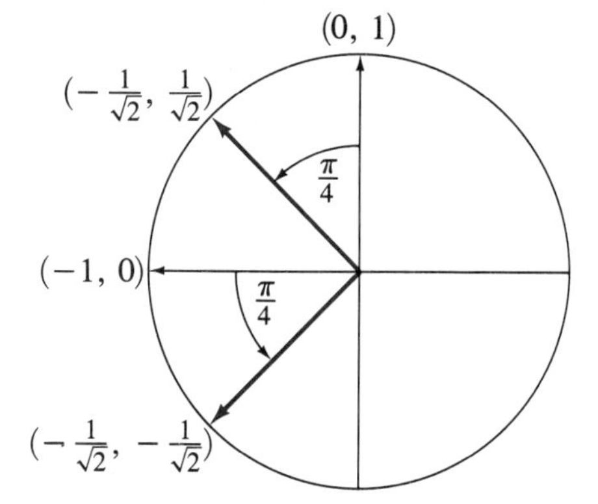

Incidentally P^T is

$$\begin{pmatrix} \cos \pi/4 & -\sin \pi/4 \\ \sin \pi/4 & \cos \pi/4 \end{pmatrix}$$

Example 7 illustrates the general form of a rotation as discussed in Chap. 2. In general, rotations have orthogonal transition matrices.

EXAMPLE 8 Recall the example of the guy wire in Fig. 3.8.3. Find the transition matrix from the standard basis for E^2.

Solution We have

$$(1, 0) = \frac{1}{\sqrt{2}}\mathbf{U}_1 + \frac{1}{\sqrt{2}}\mathbf{U}_2 \quad \text{and} \quad (0, 1) = \frac{1}{\sqrt{2}}\mathbf{U}_1 - \frac{1}{\sqrt{2}}\mathbf{U}_2$$

so that

$$P = \begin{pmatrix} 1/\sqrt{2} & 1/\sqrt{2} \\ 1/\sqrt{2} & -1/\sqrt{2} \end{pmatrix}$$

Thus, if $\mathbf{F} = (a, b)$, its normal and tangential components are given by

$$P\begin{pmatrix} a \\ b \end{pmatrix} = \begin{pmatrix} (1/\sqrt{2})(a + b) \\ (1/\sqrt{2})(a - b) \end{pmatrix} \begin{matrix} \longleftarrow \text{Normal} \\ \longleftarrow \text{Tangential} \end{matrix}$$

PROBLEMS 3.8

In Probs. 1 to 4, let $S = \{(1, 3), (-2, 1)\}$, $T = \{(1/\sqrt{2}, -1/\sqrt{2}), (1/\sqrt{2}, 1/\sqrt{2})\}$, $U = \{(-2, 1), (1, 3)\}$, and $Z = \{(\sqrt{3}/2, \frac{1}{2}), (-\frac{1}{2}, \sqrt{3}/2)\}$ be bases for E^2 with the standard inner product.

1. Find the transition matrices (*a*) from S to T, (*b*) from T to U, and (*c*) from S to U. Multiply the matrices from (*a*) and (*b*) in both ways. Is either product related to the matrix from (*c*)?

2. Let $\mathbf{x} = (1, 1)$. Find $(\mathbf{x})_S$. Using the matrices from Prob. 1, find $(\mathbf{x})_T$ and $(\mathbf{x})_U$.

3. Find the transition matrices from T to Z and from Z to T.

4. (*a*) Find the transition matrix from S to Z. (*b*) Find the transition matrix from U to Z by using interchanges in the matrix from (*a*).

In Probs. 5 to 10, let $S = \{(1, 1, 0), (0, 1, 1), (1, 0, 1)\}$, $T = \{(1/\sqrt{2}, -1/\sqrt{2}, 0), (1/\sqrt{2}, 1/\sqrt{2}, 0), (0, 0, 1)\}$, $U = \{(1, 0, 1), (0, 1, 1), (1, 1, 0)\}$, and $Z = \{(1, 0, 0), (0, 1/\sqrt{2}, -1/\sqrt{2}), (0, 1/\sqrt{2}, 1/\sqrt{2})\}$ be bases for E^3 with the standard inner product. Let $\mathbf{x} = (1, -1, 2)$.

5. Find the transition matrices (*a*) from S to T, (*b*) from T to U, and (*c*) from S to U. Multiply the matrices from (*a*) and (*b*) in both ways. Is either product related to the matrix from (*c*)?

6. Find $(\mathbf{x})_S$. Using the matrices from Prob. 1, find $(\mathbf{x})_T$ and $(\mathbf{x})_U$.

7. Find the transition matrices from T to Z and from Z to T.

8. (*a*) Find the transition matrix from S to Z. (*b*) Find the transition matrix from U to Z by using interchanges in the matrix from (*a*).

9. If

$$(\mathbf{x})_S = \begin{pmatrix} 3 \\ -2 \\ 4 \end{pmatrix}$$

what is $\mathbf{x}$?

10. If

$$(\mathbf{x})_T = \begin{pmatrix} 0 \\ 0 \\ 0 \end{pmatrix}$$

what is $\mathbf{x}$?

11. If S is any basis for V and

$$(\mathbf{x})_S = \begin{pmatrix} 0 \\ 0 \\ \vdots \\ 0 \end{pmatrix}$$

what is $\mathbf{x}$?

12. If S is any basis for V and $\mathbf{x} = \boldsymbol{\theta}$, what is $(\mathbf{x})_S$?

13. Show that if P is the transition matrix from S to T in the vector space V, then P^{-1} exists and P^{-1} is the transition matrix from T to S, by completing the following steps.

(*a*) Let $S = \{\mathbf{v}_1, \ldots, \mathbf{v}_n\}$ and $T = \{\mathbf{w}_1, \ldots, \mathbf{w}_n\}$ and Q be the transition matrix from T to S. Write QP as

$$QP = \begin{pmatrix} a_{11} & a_{12} & \cdots & a_{1n} \\ a_{21} & a_{22} & \cdots & a_{2n} \\ \cdots & \cdots & \cdots & \cdots \\ a_{n1} & a_{n2} & \cdots & a_{nn} \end{pmatrix}$$

(*b*) Multiply $(\mathbf{x})_T = P(\mathbf{x})_S$ by Q.

(*c*) Substitute the result from (*b*) into $(\mathbf{x})_S = Q(\mathbf{x})_T$ to obtain

$$(\mathbf{x})_S = QP(\mathbf{x})_S \tag{1}$$

(*d*) Let $\mathbf{x} = \mathbf{v}_1$ in Eq. (1), and show

$$\begin{pmatrix} 1 \\ 0 \\ \vdots \\ 0 \end{pmatrix} = \begin{pmatrix} a_{11} \\ a_{21} \\ \vdots \\ a_{n1} \end{pmatrix}$$

(*e*) Let $\mathbf{x} = \mathbf{v}_k$ in Eq. (1), and show

$$k\text{th place} \longrightarrow \begin{pmatrix} 0 \\ \vdots \\ 0 \\ 1 \\ 0 \\ \vdots \\ 0 \end{pmatrix} = \begin{pmatrix} a_{1k} \\ \vdots \\ a_{kk} \\ \vdots \\ a_{nk} \end{pmatrix}$$

(*f*) Conclude that $QP = I$.

14. Show that if $S = \{\mathbf{v}_1, \ldots, \mathbf{v}_n\}$ and $T = \{\mathbf{w}_1, \ldots, \mathbf{w}_n\}$ are orthonormal bases of a real vector space V, then the transition matrix from S to T is orthogonal by completing the following steps.

(*a*) Show that the transition matrix from S to T is

$$P = \begin{pmatrix} \langle \mathbf{v}_1, \mathbf{w}_1 \rangle & \langle \mathbf{v}_2, \mathbf{w}_1 \rangle & \cdots & \langle \mathbf{v}_n, \mathbf{w}_1 \rangle \\ \langle \mathbf{v}_1, \mathbf{w}_2 \rangle & \langle \mathbf{v}_2, \mathbf{w}_2 \rangle & \cdots & \langle \mathbf{v}_n, \mathbf{w}_2 \rangle \\ \cdots & \cdots & \cdots & \cdots \\ \langle \mathbf{v}_1, \mathbf{w}_n \rangle & \langle \mathbf{v}_2, \mathbf{w}_n \rangle & \cdots & \langle \mathbf{v}_n, \mathbf{w}_n \rangle \end{pmatrix}$$

(*b*) Show that the transition matrix from T to S is

$$Q = \begin{pmatrix} \langle \mathbf{w}_1, \mathbf{v}_1 \rangle & \langle \mathbf{w}_2, \mathbf{v}_1 \rangle & \cdots & \langle \mathbf{w}_n, \mathbf{v}_1 \rangle \\ \langle \mathbf{w}_1, \mathbf{v}_2 \rangle & \langle \mathbf{w}_2, \mathbf{v}_2 \rangle & \cdots & \langle \mathbf{w}_n, \mathbf{v}_2 \rangle \\ \cdots & \cdots & \cdots & \cdots \\ \langle \mathbf{w}_1, \mathbf{v}_n \rangle & \langle \mathbf{w}_2, \mathbf{v}_n \rangle & \cdots & \langle \mathbf{w}_n, \mathbf{v}_n \rangle \end{pmatrix}$$

(*c*) From Prob. 13 we know that $Q = P^{-1}$. Use (*a*) and (*b*) to show that $Q = P^T$.

(*d*) Conclude that $P^{-1} = P^T$ and so P is orthogonal.

15. Show that if P is the transition matrix from S to T, and Q is the transition matrix from T to U, where S, T, and U are bases for V, then QP is the transition matrix from S to U.

16. Let $S = \{1 + x, 1 + x^2, 1 - x^2\}$ and $T = \{1, 1 + x^2, 1 + x\}$ be bases for P_2. Find the transition matrix from S to T.

17. Show that a transition matrix for an n-dimensional vector space V is $n \times n$.

18. In $\mathbb{C}^2$ let $S = \{(i, 0), (0, i)\}$ and $T = \{(1, 0), (0, 1)\}$. Calculate the transition matrices from S to T and from T to S.

19. Consider the guy wire as shown. Compute the transition matrix from the standard basis for E^2 to $\{\mathbf{U}_1, \mathbf{U}_2\}$, where $\mathbf{U}_1$ and $\mathbf{U}_2$ have unit length.

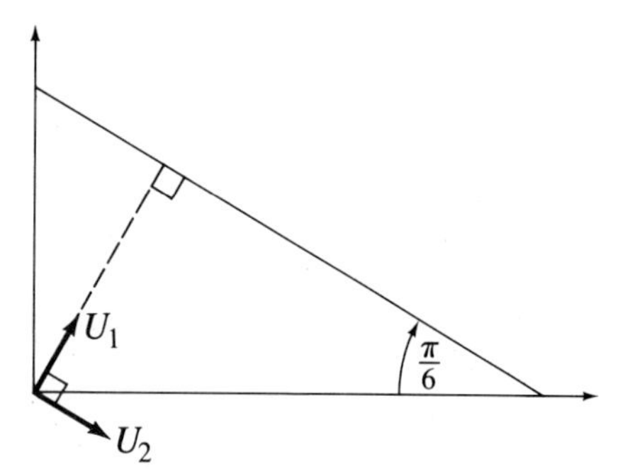

Figure P3.8.19

20. Consider the guy wire at angle α as shown. Compute the transition matrix from the standard basis for E^2 to $\{\mathbf{U}_1, \mathbf{U}_2\}$, where $\mathbf{U}_1$ and $\mathbf{U}_2$ have unit length.

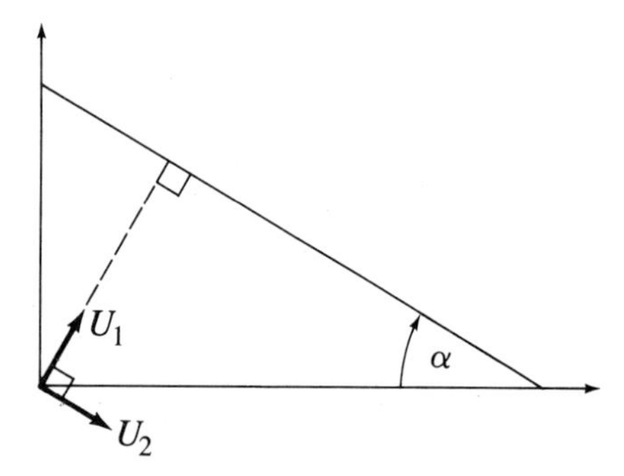

Figure P3.8.20

3.9 CALCULUS REVISITED

Vector spaces, bases, and orthogonality can be used to discuss many topics from calculus. We mention a few here.

Vector Spaces

Recall the theorem about sums and multiples of continuous functions: If f and g are continuous on an interval $[a, b]$ and c is any real number, then

$$f + g \text{ is a continuous function on } [a, b]$$

and

$$cf \text{ is a continuous function on } [a, b]$$

This theorem says that the set of functions continuous on $[a, b]$ is closed under sums and constant multiples. Thus we have a situation which may yield a vector space. With this in mind, we define $C[a, b]$ as the set of functions continuous on $[a, b]$ along with addition of functions and constant multiplication of functions defined in the usual way. To see whether $C[a, b]$ is a vector space, we must check the vector space axioms. Closure holds by the theorem stated above. The properties

$$f + g = g + f \qquad (f + g) + h = f + (g + h) \qquad c(f + g) = cf + cg$$
$$(b + c)f = bf + cf \qquad 1f = f \qquad (bc)f = b(cf)$$

are easily checked. The zero vector is the constant function (it is continuous) which is identically zero on $[a, b]$. Given f, the function $-f$ is continuous and is the additive inverse of f. Therefore, $C[a, b]$ is a vector space.

Differentiable functions are another important set of functions in calculus. If we define $C^1[a, b]$ as the set of functions f defined on $[a, b]$ with $f' \in C[a, b]$ (that is, f' is continuous) and give $C^1[a, b]$ the same operations as $C[a, b]$, then $C^1[a, b]$ is a subspace of $C[a, b]$. To see this, we need only check closure. Since a theorem from calculus tells us that the sum and constant multiples of differentiable functions are differentiable, we have the necessary closure. So $C^1[a, b]$ is a vector space in its own right as well as a subspace of $C[a, b]$.

EXAMPLE 1 Give an example of a function in $C[-1, 1]$ which is not in $C^1[-1, 1]$. This shows that $C^1[-1, 1]$ is a proper subspace of $C[-1, 1]$.

Solution A continuous function which is not differentiable will be sufficient. The function $y = |x|$ is continuous on $[-1, 1]$ but fails to have a derivative at $x = 0$.

Several functions from calculus are differentiable an infinite number of times. For example, $y = e^x$, $y = \sin x$, $y = \cos x$, and all polynomials are infinitely differentiable over all $\mathbb{R}$. Thus we can define vector spaces

$$C^2[a, b], C^3[a, b], \ldots, C^n[a, b], \ldots$$

and even

$$C^\infty[a, b]$$

DEFINITION 3.9.1 $C^n[a, b]$ is the set of all functions f defined on $[a, b]$ with $f, f', f'', \ldots, f^{(n)}$ being continuous on $[a, b]$. (Right- and left-hand derivatives are used at a and b.) With the standard operations of addition and multiplication by a constant,

$C^n[a, b]$ is a vector space. $C^\infty[a, b]$ is the set of all f in $C^n[a, b]$ for all n. $C^\infty[a, b]$ is a vector space.

EXAMPLE 2 Give an example of a function which is in $C^1[-1, 1]$ but is not in $C^2[-1, 1]$.

Solution Let $y = x^{5/3}$. Then $y' = \frac{5}{3}x^{2/3}$ and $y'' = \frac{10}{9}x^{-1/3}$. Although y' is continuous on $[-1, 1]$, y'' is not, because $y''(0)$ is undefined.

Examples 1 and 2 illustrate the subspace relationship

$$C[a, b] \supset C^1[a, b] \supset C^2[a, b] \supset \cdots \supset C^\infty[a, b]$$

among these vector spaces.

The vector space $C[a, b]$ differs from the vector spaces we have been studying in that $C[a, b]$ is *not finite-dimensional.* To see this, suppose that $\dim(C[a, b]) = n$. Then a finite basis $S = \{f_1(x), f_2(x), \ldots, f_n(x)\}$ of continuous functions in $C[a, b]$ would exist, and any set of $n + 1$ functions would have to be linearly dependent. However, $T = \{1, x, x^2, \ldots, x^n\}$ is a set of $n + 1$ functions in $C[a, b]$ which are linearly independent. So it is impossible to have $\dim(C[a, b]) = n$ for any n. Since polynomials are infinitely differentiable, the same argument shows that $C^k[a, b]$ is not finite-dimensional for any k.

Although $C[a, b]$ is not finite-dimensional, there are important finite-dimensional subspaces of $C[a, b]$.

EXAMPLE 3 Show that the set of all solutions to the equation[3] $y' = \alpha y$ is a one-dimensional subspace of $C(\mathbb{R})$.

Solution We recall that the functions[4] satisfying $y' = \alpha y$ are $y(x) = Ce^{\alpha x}$, where C is arbitrary. That is, $W = \{y \mid y(x) = Ce^{\alpha x}\}$. To show that W is a subspace of $C(\mathbb{R})$, we first note that since e^x is a continuous function, $W \subseteq C(\mathbb{R})$. Now closure of W under the operations must be shown. Since $c_1e^{\alpha x} + c_2e^{\alpha x} = (c_1 + c_2)e^{\alpha x} = c_3e^{\alpha x}$ and $c_1(c_2e^{\alpha x}) = (c_1c_2)e^{\alpha x}$, we have the closure, so W is a subspace of $C(\mathbb{R})$. A basis for W is $S = \{e^{\alpha x}\}$. Thus $\dim W = 1$.

EXAMPLE 4 Show that the set of all solutions to the equation $y'' = -\alpha^2 y$ is a two-dimensional subspace of $C(\mathbb{R})$.

Solution The solutions of the equation are of the form $A \cos \alpha x + B \sin \alpha x$. Showing

---

[3] Recall that the equation $y' = \alpha y$ is important in the study of population growth, radioactive decay, and other areas.

[4] That these are all the functions satisfying $y' = \alpha y$ is a theorem of ordinary differential equations.

that $W = \{y \mid y(x) = A \cos \alpha x + B \sin \alpha x\}$ is a subspace is done in the same way as Example 3. A basis for W is $S = \{\cos \alpha x, \sin \alpha x\}$. That S spans W is easy. For the linear independence consider $c_1 \cos \alpha x + c_2 \sin \alpha x = 0$. The equation must hold for all x. Substitute $x = 0$ to find $c_1 = 0$. Substitute $x = \pi/(2\alpha)$ to obtain $c_2 = 0$.

Linear Independence

Linear independence of a set of functions in $C[a, b]$ is not always easy to show.

EXAMPLE 5 Consider the set $S = \{\sin 2x, \sin x \cos x\}$ in $C(\mathbb{R})$. Is S a linearly independent set?

Solution We could look at

$$c_1 \sin 2x + c_2 \sin x \cos x = \boldsymbol{\theta} = 0 \tag{3.9.1}$$

as in the definition. The difficulty here is that since the vectors are functions, we do not obtain a set of linear equations with constant coefficients. The coefficients of the linear equation involve the variable x, and the equation must hold for all x. In this example, we rely on our knowledge of trigonometric identities and recall that

$$\tfrac{1}{2} \sin 2x = \sin x \cos x$$

So the set is linearly dependent.

There is another method available if the functions in a set of functions are differentiable enough times, that is, the method of the wronskian. For two functions the basic result is as follows:

Functions f and g in $C^1[a, b]$ are linearly independent if

$$\det \begin{pmatrix} f(x) & g(x) \\ f'(x) & g'(x) \end{pmatrix} \neq 0 \qquad \text{for some } x \text{ in } [a, b]$$

The determinant is called the *wronskian* of f and g and is denoted $W(f, g)$.

EXAMPLE 6 Show that $S = \{\sin x, \cos x\}$ is linearly independent in $C^1(\mathbb{R})$.

Solution We have

$$W(\sin x, \cos x) = \det\begin{pmatrix} \sin x & \cos x \\ \cos x & -\sin x \end{pmatrix} = -\sin^2 x - \cos^2 x = -1$$

Since $W(\sin x, \cos x) \neq 0$ for all x, set S is linearly independent.

Let us show why $W(f, g) \neq 0$ is sufficient for linear independence. Consider the usual equation in checking for independence

$$c_1 f(x) + c_2 g(x) = 0$$

Now we have, after differentiating the equation, the system

$$\begin{pmatrix} f(x) & g(x) \\ f'(x) & g'(x) \end{pmatrix}\begin{pmatrix} c_1 \\ c_2 \end{pmatrix} = \begin{pmatrix} 0 \\ 0 \end{pmatrix}$$

and by Cramer's rule the system has only the zero solution when

$$\det\begin{pmatrix} f(x) & g(x) \\ f'(x) & g'(x) \end{pmatrix} \neq 0$$

For larger sets of functions we have this result:

The set $\{f_1, f_2, \ldots, f_n\}$ in $C^n[a, b]$ is linearly independent if

$$W(f_1, \ldots, f_n) = \det\begin{pmatrix} f_1 & f_2 & \cdots & f_n \\ f'_1 & f'_2 & \cdots & f'_n \\ f''_1 & f''_2 & \cdots & f''_n \\ \cdots & \cdots & \cdots & \cdots \\ f_1^{(n)} & f_2^{(n)} & \cdots & f_n^{(n)} \end{pmatrix} \neq 0$$

for some x in $[a, b]$.

Orthogonality

An inner product on $C[a, b]$ can be defined by

$$\langle f, g \rangle = \int_a^b f(x)g(x)w(x)\,dx$$

where $w(x)$ is a fixed continuous function, positive on $[a, b]$; it is called a *weight function.* The simplest choice for $w(x)$ is $w(x) \equiv 1$. Once $\langle f, g \rangle$ is defined, we can say, "f is orthogonal to g if $\langle f, g \rangle = 0$" and "The norm of f is $\sqrt{\langle f, f \rangle}$," or we can make any other statement which makes sense in a vector space with inner product. We can even speak of the "angle" between two functions.

EXAMPLE 7 In $C[-1, 1]$, let $\langle f, g \rangle = \int_{-1}^{1} f(x)g(x)\,dx$. Calculate $\langle x, x^2 \rangle$, $\langle 1 + x^2, x - x^2 \rangle$, $\|1 + x\|$, and $\langle x^n, x^m \rangle$, where n is even and m is odd.

Solution

$$\langle x, x^2 \rangle = \int_{-1}^{1} x^3\, dx = \frac{x^4}{4}\bigg|_{-1}^{1} = 0$$

$$\langle 1 + x^2, x - x^2 \rangle = \int_{-1}^{1} (1 + x^2)(x - x^2)\, dx$$

$$= \int_{-1}^{1} (x - x^2 + x^3 - x^4)\, dx = -\tfrac{16}{15}$$

$$\|1 + x\| = \sqrt{\langle 1 + x, 1 + x \rangle} = \sqrt{\int_{-1}^{1} (1 + x)^2\, dx}$$

$$= \sqrt{\frac{(1 + x)^3}{3}\bigg|_{-1}^{1}} = \sqrt{\frac{8}{3}} = \frac{2\sqrt{6}}{3}$$

$$\langle x^n, x^m \rangle = \int_{-1}^{1} x^{n+m}\, dx = \frac{x^{n+m+1}}{n + m + 1}\bigg|_{-1}^{1} = 0$$

since $n + m + 1$ is even. With this inner product x^{2n+1} and x^{2m} are orthogonal.

EXAMPLE 8 In $C[0, 1]$ with inner product $\langle f, g \rangle = \int_0^1 f(x)g(x)\, dx$, find an orthonormal basis for span $\{1, x, x^2\}$.

Solution Since the Gram-Schmidt procedure works for any vector space with an inner product, we use it. We have

$$w_1(x) = 1$$

$$w_2(x) = x - \frac{\langle x, 1 \rangle}{\|1\|^2} 1$$

$$= x - \frac{\int_0^1 x\, dx}{\int_0^1 1\, dx} 1 = x - \frac{1}{2}$$

$$w_3(x) = x^2 - \frac{\langle x^2, x - \frac{1}{2} \rangle}{\|x - \frac{1}{2}\|^2}\left(x - \frac{1}{2}\right) - \frac{\langle x^2, 1 \rangle}{\|1\|^2} 1$$

$$= x^2 - \frac{\frac{1}{12}}{\frac{1}{12}}\left(x - \frac{1}{2}\right) - \frac{1}{3}$$

$$= x^2 - x + \tfrac{1}{6}$$

Finally, we normalize and obtain

$$u_1 = \frac{1}{\|1\|^2} = \frac{1}{\int_0^1 1\, dx} = 1$$

$$u_2 = \frac{x - \frac{1}{2}}{\|x - \frac{1}{2}\|} = \frac{x - \frac{1}{2}}{\sqrt{\frac{1}{12}}} = 2\sqrt{3}(x - \tfrac{1}{2})$$

$$u_3 = \frac{x^2 - x + \frac{1}{6}}{\|x^2 - x + \frac{1}{6}\|} = \frac{x^2 - x + \frac{1}{6}}{\sqrt{\int_0^1 (x^4 - 2x^3 + \frac{4}{3}x^2 - \frac{1}{3}x + \frac{1}{36})\, dx}}$$

$$= 6\sqrt{5}\,(x^2 - x + \tfrac{1}{6})$$

We still have not verified that $\langle f, g\rangle = \int_a^b f(x)g(x)w(x)\,dx$ is an inner product on $C[a, b]$. We do that now, by showing the four axioms for an inner product hold.

1.
$$\langle f, g\rangle = \int_a^b f(x)g(x)w(x)\,dx$$
$$= \int_a^b g(x)f(x)w(x)\,dx = \langle g, f\rangle$$

2.
$$\langle cf, g\rangle = \int_a^b cf(x)g(x)w(x)\,dx$$
$$= c\int_a^b f(x)g(x)w(x)\,dx = c\langle f, g\rangle$$

3.
$$\langle f + h, g\rangle = \int_a^b [f(x) + h(x)]g(x)w(x)\,dx$$
$$= \int_a^b [f(x)g(x)w(x) + h(x)g(x)w(x)]\,dx$$
$$= \int_a^b f(x)g(x)w(x)\,dx + \int_a^b h(x)g(x)w(x)\,dx$$
$$= \langle f, g\rangle + \langle h, g\rangle$$

4. **Positivity**
$$\langle f, f\rangle = \int_a^b [f(x)]^2 w(x)\,dx$$

Since $w > 0$ and $f^2(x) \geq 0$ on $[a, b]$, $\int_a^b [f(x)]^2 w(x)\,dx \geq 0$. Thus $\langle f, f\rangle \geq 0$. The definite integral of a nonnegative continuous function is zero if and only if the function is identically zero. Therefore, $\langle f, f\rangle = 0$ if and only if $f(x) \equiv 0$ on $[a, b]$.

Although the vector space structure of spaces such as $C[a, b]$ is not absolutely necessary in beginning calculus courses, it is an important tool in advanced calculus and advanced engineering mathematics. In particular, $C[-\pi, \pi]$ has an infinite orthonormal basis

$$\mathcal{O} = \left\{\frac{1}{\sqrt{2\pi}}, \frac{1}{\sqrt{\pi}}\sin x, \frac{1}{\sqrt{\pi}}\cos x, \frac{1}{\sqrt{\pi}}\sin 2x, \frac{1}{\sqrt{\pi}}\cos 2x, \ldots\right\}.$$

These wave-type functions can then be used to represent functions in $C[-\pi, \pi]$ in a definite way. This fact is the basis for Fourier analysis, which is successful in analyzing problems in heat conduction, wave propagation, and electrostatics.

PROBLEMS 3.9

1. Let $\langle f, g\rangle = \int_0^1 f(x)g(x)\,dx$ in $C[0, 1]$. For the given pairs of functions calculate $\langle f, g\rangle$.
   (*a*) $f(x) = x$, $g(x) = \sqrt{x}$
   (*b*) $f(x) = \sin \pi x$, $g(x) = \cos \pi x$

(*c*) $f(x) = e^x$, $g(x) = e^{-x}$

(*d*) $f(x) = x^2$, $g(x) = 1 - x$

2. Let $\langle f, g\rangle = \int_0^1 f(x)g(x)x^2\,dx$ in $C[0, 1]$. For the given pairs of functions calculate $\langle f, g\rangle$.

(*a*) $f(x) = 1$, $g(x) = x$

(*b*) $f(x) = 1$, $g(x) = \sqrt{x}$

(*c*) $f(x) = 1$, $g(x) = 1/(x^3 + 1)$

3. For the inner product in Prob. 1, calculate $\|f\|$ for $f(x) = x$.

4. For the inner product in Prob. 2, calculate $\|f\|$ for $f(x) = x$.

5. In $C^1[a, b]$ define

$$\langle f, g\rangle = \int_a^b [f(x)g(x) + f'(x)g'(x)]\,dx$$

Show that this is an inner product.

6. In $C^1[a, b]$ define $\langle f, g\rangle = \int_a^b f'(x)g'(x)\,dx$. Show that this is *not* an inner product.

7. In $C^1(\mathbb{R})$, determine linear independence or dependence of the following sets of functions.

(*a*) $S = \{1, x\}$

(*b*) $S = \{e^x, e^{-x}\}$

(*c*) $S = \{\sin x, 2\sin x\}$

8. Show that $\{e^x, e^{2x}\}$ is a linearly independent set in $C(\mathbb{R})$ by considering $c_1e^x + c_2e^{2x} = 0$ and putting $x = 0$ and $x = 1$.

9. Why can the wronskian not be used to determine the linear dependence or independence of $S = \{x^{2/3}, x^2\}$ in $C^1(\mathbb{R})$? Use the direct method to determine linear independence.

10. Recall the Cauchy-Schwarz and triangle inequalities

$$|\langle \mathbf{x}, \mathbf{y}\rangle| \le \|\mathbf{x}\|\,\|\mathbf{y}\| \qquad \text{and} \qquad \|\mathbf{x} + \mathbf{y}\| \le \|\mathbf{x}\| + \|\mathbf{y}\|$$

Using the inner product $\langle f, g\rangle = \int_0^1 f(x)g(x)\,dx$ in $C[0, 1]$, rewrite the inequalities above, replacing $\mathbf{x}$ and $\mathbf{y}$ by f and g, respectively.

11. Using the inner product in Prob. 1, find the angle between x and x^3. [*Remember*: If $\mathbf{u}$ and $\mathbf{v}$ are vectors, then $\cos\theta = \langle \mathbf{u}, \mathbf{v}\rangle/(\|\mathbf{u}\|\,\|\mathbf{v}\|)$.]

12. Consider $V = \text{span}\,\{1, x, x^2\}$ in $C[-1, 1]$ with inner product $\langle f, g\rangle = \int_{-1}^1 f(x)g(x)\,dx$. Find an orthonormal basis for V. The basis consists of the first three *Legendre polynomials*.

13. Show that the set of solutions to $y'' = y$ forms a subspace of $C(\mathbb{R})$. Show that the dimension is at least two.

14. Consider the subspace W of $C(0, +\infty)$ consisting of all polynomials. Define the inner product $\langle f, g\rangle = \int_0^\infty f(x)g(x)e^{-x}\,dx$. Let $V = \text{span}\,\{1, x\}$. Find an orthonormal basis for V. The basis consists of the first two *Laguerre polynomials.*

SUMMARY

Vector spaces are one of the most important structures in applied mathematics; they were defined and developed in this chapter. Some, such as E^n, are direct generalizations of two- and three-space, while others such as vector spaces of polynomials, functions, or matrices are generated by problems in calculus, differential equations, and applied mathematics. *Regardless of their appearance, finitely generated vector spaces have common attributes* such as the *defining axioms, subspace structure, bases,* and *dimension.*

The *basis problem* was posed as one of the fundamental problems of linear algebra. Its solution becomes extremely important in Chap. 5 when diagonalization of matrices is discussed. At this point its importance lies in analysis of applied problems (analysis of electron spin and resolution of forces into tangential and normal directions are two examples).

The *fundamental theorem* of this chapter guaranteed the *existence of a basis*[5] *for any finitely generated vector space V*. In fact, the proof demonstrated that a basis could be "built" from any set of vectors that spans V. For a given basis, the representation of a vector $\mathbf{x}$ in V is unique, and the coefficients in the linear combination of basis vectors for $\mathbf{x}$ are called the *coordinates of* $\mathbf{x}$ in the basis. The development of these concepts depended heavily on the concepts of *linear dependence* and *linear independence. Orthonormal bases* have some "nice" properties; we exploit them in Chap. 5.

Bases are not unique. If S and T are two bases for a vector space V and $\mathbf{x}$ is in V, then the coordinates of $\mathbf{x}$ in S and in T are related by the *transition matrix*. Transition matrices are used in Chap. 4 to aid in the representations of important functions, called linear transformations.

In calculus, the structure of the real numbers had to be developed before the concept of a function could be defined. Accordingly, because the linear functions we study have domains and ranges which are vector spaces, we are now ready to proceed to the definition of linear transformations in Chap. 4.

ADDITIONAL PROBLEMS

1. Let $V = \{(x, y, z) \text{ in } E^3 \text{ with } 3x - y + z \geq 0\}$. Is V a vector space?

2. Let $V = \{(x, y, z) \text{ in } E^3 \text{ with } x - y + z > 0\}$. Is V a vector space?

3. Let $\{\mathbf{v}_1, \mathbf{v}_2, \ldots, \mathbf{v}_n\}$ be a basis for a complex vector space V. Is $\{i\mathbf{v}_1, i\mathbf{v}_2, \ldots, i\mathbf{v}_n\}$ a basis for V?

4. Let $\{\mathbf{v}_1, \mathbf{v}_2\}$ be a basis for E^2. Let $A_{2\times 2}$ be a real matrix. Is $\{A\mathbf{v}_1, A\mathbf{v}_2\}$ a basis for E^2? Does it make any difference whether A is singular?

---

[5] Where the generating set contains nonzero vectors only.

5. Let $\{\mathbf{v}_1, \mathbf{v}_2, \ldots, \mathbf{v}_n\}$ be a basis for a complex vector space V. Is $\{c_1\mathbf{v}_1, c_2\mathbf{v}_2, \ldots, c_n\mathbf{v}_n\}$ a basis for V? Assume $c_k \neq 0$ for $k = 1, 2, \ldots, n$.

6. If $\{(a, b), (c, d)\}$ is to be an orthonormal basis for E^2, what are c and d in terms of a and b? What must be true about

$$\det\begin{pmatrix} a & c \\ b & d \end{pmatrix}$$

7. Let A be an $m \times n$ matrix, $m \neq n$. Let $P = A(A^TA)^{-1}A^T$. Show that P is symmetric and that $P^2 = P$.

8. Find a first column that will make

$$\begin{pmatrix} & \frac{-1}{\sqrt{6}} & \frac{1}{\sqrt{3}} \\ & \frac{-1}{\sqrt{6}} & \frac{1}{\sqrt{3}} \\ & \frac{2}{\sqrt{6}} & \frac{1}{\sqrt{3}} \end{pmatrix}$$

an orthogonal matrix.

9. Is the set of all orthogonal $n \times n$ matrices a subspace of $\mathscr{M}_{nn}$?

10. Let A be a fixed $n \times n$ matrix. Let V be the set of all matrices of the form $P^{-1}AP$, where P runs through the set of all $n \times n$ invertible matrices. Is V a subspace of $\mathscr{M}_{nn}$?

11. Let V be the set of solutions of $A_{n\times n}X_{n\times 1} = cX_{n\times 1}$, where c is a complex number. Give V the operations of $\mathscr{C}_{n1}$. Is V a vector space?

12. Show that if A and B are $n \times n$ hermitian matrices, then $i(AB - BA)$ is hermitian.

13. Approximation of a given function by other functions is a common tool of applied mathematics. Let the given function be $g(x)$, x in $[a, b]$. To carry out the approximation, one must set up a finite-dimensional vector space V of functions defined on $[a, b]$ and make sure the vector space has an inner product (which generates a norm). The function g must have $\|g\|$ a finite number. Then if $S = \{f_1, f_2, \ldots, f_n\}$ is a basis for V, we try to find constants $c_1, c_2, \ldots, c_n$ so that

$$\|(c_1f_1 + c_2f_2 + \cdots + c_nf_n) - g\|$$

is as small as possible. This process of minimization is simplified if S is an orthonormal basis, for then the best approximation is

$$\langle g, f_1\rangle f_1 + \cdots + \langle g, f_n\rangle f_n$$

Approximate $g(x) = e^x$ by a quadratic polynomial. Use the vector space $\mathscr{P}_2$ with inner product defined by

$$\langle a_1x^2 + a_2x + a_3, b_1x^2 + b_2x + b_3\rangle = a_1b_1 + a_2b_2 + a_3b_3$$

14. Suppose in a real vector space V with basis $S = \{\mathbf{v}_1, \mathbf{v}_2, \ldots, \mathbf{v}_n\}$ we define an inner product for which the basis is not orthogonal. Define $G_{n\times n} = (\langle \mathbf{v}_i, \mathbf{v}_j\rangle)_{n\times n}$. Show that $G \neq I$. Show that G is symmetric. In mathematical physics, G is called the *metric* of the vector space.

15. Regarding Prob. 14, show for vectors $\mathbf{x}$ and $\mathbf{y}$ in V, with $\mathbf{x} = a_1\mathbf{v}_1 + \cdots + a_n\mathbf{v}_n$ and $\mathbf{y} = b_1\mathbf{v}_1 + \cdots + b_n\mathbf{v}_n$, that

$$\langle \mathbf{x}, \mathbf{y}\rangle = (\mathbf{x})_S{}^T G(\mathbf{y})_S$$

16. Use the result of Prob. 15 to show that for any nonzero vector X in E^n

$$X^T G X > 0$$

This means that the metric of a vector space is a *positive definite* matrix.

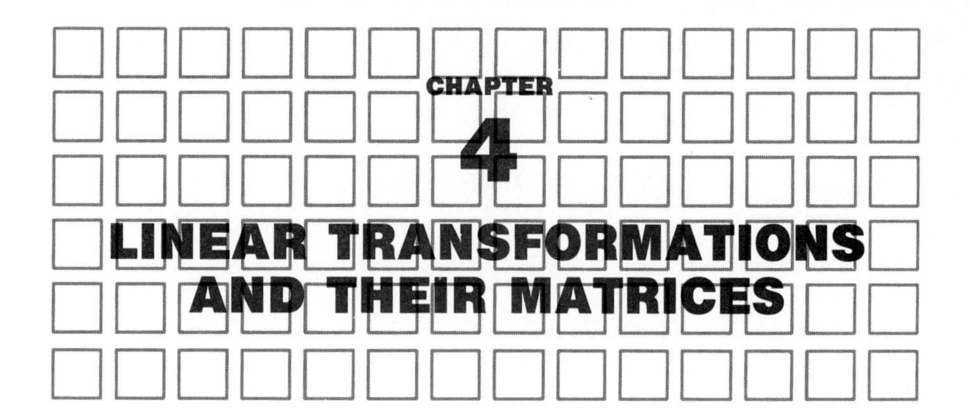

4.1 LINEAR TRANSFORMATIONS

The central objective of linear algebra is the analysis of linear functions defined on a finite-dimensional vector space. For example, analysis of the shear transformation is a problem of this sort. First we define the concept of a linear function or transformation.

DEFINITION 4.1.1 Let V and W be real vector spaces (their dimensions can be different), and let T be a function with domain V and range in W (written $T: V \to W$). We say T is a *linear transformation* if

(*a*) For all $\mathbf{x}, \mathbf{y} \in V$, $T(\mathbf{x} + \mathbf{y}) = T(\mathbf{x}) + T(\mathbf{y})$($T$ is *additive*).

(*b*) For all $\mathbf{x} \in V$, $r \in \mathbb{R}$, $T(r\mathbf{x}) = rT(\mathbf{x})$($T$ is *homogeneous*).

If V and W are complex vector spaces, the definition is the same except in (*b*), $r \in \mathbb{C}$. If $V = W$, then T can be called a *linear operator*.

EXAMPLE 1 Let $V = W = E^1$. Define $T(x) = mx$, where m is a fixed real number. Show that T is a linear transformation.

Solution We must show that T is additive and homogeneous. For the additivity, we let x and y be in E^1 and calculate

$$T(x + y) = m(x + y) = mx + my$$
$$T(x) + T(y) = mx + my$$

Since $T(x + y) = T(x) + T(y)$, we know that T is additive. Also T is homogeneous since

$$T(rx) = m(rx) = (mr)x = r(mx) = rT(x)$$

Thus T is a linear transformation.

EXAMPLE 2 Let $V = W = E^1$. For $x \in V$, define $F(x) = mx + b$, where m and b are real numbers and $b \neq 0$. Show that F is *not* a linear transformation.

Solution First we check additivity, noting that $F(\cdot) = m(\cdot) + b$:

$$F(x + y) = m(x + y) + b = mx + my + b$$

However,

$$F(x) + F(y) = (mx + b) + (my + b) = mx + my + 2b$$

Since $b \neq 0$, $2b \neq b$ so $F(x + y) \neq F(x) + F(y)$ for all $x, y \in V$, and F is not linear.

EXAMPLE 3 Let $V = \mathscr{P}_n$ and $W = \mathscr{P}_{n-1}$, and define, for f in V, $T: V \to W$ by $(T(f)) = f'(x)$ x in $\mathbb{R}$. That is, T is differentation. From calculus, we know that for differentiable functions f and g, $(f + g)' = f' + g'$ and $(rf)' = rf'$, so T is linear.

EXAMPLE 4 Let $V = \{$real-valued functions defined and continuous on $[a, b]\} = C[a, b]$. Let $W = E^1$ and define $T: V \to W$ by $T(f) = \int_a^b f(x)\,dx$. Then T is linear because from calculus we know that for integrable functions f and g,

$$\int_a^b f(x) + g(x)\,dx = \int_a^b f(x)\,dx + \int_a^b g(x)\,dx \text{ and } \int_a^b rf(x)\,dx = r\int_a^b f(x)\,dx, \qquad \text{for } r \text{ in } \mathbb{R}$$

EXAMPLE 5 Let $V = \mathscr{C}_{nn}$ and let $W = \mathbb{C}^1$. Define $T: V \to W$ by $T(A) = \operatorname{tr} A$, for A in $\mathscr{C}_{nn}$. So T is linear by properties of the trace of a matrix.

EXAMPLE 6 Let $V = \mathscr{M}_{mn}$ and $W = \mathscr{M}_{nm}$, and define $T(A) = A^T$ for A in $\mathscr{M}_{mn}$. Then T is linear by properties of the transpose.

In Examples 1 and 2, the functions T and F have graphs as straight lines, yet in Example 2 we found F was not linear. The difference between T and F is in the constant term. If $b = 0$, we have linearity; if not, we do not have linearity. In examples 3 through 6, $T(\boldsymbol{\theta}) = \boldsymbol{\theta}$. This gives us a clue to the first property of linear transformations.

THEOREM 4.1.1 Let V and W be vector spaces. If $T: V \to W$ is a linear transformation, then

$T(\boldsymbol{\theta}_V) = \boldsymbol{\theta}_W$. (The subscripts emphasize the vector space that the zero vector comes from.)

Proof Since $\boldsymbol{\theta}_V + \boldsymbol{\theta}_V = \boldsymbol{\theta}_V$,

$$\underbrace{T(\boldsymbol{\theta}_V)}_{\text{In } W} = T(\boldsymbol{\theta}_V + \boldsymbol{\theta}_V) \underset{\text{By additivity}}{=} \underbrace{T(\boldsymbol{\theta}_V) + T(\boldsymbol{\theta}_V)}_{\text{In } W}$$

and so

$$T(\boldsymbol{\theta}_V) = T(\boldsymbol{\theta}_V) + T(\boldsymbol{\theta}_V)$$

By uniqueness of $\boldsymbol{\theta}_W$ in W, the only way the last equation can hold is if $T(\boldsymbol{\theta}_V) = \boldsymbol{\theta}_W$.

This theorem can sometimes be used to show transformations are nonlinear. A logical consequence of the theorem is

If $T(\boldsymbol{\theta}_V) \neq \boldsymbol{\theta}_W$, then T is not linear.

EXAMPLE 7 Show that $T: E^2 \to E^2$, defined by

$$T((x_1, x_2)) = (x_1 + x_2, x_1 - x_2 + 1)$$

is not linear.

Solution In E^2, $T(\boldsymbol{\theta}) = T((0, 0)) = (0, 1) \neq \boldsymbol{\theta}$. Therefore, T is not linear.

EXAMPLE 8 Let $T: E^2 \to E^1$ be defined by

$$T((x_1, x_2)) = {x_1}^2 + {x_2}^2$$

Show that T is not linear even though $T(\boldsymbol{\theta}) = \boldsymbol{\theta}$.

Solution We have $T(\boldsymbol{\theta}) = T((0, 0)) = 0^2 + 0^2 = 0$, which is the zero of E^1. This allows no conclusion; the definition of linearity must be used. To check additivity we calculate

$$\begin{aligned} T(\mathbf{x} + \mathbf{y}) &= T((x_1, x_2) + (y_1, y_2)) \\ &= T((x_1 + y_1, x_2 + y_2)) = (x_1 + y_2)^2 + (x_2 + y_2)^2 \\ &= {x_1}^2 + 2x_1y_1 + {y_1}^2 + {x_2}^2 + 2x_2y_2 + {y_2}^2 \end{aligned}$$

and

$$T(\mathbf{x}) + T(\mathbf{y}) = T((x_1, x_2)) + T((y_1, y_2)) = {x_1}^2 + {x_2}^2 + {y_1}^2 + {y_2}^2$$

Since $T(\mathbf{x} + \mathbf{y}) \neq T(\mathbf{x}) + T(\mathbf{y})$, we know that T is not linear. In most cases, to determine linearity or nonlinearity of a transformation, we use the definition.

EXAMPLE 9 Show that the following transformations are linear.

(*a*) $T: E^3 \to E^3$ defined by

$$T((x_1, x_2, x_3)) = (x_1 + x_2, x_2 + x_3, x_3 + x_1)$$

(*b*) $T: E^3 \to E^3$ defined by

$$T((x_1, x_2, x_3)) = \mathbf{v} \times (x_1, x_2, x_3)$$

where $\mathbf{v}$ is a fixed vector in E^3

(*c*) $T: E^3 \to E^1$ defined by

$$T((x_1, x_2, x_3)) = ax_1 + bx_2 + cx_3$$

where a, b, and c are fixed real numbers

(*d*) $T: \mathcal{M}_{22} \to \mathcal{M}_{22}$ defined by

$$T\left(\begin{pmatrix} a & b \\ c & d \end{pmatrix}\right) = \begin{pmatrix} 1 & -1 \\ 2 & 4 \end{pmatrix}\begin{pmatrix} a & b \\ c & d \end{pmatrix}$$

(*e*) $T: \mathcal{P}_1 \to \mathcal{P}_2$ defined by

$$T(ax + b) = \frac{ax^2}{2} + bx$$

(*f*) $T: \mathbb{C}^2 \to \mathbb{C}^2$ defined by

$$T((z_1, z_2)) = (z_1 + z_2, z_1 - 2z_2)$$

Solution Parts (*a*) through (*e*) are left to the problems.

(*f*)
$$\begin{aligned} T((z_1, z_2) + (u_1, u_2)) &= T(z_1 + u_1, z_2 + u_2) \\ &= (z_1 + u_1 + z_2 + u_2, z_1 + u_1 - 2z_2 - 2u_2) \\ &= (z_1 + z_2, z_1 - 2z_2) + (u_1 + u_2, u_1 - 2u_2) \\ &= T(z_1, z_2) + T(u_1, u_2) \\ T(c(z_1, z_2)) = T(cz_1, cz_2) &= (cz_1 + cz_2, cz_1 - 2cz_2) \\ &= c(z_1 + z_2, z_1 - 2z_2) \\ &= cT(z_1, z_2) \end{aligned}$$

Thus T is linear.

EXAMPLE 10 Show that $T: \mathscr{C}_{22} \to \mathscr{C}_{22}$ defined by $T(A) = \bar{A}$ is not linear.

Solution We know that $T(cA) = \overline{cA} = \bar{c}\bar{A} = \bar{c}T(A) \neq cT(A)$ unless $c \in \mathbb{R}$, but c can have a nonzero imaginary part. So T is not linear. [However, T is called *conjugate linear* because $T(cA) = \bar{c}T(A)$ and $T(A + B) = T(A) + T(B)$.]

EXAMPLE 11 Let $V = \mathscr{M}_{n1}$ and $W = \mathscr{M}_{m1}$. Let M be an $m \times n$ real matrix. Define $T: V \to W$ by

$$T(X) = MX$$

T is linear because by matrix algebra

$$T(X + Y) = M(X + Y) = MX + MY$$
$$T(cX) = M(cX) = c(MX)$$

EXAMPLE 12 Let $V = \mathscr{C}_{n1}$ and $W = \mathscr{C}_{m1}$ and let Z be an $m \times n$ matrix from $\mathscr{C}_{mn}$. Define $T: V \to W$ by $T(X) = ZX$. Then T is linear because by matrix algebra

$$Z(X + Y) = ZX + ZY$$
$$Z(cX) = c(ZX)$$

Some special linear transformations must be noted for future use. The *zero* transformation $T_{\boldsymbol{\theta}}$ from V to W is defined as

$$T(\mathbf{x}) = \boldsymbol{\theta}_W \qquad \text{for all } \mathbf{x} \text{ in } V$$

The *identity* transformation I from V to V is defined as

$$I(\mathbf{x}) = \mathbf{x} \qquad \text{for all } \mathbf{x} \text{ in } V$$

The *contraction* transformation T_α from V to V is

$$T_\alpha(\mathbf{x}) = \alpha\mathbf{x} \qquad 0 < \alpha < 1, \text{ for all } \mathbf{x} \in V$$

The *dilation* transformation T_β from V to V is

$$T_\beta \mathbf{x} = \beta\mathbf{x} \qquad 1 < \beta, \text{ for all } \mathbf{x} \in V$$

Verification that these are linear transformations is left to the problems.

Although several examples of linear transformations have now been given, we have not yet begun to analyze linear transformations. In algebra, analysis of functions was done with graphs of the functions. *In our present situation we must usually be satisfied without the types of graphs we drew in algebra.* Usually we draw "graphs," as indicated in Fig. 4.1.1, whenever possible. Ordinarily this can be done only when V and W are versions of E^n, $n = 1, 2$, or 3. Other cases require considerable imagination. Consider Example 13.

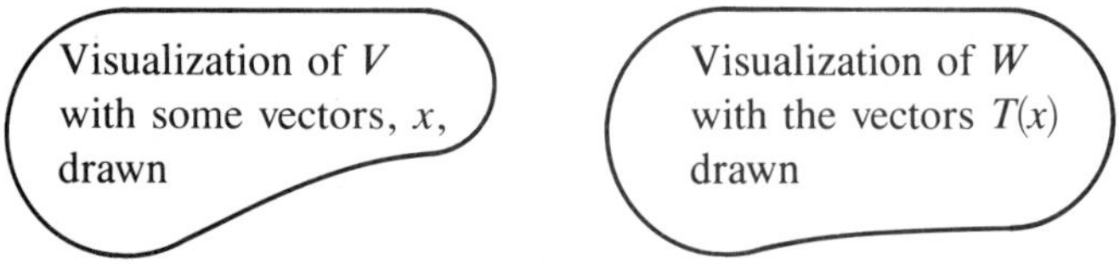

Figure 4.1.1
Graphing $T: V \to W$.

Visualization of V with some vectors, x, drawn

Visualization of W with the vectors $T(x)$ drawn

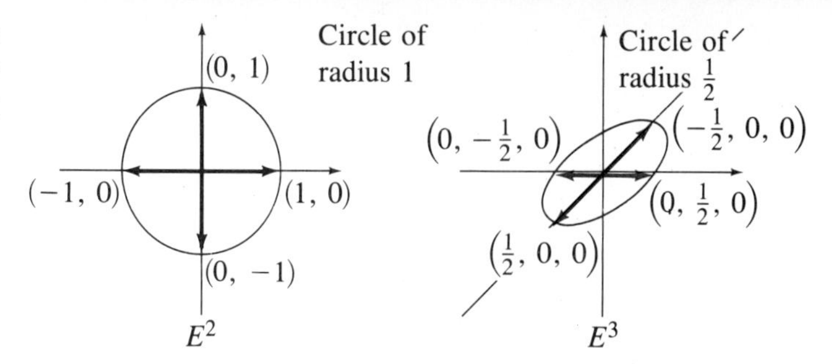

Figure 4.1.2 Graph of $T:E^2 + E^3$, from Example 13.

EXAMPLE 13 "Graph" the transformation $T:E^2 \to E^3$, defined by $T(x_1, x_2) = \frac{1}{2}(x_1, x_2, 0)$.

Solution The visualizations of E^2 and E^3 as well as some special vectors are shown in Fig. 4.1.2. The image of these vectors after T acts on them is also shown in that figure. If we put more vectors of length 1 in the circle in Fig. 4.1.2*a*, the terminal points of the images lie on the circle of radius $\frac{1}{2}$, as in Fig. 4.2.2*b*. This supports our intuitive feeling that T "shrinks" all vectors in the domain, much like a contraction transformation.

Since graphs are not simple for linear transformations, we must be able to analyze them without graphs as well. Unless specified otherwise, all vector spaces from now on are assumed to be finite-dimensional. One of the basic tools for the analysis of linear transformations is the following:

> Kernel problem
>
> Given $T:V \to W$, find all $\mathbf{x}$ in V such that $T(\mathbf{x}) = \boldsymbol{\theta}$. The set of all such $\mathbf{x}$ is called the *kernel of* T and written ker T.

Roughly speaking, the kernel problem is very much like the problem from algebra of solving the equation $f(x) = 0$, for example, solving $x^2 - 2x - 3 = 0$. In algebra this problem is solved by factoring or using the quadratic formula. In linear algebra the solution to the kernel problem many times reduces to solving m equations in n unknowns (the "first basic problem of linear algebra").

EXAMPLE 14 Find ker T, where $T:E^3 \to E^2$ is defined by $T((x_1, x_2, x_3)) = (x_1 + x_2, x_2 - x_3)$.

Solution Since $\ker T = \{\mathbf{x} \mid T(\mathbf{x}) = \boldsymbol{\theta}\}$, we must solve $T((x_1, x_2, x_3)) = (0, 0)$, that is,

$$(x_1 + x_2, x_2 - x_3) = (0, 0)$$

The resulting equations are

$$x_1 + x_2 = 0$$
$$x_2 - x_3 = 0$$

which have solution $(-k, k, k)$. Therefore

$$\ker T = \{\mathbf{v} \in E^3 \mid \mathbf{v} = k(-1, 1, 1)\} = \text{span}\,\{(-1, 1, 1)\}$$

In example 14 the kernel of the given linear transformation was a subspace of the domain. In fact, a basis for ker T was $\{(-1, 1, 1)\}$. The kernel of a linear transformation is always a vector space.

THEOREM 4.1.2 Let V and W be vector spaces, and let $T\colon V \to W$ be a linear transformation. The set ker T is a subspace of V.

Proof The kernel of T is nonempty because $T(\boldsymbol{\theta}) = \boldsymbol{\theta}$. We need to show that ker T is closed under addition and scalar multiplication. Recall that $\mathbf{x} \in \ker T$ if and only if $T\mathbf{x} = \boldsymbol{\theta}$. Let $\mathbf{x}$ and $\mathbf{y}$ be in ker T, and let c be a number. By the linearity of T,

$$T(\mathbf{x} + \mathbf{y}) = T(\mathbf{x}) + T(\mathbf{y}) = \boldsymbol{\theta} + \boldsymbol{\theta} = \boldsymbol{\theta}$$

and

$$T(c\mathbf{x}) = cT(\mathbf{x}) = c\boldsymbol{\theta} = \boldsymbol{\theta}$$

so $\mathbf{x} + \mathbf{y} \in \ker T$ and $c\mathbf{x} \in \ker T$. Thus ker T is a subspace of V.

Since ker T is a subspace of V, it has dimension. The dimension of ker T is called the *nullity* of T. Thus for the linear transformation in Example 14 the nullity is 1. We write this

$$\eta(T) = 1$$

EXAMPLE 15 Calculate $\eta(T)$ for the linear transformation $T\colon E^3 \to E^2$ defined by

$$T((a, b, c)) = (a + 2b + c, \; -a + 3b + c)$$

Find a basis for ker T.

Solution We must find the set of all vectors (a, b, c) in E^3 that $T(a, b, c) = (0, 0)$. That is, the equation

$$\begin{pmatrix} a + 2b + c \\ -a + 3b + c \end{pmatrix} = \begin{pmatrix} 0 \\ 0 \end{pmatrix}$$

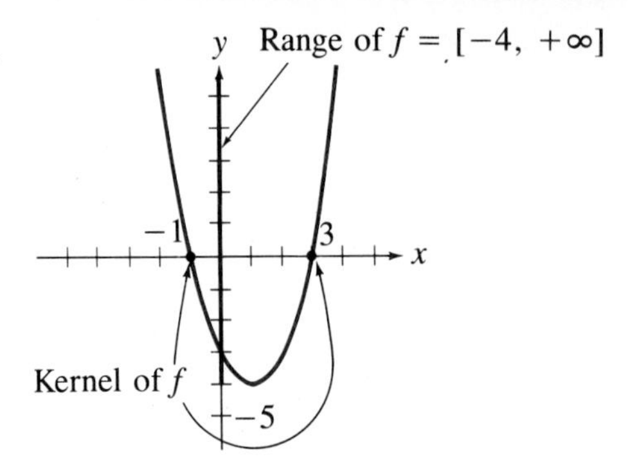

Figure 4.1.3
Kernel and range of $y = x^2 - 2x - 3$.

must be solved. The solution is

$$\begin{pmatrix} a \\ b \\ c \end{pmatrix} = \begin{pmatrix} -k \\ -2k \\ 5k \end{pmatrix}$$

and $\ker T = \text{span}\{(-1, -2, 5)\}$. Therefore $\dim(\ker T) = 1$, so $\eta(T) = 1$. A basis is $\{(-1, -2, 5)\}$.

To continue the analysis of linear transformations, we consider the *range* of T. In algebra finding the range of a function f is important in graphing $y = f(x)$. For example, $y = x^2 - 2x - 3$ has range $\{y | -4 \le y \le \infty\}$. The solutions of $x^2 - 2x - 3 = 0$ are $x = 3$ and $x = -1$. (That is, the kernel of f is $\{-1, 3\}$.) All this information is shown in Fig. 4.1.3. The range of a linear transformation cannot always be used to obtain a graph of T, but it is quite useful in other ways.

DEFINITION 4.1.2 Let $T: V \to W$ be a linear transformation. The *range of* T is the set of all possible $\mathbf{v}$ in W such that $\mathbf{y} = T(\mathbf{x})$ for some $\mathbf{x}$ in V. The range of T is written range T. The range of T is a subspace of W (see the problems).

EXAMPLE 16 Define T from E^3 to E^3 by $T((a, b, c)) = (a - b + c, 2a + b - c, -a - 2b + 2c)$. Determine range T and dim(range T). Find two vectors in range T and two vectors not in range T. Find a basis for range T. Find ker T. Graph ker T and range T. Attempt a graph of T.

Solution Let $\mathbf{y} = (y_1, y_2, y_3)$ be in range T. Thus $\mathbf{y} = T((a, b, c))$ for some vector (a, b, c) in E^3. That is, the equation $\mathbf{y} = T((a, b, c))$ *must be consistent*. We reduce the equations and see what conditions the consistency forces. The equations are

$$\begin{aligned} a - b + c &= y_1 \\ 2a + b - c &= y_2 \\ -a - 2b + 2c &= y_3 \end{aligned}$$

and they reduce to

$$\begin{aligned} a - b + \quad c &= \quad y_1 \\ 3b - 3c &= \quad y_2 - 2y_1 \\ 0 &= -y_1 + y_2 + y_3 \end{aligned}$$

So if $\mathbf{y} = (y_1, y_2, y_3)$ is to be the range T, then $-y_1 + y_2 + y_3 = 0$. That is,

$$\text{range } T = \{(y_1, y_2, y_3) | y_1 = y_2 + y_3\}$$

The condition on y_1, y_2, and y_3 gives a criterion for inclusion in range T. Some vectors in range T are $(-2, -1, -1)$ and $(0, -1, 1)$. Some vectors not in range T are $(1, 1, 1)$ and $(1, 0, 0)$. The dimension of range T is 2, since the equation $-y_1 + y_2 + y_3 = 0$ allows the assignment of arbitrary values to any *two* of the values of y_k.

To obtain a basis, we can use $(-2, -1, -1)$ and $(0, -1, 1)$ as above, since they are linearly independent in the range and dim(range T) = 2. In fact, any two linearly independent vectors in range T form a basis for range T. The kernel of T is found by setting $y_1 = y_2 = y_3 = 0$ in the linear equations above. We obtain ker $T = \text{span}\{(0, 1, 1)\}$. Graphs are shown in Fig. 4.1.4.

The dimension of the range of a linear transformation T is called the *rank* of T and written $\mathscr{R}(T)$. That is,

$$\mathscr{R}(T) = \dim(\text{range } T)$$

The rank and nullity of a linear transformation are related to each other by the equation

$$\text{rank } T + \text{nullity } T = \dim(\text{domain})$$

This is the result of the following basic theorem, one of the most important in linear algebra:

THEOREM 4.1.3

If $T: V \to W$ is a linear transformation and $\dim V = n$, then

$$\mathscr{R}(T) + \eta(T) = n \tag{4.1.1}$$

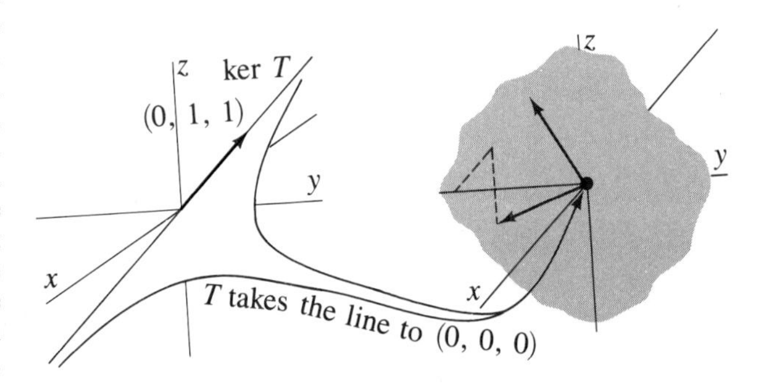

Figure 4.1.4 Graph of the transformation in Example 16. The kernel of T is the line spanned by $(0, 1, 1)$. The range, shaded in color, on the right, is the plane spanned by $(0, -1, 1)$ and $(-2, -1, -1)$. The plane is tilted so that the upper right part is coming out of the page. It has been truncated in the graph.

Before proving this theorem, we consider an example of its use.

EXAMPLE 17 Find the nullity of the linear transformation in Example 16.

Solution We had $T: E^3 \to E^3$ and found $\mathscr{R}(T) = 2$. Since $\dim V = \dim E^3 = 3$, Eq. (4.1.1) leads to

$$2 + \eta(T) = 3$$

Therefore $\eta(T) = 1$

Proof of Theorem 4.1.3 Since ker T and range T are vector spaces, $\mathscr{R}(T)$ and $\eta(T)$ are defined. We consider three cases: $\eta(T) = 0$, $\eta(T) = n$, and $1 \leq \eta(T) \leq n - 1$.

Case 1: $\eta(T) = 0$. Suppose $\mathscr{R}(T) = k < n$. That is, suppose that $\eta(T) + \mathscr{R}(T) < n$. We will obtain a contradiction. Since $\mathscr{R}(T) = k$, any set of more than k vectors in range T is linearly dependent. Let $\{\mathbf{v}_1, \ldots, \mathbf{v}_n\}$ be a basis for V. Since $k < n$, $\{T(\mathbf{v}_1), \ldots, T(\mathbf{v}_n)\}$ must be linearly dependent and so there exist $c_1, \ldots, c_n$, not all zero, with

$$c_1 T(\mathbf{v}_1) + \cdots + c_n T(\mathbf{v}_n) = \boldsymbol{\theta}$$

Thus by linearity $T(c_1\mathbf{v}_1 + \cdots + c_n\mathbf{v}_n) = \boldsymbol{\theta}$ and $c_1\mathbf{v}_1 + \cdots + c_n\mathbf{v}_n \in \ker T$. Since $\ker T = \{\boldsymbol{\theta}\}$ and not all the c_i's are zero, we have $\{\mathbf{v}_1, \ldots, \mathbf{v}_n\}$ being linearly dependent, which is a contradiction. Therefore, $\mathscr{R}(T) = n$ and $\eta(T) + \mathscr{R}(T) = 0 + n = n$.

Case 2: $\eta(T) = n$. Since ker T is a subspace of V and dim (ker T) = dim V, we actually have ker $T = V$ and $T(\mathbf{x}) = \boldsymbol{\theta}$ for all $\mathbf{x} \in V$. Therefore range $T = \{\boldsymbol{\theta}\}$ and dim (range T) = 0. Thus $\mathscr{R}(T) + \eta(T) = 0 + n = n$.

Case 3: $\eta(T) = k$, $1 \leq k \leq n - 1$. Let $B = \{\mathbf{v}_1, \ldots, \mathbf{v}_k\}$ be a basis for ker T. By a previous result, B can be extended to a basis $S = \{\mathbf{v}_1, \ldots, \mathbf{v}_k, \mathbf{u}_{k+1}, \ldots, \mathbf{u}_n\}$ of V, since dim $V = n$. We will show that $\mathscr{T} = \{T(\mathbf{u}_{k+1}), \ldots, T(\mathbf{u}_n)\}$ is a basis for range T. Then we will have $\mathscr{R}(T) = n - k$ and

$$\mathscr{R}(T) + \eta(T) = (n - k) + k = n$$

$\mathscr{T}$ Is Linearly Independent Consider

$$c_{k+1} T(\mathbf{u}_{k+1}) + \cdots + c_n T(\mathbf{u}_n) = \boldsymbol{\theta}$$

By the linearity of T,

$$T(c_{k+1}\mathbf{u}_{k+1} + \cdots + c_n\mathbf{u}_n) = \boldsymbol{\theta}$$

and so $c_{k+1}\mathbf{u}_{k+1} + \cdots + c_n\mathbf{u}_n$ is in ker T. So there exist $c_1, \ldots, c_k$ with

$$c_1\mathbf{v}_1 + \cdots + c_k\mathbf{v}_k = c_{k+1}\mathbf{v}_{k+1} + \cdots + c_n\mathbf{u}_n$$

That is,

$$c_1\mathbf{v}_1 + \cdots + c_k\mathbf{v}_k - c_{k+1}\mathbf{u}_{k+1} - \cdots - c_n\mathbf{u}_n = \boldsymbol{\theta}$$

and since S is a basis for V, we know that $c_1 = \cdots = c_k = c_{k+1} = \cdots = c_n = 0$. Thus $\mathscr{T}$ is linearly independent.

$\mathscr{T}$ Spans range T Let $\mathbf{y} \in$ range T, so that $\mathbf{y} = T(\mathbf{x})$ for some $\mathbf{x} \in \mathbf{V}$. Since S is a basis for V, we can write $\mathbf{x} = c_1\mathbf{v}_1 + \cdots + c_k\mathbf{v}_k + c_{k+1}\mathbf{u}_{k+1} + \cdots + c_n\mathbf{u}_n$, and so

$$\begin{aligned}\mathbf{y} = T(\mathbf{x}) &= T(c_1\mathbf{v}_1 + \cdots + c_k\mathbf{v}_k) + c_{k+1}T(\mathbf{u}_{k+1}) + \cdots + c_nT(\mathbf{u}_n) \\ &= \qquad\qquad \boldsymbol{\theta} \qquad\qquad + c_{k+1}T(\mathbf{u}_{k+1}) + \cdots + c_nT(\mathbf{u}_n)\end{aligned}$$

Thus $\mathbf{y} \in$ span $\mathscr{T}$.

EXAMPLE 18 "Graph" $T: E^3 \to E^3$, defined by $T((x_1, x_2, x_3)) = (-x_1 + x_2 + x_3, 2x_1 - x_2, x_1 + x_2 + 3x_3)$, indicating ker T and range T.

Solution Solving $T(\mathbf{x}) = \mathbf{y}$, we find

$$\left(\begin{array}{ccc|c} -1 & 1 & 1 & y_1 \\ 0 & 1 & 2 & y_2 + 2y_1 \\ 0 & 0 & 0 & y_3 - 2y_2 - 3y_1 \end{array}\right) \tag{4.1.2}$$

To determine ker T, set $\mathbf{y} = \boldsymbol{\theta}$. The solution of the resulting equations is $\mathbf{x} = (-k, -2k, k)$, so ker $T = \text{span}\ \{(-1, -2, 1)\}$. From Eq. (4.1.2) we see that range $T = \{\mathbf{y} \mid y_3 - 2y_2 - 3y_1 = 0\} = \{\mathbf{y} \mid \mathbf{y} = (s, t, 3s + 2t)\} = \text{span}\ \{(1, 0, 3), (0, 1, 2)\}$. A graph is shown in Fig. 4.1.5.

Two more ways to view the action of a linear transformation are to determine the images under T of geometric figures such as squares and circles.

Figure 4.1.5
Graph of T in Example 18.

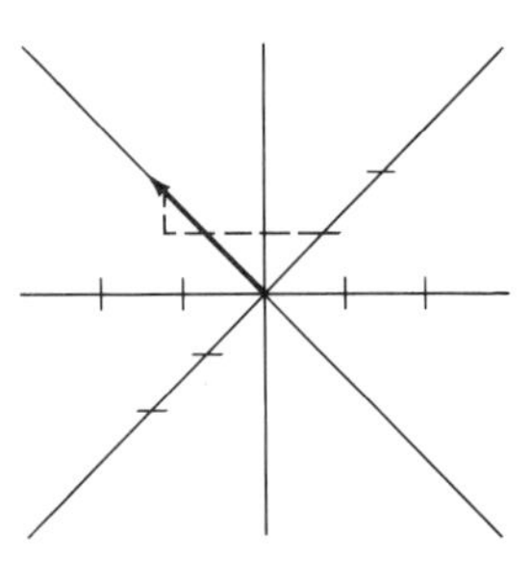

ker T = line containing $(-1, -2, 1)$

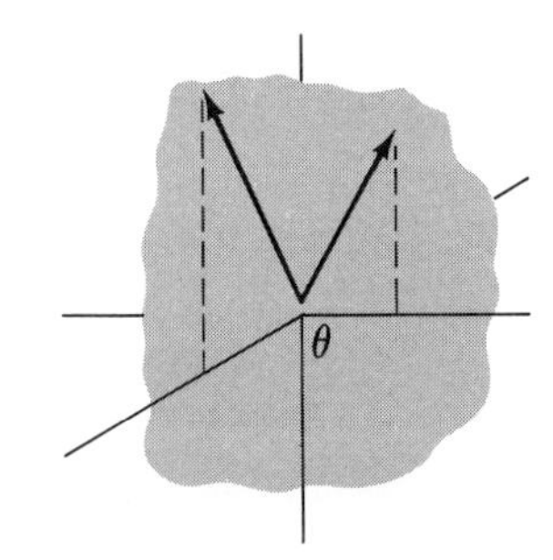

Range T = plane containing $(1, 0, 3)$ and $(0, 1, 2)$

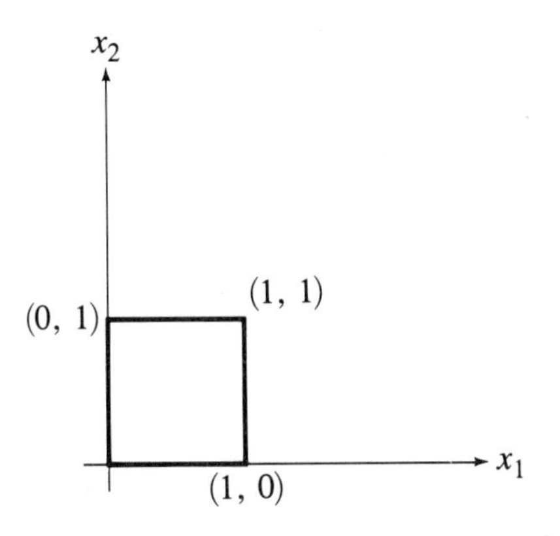

Figure 4.1.6
Unit square.

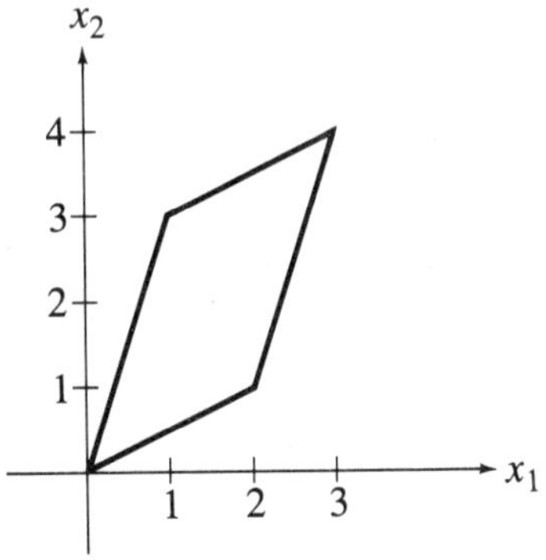

Figure 4.1.7
Image of unit square from Fig. 4.1.6.

EXAMPLE 19 It can be shown (see the problems) that if $T:\mathscr{M}_{21} \to \mathscr{M}_{21}$ is a linear transformation defined by

$$T\left(\begin{pmatrix} x_1 \\ x_2 \end{pmatrix}\right) = A\begin{pmatrix} x_1 \\ x_2 \end{pmatrix}$$

where A is a nonsingular matrix, then the image of a straight-line segment from P to Q in E^2 will be a straight-line segment from $T(P)$ to $T(Q)$. Let T be defined as

$$T\left(\begin{pmatrix} x_1 \\ x_2 \end{pmatrix}\right) = \begin{pmatrix} 1 & 2 \\ 3 & 1 \end{pmatrix}\begin{pmatrix} x_1 \\ x_2 \end{pmatrix}$$

Find the image under T of the "unit square" shown in Fig. 4.1.6.

Solution We find the images of the vertices. Since each side of the square is a straight-line segment, the image of the square will be the figure generated by joining the images of the vertices with a straight-line segment. Now

$$T\left(\begin{pmatrix} 0 \\ 0 \end{pmatrix}\right) = \begin{pmatrix} 0 \\ 0 \end{pmatrix} \qquad T\left(\begin{pmatrix} 1 \\ 0 \end{pmatrix}\right) = \begin{pmatrix} 1 \\ 3 \end{pmatrix}$$

$$T\left(\begin{pmatrix} 0 \\ 1 \end{pmatrix}\right) = \begin{pmatrix} 2 \\ 1 \end{pmatrix} \qquad T\left(\begin{pmatrix} 1 \\ 1 \end{pmatrix}\right) = \begin{pmatrix} 3 \\ 4 \end{pmatrix}$$

Therefore the image is the parallelogram shown in Fig. 4.1.7. Note that points are associated with terminal points of the vectors naturally associated with the elements of $\mathscr{M}_{21}$.

EXAMPLE 20 Show that $T:\mathscr{M}_{21} \to \mathscr{M}_{21}$ defined by

$$T\left(\begin{pmatrix} x_1 \\ x_2 \end{pmatrix}\right) = \begin{pmatrix} 2 & 0 \\ 0 & 3 \end{pmatrix}\begin{pmatrix} x_1 \\ x_2 \end{pmatrix}$$

transforms the unit circle $x_1^2 + x_2^2 = 1$ to an ellipse.

Solution Since the image of

$$\begin{pmatrix} x_1 \\ x_2 \end{pmatrix}$$

under T is

$$\begin{pmatrix} y_1 \\ y_2 \end{pmatrix} = \begin{pmatrix} 2x_1 \\ 3x_2 \end{pmatrix}$$

we have

$$\frac{y_1^2}{4} + \frac{y_2^2}{9} = \frac{4x_1^2}{4} + \frac{9x_2^2}{9} = x_1^2 + x_2^2 = 1$$

Therefore the image of the circle is an ellipse. This action of T is shown in Fig. 4.1.8.

An interpretation of Fig. 4.1.8 is that T dilates with constant 3 in the x_2 direction and constant 2 in the x_1 direction.

EXAMPLE 21 Show that $T: \mathscr{M}_{31} \to \mathscr{M}_{31}$ defined by

$$\begin{pmatrix} y_1 \\ y_2 \\ y_3 \end{pmatrix} = T\left(\begin{pmatrix} x_1 \\ x_2 \\ x_3 \end{pmatrix}\right) = \begin{pmatrix} a & 0 & 0 \\ 0 & b & 0 \\ 0 & 0 & c \end{pmatrix}\begin{pmatrix} x_1 \\ x_2 \\ x_3 \end{pmatrix} \qquad a, b, c > 0$$

transforms the sphere $x_1^2 + x_2^2 + x_3^2 = R^2$ to an ellipsoid.

Figure 4.1.8
T acting on the unit circle.

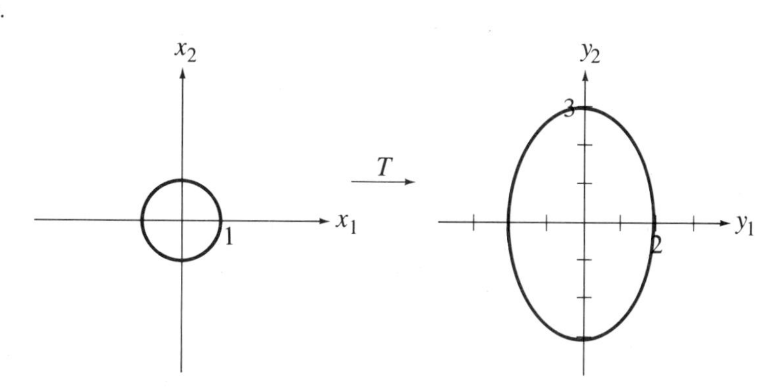

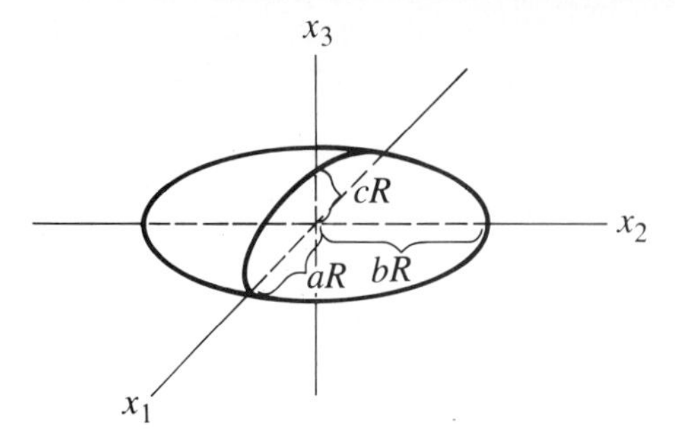

Figure 4.1.9 Image of sphere in Example 21.

Solution The image of

$$\begin{pmatrix} x_1 \\ x_2 \\ x_3 \end{pmatrix} \quad \text{is} \quad \begin{pmatrix} ax_1 \\ bx_2 \\ cx_3 \end{pmatrix}$$

and

$$\frac{(ax_1)^2}{a^2} + \frac{(bx_2)^2}{b^2} + \frac{(cx_3)^2}{c^2} = x_1{}^2 + x_2{}^2 + x_3{}^2 = R^2$$

Division by R^2 yields

$$\frac{y_1{}^2}{(aR)^2} + \frac{y_2{}^2}{(bR)^2} + \frac{y_3{}^2}{(cR)^2} = 1$$

which is an equation for the ellipsoid shown in Fig. 4.1.9.

Important examples of linear transformations exist which cannot be analyzed geometrically except in some generalized way. One example is $T:\mathscr{P}_n \to \mathscr{P}_{n+1}$ defined by

$$T(a_0 + a_1x + \cdots + a_nx^n) = a_0x + a_1x^2 + \cdots + a_nx^{n+1}$$

That is, for f in $\mathscr{P}_n$, $T(f)$ is the function obtained at each x by multiplying $f(x)$ by x. That is,

$$(T(f))(x) = xf(x)$$

This linear transformation is a special case of the *coordinate operator* in quantum mechanics.

If we allow complex vector spaces and consider the set of nth-degree polynomials with complex coefficients with the same operations as $\mathscr{P}_n$, we will have a vector space $\mathscr{P}_n{}^{\mathbb{C}}$. We can define a transformation

$$T(a_0 + a_1x + a_2x^2 + \cdots + a_nx^n) = -i(a_1 + 2a_2x + \cdots + na_nx^{n-1})$$

that is,

$$(T(f))(x) = -i\frac{d}{dx}(f(x))$$

Now T is linear, and the rule for T is the same as the rule for the *momentum operator* in quantum mechanics. However, the momentum operator has a different domain.

PROBLEMS 4.1

1. Determine whether the following transformations $T: E^3 \to E^3$ are linear.
   (*a*) $T((x_1, x_2, x_3)) = (x_1, x_1 - x_2, x_2 + x_3)$
   (*b*) $T((x_1, x_2, x_3)) = (x_1, x_2, x_2 x_3)$
   (*c*) $T((x_1, x_2, x_3)) = (x_1, 0, 0)$
   (*d*) $T((x_1, x_2, x_3)) = (1, 0, 0)$
   (*e*) $T((x_1, x_2, x_3)) = (3x_1 + 2x_2, x_3, |x_2|)$
   (*f*) $T((x_1, x_2, x_3)) = (x_1 - x_2, x_1 + x_2, x_3)$
2. Determine whether the following transformations defined on $\mathscr{M}_{22}$ are linear.
   (*a*) $T(A) = A^T$
   (*b*) $T(A) = A + A^T$
   (*c*) $T(A) = A^T A$
   (*d*) $T(A) = A + I$ (where I is the identity matrix)
   (*e*) $T(A) = kA$ (where k is a real number), $k \neq 0$
   (*f*) $T(A) = A^2$
   (*g*) $T(A) = \det A$
   (*h*) $T(A) = \text{tr} A$ (trace of A = sum of the diagonal elements)
3. Determine whether the following transformations $T: \mathscr{P}_2 \to \mathscr{P}_2$ are linear.
   (*a*) $T(a + bx + cx^2) = ax + bx^2$
   (*b*) $T(a + bx + cx^2) = x - c$
   (*c*) $T(a + bx + cx^2) = 2$
   (*d*) $T(a + bx + cx^2) = \dfrac{1}{a + bx + cx^2}$
   (*e*) $T(a + bx + cx^2) = (a - b) + (b + c)x + (a - c)x^2$
4. Determine whether the following transformations $\mathscr{C}_{22} \to \mathscr{C}_{22}$ are linear.
   (*a*) $T(A) = A^*$ (*b*) $T(A) = A^* A$
   (*c*) $T(A) = A + \bar{A}$ (*d*) $T(A) = iA$
5. For each *linear* transformation from Prob. 1, determine ker T and range T, find bases for ker T and range T, and verify the equation $\eta(T) + \mathscr{R}(T) =$ dim (domain T). Sketch a "graph" of T as in Example 13 or 19.
6. For each *linear* transformation from Prob. 2, determine ker T and range T, find bases for ker T and range T, and verify the equation $\eta(T) + \mathscr{R}(T) =$ dim (domain T).

7. For each *linear* transformation from Prob. 3, determine ker T and range T, find bases for ker T and range T, and verify the equation $\eta(T) + \mathscr{R}(T) = \dim(\text{domain } T)$.

8. Show that the zero transformation is linear.

9. Show that the identity transformation is linear.

10. Show that a contraction operator is linear.

11. Show that a dilation transformation is linear.

12. Show that a linear transformation $T: E^2 \to E^2$ defined by

$$T\left(\begin{pmatrix} x_1 \\ x_2 \end{pmatrix}\right) = \begin{pmatrix} a & 0 \\ 0 & b \end{pmatrix}\begin{pmatrix} x_1 \\ x_2 \end{pmatrix}$$

transforms the circle ${x_1}^2 + {x_2}^2 = 1$ to the ellipse ${y_1}^2/a^2 + {y_2}^2/b^2 = 1$.

13. Draw the image under T of the unit square in E^2 for T defined by

(*a*)
$$T\left(\begin{pmatrix} x_1 \\ x_2 \end{pmatrix}\right) = \begin{pmatrix} \frac{1}{\sqrt{2}} & -\frac{1}{\sqrt{2}} \\ \frac{1}{\sqrt{2}} & \frac{1}{\sqrt{2}} \end{pmatrix}\begin{pmatrix} x_1 \\ x_2 \end{pmatrix}$$

(*b*)
$$T\begin{pmatrix} x_1 \\ x_2 \end{pmatrix} = \begin{pmatrix} x_1 - x_2 \\ 2x_1 + 3x_2 \end{pmatrix}$$

14. Show that the additivity condition $T(\mathbf{x} + \mathbf{y}) = T(\mathbf{x}) + T(\mathbf{y})$ implies that $T(n\mathbf{x}) = nT(\mathbf{x})$ for any positive integer n.

15. Show that the additivity condition $T(\mathbf{x} + \mathbf{y}) = T(\mathbf{x}) + T(\mathbf{y})$ implies that $T((p/q)\mathbf{x}) = (p/q)T(\mathbf{x})$, where p and q are positive integers. [*Hint*: By Prob. 14, $pT(\mathbf{x}) = T(p\mathbf{x}) = T(q(p/q)\mathbf{x})$.]

16. Show that this definition is equivalent to Definition 4.1.1:

> Let V and W be vector spaces, and let T be a function with domain V and range in W. Then T is a linear transformation if for all $a, b \in \mathbb{R}$, $\mathbf{x}, \mathbf{y} \in V$, $T(a\mathbf{x} + b\mathbf{y}) = aT(\mathbf{x}) + bT(\mathbf{y})$.

17. Let $T: V \to W$ be a linear transformation. Show that range T is a subspace of W.

18. Complete the details of Example 9*a* through *e*.

19. Verify that the coordinate operator as defined in this section is linear.

20. Verify that the momentum operator as defined in this section is linear.

21. Let $T: E^3 \to E^3$ be linear, and suppose that $T((1, 0, 1)) = (1, -1, 3)$ and $T((2, 1, 0)) = (0, 2, 1)$. Determine $T((8, 3, 2))$. [*Hint*: write $(8, 3, 2)$ as a linear combination of $(1, 0, 1)$ and $(2, 1, 0)$, and use the linearity of T.]

22. Regarding T as in Prob. 21, calculate $T((1, 2, -3))$ and $T((4, -4, 12))$.

23. Regarding T as in Prob. 21, why can $T((3, 0, 4))$ not be calculated from the information given?

24. Define $T:\mathscr{P}_2 \to \mathscr{M}_{22}$ for each f in $\mathscr{P}_2$ by

$$T(f) = \begin{pmatrix} f(1) & f(0) \\ f(0) & f(2) \end{pmatrix}$$

Is T a linear transformation?

25. Define $T:\mathscr{M}_{21} \to \mathscr{M}_{21}$ by

$$T\left(\begin{pmatrix} a \\ b \end{pmatrix}\right) = M\begin{pmatrix} a \\ b \end{pmatrix}$$

where M is an invertible 2×2 real matrix. Show that the image of a line under T is again a line. (*Hint*: Describe a line in the domain by its *vector equation*; then use the linearity of T.)

4.2 MATRIX REPRESENTATION PROBLEM FOR LINEAR TRANSFORMATIONS

The methods of Sec. 4.1 can be used to discuss systems of linear equations. For a set of equations

$$A_{m \times n} X_{n \times 1} = B_{m \times 1}$$

the matrix on the left-hand side represents a linear transformation from $\mathscr{M}_{n1}$ to $\mathscr{M}_{m1}$ because, by the laws of matrix algebra, $A(cX + dY) = cAX + dAY$. The kernel of the linear transformation is the solution set for the homogeneous equation $AX = \boldsymbol{\theta}$. The range of the linear transformation is the set of all vectors B for which $AX = B$ has a solution. The "rank-kernel equation" from theorem 4.1.3 of Sec. 4.1 means that

$$\dim\,(\text{solution space of } AX = \boldsymbol{\theta}) + \dim\,(\text{range}) = n$$

However from previous chapters we know that the dimension of the solution space is the number of zero rows of the reduced row echelon form of A. So the last equation can be rewritten

$$(\dim \text{ of solution space of } AX = \boldsymbol{\theta}) + (\text{rank } A) = n = \text{no. of columns of } A$$

EXAMPLE 1 Find the dimension of the solution space of $AX = \boldsymbol{\theta}$, where

$$A = \begin{pmatrix} 1 & 2 & -1 & 2 \\ 3 & 1 & 2 & 2 \\ 2 & -1 & 3 & 0 \\ 1 & -2 & 4 & -2 \end{pmatrix}$$

Solution Since $n = 4$, if we find rank A, then $4 -$ rank A is the number we desire. Since rank $A =$ row rank A, we row-reduce A:

$$\begin{pmatrix} 1 & 2 & -1 & 2 \\ 3 & 1 & 2 & 2 \\ 2 & -1 & 3 & 0 \\ 1 & -3 & 4 & -2 \end{pmatrix} \longrightarrow \begin{pmatrix} 1 & 2 & -1 & 2 \\ 0 & -5 & 5 & -4 \\ 0 & 0 & 0 & 0 \\ 0 & 0 & 0 & 0 \end{pmatrix}$$

to find rank $A = 2$. Therefore the dimension of the solution space of $AX = \boldsymbol{\theta}$ is 2. Notice that this is also the number of unknowns which can be arbitrarily set. Thus the terminology *2 degrees of freedom.*

Example 1 illustrates this general principle:

If T is a linear transformation generated by a matrix A, then $\eta(T)$ and $\mathscr{R}(T)$ can be found by row-reducing matrix A. That is, information about a linear transformation can be gained by analyzing a matrix.

For this reason (and others which appear later), representation of a linear transformation by a matrix is important. Thus we come to the third basic problem of linear algebra.

Third Basic Problem of Linear Algebra

Given a linear transformation $T: V \to W$, where $\dim V = n$ and $\dim W = m$, find an $m \times n$ matrix A which "represents" T.

Before stating precisely what the word *represents* means, we consider some simple examples.

EXAMPLE 2 Consider the identity transformation $T: E^3 \to E^3$, defined by $T(\mathbf{x}) = \mathbf{x}$. Let $X = (\mathbf{x})_{\mathscr{E}}$, where $\mathscr{E}$ is the standard ordered basis $\{(1, 0, 0), (0, 1, 0), (0, 0, 1)\}$. Then

$$X = \begin{pmatrix} x_1 \\ x_2 \\ x_3 \end{pmatrix}$$

when $\mathbf{x} = (x_1, x_2, x_3)$. So

$$I_{3 \times 3}X = \begin{pmatrix} 1 & 0 & 0 \\ 0 & 1 & 0 \\ 0 & 0 & 1 \end{pmatrix} \begin{pmatrix} x_1 \\ x_2 \\ x_2 \end{pmatrix} = \begin{pmatrix} x_1 \\ x_2 \\ x_3 \end{pmatrix} = X$$

The action of the *identity transformation* is represented by matrix multiplication of coordinate matrices by the *identity matrix* I.

EXAMPLE 3 Consider the projection $P: E^3 \to E^3$ defined by $P(x_1, x_2, x_3) = (x_1, x_2, 0)$. For $\mathscr{E}$ as in Example 2,

$$MX = \begin{pmatrix} 1 & 0 & 0 \\ 0 & 1 & 0 \\ 0 & 0 & 0 \end{pmatrix} \begin{pmatrix} x_1 \\ x_2 \\ x_3 \end{pmatrix} = \begin{pmatrix} x_1 \\ x_2 \\ 0 \end{pmatrix}$$

we see that the action of P is represented by matrix multiplication by

$$M = \begin{pmatrix} 1 & 0 & 0 \\ 0 & 1 & 0 \\ 0 & 0 & 0 \end{pmatrix}$$

Note that $P(P(X)) = P(X)$; also $MM = M$. We have the projection represented by an idempotent matrix.

EXAMPLE 4 Consider differentiation $D: \mathscr{P}_1 \to \mathscr{P}_1$ defined by $D(a + bx) = b$. If we use the standard ordered basis $\mathscr{E} = \{1, x\}$, then

$$(a + bx)_{\mathscr{E}} = \begin{pmatrix} a \\ b \end{pmatrix}$$

and we can write

$$D: \begin{pmatrix} a \\ b \end{pmatrix} \to \begin{pmatrix} b \\ 0 \end{pmatrix}$$

Now the matrix

$$M = \begin{pmatrix} 0 & 1 \\ 0 & 0 \end{pmatrix}$$

satisfies

$$M \begin{pmatrix} a \\ b \end{pmatrix} = \begin{pmatrix} 0 & 1 \\ 0 & 0 \end{pmatrix} \begin{pmatrix} a \\ b \end{pmatrix} = \begin{pmatrix} b \\ 0 \end{pmatrix}$$

that is, the action of D is represented by multiplication by M. We note that

$$D(D(a + bx)) = D(b + 0x) = 0 + 0x = \boldsymbol{\theta}$$

and

$$MM = \begin{pmatrix} 0 & 1 \\ 0 & 0 \end{pmatrix} \begin{pmatrix} 0 & 1 \\ 0 & 0 \end{pmatrix} = \begin{pmatrix} 0 & 0 \\ 0 & 0 \end{pmatrix}$$

The transformation D is represented by a matrix which is nilpotent of exponent 2. This just means that the second derivative of a first-degree polynomial is zero.

Examples 2 to 4 depend on the fact that we used the standard basis to represent the vectors in each vector space. We will see that in general the *representing matrix depends on the bases used for the domain and range.*

Solution of Representation Problem The basic principle which leads to the solution of the basis problem for a linear transformation $T: V \to W$ is as follows:

> If $\dim V = n$ and $\{\mathbf{v}_1, \mathbf{v}_2, \ldots, \mathbf{v}_n\}$ is a basis for V, then the range of T is completely describable in terms of the images $T(\mathbf{v}_1), \ldots, T(\mathbf{v}_n)$ of the basis vectors.

To see this, let $\mathbf{x}$ be any vector in V. There exist constants $c_1, \ldots, c_n$ such that $\mathbf{x} = c_1\mathbf{v}_1 + c_2\mathbf{v}_2 + \cdots + c_n\mathbf{v}_n$. Therefore, $T(\mathbf{x}) = c_1 T(\mathbf{v}_1) + c_2 T(\mathbf{v}_2) + \cdots + c_n T(\mathbf{v}_n)$ and we see that every element $T(\mathbf{x})$ in the range is a linear combination of the images of basis elements. That is $T(\mathbf{x}) \in \text{span}\{T(\mathbf{v}_1), \ldots, T(\mathbf{v}_n)\}$.

The setup and procedure for solving the representation problem are as follows:

> Suppose $\dim V = n$, $\mathscr{S} = \{\mathbf{v}_1, \ldots, \mathbf{v}_n\}$ is an ordered basis for V, and suppose $\dim W = m$ and $\mathscr{T} = \{\mathbf{w}_1, \ldots, \mathbf{w}_m\}$ is an ordered basis for W.
>
> **1.** Calculate $T(\mathbf{v}_1), T(\mathbf{v}_2), \ldots, T(\mathbf{v}_n)$.
> **2.** Find the coordinate vectors $(T(\mathbf{v}_1))_{\mathscr{T}}, (T(\mathbf{v}_2))_{\mathscr{T}}, \ldots, (T(\mathbf{v}_m))_{\mathscr{T}}$.
> **3.** Write the matrix with columns as the column vectors calculated in Step 2:
>
> $$M = ((T(\mathbf{v}_1))_{\mathscr{T}} (T(\mathbf{v}_2))_{\mathscr{T}} \cdots (T(\mathbf{v}_n))_{\mathscr{T}})$$

The $m \times n$ matrix M represents T, as indicated in Fig. 4.2.1. Whenever necessary, we write M_T to denote the matrix of T. The diagram gives the content of the theorem that we will state for the solution of the representation problem. Before spelling out the theorem, we consider several examples. In these examples, we write $(V, \mathscr{S})$ to indicate the vector space V with basis $\mathscr{S}$.

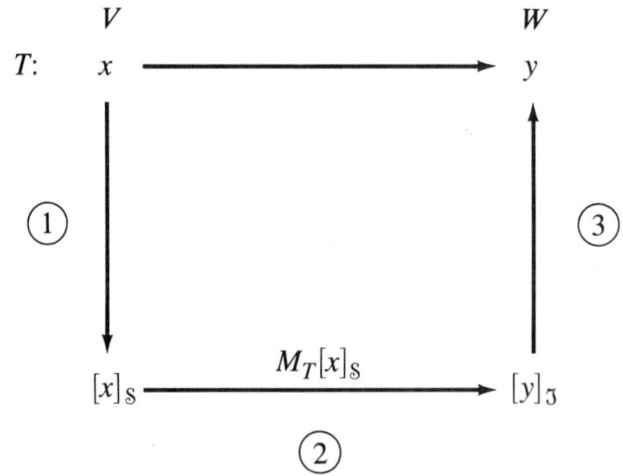

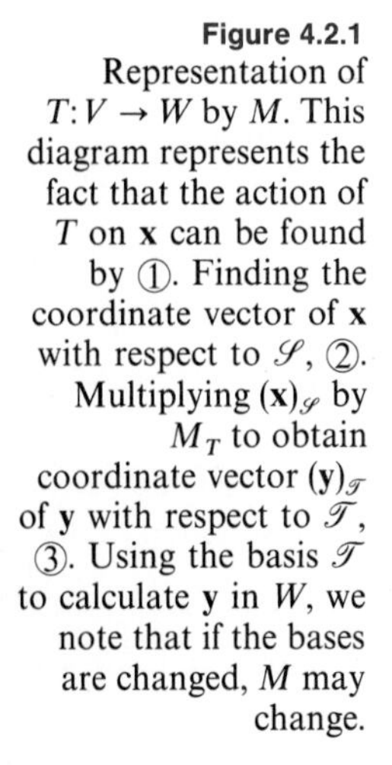
Figure 4.2.1 Representation of $T: V \to W$ by M. This diagram represents the fact that the action of T on $\mathbf{x}$ can be found by ①. Finding the coordinate vector of $\mathbf{x}$ with respect to $\mathscr{S}$, ②. Multiplying $(\mathbf{x})_{\mathscr{S}}$ by M_T to obtain coordinate vector $(\mathbf{y})_{\mathscr{T}}$ of $\mathbf{y}$ with respect to $\mathscr{T}$, ③. Using the basis $\mathscr{T}$ to calculate $\mathbf{y}$ in W, we note that if the bases are changed, M may change.

EXAMPLE 5 Let $T:(E^2, \mathscr{S}) \to (E^2, \mathscr{T})$ be defined by $T((x_1, x_2)) = (x_1 + 2x_2, x_1 - x_2)$. Find the matrix M representing T when

(*a*) $\mathscr{S} = \mathscr{T} = \{\mathbf{e}_1, \mathbf{e}_2\}$, the standard basis

(*b*) $\mathscr{S} = \mathscr{T} = \{(1, 2), (3, -1)\}$

(*c*) $\mathscr{S} = \mathscr{T} = \{(3, -1), (1, 2)\}$

(*d*) $\mathscr{S} = \{(1, -1), (1, 1)\}$, $\mathscr{T} = \{(1, 0), (0, -1)\}$

(*e*) $\mathscr{S} = \{(1, -1), (1, 1)\}$, $\mathscr{T} = \{(0, -1), (1, 0)\}$

In each case, calculate $T((3, 2))$ directly and by using M.

Solution (*a*) $T(\mathbf{e}_1) = (1, 1) = 1(1, 0) + 1(0, 1)$ so $(T(\mathbf{e}_1))_{\mathscr{T}} = \begin{pmatrix} 1 \\ 1 \end{pmatrix}$

$T(\mathbf{e}_2) = (2, -1) = 2(1, 0) + (-1)(0, 1)$ so $(T(\mathbf{e}_2))_{\mathscr{T}} = \begin{pmatrix} 2 \\ -1 \end{pmatrix}$

The matrix is

$$M = \begin{pmatrix} 1 & 2 \\ 1 & -1 \end{pmatrix}$$

Now $T((3, 2)) = (7, 1)$ from the definition of T. But

$$((3, 2))_{\mathscr{S}} = \begin{pmatrix} 3 \\ 2 \end{pmatrix}$$

so

$$(T(3, 2))_{\mathscr{T}} = M\begin{pmatrix} 3 \\ 2 \end{pmatrix} = \begin{pmatrix} 1 & 2 \\ 1 & -1 \end{pmatrix}\begin{pmatrix} 3 \\ 2 \end{pmatrix} = \begin{pmatrix} 7 \\ 1 \end{pmatrix}$$

and finally,

$$\begin{pmatrix} 7 \\ 1 \end{pmatrix} \longrightarrow 7(1, 0) + 1(0, 1) = (7, 1)$$

In this case, since $\mathscr{S} = \mathscr{T}$, we say that M is the *matrix of T with respect to* $\mathscr{S}$. Also, since $\mathscr{S} = \mathscr{T} =$ the standard basis, M is called the *standard matrix of T*.

(*b*) We work as in (*a*):

$T((1, 2)) = (5, -1) = \frac{2}{7}(1, 2) + \frac{11}{7}(3, -1)$ so $(T(1, 2))_{\mathscr{T}} = \begin{pmatrix} \frac{2}{7} \\ \frac{11}{7} \end{pmatrix}$

$T((3, -1)) = (1, 4) = \frac{13}{7}(1, 2) + (-\frac{2}{7})(3, -1)$ so $(T(3, -1))_{\mathscr{T}} = \begin{pmatrix} \frac{13}{7} \\ -\frac{2}{7} \end{pmatrix}$

So we have the matrix of T with respect to $\mathscr{S}$:

$$M = \begin{pmatrix} \frac{2}{7} & \frac{13}{7} \\ \frac{11}{7} & -\frac{2}{7} \end{pmatrix}$$

From the definition of T, $T(3, 2) = (7, 1)$; since

$$((3, 2))_{\mathscr{S}} = \begin{pmatrix} \frac{9}{7} \\ \frac{4}{7} \end{pmatrix}$$

using the matrix M we have

$$(T(3, 2))_{\mathscr{T}} = M\begin{pmatrix} \frac{9}{7} \\ \frac{4}{7} \end{pmatrix} = \begin{pmatrix} \frac{2}{7} & \frac{13}{7} \\ \frac{11}{7} & -\frac{2}{7} \end{pmatrix}\begin{pmatrix} \frac{9}{7} \\ \frac{4}{7} \end{pmatrix} = \begin{pmatrix} \frac{10}{7} \\ \frac{13}{7} \end{pmatrix}$$

Finally

$$\begin{pmatrix} \frac{10}{7} \\ \frac{13}{7} \end{pmatrix} \longrightarrow \tfrac{10}{7}(1, 2) + \tfrac{13}{7}(3, -1) = (7, 1)$$

(*c*) The case differs from (*b*) in that the order of $\mathscr{S}$ has been reversed. The calculations, however, are similar:

$$T((3, -1)) = (1, 4) = -\tfrac{2}{7}(3, -1) + \tfrac{13}{7}(1, 2) \qquad \text{so} \qquad (T(3, -1))_{\mathscr{T}} = \begin{pmatrix} -\frac{2}{7} \\ \frac{13}{7} \end{pmatrix}$$

$$T((1, 2)) = (5, -1) = \tfrac{11}{7}(3, -1) + \tfrac{2}{7}(1, 2) \qquad \text{so} \qquad (T(1, 2))_{\mathscr{T}} = \begin{pmatrix} \frac{11}{7} \\ \frac{2}{7} \end{pmatrix}$$

Therefore the matrix of T with respect to $\mathscr{S}$ is

$$M = \begin{pmatrix} -\frac{2}{7} & \frac{11}{7} \\ \frac{13}{7} & \frac{2}{7} \end{pmatrix}.$$

Note that this matrix differs from the representing matrix in (*b*) in that an interchange of rows and columns has occurred. Calculation of $T(3, 2)$ is left to the reader.

(*d*) In this case $\mathscr{S} \neq \mathscr{T}$, but our usual procedure can be used.

$$T((1, -1)) = (-1, 2) = -1(1, 0) + (-2)(0, -1) \quad \text{so} \quad (T(1, -1))_{\mathscr{T}} = \begin{pmatrix} -1 \\ -2 \end{pmatrix}$$

$$T((1, 1)) = (3, 0) = 3(1, 0) + 0(0, -1) \qquad \text{so} \qquad (T(1, 1))_{\mathscr{T}} = \begin{pmatrix} 3 \\ 0 \end{pmatrix}$$

The *matrix of T with respect to* $\mathscr{S}$ and $\mathscr{T}$ is

$$M = \begin{pmatrix} -1 & 3 \\ -2 & 0 \end{pmatrix}$$

Now to calculate $T(3, 2)$ in two ways, $T(3, 2) = (7, 1)$ by definition, but we also have

$$((3, 2))_{\mathscr{S}} = \begin{pmatrix} \frac{1}{2} \\ \frac{5}{2} \end{pmatrix}$$

and

$$M\begin{pmatrix} \frac{1}{2} \\ \frac{5}{2} \end{pmatrix} = \begin{pmatrix} 7 \\ -1 \end{pmatrix} \longrightarrow 7(1, 0) + (-1)(0, -1) = (7, 1)$$

(*e*) The details of this case are left to the reader. The matrix in this case is

$$M = \begin{pmatrix} -2 & 0 \\ -1 & 3 \end{pmatrix}$$

Note that the difference between (*d*) and (*e*) is that the order of the basis in the range is reversed. This resulted in an interchange of rows in the representing matrix.

Having solved some examples, we now state and prove the theorem which furnishes our procedure for the solution of the representation problem.

THEOREM 4.2.1 (Solution of the representation problem) Let $T: V \to W$, where $\dim V = n$ and $\dim W = m$, be a linear transformation. Let $\mathscr{S} = \{\mathbf{v}_1, \ldots, \mathbf{v}_n\}$ and $\mathscr{T} = \{\mathbf{w}_1, \ldots, \mathbf{w}_m\}$ be bases for V and W, respectively. There exists an $m \times n$ matrix M (unique to the ordered bases $\mathscr{S}$ and $\mathscr{T}$) with the property that for any $\mathbf{x} \in V$, $(T(\mathbf{x}))_{\mathscr{T}} = M(\mathbf{x})_{\mathscr{S}}$.

Proof Let $\mathbf{x} \in V$. Because $\mathscr{S}$ is a basis, there exists a unique set $\{c_1, \ldots, c_n\}$ of constants such that $\mathbf{x} = c_1\mathbf{v}_1 + c_2\mathbf{v}_2 + \cdots + c_n\mathbf{v}_n$. Now $T(\mathbf{x}) = c_1(T(\mathbf{v}_1)) + c_2(T(\mathbf{v}_2)) + \cdots + c_n(T(\mathbf{v}_n))$ by the linearity of T. For each k $(1 \le k \le n)$, $T(\mathbf{v}_k)$ is in W and can be represented by the basis elements for W:

$$T(\mathbf{v}_k) = a_{1k}\mathbf{w}_1 + a_{2k}\mathbf{w}_2 + \cdots + a_{mk}\mathbf{w}_m$$

Therefore,

$$\begin{aligned} T(\mathbf{x}) = {} & c_1(a_{11}\mathbf{w}_1 + a_{21}\mathbf{w}_2 + \cdots + a_{m1}\mathbf{w}_m) \\ & + c_2(a_{12}\mathbf{w}_1 + a_{22}\mathbf{w}_2 + \cdots + a_{m2}\mathbf{w}_m) + \cdots \\ & + c_n(a_{1n}\mathbf{w}_1 + a_{2n}\mathbf{w}_2 + \cdots + a_{mn}\mathbf{w}_m) \end{aligned}$$

and after collecting terms we have

$$\begin{aligned} T(\mathbf{x}) = {} & (a_{11}c_1 + a_{12}c_2 + \cdots + a_{1n}c_n)\mathbf{w}_1 \\ & + (a_{21}c_1 + a_{22}c_2 + \cdots + a_{2n}c_n)\mathbf{w}_2 + \cdots \\ & + (a_{m1}c_1 + a_{m2}c_2 + \cdots + a_{mn}c_n)\mathbf{w}_m \end{aligned}$$

The coefficients of $\mathbf{w}_1, \ldots, \mathbf{w}_m$ in the last expression are exactly the row-column products from

$$M\begin{pmatrix} c_1 \\ \vdots \\ c_n \end{pmatrix} = \begin{pmatrix} a_{11} & a_{12} & \cdots & a_{1n} \\ a_{21} & a_{22} & \cdots & a_{2n} \\ \cdots & \cdots & \cdots & \cdots \\ a_{m1} & a_{m2} & \cdots & a_{mn} \end{pmatrix}\begin{pmatrix} c_1 \\ c_2 \\ \vdots \\ c_n \end{pmatrix}$$

$$(T(\mathbf{v}_1))_{\mathscr{T}} \quad (T(\mathbf{v}_2))_{\mathscr{T}} \cdots (T(\mathbf{v}_n))_{\mathscr{T}}$$

Therefore the $m \times n$ matrix M, the columns of which are the coordinate vectors of $T(\mathbf{v}_1), \ldots, T(\mathbf{v}_n)$, is the desired matrix. The matrix is unique to the pair of bases since coordinate vectors are unique in a given basis.

EXAMPLE 6 Let $T:\mathscr{P}_1 \to \mathscr{P}_2$ be defined by $T(a + bx) = ax + (b/2)x^2$. Give $\mathscr{P}_1$ and $\mathscr{P}_2$ the standard bases $\mathscr{S} = \{1, x\}$ and $\mathscr{T} = \{1, x, x^2\}$, respectively. Find the matrix of T with respect to these bases. Do the same for $L:\mathscr{P}_2 \to \mathscr{P}_1$ defined by $L(a + bx + cx^2) = b + 2cx$.

Solution Now $T(1) = x$, so

$$(T(1))_{\mathscr{T}} = \begin{pmatrix} 0 \\ 1 \\ 0 \end{pmatrix}$$

Likewise, $T(x) = \frac{1}{2}x^2$, so

$$(T(x))_{\mathscr{T}} = \begin{pmatrix} 0 \\ 0 \\ \frac{1}{2} \end{pmatrix}$$

Therefore the matrix M_T representing T is

$$M_T = \begin{pmatrix} 0 & 0 \\ 1 & 0 \\ 0 & \frac{1}{2} \end{pmatrix}$$

For the transformation L,

$$L(1) = 0 \qquad L(x) = 1 \qquad L(x^2) = 2x$$

Thus

$$(L(1))_{\mathscr{S}} = \begin{pmatrix} 0 \\ 0 \end{pmatrix} \qquad (L(x))_{\mathscr{S}} = \begin{pmatrix} 1 \\ 0 \end{pmatrix} \qquad (L(x^2))_{\mathscr{S}} = \begin{pmatrix} 0 \\ 2 \end{pmatrix}$$

and the matrix M_L representing L is

$$M_L = \begin{pmatrix} 0 & 1 & 0 \\ 0 & 0 & 2 \end{pmatrix}$$

Note that

$$M_L M_T = \begin{pmatrix} 1 & 0 \\ 0 & 1 \end{pmatrix} = I_2$$

so that we could call M_L a *left inverse* of M_T. However,

$$M_T M_L = \begin{pmatrix} 0 & 0 & 0 \\ 0 & 1 & 0 \\ 0 & 0 & 1 \end{pmatrix} \neq I_3$$

and M_T is *not* a left inverse of M_L. Note that T is just antidifferentiation with arbitrary constant set to zero. When we antidifferentiate and then differentiate, we get the original function back. This is reflected by $M_L M_T = I$.

EXAMPLE 7 Let $T:\mathscr{M}_{22} \to \mathscr{M}_{22}$ be defined by $T(A) = A - A^T$. Give $\mathscr{M}_{22}$ the standard basis

$$\mathscr{S} = \left\{\begin{pmatrix}1 & 0\\0 & 0\end{pmatrix}, \begin{pmatrix}0 & 1\\0 & 0\end{pmatrix}, \begin{pmatrix}0 & 0\\1 & 0\end{pmatrix}, \begin{pmatrix}0 & 0\\0 & 1\end{pmatrix}\right\} = \{\mathbf{e}_1, \mathbf{e}_2, \mathbf{e}_3, \mathbf{e}_4\}$$

and find the matrix for T with respect to $\mathscr{S}$.

Solution First we calculate the images of the basis vectors:

$$T\left(\begin{pmatrix}1 & 0\\0 & 0\end{pmatrix}\right) = \begin{pmatrix}0 & 0\\0 & 0\end{pmatrix} = 0\mathbf{e}_1 + 0\mathbf{e}_2 + 0\mathbf{e}_3 + 0\mathbf{e}_4$$

$$T\left(\begin{pmatrix}0 & 1\\0 & 0\end{pmatrix}\right) = \begin{pmatrix}0 & 1\\-1 & 0\end{pmatrix} = 0\mathbf{e}_1 + \mathbf{e}_2 - \mathbf{e}_3 + 0\mathbf{e}_4$$

$$T\left(\begin{pmatrix}0 & 0\\1 & 0\end{pmatrix}\right) = \begin{pmatrix}0 & -1\\1 & 0\end{pmatrix} = 0\mathbf{e}_1 - \mathbf{e}_2 + \mathbf{e}_3 + 0\mathbf{e}_4$$

$$T\left(\begin{pmatrix}0 & 0\\0 & 1\end{pmatrix}\right) = \begin{pmatrix}0 & 0\\0 & 0\end{pmatrix} = 0\mathbf{e}_1 + 0\mathbf{e}_2 + 0\mathbf{e}_3 + 0\mathbf{e}_4$$

Therefore,

$$M = \begin{pmatrix}0 & 0 & 0 & 0\\0 & 1 & -1 & 0\\0 & -1 & 1 & 0\\0 & 0 & 0 & 0\end{pmatrix}$$

EXAMPLE 8 Define $T:\mathbb{C}^2$ to $\mathbb{C}^2$ by $T((z_1, z_2)) = (iz_1, (1 + i)z_2 - z_1)$. Let $\mathbb{C}^2$ have the basis $\mathscr{S} = \{(i, 0), (0, 1)\}$. Calculate M_T.

Solution

$$T(i, 0) = (-1, -i) = i(i, 0) + (-i)(0, 1)$$
$$T(0, 1) = (0, 1 + i) = 0(i, 0) + (1 + i)(0, 1)$$

Therefore

$$M_T = \begin{pmatrix}i & 0\\-i & 1 + i\end{pmatrix}$$

Some Algebra of Linear Transformations Let V and W be vector spaces, and let L and T be linear transformations from V to W. We can define the scalar multiple rL of L and the sum $L + T$ of L and T as linear transformations from V to W by the rules

$$(rL)(v) = r(L(v)) \qquad r \text{ a number, } v \text{ in } V$$
$$(L + T)(v) = L(v) + T(v) \qquad v \text{ in } V$$

If M_L and M_T are the representing matrices with respect to bases $\mathscr{S}$ and $\mathscr{T}$, respectively, then rM_L represents rL and $M_L + M_T$ represents $L + T$. This can be shown by the method of proof in Theorem 4.2.1.

EXAMPLE 9 Consider T as defined in Example 5a, and define $L: E^2 \to E^2$ by $L((x_1, x_2)) = (x_2, x_1)$. The standard matrices for T and L are

$$M_T = \begin{pmatrix} 1 & 2 \\ 1 & -1 \end{pmatrix} \qquad M_L = \begin{pmatrix} 0 & 1 \\ 1 & 0 \end{pmatrix}$$

So $T + L$ is defined as

$$\begin{aligned}(T + L)((x_1, x_2)) &= T((x_1, x_2)) + L((x_1, x_2)) \\ &= (x_1 + 2x_2, x_1 - x_2) + (x_2, x_1) \\ &= (x_1 + 3x_2, 2x_1 - x_2)\end{aligned}$$

and rT is defined as

$$\begin{aligned}(rT)((x_1, x_2)) = r(T((x_1, x_2))) &= r(x_1 + 2x_2, x_1 - x_2) \\ &= (rx_1 + 2rx_2, rx_1 - rx_2)\end{aligned}$$

The standard matrix for $T + L$ is

$$\begin{pmatrix} 1 & 3 \\ 2 & -1 \end{pmatrix}$$

and the standard matrix for rT is

$$\begin{pmatrix} r & 2r \\ r & -r \end{pmatrix}$$

Direct calculation shows that

$$\begin{aligned}M_{T+L} &= M_T + M_L \\ M_{rL} &= rM_L\end{aligned}$$

PROBLEMS 4.2

1. For the following sets of homogeneous equations $AX = 0$, find rank A, dim (solution space), and verify that

$$\dim(\text{solution space}) + \text{rank } A = \text{no. of columns of } A$$

(a) $A = \begin{pmatrix} 1 & 2 & 3 \\ 4 & 5 & 6 \\ 7 & 8 & 9 \end{pmatrix}$ $\qquad$ (b) $A = \begin{pmatrix} 1 & -1 & 2 \\ 0 & 1 & 1 \\ 1 & 0 & 3 \end{pmatrix}$

(c) $A = \begin{pmatrix} 1 & 2 & -1 & 0 \\ 4 & -2 & 3 & 1 \\ 6 & 2 & 1 & 1 \\ 3 & -4 & 4 & 1 \end{pmatrix}$

2. For the following transformations $T: V \to W$, find the matrix of T, *assuming*

the standard basis in both V and *W.*

(a) $T: E^2 \to E^2$, $\quad T((x_1, x_2)) = (3x_1, 2x_1 - 5x_2)$

(b) $T: \mathscr{P}_1 \to E^2$, $\quad T(ax + b) = (3a, 2a - 5b)$ for $\mathscr{P}_1$, $S = \{1, x\}$

(c) $T: E^2 \to E^3$, $\quad T((x_1, x_2)) = (x_1, x_1 + x_2, 3x_1 - x_2)$

(d) $T: \mathscr{P}_2 \to \mathscr{P}_1 \longrightarrow T(ax^2 + bx + c) = cx + b$ for $\mathscr{P}_2$, $S = \{1, x, x^2\}$

(e) $T: \mathscr{M}_{22} \to \mathscr{M}_{22}$, $\quad T\left(\begin{pmatrix} a & b \\ c & d \end{pmatrix}\right) = \begin{pmatrix} c & d \\ a & b \end{pmatrix}$

(f) $T: \mathscr{M}_{22} \to \mathscr{M}_{21}$, $\quad T\left(\begin{pmatrix} a & b \\ c & d \end{pmatrix}\right) = \begin{pmatrix} a \\ b \end{pmatrix}$

3. Let $T: E^3 \to E^2$ be defined by $T((x_1, x_2, x_3)) = (x_1 - x_2 + x_3, x_2 - x_3)$, let E^3 have the standard basis, and let E^2 have the basis $\mathscr{S} = \{(1, 1), (1, -1)\}$.

(a) Find the matrix of T with respect to these bases.

(b) Calculate $T((1, -1, 2))$ directly and by using the matrix of T.

4. Let $T: E^3 \to E^3$ be defined by $T((x_1, x_2, x_3)) = (x_1 + x_2, x_2 + x_3, x_3 + x_1)$, and let E^3 have the basis $\mathscr{S} = \{(1, 1, 0), (0, 1, 1), (1, 0, 1)\}$.

(a) Find the matrix of T with respect to $\mathscr{S}$.

(b) Calculate $T((1, 1, 1))$ directly and by using the matrix of T.

5. Let $T: E^2 \to E^2$ be a linear transformation with the property $T(1, 1) = (1, 0)$ and $T(1, -1) = (0, 1)$. Find a matrix representation for T. [*Hint:* For the domain use $\mathscr{S} = \{(1, 1), (1, -1)\}$ as a basis.]

6. Define $L: E^n \to E^n$ by $L((x_1, x_2, x_3, \ldots, x_n)) = (x_2, x_3, \ldots, x_n, 0)$ and define $R: E^n \to E^n$ by $R((x_1, x_2, x_3, \ldots, x_n)) = (0, x_1, x_2, x_3, \ldots, x_{n-1})$. Find the matrices for L and R with respect to the standard basis.

7. Show that the matrix representing the zero transformation is the zero matrix regardless of basis.

8. Show that a contraction or dilation transformation from V to V has a diagonal matrix representation regardless of the basis given to V (same basis in domain and range).

9. Let T be defined by $T: \mathscr{P}_n \to \mathscr{P}_{n+1}$ as the coordinate operator

$$(T(f))(x) = xf(x)$$

Show that the standard matrix of T is

$$\begin{pmatrix} 0 & 0 & 0 & \cdots & 0 \\ 1 & 0 & 0 & & 0 \\ 0 & 1 & \ddots & \ddots & \vdots \\ \vdots & & \ddots & \ddots & 0 \\ 0 & \cdots & 0 & 1 & 0 \\ 0 & & \cdots & 0 & 1 \end{pmatrix}_{(n+2) \times (n+1)}$$

10. Let T be the momentum operator defined at the end of Sec. 4.1. Given that $\mathscr{P}_n^{\mathbb{C}}$ has the standard ordered basis $\{1, x, \ldots, x^n\}$, show that the standard matrix of T is

$$\begin{pmatrix} 0 & -i & 0 & 0 & 0 & \cdots & 0 \\ 0 & 0 & -2i & 0 & 0 & \cdots & 0 \\ 0 & 0 & 0 & -3i & 0 & \cdots & 0 \\ \vdots & \vdots & & \ddots & \ddots & & \vdots \\ & & & & & & 0 \\ 0 & 0 & & \cdots & 0 & \cdots & -ni \end{pmatrix}_{n \times (n+1)}$$

11. For the following transformations $T: V \to W$, find the standard matrix of T.

(*a*) $T: \mathbb{C}^2 \to \mathbb{C}^2$, $T((z_1, z_2)) = (z_1 + z_2, iz_2)$

(*b*) $T: \mathbb{C}^3 \to \mathbb{C}^3$, $T((z_1, z_2, z_3)) = (iz_2, iz_3, 0)$

(*c*) $T: \mathscr{C}_{22} \to \mathscr{C}_{22}$, $T(A) = A + iA^T$

4.3 SIMILAR MATRICES AND CHANGE OF BASIS

The purpose of a matrix representation M for a linear transformation T is to enable us to analyze T by working with M. If M is easy to work with, we have gained an advantage; if not, we have no advantage. Since different bases lead to different matrices, the "right" choice of basis to obtain a simple matrix M is important. The right choice of basis is not obvious, as Example 1 shows.

EXAMPLE 1 Show that $T: E^2 \to E^2$ defined by $T(x_1, x_2) = (x_1 + 6x_2, 3x_1 + 4x_2)$ has standard matrix

$$\begin{pmatrix} 1 & 6 \\ 3 & 4 \end{pmatrix}$$

Then show that, with respect to the basis $\mathscr{T} = \{(2, -1), (1, 1)\}$, T has a diagonal matrix representation.

Solution For the standard matrix we have

$$T((1, 0)) = (1, 3) = 1(1, 0) + 3(0, 1) \quad \text{so} \quad (T((1, 0)))_{\text{std}} = \begin{pmatrix} 1 \\ 3 \end{pmatrix}$$

$$T((0, 1)) = (6, 4) = 6(1, 0) + 4(0, 1) \quad \text{so} \quad (T((0, 1)))_{\text{std}} = \begin{pmatrix} 6 \\ 4 \end{pmatrix}$$

and

$$M_{\text{std}} = \begin{pmatrix} 1 & 6 \\ 3 & 4 \end{pmatrix}$$

But with respect to $\mathscr{T}$,

$$T((2, -1)) = (-4, 2) = -2(2, -1) + 0(1, 1) \quad \text{so} \quad [T((1, 0))]_{\mathscr{T}} = \begin{pmatrix} -2 \\ 0 \end{pmatrix}$$

$$T((1, 1)) = (7, 7) = 0(2, -1) + 7(1, 1) \quad \text{so} \quad [T((1, 1))]_{\mathscr{T}} = \begin{pmatrix} 0 \\ 7 \end{pmatrix}$$

and the matrix with respect to $\mathscr{T}$ is

$$M = \begin{pmatrix} -2 & 0 \\ 0 & 7 \end{pmatrix}$$

We recall from matrix algebra that diagonal matrices are easy to work with for certain operations: inversion, determinants, and multiplication, to name three. As the dimension of the vector spaces (and size of the matrices) grows, this is even more the case. We need, then, to find a way of getting the simplest possible matrix to represent a transformation T. To solve this problem (a solution is presented in Chap. 5), we must discover how to relate different matrix representations for the given linear transformation. We restrict our attention to the case $V = W$ with the same basis in V and W. This case occurs most frequently in applications.

To discover the relationship, suppose that $M_{(\mathscr{S})}$ is the matrix representing $T:(V, \mathscr{S}) \to (V, \mathscr{S})$ and that $M_{(\mathscr{T})}$ represents $T:(V, \mathscr{T}) \to (V, \mathscr{T})$. Let P be the transition matrix from basis $\mathscr{T}$ to basis $\mathscr{S}$, so that for any x in V,

$$\begin{aligned} M_{(\mathscr{T})}(\mathbf{x})_{\mathscr{T}} = (T\mathbf{x})_{\mathscr{T}} = P^{-1}(T\mathbf{x})_{\mathscr{S}} &= P^{-1}(M_{(\mathscr{S})}(\mathbf{x})_{\mathscr{S}}) \\ &= P^{-1}M_{(\mathscr{S})}I(\mathbf{x})_{\mathscr{S}} \\ &= P^{-1}M_{\mathscr{S}}(PP^{-1})(\mathbf{x})_{\mathscr{S}} \\ &= (P^{-1}M_{(\mathscr{S})}P)(P^{-1}(\mathbf{x})_{\mathscr{S}}) \\ &= (P^{-1}M_{(\mathscr{S})}P)(\mathbf{x})_{\mathscr{T}} \end{aligned}$$

Therefore

$$M_{(\mathscr{T})}(\mathbf{x})_{\mathscr{T}} = (P^{-1}M_{(\mathscr{S})}P)(\mathbf{x})_{\mathscr{T}} \qquad \text{for all } \mathbf{x} \text{ in } V$$

so that $M_{(\mathscr{T})} = P^{-1}M_{(\mathscr{S})}P$. These equations actually give a proof of the basic result.

THEOREM 4.3.1 Let $T: V \to V$ be a linear transformation with matrix $M_{(\mathscr{S})}$ with respect to a basis $\mathscr{S}$ and with matrix $M_{(\mathscr{T})}$ with respect to a basis $\mathscr{T}$. If P is the transition matrix from basis $\mathscr{T}$ to basis $\mathscr{S}$, then

$$M_{(\mathscr{T})} = P^{-1}M_{(\mathscr{S})}P$$

The relation $M_{(\mathscr{T})} = P^{-1}M_{(\mathscr{S})}P$ is important enough to be given a name.

DEFINITION 4.3.1 Two $n \times n$ matrices A and B are *similar* if there exists an invertible matrix P such that $B = P^{-1}AP$.

Note that the definition in no way tells us how to find the *similarity transform* P. In the case of two representation matrices for a linear transformation, P is a transition matrix from one basis to another, as we saw in Theorem 4.3.1.

An important restatement of Theorem 4.3.1 is as follows:

> Let $T: V \to V$ be a linear transformation. Any two representing matrices of T are similar.

EXAMPLE 2 In Example 1, denote the standard basis by $\mathscr{S}$. Illustrate Theorem 4.3.1 for T, $\mathscr{S}$, and $\mathscr{T}$, as given in Example 1.

Solution The standard matrix, as before, is

$$M_{(\mathscr{S})} = \begin{pmatrix} 1 & 6 \\ 3 & 4 \end{pmatrix}$$

Now calculate the transition matrix from $\mathscr{T}$ to $\mathscr{S}$:

$$(2, -1) = 2(1, 0) + (-1)(0, 1)$$

$$(1, 1) = 1(1, 0) + 1(0, 1)$$

$$P = \begin{pmatrix} 2 & 1 \\ -1 & 1 \end{pmatrix} \Rightarrow P^{-1} = \begin{pmatrix} \frac{1}{3} & -\frac{1}{3} \\ \frac{1}{3} & \frac{2}{3} \end{pmatrix}$$

Then

$$\begin{aligned} P^{-1}M_{(\mathscr{S})}P = P^{-1}\begin{pmatrix} 1 & 6 \\ 3 & 4 \end{pmatrix}P &= \begin{pmatrix} \frac{1}{3} & -\frac{1}{3} \\ \frac{1}{3} & \frac{2}{3} \end{pmatrix}\begin{pmatrix} 1 & 6 \\ 3 & 4 \end{pmatrix}\begin{pmatrix} 2 & 1 \\ -1 & 1 \end{pmatrix} \\ &= \begin{pmatrix} \frac{1}{3} & -\frac{1}{3} \\ \frac{1}{3} & \frac{2}{3} \end{pmatrix}\begin{pmatrix} -4 & 7 \\ 2 & 7 \end{pmatrix} \\ &= \begin{pmatrix} -2 & 0 \\ 0 & 7 \end{pmatrix} = M_{(\mathscr{T})} \end{aligned}$$

One way to remember how to relate matrices with respect to $\mathscr{S}$ and $\mathscr{T}$ is to use the diagram

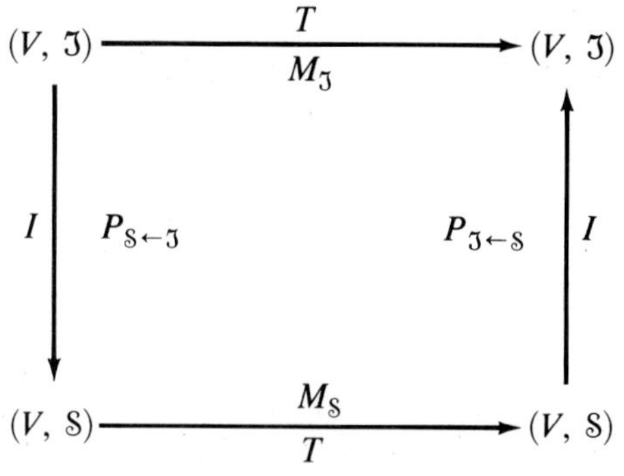

$(V, \mathfrak{T}) \xrightarrow[M_{\mathfrak{T}}]{T} (V, \mathfrak{T})$

I $P_{\mathcal{S} \leftarrow \mathfrak{T}}$ (down) $\qquad$ $P_{\mathfrak{T} \leftarrow \mathcal{S}}$ I (up)

$(V, \mathcal{S}) \xrightarrow[T]{M_{\mathcal{S}}} (V, \mathcal{S})$

The point is that the transition matrix is just the matrix which represents the identity transformation. The notation $P_{\mathscr{A}\leftarrow\mathscr{B}}$ indicates the transition matrix from $\mathscr{B}$ to $\mathscr{A}$. Note that

$$M_{(\mathscr{T})} = P_{\mathscr{T}\leftarrow\mathscr{S}} M_{(\mathscr{S})} \; P_{\mathscr{S}\leftarrow\mathscr{T}}$$

All the same

and remember that $P_{\mathscr{T}\leftarrow\mathscr{S}} = (P_{\mathscr{S}\leftarrow\mathscr{T}})^{-1}$

To exploit fully the result of Theorem 4.3.1, we will later use some properties of similar matrices.

THEOREM 4.3.2 The following statements regarding similarity are true. In all cases the matrices are $n \times n$.

(*a*) Matrix A is similar to A.

(*b*) If A is similar to B, then B is similar to A.

(*c*) If A is similar to B and B is similar to C, then A is similar to C.

(*d*) If A is similar to B, then $\det A = \det B$.

(*e*) If A is similar to B, then $\operatorname{tr} A = \operatorname{tr} B$.

(*f*) If A is similar to B, then A^m is similar to B^m for any positive integer m.

(*g*) If A is similar to B, then A is invertible if and only if B is invertible. In that case A^{-1} is similar to B^{-1}.

[*Note*: Parts (*d*), (*e*), and (*g*) state, respectively, that the determinant, trace, and invertibility are *invariant under similarity*.]

Proof (*a*) Since $A = IAI = I^{-1}AI$, A is similar to A.

(*b*) If A is similar to B, then $B = P^{-1}AP$. Thus $PBP^{-1} = P(P^{-1}AP)P^{-1} = A$. Thus $A = (P^{-1})^{-1}B(P^{-1})$. Calling P^{-1} by the name $\mathscr{P}$, we have $A = \mathscr{P}^{-1}B\mathscr{P}$, and B is similar to A.

(*c*) If A is similar to B and B is similar to C, we have $B = P^{-1}AP$ and $C = Q^{-1}BQ$, where P and Q are, in general, not equal. Now we have

$$C = Q^{-1}(B)Q = Q^{-1}(P^{-1}AP)Q = (Q^{-1}P^{-1})A(PQ) = (PQ)^{-1}A(PQ)$$

Since PQ is invertible, C is similar to A.

(*d*) If $B = P^{-1}AP$, then

$$\begin{aligned}\det B &= \det(P^{-1}AP)\\ &= (\det P^{-1})(\det A)(\det P) = (\det A)(\det P^{-1})(\det P)\\ &= (\det A)\left(\frac{1}{\det P}\right)(\det P) = \det A\end{aligned}$$

(*e*) If $B = P^{-1}AP$, since $\operatorname{tr}(AB) = \operatorname{tr}(BA)$ we have $\operatorname{tr} B = \operatorname{tr}(P^{-1}AP) = \operatorname{tr}((P^{-1}A)P) = \operatorname{tr}(P(P^{-1}A)) = \operatorname{tr} A$.

(*f*) If $B = P^{-1}AP$, then

$$\begin{aligned}B^2 = (P^{-1}AP)^2 &= (P^{-1}AP)(P^{-1}AP)\\ &= (P^{-1}A)(PP^{-1})(AP)\\ &= P^{-1}A^2P\end{aligned}$$

Thus B^2 is similar to A^2 and has the same similarity transform P.

Now we use induction. Suppose for the induction hypothesis that A^k is similar to B^k with $B^k = P^{-1}A^kP$. Now we show that B^{k+1} is similar to A^{k+1}. By the induction hypothesis

$$B^k = P^{-1}A^kP$$

Now

$$\begin{aligned}B^{k+1} &= BP^{-1}A^kP = (P^{-1}AP)P^{-1}A^kP\\ &= P^{-1}AIA^kP\\ &= P^{-1}A^{k+1}P\end{aligned}$$

Therefore B^{k+1} is similar to A^{k+1} with similarity transform P. By the principle of mathematical induction, (*f*) is proved.

(*g*) Since $\det A = \det B$, we know that $\det A \neq 0$ if and only if $\det B \neq 0$. For the second part note that

$$A = P^{-1}BP,\ A^{-1} = (P^{-1}BP)^{-1} = P^{-1}B^{-1}(P^{-1})^{-1} = P^{-1}B^{-1}P.$$

Parts (*d*) and (*e*) are illustrated by the similar matrices

$$\begin{pmatrix}1 & 6\\ 3 & 4\end{pmatrix} \quad \text{and} \quad \begin{pmatrix}-2 & 0\\ 0 & 7\end{pmatrix}$$

from Example 1. Both matrices have trace 5 and determinant -14.

The problem of determining whether two given matrices are similar is generally difficult. But parts (*d*) and (*e*) can be used to rule out similarity: If $\operatorname{tr} A \neq \operatorname{tr} B$ or $\det A \neq \det B$, then A cannot be similar to B.

EXAMPLE 3 For the following pairs of matrices, determine whether A is similar to B.

(*a*) $A = \begin{pmatrix}1 & -1\\ 2 & 3\end{pmatrix} \qquad B = \begin{pmatrix}2 & 1\\ 2 & 2\end{pmatrix}$

(*b*)
$$A = \begin{pmatrix} 2 & 6 & 2 \\ 5 & 1 & -1 \\ 4 & 1 & 3 \end{pmatrix} \qquad B = \begin{pmatrix} 1 & 1 & 3 \\ 0 & 1 & 2 \\ 2 & 2 & 5 \end{pmatrix}$$

(*c*)
$$A = \begin{pmatrix} 1 & 1 \\ 0 & 1 \end{pmatrix} \qquad B = \begin{pmatrix} 1 & 0 \\ 0 & 1 \end{pmatrix}$$

Solution (*a*) Although $\operatorname{tr} A = \operatorname{tr} B$, $\det A \neq \det B$, so A is not similar to B.

(*b*) Since $\operatorname{tr} A \neq \operatorname{tr} B$, matrix A is not similar to matrix B.

(*c*) In this case, the traces and determinants of A and B coincide, so similarity cannot be easily ruled out. Since A and B are small, we can check the equation

$$B = P^{-1}AP$$

If this is to hold, we must have

$$PB = AP$$

for a nonsingular matrix p, Let

$$P = \begin{pmatrix} a & b \\ c & d \end{pmatrix}$$

Then the required equation is

$$\begin{pmatrix} a & b \\ c & d \end{pmatrix}\begin{pmatrix} 1 & 0 \\ 0 & 1 \end{pmatrix} = \begin{pmatrix} 1 & 1 \\ 0 & 1 \end{pmatrix}\begin{pmatrix} a & b \\ c & d \end{pmatrix}$$

which leads to

$$\begin{aligned} a &= a + c \Rightarrow 0 = c \\ b &= b + d \Rightarrow 0 = d \\ c &= c \qquad \Rightarrow c = c \\ d &= d \qquad \Rightarrow d = d \end{aligned}$$

Therefore a and b are arbitrary, and $c = d = 0$ which leads to

$$P = \begin{pmatrix} a & b \\ 0 & 0 \end{pmatrix}$$

which is singular. Therefore A is not similar to B. For large matrices, this method of solution is not practical.

PROBLEMS 4.3 In Probs. 1 to 9 a linear transformation $T: V \to V$ and bases $\mathscr{S}$ and $\mathscr{T}$ are given.

(*a*) Find the matrix of T with respect to $\mathscr{S}$.

(*b*) Find the matrix of T with respect to $\mathcal{T}$ by using the transition matrix.
(*c*) Find the matrix of T with respect to $\mathcal{T}$ directly.

1. $T: E^2 \to E^2$, $T((x_1, x_2)) = (x_1, 0)$
$\mathcal{S}$ = standard basis $\quad \mathcal{T} = \{(1, 1), (1, -1)\}$

2. $T: E^2 \to E^2$, $T((x_1, x_2)) = (2x_1, 2x_2)$
$\mathcal{S}$ = standard basis $\quad \mathcal{T} = \{(1, 1), (1, -1)\}$
(This is a dilation. Compare the result of this problem with Prob. 8 of the last section.)

3. $T: E^2 \to E^2$, $T((x_1, x_2)) = (x_1 + x_2, 2x_1 - 3x_2)$
$\mathcal{S}$ = standard basis $\quad \mathcal{T} = \{(1, 1), (1, -1)\}$

4. T as in Prob. 3
$\mathcal{S} = \{(1, 1), (1, -1)\} \quad \mathcal{T} = \{(1, 1), (1, 2)\}$

5. $T: E^2 \to E^2$, where T is rotation through $\pi/4$ radians counterclockwise.
$\mathcal{S}$ = standard basis $\quad \mathcal{T} = \{(1, 1), (1, 2)\}$

6. $T: E^3 \to E^3$, $T((x_1, x_2, x_3)) = (x_1, x_1 + x_2, x_2 - x_3)$
$\mathcal{S}$ = standard basis $\quad \mathcal{T} = \{(1, 1, 0), (-1, 1, 0), (0, 0, 1)\}$

7. $T: E^3 \to E^3$, $T((x_1, x_2, x_3)) = (x_1 + 2x_2, x_1 + x_2 + x_3, x_3)$
$\mathcal{S}$ = standard basis $\quad \mathcal{T} = \{(1, 1, 0), (0, 1, 1), (1, 0, 1)\}$

8. $T: \mathcal{P}_1 \to \mathcal{P}_1$, $T(a + bx) = a + b + (2a - 3b)x$
$\mathcal{S}$ = standard basis $\quad \mathcal{T} = \{1 + x, 1 - x\}$

9. $T: \mathcal{M}_{22} \to \mathcal{M}_{22}$, $T(A) = A^T$

$\mathcal{S}$ = standard basis $\quad \mathcal{T} = \left\{\begin{pmatrix} 1 & 1 \\ 0 & 0 \end{pmatrix}, \begin{pmatrix} 0 & 0 \\ 1 & 1 \end{pmatrix}, \begin{pmatrix} 1 & 0 \\ 1 & 0 \end{pmatrix}, \begin{pmatrix} 1 & 0 \\ 0 & 1 \end{pmatrix}\right\}$

10. Show that if A is similar to B and A is invertible, then A^{-k} is similar to B^{-k} for $k = 1, 2, \ldots$.

11. Pairs of matrices A and B are given. In each case show that A and B are not similar.

(*a*) $A = \begin{pmatrix} 1 & 0 \\ 2 & 3 \end{pmatrix} \qquad B = \begin{pmatrix} 3 & 1 \\ 3 & 2 \end{pmatrix}$

(*b*) $A = \begin{pmatrix} 9 & 3 & 7 \\ 0 & 5 & 6 \\ 0 & 0 & 0 \end{pmatrix} \qquad B = \begin{pmatrix} 1 & 2 & 3 \\ 4 & 5 & 6 \\ 7 & 8 & 9 \end{pmatrix}$

(*c*) $A = \begin{pmatrix} 2 & 0 \\ 1 & 2 \end{pmatrix} \qquad B = \begin{pmatrix} 2 & 0 \\ 0 & 2 \end{pmatrix}$

12. Show that if A and B are similar matrices, then rank A = rank B. (*Hint*: Use the fact that similar matrices represent the same linear transformation.)

13. Compare the results of Probs. 3 and 8. Is there a reasonable identification of the transformations (and V) in these problems?

14. Let M be an $m \times n$ matrix. Let V be an n-dimensional vector space with basis S, and let W be an m-dimensional vector space with basis T. Let $\mathbf{x}$ be in V. Define a transformation L from V to W by this rule: L takes $\mathbf{x}$ in V to $\mathbf{y}$ in W as follows:

1. Replace x by $(\mathbf{x})_S$, the coordinate matrix of x.
2. Calculate $M\,(\mathbf{x})_S$.
3. Let y be the vector in W with $(\mathbf{y})_T = M(\mathbf{x})_S$.

Show that L so defined is linear. This means that any matrix generates a linear transformation.

4.4 INVERTIBLE TRANSFORMATIONS AND CLASSIFYING TRANSFORMATIONS

One of the most important problems in applied mathematics is to solve

$$T(\mathbf{x}) = \mathbf{y} \tag{4.4.1}$$

where T: $V \to V$ is a linear transformation, V is a vector space, $\mathbf{y}$ is given, and $\mathbf{x}$ is to be found. If V is given a basis $\mathscr{S} = \{\mathbf{v}_1, \ldots, \mathbf{v}_n\}$ and M is the matrix of T with respect to $\mathscr{S}$, then the corresponding matrix problem is

$$M(\mathbf{x})_{\mathscr{S}} = (\mathbf{y})_{\mathscr{S}} \tag{4.4.2}$$

which is the first fundamental problem of linear algebra. So we can see the importance of the representation problem in reducing transformation equations in applied mathematics to matrix equations. If we solve Eq. (4.4.2), we have essentially inverted the transformation T. This concept must now be developed.

DEFINITION 4.4.1 Let $T: V \to W$ and $L: W \to Z$, where V, W, and Z are vector spaces. The composition of L and T, denoted $L \circ T$, is defined for $\mathbf{x} \in V$ by

$$(L \circ T)(\mathbf{x}) = L(T(\mathbf{x}))$$

Pictorially the situation in Definition 4.4.1 is shown in Fig. 4.4.1. It is not hard to define longer strings of compositions $L(T(U(\mathbf{x}))$ and to see that $((L \circ T) \circ U)(\mathbf{x}) = (L \circ (T \circ U))(\mathbf{x})$.

EXAMPLE 1 Let $T: E^2 \to E^3$ and $L: E^3 \to E^2$ be defined by $T((x_1, x_2)) = (x_1, x_1 + x_2, x_2)$ and $L((x_1, x_2, x_3)) = (x_1 - x_2, x_2 - x_3)$. Calculate $(L \circ T)((x_1, x_2))$. Show that $L \circ T$ is a linear transformation. Determine the standard matrices for L, T, and

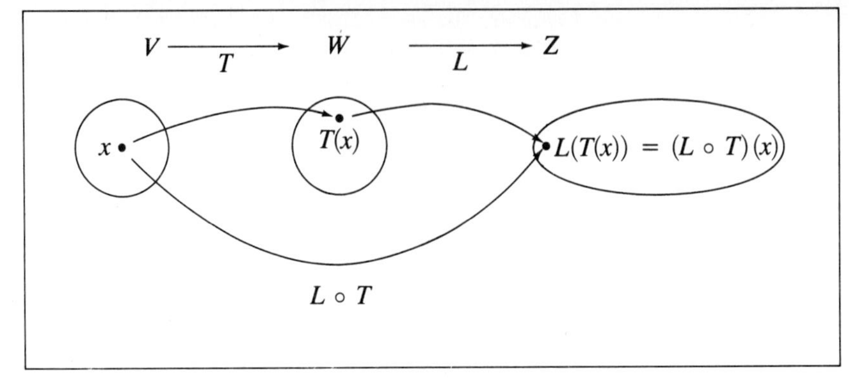

Figure 4.4.1 Composition of transformations.

$L \circ T$. Show that the matrix of $L \circ T$ is the product of the matrix of L and the matrix of T.

Solution By Definition, $(L \circ T)((x_1, x_2)) = L(T(x_1, x_2)) = L((x_1, x_1 + x_2, x_2)) = (x_1 - (x_1 + x_2), (x_1 + x_2) - x_2) = (-x_2, x_1)$. Now let (x_1, x_2) and (y_1, y_2) be two vectors in E^2. We have $(L \circ T)((x_1, x_2) + (y_1, y_2)) = (L \circ T)((x_1 + y_1, x_2 + y_2)) = (-(x_2 + y_2), x_1 + y_1)$. Also $(L \circ T)((x_1, x_2)) + (L \circ T)((y_1, y_2)) = (-x_2, x_1) + (-y_2, y_1) = (-x_2 - y_2, x_1 + y_1)$. It is easy to show that $(L \circ T)(r(x_1, x_2)) = r(L \circ T)((x_1, x_2))$; therefore, $L \circ T$ is linear. By the methods of Sec. 4.3 we have the following:

TRANSFORMATION	STANDARD MATRIX
L	$M_L = \begin{pmatrix} 1 & -1 & 0 \\ 0 & 1 & -1 \end{pmatrix}$
T	$M_T = \begin{pmatrix} 1 & 0 \\ 1 & 1 \\ 0 & 1 \end{pmatrix}$
$L \circ T$	$M_{L \circ T} = \begin{pmatrix} 0 & -1 \\ 1 & 0 \end{pmatrix}$

Furthermore,

$$M_L M_T = \begin{pmatrix} 1 & -1 & 0 \\ 0 & 1 & -1 \end{pmatrix} \begin{pmatrix} 1 & 0 \\ 1 & 1 \\ 0 & 1 \end{pmatrix} = \begin{pmatrix} 0 & -1 \\ 1 & 0 \end{pmatrix} = M_{L \circ T}$$

Note that $M_T M_L \neq M_{L \circ T}$. This reflects something we already know: Matrix multiplication is, in general, not commutative.

Example 1 illustrates the following theorem.

THEOREM 4.4.1 If $T:(V, \mathscr{S}) \to (W, \mathscr{T})$ and $L:(W, \mathscr{T}) \to (Z, \mathscr{B})$ are linear, then $(L \circ T):(V, \mathscr{S}) \to (Z, \mathscr{B})$ is linear. If M_T, M_L, and $M_{L \circ T}$ are the matrices representing T, L, and $L \circ T$, respectively, then

$$M_{L \circ T} = M_L M_T$$

Proof For the linearity, let $\mathbf{u}$ and $\mathbf{v}$ be vectors in V, and let r and s be numbers. By the linearity of L and T separately, $(L \circ T)(r\mathbf{u} + s\mathbf{v}) = L(T(r\mathbf{u} + s\mathbf{v})) = L(rT(\mathbf{u}) + sT(\mathbf{v})) = rL(T(\mathbf{u})) + sL(T(\mathbf{v})) = r(L \circ T)(\mathbf{u}) + s(L \circ T)(\mathbf{v})$. Thus $L \circ T$ is linear. To show the matrix representation, let $\mathbf{x} \in V$, $T(\mathbf{x}) = \mathbf{y}$, and $L(T(\mathbf{x})) = \mathbf{z}$. Now $(\mathbf{y})_{\mathscr{T}} = M_T(\mathbf{x})_{\mathscr{S}}$ and $(\mathbf{z})_{\mathscr{B}} = M_L(\mathbf{y})_{\mathscr{T}}$. Therefore $(\mathbf{z})_{\mathscr{B}} = M_L(M_T(\mathbf{x})_{\mathscr{S}}) = (M_L M_T)(\mathbf{x})_{\mathscr{S}}$. But we also have

$$(\mathbf{z})_{\mathscr{B}} = M_{L \circ T}(\mathbf{x})_{\mathscr{S}}$$

By the uniqueness of matrix representation,

$$M_{L \circ T} = M_L M_T$$

The idea of the composition of transformation is set; we can define the inverse of a transformation.

DEFINITION 4.4.2 Let $T: V \to V$ be a linear transformation. The (two-sided) inverse of T is a transformation $T^{-1}: V \to V$ for which

$$(T^{-1} \circ T)(\mathbf{x}) = \mathbf{x} \qquad \text{for all } \mathbf{x} \in V$$

and

$$(T \circ T^{-1})(\mathbf{x}) = \mathbf{x} \qquad \text{for all } \mathbf{x} \in V$$

If T^{-1} exists, then T is called *invertible.*

We note that T^{-1} is linear. In fact, $T^{-1}(\mathbf{x} + \mathbf{y}) = T^{-1}(T(T^{-1}(\mathbf{x})) + T(T^{-1}(\mathbf{y}))) = T^{-1}(T(T^{-1}(\mathbf{x}) + T^{-1}(\mathbf{y}))) = (T^{-1}T)(T^{-1}(\mathbf{x}) + T^{-1}(\mathbf{y})) = T^{-1}(\mathbf{x}) + T^{-1}(\mathbf{y})$ and $T^{-1}(rx) = T^{-1}(rT(T^{-1}(\mathbf{x}))) = T^{-1}(T(rT^{-1}(\mathbf{x}))) = (T^{-1}T)(rT^{-1}(\mathbf{x})) = rT^{-1}(\mathbf{x})$. Other important properties of inverses are contained in Theorem 4.4.2.

THEOREM 4.4.2 Let $T:(V, \mathscr{S}) \to (V, \mathscr{S})$ be a linear transformation.

(*a*) Let M be the matrix of T with respect to $\mathscr{S}$. Then T is invertible if and only if M^{-1} exists. Moreover, the matrix for T^{-1} in this case is precisely M^{-1}.

(*b*) T is invertible if and only if $\eta(T) = 0$ [that is, dim (ker T) = 0].

(*c*) T is invertible if and only if $T(\mathbf{x}) = T(\mathbf{y})$ implies that $\mathbf{x} = \mathbf{y}$ for all $\mathbf{x}, \mathbf{y} \in V$. (That is, T is a one-to-one function.)

Proof (*a*) ($\Rightarrow$) Since $(T \circ T^{-1})(\mathbf{x}) = \mathbf{x}$, we have

$$(\text{matrix of } T \text{ times matrix of } T^{-1})\ (\mathbf{x})_s = (\mathbf{x})_s$$

If $M_{T^{-1}}$ denotes the matrix for T^{-1}, the last equation is

$$MM_{T^{-1}}(\mathbf{x})_s = (\mathbf{x})_s$$

which means that $MM_{T^{-1}} = I$. Therefore $M_{T^{-1}} = M^{-1}$.

($\Leftarrow$) Suppose M^{-1} exists. Since it represents a linear transformation L and $MM^{-1} = M^{-1}M = I$, we know that $(L \circ T)(\mathbf{x}) = \mathbf{x}$ and $(T \circ L)(\mathbf{x}) = \mathbf{x}$. Thus $L = T^{-1}$, and T is invertible.

(*b*) Let dim $V = n$ and M be the matrix of T.

($\Rightarrow$) Suppose T is invertible and $\eta(T) > 0$. Then since $\eta(T) + \text{rank } T = n$, rank $T < n$. Therefore rank $M < n$ and M^{-1} does not exist, a contradiction since if T^{-1} exists, M^{-1} must exist.[1]

($\Leftarrow$) If $\eta(T) = 0$, then since

$$\eta(T) + \mathscr{R}(T) = n$$

we have $\mathscr{R}(T) = \text{rank } T = n$. Thus rank $M = n$ and M exists. Reasoning as in (*a*) now, we see that T^{-1} must exist.

(*c*) ($\Rightarrow$) Suppose T is invertible and there exist $\mathbf{x}'$ and $\mathbf{y}'$ in V with $\mathbf{x}' \neq \mathbf{y}'$ and $T(\mathbf{x}') = T(\mathbf{y}')$. Then $\mathbf{x}' - \mathbf{y}' \neq \boldsymbol{\theta}$ and $T(\mathbf{x}' - \mathbf{y}') = \boldsymbol{\theta}$. But then dim (ker T) $\neq$ 0. That is, $\eta(T) \neq 0$, which contradicts (*b*).

($\Leftarrow$) Suppose $T(\mathbf{x}) = T(\mathbf{y})$ implies that $\mathbf{x} = \mathbf{y}$ for all $\mathbf{x}$, $\mathbf{y}$ in V and that T is not invertible. Then by (*b*) dim ker $T \neq 0$. Therefore ker T contains at least $\boldsymbol{\theta}$ and some $\mathbf{z} \neq \boldsymbol{\theta}$. Now since ker T is a subspace, $\mathbf{z} - \boldsymbol{\theta} \in$ ker T. Thus $T(\mathbf{z} - \boldsymbol{\theta}) = T(\mathbf{z}) - T(\boldsymbol{\theta}) = \boldsymbol{\theta}$. So we have $T(\mathbf{z}) = T(\boldsymbol{\theta})$, but $\mathbf{z} \neq \boldsymbol{\theta}$, which contradicts our assumption.

Part (*a*) of Theorem 4.4.2 tells us that we can determine the invertibility of a transformation by determining the invertibility of any representing matrix. This is so because if A and B are any two representing matrices, then they are similar: $A = P^{-1}BP$. Now since det A = det B, matrix A is invertible if and only if matrix B is invertible. So if one representing matrix is invertible, all are.

EXAMPLE 2 Define $T: E^3 \to E^3$ by $T(\mathbf{x}) = T((x_1, x_2, x_3)) = (bx_3 - cx_2, cx_1 - ax_3, ax_2 - bx_1)$. This transformation maps (x_1, x_2, x_3) to $(a, b, c) \times (x_1, x_2, x_3)$, the cross product of the fixed vector (a, b, c) with $\mathbf{x}$. Show that T is not invertible, thereby showing that we cannot determine a vector from its cross product with a known vector (a, b, c).

Solution We will find a matrix M representing T and show that M^{-1} does not exist. Since invertibility is preserved by similarity, we may use any representing matrix to determine invertibility. We use the standard matrix since it is easiest to compute. We have

$$T(1, 0, 0) = (0, c, -b)$$

$$T(0, 1, 0) = (-c, 0, a)$$

$$T(0, 0, 1) = (b, -a, 0)$$

[1] Prob. 12 of Sec. 4.3 shows that rank T is defined as rank M_T.

so the matrix is

$$M = \begin{pmatrix} 0 & -c & b \\ c & 0 & -a \\ -b & a & 0 \end{pmatrix}$$

Note that M is antisymmetric. Now $\det M = abc - abc = 0$, so M^{-1} does not exist. Therefore T is not invertible.

EXAMPLE 3 (Shear revisited) Imagine a cube of gelatin held between the hands as viewed from the side in Fig. 4.4.2*a*. Suppose the upper hand is moved to the right k units (k is small). Then the height of the gelatin will not change much, and the side view will look like a parallelogram. This action in mechanics leads to shear. In Fig. 4.4.2*b*, some vectors are imposed on the face of the gelatin. *Assuming that the shear is a linear transformation* S, find a matrix representing S and show that S is invertible. (Invertibility is reasonable: To "undo" the shear, we can just move the top hand back to its original position.)

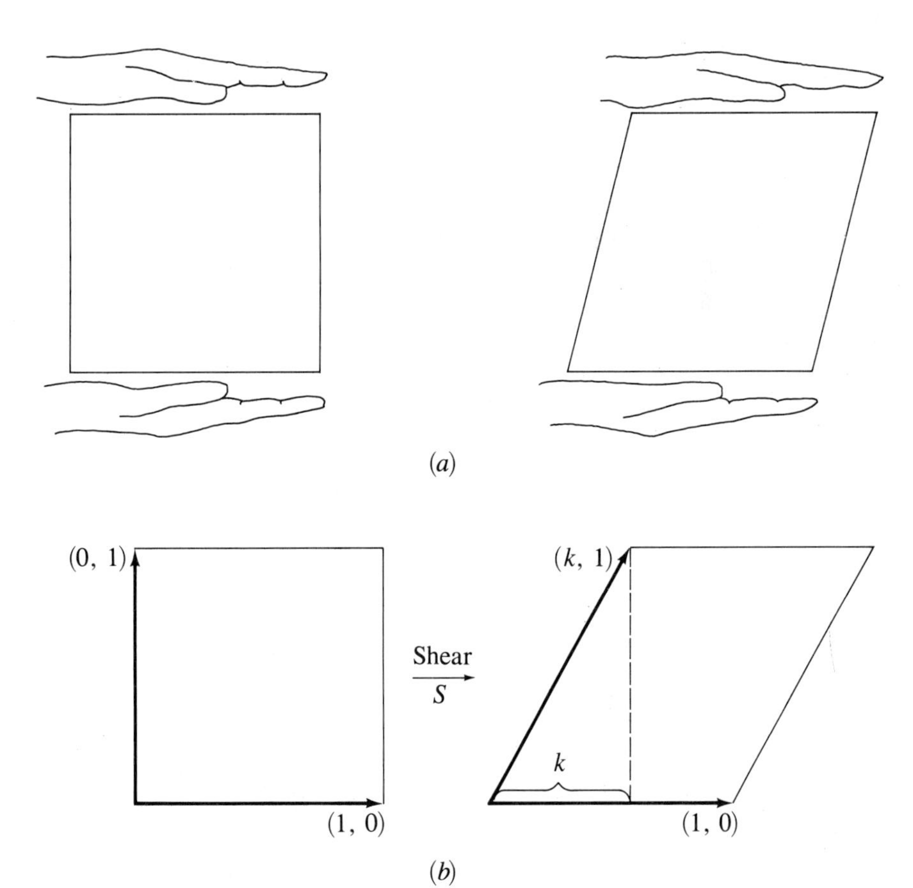

Figure 4.4.2 (*a*) Shear. (*b*) Shear with vectors.

Solution As stated before, the basic principle for finding matrix representations is that a linear transformation is determined by its action or basis elements. Since $\mathscr{S} = \{(1, 0), (0, 1)\}$ is a basis for E^2, the actions of S on $(1, 0)$ and $(0, 1)$ can be used:

$$S((1, 0)) = (1, 0) = 1(1, 0) + 0(0, 1)$$
$$S((0, 1)) = (k, 1) = k(1, 0) + 1(0, 1)$$

Therefore the matrix M of S with respect to $\mathscr{S}$ is

$$M = \begin{pmatrix} 1 & k \\ 0 & 1 \end{pmatrix}$$

Matrix M is invertible; therefore by Theorem 4.4.2*a*, S is invertible. Note that

$$M^{-1} = \begin{pmatrix} 1 & -k \\ 0 & 1 \end{pmatrix}$$

which represents shear with the upper face of the cube moving $-k$ units to the right, which means k units to the left. This, of course, makes sense: To undo moving the upper hand to the right, simply move it to the left the same distance.

EXAMPLE 4 Consider a square as shown in Fig. 4.4.3. Let R be a counterclockwise rotation about c of 90°. Show, using matrices, that four successive applications of R give the identity transformation.

Solution We find the standard matrix of R and calculate its fourth power. Since $R(1, 0) = (0, 1) = 0(1, 0) + 1(0, 1)$ and $R(0, 1) = (-1, 0) = -1(1, 0) + 0(0, 1)$, the standard matrix is

$$M = \begin{pmatrix} 0 & -1 \\ 1 & 0 \end{pmatrix}$$

Now

$$M^2 = \begin{pmatrix} -1 & 0 \\ 0 & -1 \end{pmatrix} \quad \text{and} \quad M^4 = M^2M^2 = I$$

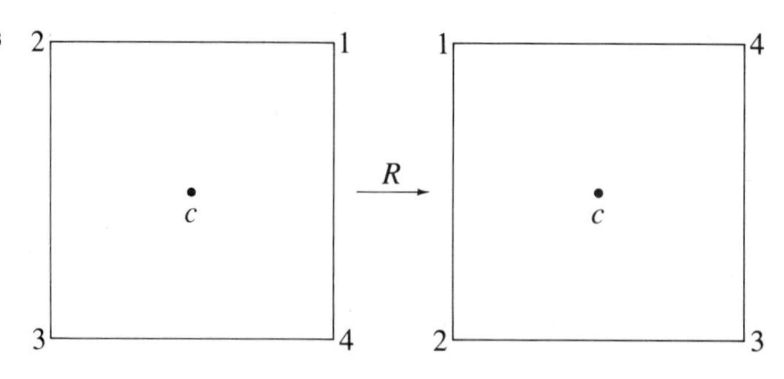

Figure 4.4.3 Rotating square.

Classifying Linear Transformations We have seen that if a matrix A is invertible and a matrix B is similar to A, then B is invertible also. That is, invertibility is preserved under similarity. As a result of Theorem 4.4.2*a*, we say that a linear transformation T is invertible if any matrix representation of T is an invertible matrix. Because other properties of matrices are preserved under similarity, we make the following definition.

DEFINITION 4.4.3 Let $\mathscr{P}$ be a property of matrices which is preserved under similarity. We classify a linear transformation $T: V \to V$ as having property $\mathscr{P}$ if in some basis the matrix representing T has property $\mathscr{P}$.

EXAMPLE 5 Show that the property of idempotency is preserved under similarity.

Solution Let A be idempotent (that is, $A^2 = A$), and let B be similar to $A: B = P^{-1}AP$. Now $B^2 = (P^{-1}AP)(P^{-1}AP) = P^{-1}AIAP = P^{-1}A^2P = P^{-1}AP = B$. Therefore B is idempotent.

EXAMPLE 6 Show that the linear transformation $T: E^3 \to E^3$ of projection $T((x_1, x_2, x_3)) = (x_1, 0, x_3)$ is idempotent.

Solution By Example 5, idempotency is preserved under similarity. All we need to do is to find a matrix representation of T which is an idempotent matrix. We look at the standard matrix, which is easiest to compute. It is

$$M = \begin{pmatrix} 1 & 0 & 0 \\ 0 & 0 & 0 \\ 0 & 0 & 1 \end{pmatrix}$$

Now $M^2 = M$, so T is idempotent.

Another property preserved under similarity is nilpotency.

EXAMPLE 7 Show that nilpotency is preserved under similarity. Then show that the *left-shift* linear transformation $T: E^4 \to E^4$ defined by $T(x_1, x_2, x_3, x_4) = (x_2, x_3, x_4, 0)$ is a nilpotent transformation.

Solution Let A be nilpotent of exponent k, so that $A^k = 0$. If B is similar to A, then $B^k = P^{-1}A^kP$ by Theorem 4.3.2*f*. Thus $B^k = P^{-1}0P = 0$, and B is nilpotent of exponent k. The standard matrix for T is

$$M = \begin{pmatrix} 0 & 1 & 0 & 0 \\ 0 & 0 & 1 & 0 \\ 0 & 0 & 0 & 1 \\ 0 & 0 & 0 & 0 \end{pmatrix}$$

The powers of M are

$$M^2 = \begin{pmatrix} 0 & 0 & 1 & 0 \\ 0 & 0 & 0 & 1 \\ 0 & 0 & 0 & 0 \\ 0 & 0 & 0 & 0 \end{pmatrix} \qquad M^3 = \begin{pmatrix} 0 & 0 & 0 & 1 \\ 0 & 0 & 0 & 0 \\ 0 & 0 & 0 & 0 \\ 0 & 0 & 0 & 0 \end{pmatrix} \qquad M^4 = 0$$

Therefore T is nilpotent of exponent 4. So four successive applications of T produce the zero vector. Note that T is not invertible because M is not invertible. The noninvertibility makes sense because when we shift left, we lose entirely the information from the first component.

Linear Equations (Reprise) As we said at the beginning of this section, the problem of solving

$$L(\mathbf{x}) = \mathbf{y}$$

for $\mathbf{x}$ in V, given the linear transformation $L: V \to V$ and $\mathbf{y}$ in V, is a generalization of the first basic problem of linear algebra. When V is finite-dimensional, the problem reduces to the first basic problem of solving linear equations once a basis is assigned to V and a matrix representing L is found. In this case the equation $L(\mathbf{x}) = \mathbf{y}$ is uniquely solvable if and only if M_L is invertible. When M_L is not invertible, dim (ker L) $\neq 0$ and the general solution is of the form $\mathbf{p} + \mathbf{h}$ where $\mathbf{p}$ is a particular solution of $L(\mathbf{x}) = \mathbf{y}$ and $\mathbf{h}$ is a solution of the associated homogeneous problem $L(\mathbf{x}) = \boldsymbol{\theta}$. (See Theorem 1.5.5.) To illustrate this, consider $L: \mathscr{P}_2 \to \mathscr{P}_2$, where L is differentiation. We wish to solve $L(f) = g$. The standard matrix for L is

$$M = \begin{pmatrix} 0 & 1 & 0 \\ 0 & 0 & 2 \\ 0 & 0 & 0 \end{pmatrix}$$

The matrix is not invertible, so we cannot expect a unique solution. Let

$$(g)_{\mathscr{S}} = \begin{pmatrix} a \\ b \\ c \end{pmatrix}$$

Then

$$M(f)_{\mathscr{S}} = \begin{pmatrix} a \\ b \\ c \end{pmatrix}$$

can have a solution if and only if $c = 0$. When $c \neq 0$, g is not in the range of L. Now when $c = 0$, g is in the range of L, and using the matrix equation

$$M(f)_{\mathscr{S}} = (g)_{\mathscr{S}}$$

we find

$$(f)_{\mathscr{S}} = \begin{pmatrix} k \\ a \\ b/2 \end{pmatrix}$$

where k is arbitrary. That is, $f(x) = k + ax + bx^2/2$, where the constant k is arbitrary (remember antiderivatives and the arbitrary "constant of integration"?). Note that $p(x) = ax + bx^2/2$ is a particular solution of $L(f) = g$, and $h(x) = k$ is the most general solution to $L(f) = \boldsymbol{\theta}$. Thus the general solution, which exists only when g is not a quadratic ($c = 0$), is of the form $p + h$, where h is in the kernel of L.

For a pure matrix problem consider

$$\begin{aligned} x + y - z + w &= 1 \\ 2x + y - w &= -2 \\ x + z - 2w &= -3 \\ y - 2z + 3w &= 4 \end{aligned}$$

The equations are equivalent to

$$\left(\begin{array}{cccc|c} 1 & 0 & 1 & -2 & -3 \\ 0 & 1 & -2 & 3 & 4 \\ 0 & 0 & 0 & 0 & 0 \\ 0 & 0 & 0 & 0 & 0 \end{array}\right)$$

Thus the solution exists but is not unique. A particular solution is obtained by putting $z = 1$ and $w = 1$, which leads to $y = 3$ and $x = -2$. Looking at the associated homogeneous system

$$\left(\begin{array}{cccc|c} 1 & 0 & 1 & -2 & 0 \\ 0 & 1 & -2 & 3 & 0 \\ 0 & 0 & 0 & 0 & 0 \\ 0 & 0 & 0 & 0 & 0 \end{array}\right)$$

we find a solution

$$\mathbf{h} = \begin{pmatrix} 2s - r \\ 2r - 3s \\ r \\ s \end{pmatrix}$$

Thus the most general solution is

$$\mathbf{p} + \mathbf{h} = \begin{pmatrix} -2 \\ 3 \\ 1 \\ 1 \end{pmatrix} + \begin{pmatrix} 2s - r \\ 2r - 3s \\ r \\ s \end{pmatrix}$$

The preceding examples illustrate the following theorem, which is stated without proof.

THEOREM 4.4.3 Let V be a vector space with $\dim V \geq 1$, and consider the linear equation

$$L(\mathbf{x}) = \mathbf{y}$$

where $L: V \to V$ is linear, dim (range L) ≥ 1, and $\mathbf{y}$ is in V. If $\mathbf{y}$ is in the range of L, then either (*a*) the equation has a unique solution or (*b*) the equation has an infinite number of solutions which are of the form $\mathbf{p} + \mathbf{h}$, where $\mathbf{h}$ is the general solution to $L(\mathbf{x}) = \boldsymbol{\theta}$ and p is any particular solution of $L(\mathbf{x}) = \mathbf{y}$. If $\mathbf{y}$ is not in the range of L, then the equation has no solution.

Finally let us consider a case which can arise in applications: attempting to solve $L(\mathbf{x}) = \mathbf{y}$ when $\mathbf{y}$ is not in the range of L: Theorem 4.4.3 tells us that a solution in the usual sense cannot be expected but if an inner product can be defined on V, we could call a vector $\tilde{\mathbf{x}}$ an *approximate solution* if

$$\|L(\tilde{\mathbf{x}}) - \mathbf{y}\|^2 \leq \|L(\mathbf{x}) - \mathbf{y}\|^2 \qquad \text{for all } \mathbf{x} \text{ in } V$$

Consider, the problem of solving

$$x_1 + x_2 = 2$$
$$x_1 + x_2 = 1$$

which has no solution. Defining $L: \mathscr{M}_{21} \to \mathscr{M}_{21}$ by

$$L\left(\begin{pmatrix} x_1 \\ x_2 \end{pmatrix}\right) = \begin{pmatrix} 1 & 1 \\ 1 & 1 \end{pmatrix}\begin{pmatrix} x_1 \\ x_2 \end{pmatrix}$$

we see that

$$\begin{pmatrix} 2 \\ 1 \end{pmatrix}$$

the right-hand side of the equations, is not in the range of L. Using the standard inner product on $\mathscr{M}_{21}$, we have

$$\left\| L\left(\begin{pmatrix} x_1 \\ x_2 \end{pmatrix}\right) - \begin{pmatrix} 2 \\ 1 \end{pmatrix} \right\|^2 = (x_1 + x_2 - 2)^2 + (x_1 + x_2 - 1)^2$$

Methods of calculus show that this last expression is minimized if $x_1 + x_2 = \frac{3}{2}$; this does not yield a unique approximate solution. One way to avoid this problem is to require that the approximate solution be of *minimum norm* (in this

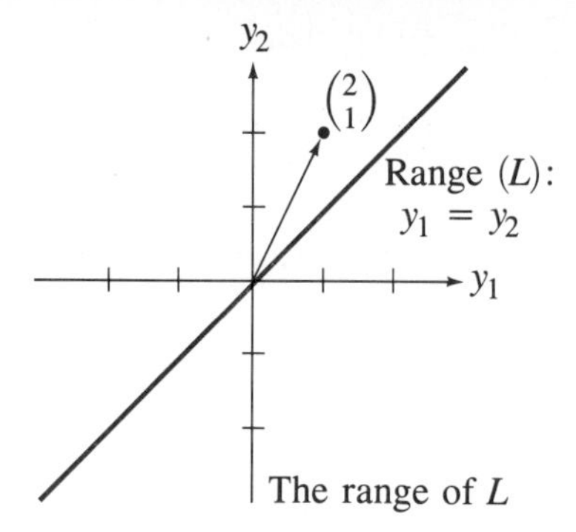

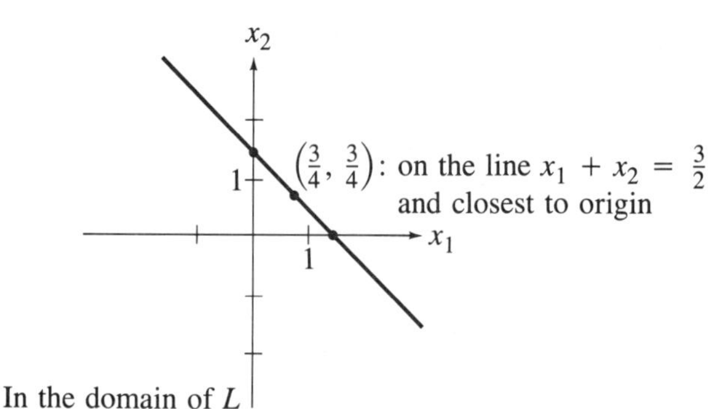

Figure 4.4.4 Minimum norm approximate solution.

case, the one closest to the origin). With this requirement we obtain $\tilde{x} = (\frac{3}{4}, \frac{3}{4})$ (see Fig. 4.4.4). Finding minimum norm solutions leads to the concepts of *generalized inverses* and *regularization*, which the interested reader can find in texts such as *Regression and the Moore-Penrose Pseudoinverse*, by A. E. Albert (Academic Press, New York, 1972).

PROBLEMS 4.4

In Probs. 1 to 6, a linear transformation and a property or properties (similarity preserved) are given. Determine whether the given transformation has the given property.

1. $T: E^3 \to E^3$, $T((x_1, x_2, x_3)) = (x_1, x_2, 0)$; invertibility, idempotency
2. $T: \mathscr{P}_2 \to \mathscr{P}_2$, $T(ax^2 + bx + c) = 2ax + b$; invertibility, nilpotency
3. $T: \mathscr{M}_{22} \to \mathscr{M}_{22}$, $T(A) = A^T$; invertibility, idempotency
4. $T: E^2 \to E^2$, T is rotation by 180°; invertibility
5. $T: E^3 \to E^3$, $T((x_1, x_2, x_3)) = (x_1, x_1 + x_2, x_1 + x_2 + x_3)$; invertibility, nilpotency
6. $T: \mathscr{P}_1 \to \mathscr{P}_1$, $T(a + bx) = b + ax$; invertibility, idempotency

7. Consider an equilateral triangle

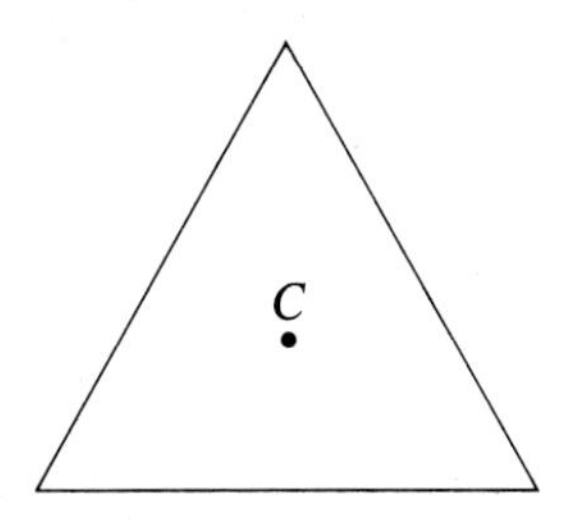

Figure P4.4.7

with center C. Let R be a counterclockwise rotation about C of 120°. Show that three successive applications of R yield the identity transformation. (Use matrices.)

8. Let T be a counterclockwise rotation about C of 240°. Show that $T \circ T \circ T = I$.

9. Consider $T: \mathbb{C}^n \to \mathbb{C}^n$ defined by

$$T((z_1, z_2, \ldots, z_n)) = (0, z_1, z_2, \ldots, z_{n-1})$$

(T is called a *right shift*.) Is T invertible? Is T nilpotent? Is T idempotent?

10. Consider $T: \mathbb{C}^n \to \mathbb{C}^n$ defined by

$$T((z_1, z_2, \ldots, z_n)) = (z_2, z_3, \ldots, z_n, z_1)$$

Is T invertible? Is T nilpotent? Is T idempotent?

11. If A is an involutory matrix ($A^2 = I$) and B is similar to A, is B involutory?

12. If A is an orthogonal matrix and B is similar to A, is B orthogonal?

13. If A is a symmetric matrix, P is orthogonal, and $B = P^T AP$, is B symmetric?

14. Let us say that a matrix $A_{n \times n}$ is diagonalizable if it is similar to a diagonal matrix D. If B is similar to A, is B similar to a diagonal matrix?

15. Let L be an invertible linear operator from V to V, and let $\mathcal{S} = \{\mathbf{v}_1, \mathbf{v}_2, \ldots, \mathbf{v}_k\}$ be a linearly independent set in V. Show that the set $\{L(\mathbf{v}_1), L(\mathbf{v}_2), \ldots, L(\mathbf{v}_k)\}$ is linearly independent in the range of L.

4.5 CALCULUS REVISITED

Two of the most important linear transformations in applied mathematics are differentiation and definite integration, which are studied in calculus. The familiar rules[2]

[2] Which, of course, come from theorems in which f and g must satisfy certain conditions.

$$\frac{d}{dx}(f(x)+g(x)) = \frac{d}{dx}f(x) + \frac{d}{dx}g(x)$$

$$\frac{d}{dx}(cf(x)) = c\frac{d}{dx}f(x)$$

$$\int_a^b [f(x)+g(x)]\,dx = \int_a^b f(x)\,dx + \int_a^b g(x)\,dx$$

$$\int_a^b cf(x)\,dx = c\int_a^b f(x)\,dx$$

are among the most useful for calculating derivatives and integrals. These rules are just statements of linearity. In fact, if we denote differentiation by D, the first two rules above are just

$$D(f+g) = D(f) + D(g) \qquad \text{and} \qquad D(cf) = cD(f)$$

EXAMPLE 1 Let D be the transformation of differentiation. Show that $D:\mathscr{P}_3 \to \mathscr{P}_3$ is not invertible. Give $\mathscr{P}_3$ the standard basis $\mathscr{S} = \{1, x, x^2, x^3\}$. Find the matrix of D with respect to $\mathscr{S}$. Show that D is nilpotent.

Solution To show that D is not invertible, we show that dim (ker D) $\neq 0$. Now ker D is the set of polynomials f for which $D(f) = 0$ (the zero function). That is, the kernel is the set of all polynomials which have derivative zero. From calculus we know that this is the set of constant functions. Thus ker D = span 1, and $\eta(D) = 1$. Therefore D is not invertible. To find the matrix, we calculate

$$\begin{aligned} D(x^3) &= 3x^2 = 0x^3 + 3x^2 + 0x + 0 \\ D(x^2) &= 2x = 0x^3 + 0x^2 + 2x + 0 \\ D(x) &= 1 = 0x^3 + 0x^2 + 0x + 1 \\ D(1) &= 0 = 0x^3 + 0x^2 + 0x + 0 \end{aligned}$$

to find

$$M = \begin{pmatrix} 0 & 0 & 0 & 0 \\ 3 & 0 & 0 & 0 \\ 0 & 2 & 0 & 0 \\ 0 & 0 & 1 & 0 \end{pmatrix}$$

Notice that M is not invertible, as we would expect since D is not invertible. Now since $M^4 = 0$, matrix M is nilpotent and thus D is nilpotent of order 4. This is just a linear algebra way of saying that if we differentiate a cubic 4 times or more, we obtain the zero function.

Some calculus students would object to the statement "D is not invertible" because "Everyone knows integration and differentiation are just opposite operations." We must be careful since the process of antidifferentiation involves the addition of an arbitrary constant. If we let A denote antidifferentiation, then, for example, $A(2x) = x^2 + c$, where c is an arbitrary constant. If A were to be an inverse to D, we would have $A(D(x^2)) = x^2$; however, $A(D(x^2)) = A(2x) = x^2 + c$, which is equal to x^2 only if $c = 0$. So in general $A \circ D \neq I$: therefore, A and D are not inverses. The next example shows, however, that for some vector spaces antidifferentiation *is* the inverse of differentiation.

EXAMPLE 2 Let V be the subspace of $\mathscr{P}_3$ defined by $V = \{f(x) \text{ in } \mathscr{P}_3 | f(0) = 0\}$. Show that $\dim V = 3$. Show that the range of D with domain restricted to V is $\mathscr{P}_2$. Define $A_0: \mathscr{P}_2 \to V$ as antidifferentiation with arbitrary constant zero. Show that $(A_0 \circ D)(f) = f = I(f)$ for any f in $\mathscr{P}_3$.

Solution A natural basis for V is $\mathscr{S} = \{x, x^2, x^3\}$ since any cubic $p(x) = ax^3 + bx^2 + cx + d$ has the property $p(0) = 0$ if and only if $d = 0$. Therefore $\dim V = 3$. Now we calculate the range of D. The range of D is spanned by the images under D of the basis vectors for V; we have

$$\begin{aligned} D(x) &= 1 \\ D(x^2) &= 2x \\ D(x^3) &= 3x^2 \end{aligned}$$

so that range $D = \text{span}\,\{1, 2x, 3x^2\}$. The range of D is all linear combinations of $1, 2x$, and $3x^2$ which is $\mathscr{P}_2$. To show that $A_0 \circ D = I$, we calculate the matrices for A_0 and D, using the natural bases for V and $\mathscr{P}_2$. We find

$$\begin{aligned} D(x) &= 1 + 0x + 0x^2 \\ D(x^2) &= 0 + 2x + 0x^2 \\ D(x^3) &= 0 + 0x + 3x^2 \end{aligned}$$

and

$$\begin{aligned} A_0(1) &= 1x + 0x^2 + 0x^3 \\ A_0(x) &= 0x + \tfrac{1}{2}x^2 + 0x^3 \\ A_0(x_2) &= 0x + 0x^2 + \tfrac{1}{3}x^3 \end{aligned}$$

Therefore

$$M_D = \begin{pmatrix} 1 & 0 & 0 \\ 0 & 2 & 0 \\ 0 & 0 & 3 \end{pmatrix} \qquad M_{A_0} = \begin{pmatrix} 1 & 0 & 0 \\ 0 & \frac{1}{2} & 0 \\ 0 & 0 & \frac{1}{3} \end{pmatrix} \qquad M_{A_0} M_D = I$$

Also we see that $M_D M_{A_0} = I$, so that $D \circ A = I$. In this example we have forced the arbitrary constant of antidifferentiation to be zero.

Example 2 shows the general principle that to "invert" differentiation, we need some conditions on the domain of D. In Example 2 the condition was

$f(0) = 0$. In general, such conditions can be called *initial conditions*; they arise in problems in the area of mathematics known as *differential equations*. In some textbooks the solving of a differential equation is referred to as *integrating* the equation. This terminology comes up because of the inverse relation we have just seen.

EXAMPLE 3 Show that the transformation $\mathscr{A}$ defined on $C[0, 1]$ by

$$(\mathscr{A}(f))(x) \equiv \int_{x_0}^{x} f(t)\,dt \qquad \text{for all } x \text{ in } [0, 1]$$

is linear.

Solution Let f and g be functions in $C[0, 1]$. We have

$$\begin{aligned}(\mathscr{A}(f + g))(x) &\equiv \int_{x_0}^{x} [f(t) + g(t)]\,dt \\ &= \int_{x_0}^{x} f(t)\,dt + \int_{x_0}^{x} g(t)\,dt \\ &= (\mathscr{A}(f))(x) + (\mathscr{A}(g))(x)\end{aligned}$$

and

$$\begin{aligned}(\mathscr{A}(cf))(x) &\equiv \int_{x_0}^{x} cf(t)\,dt \\ &= c\int_{x_0}^{x} f(t)\,dt \\ &= (c\mathscr{A}(f))(x)\end{aligned}$$

and $\mathscr{A}$ is linear. Notice that $(\mathscr{A}(f))(x_0) = 0$ for all f.

The fundamental theorem of calculus may be stated in this form:

If $f'(x)$ is continuous on $[0, 1]$, then for a fixed $x_0 \in [0, 1]$ and any $x \in [0, 1]$,

$$\int_{x_0}^{x} f'(t)\,dt = f(x) - f(x_0)$$

Using our notation of D and $\mathscr{A}$, we see that the last equation is

$$\mathscr{A}(D(f))(x) = f(x) - f(x_0)$$

Therefore $\mathscr{A}$ is the inverse of D if and only if $f(x_0) = 0$. Again we see an extra condition to guarantee invertibility of differentiation.

To illustrate this last point, we consider a simple differential equation: Find a function $y(x)$, in $C^1(\mathbb{R})$, which satisfies

$$y'(x) = y(x) \qquad x \in \mathbb{R}$$

Now we know that *any* function of the form $y(x) = Ce^x$ satisfies this equation. Since C is arbitrary, we have not *uniquely solved* the problem. However, if we require further a condition such as $y(0) = 3$, we find $y(x) = 3e^x$ as the only[3] solution of the problem. Unique solvability of a differential equation and invertibility of the differentiation operator are closely connected.

Linear transformations arise in several-variable calculus also. Recall the gradient: Let $f: \mathbb{R}^3 \to \mathbb{R}$ with $w = f(x, y, z)$. We have

$$(\text{grad } f)(x, y, z) = \left(\frac{\partial f}{\partial x}(x, y, z), \frac{\partial f}{\partial y}(x, y, z), \frac{\partial f}{\partial z}(x, y, z)\right)$$

and the properties

$$\text{grad } (f + g) = \text{grad } f + \text{grad } g$$
$$\text{grad } (cf) = c(\text{grad } f)$$

are not hard to see. Therefore the gradient operation is a linear transformation from the space of all real-valued functions of three variables with continuous derivatives to the vector space of ordered triples of continuous functions of three variables.

The *Jacobian* is another linear operator which is studied in several-variable calculus. It is generated by a matrix. If **f** is a function defined by

$$\mathbf{f}(x, y) = (g(x, y), h(x, y))$$

then the Jacobian of **f** is the matrix of functions

$$J(x, y) = \begin{pmatrix} \dfrac{\partial g(x, y)}{\partial x} & \dfrac{\partial g(x, y)}{\partial y} \\ \dfrac{\partial h(x, y)}{\partial x} & \dfrac{\partial h(x, y)}{\partial y} \end{pmatrix}$$

At each point (x_0, y_0), $J(x_0, y_0)$ is a fixed 2×2 matrix. If $J(x_0, y_0)$ is an invertible matrix, then **f** is invertible in some neighborhood of (x_0, y_0). In this way the local invertibility of a *nonlinear* function **f** is studied by determining the invertibility of an associated *linear* transformation (generated by the Jacobian).

EXAMPLE 4 Consider the nonlinear transformation $\mathbf{f}: E^2 \to E^2$ defined by $\mathbf{f}(x, y) = (x \cos y, x \sin y)$. Find the Jacobian of **f**. Where is **f** locally invertible?

---

[3] This is proved in differential equations courses.

Solution

$$J(x, y) = \begin{pmatrix} \dfrac{\partial(x \cos y)}{\partial x} & \dfrac{\partial(x \cos y)}{\partial y} \\ \dfrac{\partial(x \sin y)}{\partial x} & \dfrac{\partial(x \sin y)}{\partial y} \end{pmatrix}$$

$$= \begin{pmatrix} \cos y & x \sin y \\ \sin y & x \cos y \end{pmatrix}$$

This matrix is invertible if and only if

$$\det J(x, y) \neq 0$$

That is, $x(\cos^2 y + \sin^2 y) = x \neq 0$. Therefore we say **f** is *locally invertible* in a neighborhood of any point (x_0, y_0) with $x_0 \neq 0$.

These examples illustrate the fact that the operations of differentiation and integration are the source of many linear transformations in applied mathematics.

PROBLEMS 4.5

1. Let V be the set of all functions $f(x)$ for which $\lim_{x \to a} f(x)$ exists. Show that V, with the usual definitions of addition and scalar multiplication for functions, is a vector space. How does the vector space structure depend on the linearity of the transformation $L_a: V \to \mathbb{R}$ defined by $L_a(f) = \lim_{x \to a} f(x)$?

2. Let f be a fixed function in $C[0, \pi]$. Define $L_f: C[0, \pi] \to \mathbb{R}$ by $L_f(g) = \int_0^\pi f(x)g(x)\,dx$. Show that L_f is a linear operator.

3. Consider the linear operator in Prob. 2. The kernel of L_f is the set of all functions g orthogonal to f [where the dot product is $\langle f, g \rangle = \int_0^\pi f(x)g(x)\,dx$]. Show that if $f(x) = \sin x$, then $g(x) = \sin nx$, $n \neq 1$, n a positive integer, is in ker L_f.

4. Let $\mathbf{f}(x, y) = (x^2, y^2)$ be a nonlinear operator mapping E^2 into E^2. Use the Jacobian to determine the local invertibility of **f**.

5. Show that $D^2: \mathscr{P}_3 \to \mathscr{P}_3$ defined by $D^2(p(x)) = p''(x)$ is a linear transformation. Find the standard matrix of D^2. Show that this matrix is the square of the matrix for D found in Example 1.

6. Let V be the set of all functions $f: \mathbb{R} \to \mathbb{R}$ with Maclaurin series convergent to the function for all x. Let

$$f(x) = a_0 + a_1 x + a_2 x^2 + \cdots + a_n x^n + \cdots$$
$$g(x) = b_0 + b_1 x + b_2 x^2 + \cdots + b_n x^n + \cdots$$

and define

$$(f + g)(x) = (a_0 + b_0) + (a_1 + b_1)x + (a_2 + b_2)x^2 + \cdots + (a_n + b_n)x^n + \cdots$$
$$(cf)(x) = ca_0 + ca_1x + ca_2x^2 + \cdots + ca_nx^n + \cdots$$

Show that V with these operations is a vector space (recall theorems about convergent power series).

7. Consider the vector space V from Prob. 6. From calculus we know that if $f \in V$, and f is differentiable on $\mathbb{R}$ then

$$f'(x) = a_1 + 2a_2x + 3a_3x^2 + \cdots + na_nx^{n-1} + \cdots$$

Therefore, $D: V \to V$. What would be an inverse operator to D?

8. Regarding Prob. 7, show how we might consider the "infinite matrix"

$$M = \begin{pmatrix} 0 & 1 & 0 & 0 & \cdots & & & \\ 0 & 0 & 2 & 0 & \cdots & & & \\ 0 & 0 & 0 & 3 & \cdots & & & \\ 0 & 0 & 0 & 0 & \ddots & & & \\ \vdots & \vdots & \vdots & \vdots & & n & 0 & \\ \vdots & \vdots & \vdots & \vdots & 0 & 0 & \ddots & \cdots \\ \vdots & \vdots & \vdots & \vdots & \vdots & & & \ddots \end{pmatrix}$$

to be the matrix of D. [*Hint*: What would be a good choice for a "basis" of V?]

9. Recall that for $\mathbf{f}: E^3 \to E^3$ the curl of $\mathbf{f}$ can be defined, when $\mathbf{f}(x, y, z) = \langle f_1(x, y, z), f_2(x, y, z), f_3(x, y, z) \rangle$, as

$$\text{curl } \mathbf{f} = \text{"det"} \begin{pmatrix} \mathbf{i} & \mathbf{j} & \mathbf{k} \\ \dfrac{\partial}{\partial x} & \dfrac{\partial}{\partial y} & \dfrac{\partial}{\partial z} \\ f_1 & f_2 & f_3 \end{pmatrix}$$

Show that curl is a linear transformation from the space of ordered triples of continuously differentiable functions of three variables to the space of ordered triples of continuous functions of three variables.

10. For a function $\mathbf{f}: E^3 \to E^3$ given by

$$\mathbf{f}(x, y, z) = (f_1(x, y, z), f_2(x, y, z), f_3(x, y, z))$$

the divergence of f is defined by

$$\text{div } \mathbf{f} = \frac{\partial f_1}{\partial x} + \frac{\partial f_2}{\partial y} + \frac{\partial f_3}{\partial z}$$

Show that div is a linear transformation from the space of ordered triples of continuously differentiable functions of three variables to the space of real-valued continuous functions of three variables.

11. Let V be the space of all real-valued functions $f:[0, 1] \to \mathbb{R}$, along with the standard operations of addition and scalar multiplication.

(*a*) Show that $T: V \to V$ defined by

$$T(f)(x) = xf(x) \qquad \text{for all } x \text{ in } [0, 1]$$

is linear.

(*b*) Show that g defined by

$$g(x) = \begin{cases} 1 & x = 0 \\ 0 & x \neq 0 \end{cases}$$

is in ker T.

(*c*) Is T invertible?

12. Let $V = C[0, 1]$, and define T as in Prob. 11*a*. Is T invertible with this domain? (Check the kernel.)

SUMMARY

Linear transformations, central to applied mathematics, were defined and then analyzed by considering the *kernel* and *range* of the transformation as well as the *matrix representing the transformation*. To find the matrix representing a transformation is the *third basic problem of linear algebra.*

Linear transformations from a vector space to itself *can be classified* in a definite way because certain properties of matrices are invariant under similarity (invertibility, nilpotency, and idempotency are three) and *all representing matrices of a transformation from* $(V, \mathscr{S})$ *to* $(V, \mathscr{S})$ *are similar*. The similarity transform P in these cases is simply a transition matrix, as defined in Chap. 3.

Having the ideas of matrix algebra, linear transformations, and vector spaces in our repertoire, we are ready to tackle two extremely important problems of linear algebra: the eigenvalue-eigenvector problem and the diagonalization problem. In Chap. 5, determinants, inverses, systems of homogeneous equations, vector spaces, bases, dimension, linear dependence, linear independence, similarity, and orthogonality are all used to solve these problems. Thus the results from the first four chapters will at last come together to solve important theoretical and practical problems.

ADDITIONAL PROBLEMS

1. Define $T: \mathscr{M}_{nn} \to \mathscr{M}_{nn}$ by $T(A) = A - A^T$. Show that T is linear. Describe the kernel of T.

2. Define $L: \mathscr{M}_{nn} \to \mathscr{M}_{nn}$ by $L(A) = A + A^T$. Show that L is linear. Describe the kernel of L.

3. Compare the dimension of the vector space of $n \times n$ symmetric matrices and the dimension of the vector space of $n \times n$ upper triangular matrices.

4. Let A be an $n \times n$ invertible matrix. Define $T:\mathcal{M}_{nn} \to \mathcal{M}_{nn}$ by $T(B) = A^{-1}BA$. Is T linear? Is T one-to-one?

5. Let $P \neq 0$ be a matrix with $P^2 = P$, and define $T:\mathcal{M}_{nn} \to \mathcal{M}_{nn}$ by $T(A) = PA$. Show that T is linear. Is T one-to-one?

6. Let c be in $\mathbb{C}$ and A be in $\mathscr{C}_{nn}$. Define $T:\mathscr{C}_{n1} \to \mathscr{C}_{n1}$ by $T(X) = AX - cIX$. Show that T is linear.

7. Let A be a square matrix. Consider the matrix $B = I + A$. Multiply $B[I - A + A^2 - A^3 + \cdots + (-1)^n A^n]$ for different values of n. If A is nilpotent of exponent k, show how the sum in the brackets can be used to compute $(I + A)^{-1}$.

8. Let A be a square matrix such that A^n tends to the zero matrix as n increases without bound. How can you "construct" $(I + A)^{-1}$ by using a sum such as in Prob. 7?

9. Describe the linear transformation $L \circ T$, where L and T come from Probs. 1 and 2. Is $L \circ T = T \circ L$?

10. Define $N:\mathcal{M}_{nn} \to \mathcal{M}_{nn}$ by $N(A) = A^T$. Using L and T from Probs. 1 and 2, calculate $T \circ N$ and $L \circ N$. Compare $T \circ N$ and $N \circ T$. Compare $L \circ N$ and $N \circ L$.

11. Linear transformations can be indexed by a variable. For example, if a particle is rotating about the origin of the plane with constant angular velocity ω, the position of the particle is defined by

$$\begin{pmatrix} x(t) \\ y(t) \end{pmatrix} = \begin{pmatrix} \cos \omega t & -\sin \omega t \\ \sin \omega t & \cos \omega t \end{pmatrix} \begin{pmatrix} x_0 \\ y_0 \end{pmatrix}$$

where

$$\begin{pmatrix} x_0 \\ y_0 \end{pmatrix}$$

is the initial position vector. Notice that the 2×2 matrix $A(t)$ in the equation is a matrix function. Show that for all t, $A(t)$ is invertible. Show that for all t, $A(t)$ is orthogonal.

12. The matrix

$$A = \begin{pmatrix} b_1 & b_2 & b_3 & \cdots & b_n \\ 1 - d_1 & 0 & 0 & \cdots & 0 \\ 0 & 1 - d_2 & 0 & \cdots & 0 \\ 0 & 0 & 1 - d_3 & \cdots & 0 \\ \vdots & \vdots & \vdots & \ddots & \vdots \\ 0 & 0 & 0 & \cdots \; 1 - d_{n-1} & 0 \end{pmatrix}$$

is used in difference equations for determining age distributions in populations; that is, determining the numbers of individuals in different age brackets. The matrix represents a linear transformation from the state of age distributions at one observation time to the next. The b_k are birth rates, and the d_k are death rates for the kth age bracket. Calculate det A for various values of n. Can you generate a formula for det A for arbitrary n?

13. In mechanics, when a body is subjected to forces, some change in relative positions of particles in the body may result. For example, when a rod is bent, the relative positions of points on the curved surface change. This change in relative position is known as *strain*. The term *homogeneous strain* refers to strain in which the new coordinates of a material point, given by Y in $\mathcal{M}_{31}$, are related to the old coordinates, given by X in $\mathcal{M}_{31}$, by $Y = AX$, where A is a 3×3 real matrix. Show that for homogeneous strain straight lines remain straight.

14. Show that in the case of homogeneous strain, parallel straight lines remain parallel.

15. Newton's method for solving $f(x) = 0$ can be extended to the problem of solving

$$f(x, y) = 0 \qquad g(x, y) = 0$$

Recall that the iterative step for Newton's method is

$$x_{n+1} = x_n - \frac{f(x_n)}{f'(x_n)}$$

For the case of the two equations involving $f(x, y)$ and $g(x, y)$, we have

$$\begin{pmatrix} x_{n+1} \\ y_{n+1} \end{pmatrix} = \begin{pmatrix} x_n \\ y_n \end{pmatrix} - \begin{pmatrix} \dfrac{\partial f(x_n, y_n)}{\partial x} & \dfrac{\partial f(x_n, y_n)}{\partial y} \\ \dfrac{\partial g(x_n, y_n)}{\partial x} & \dfrac{\partial g(x_n, y_n)}{\partial y} \end{pmatrix}^{-1} \begin{pmatrix} f(x_n, y_n) \\ g(x_n, y_n) \end{pmatrix}$$

Apply this to

$$x^2 + y^2 - 2 = 0$$
$$x - y = 0$$

and start with

$$\begin{pmatrix} x_1 \\ y_1 \end{pmatrix} = \begin{pmatrix} 1 \\ 0 \end{pmatrix}$$

Complete three steps of the iteration. Actual solutions are

$$\begin{pmatrix} 1 \\ 1 \end{pmatrix} \quad \text{and} \quad \begin{pmatrix} -1 \\ -1 \end{pmatrix}$$

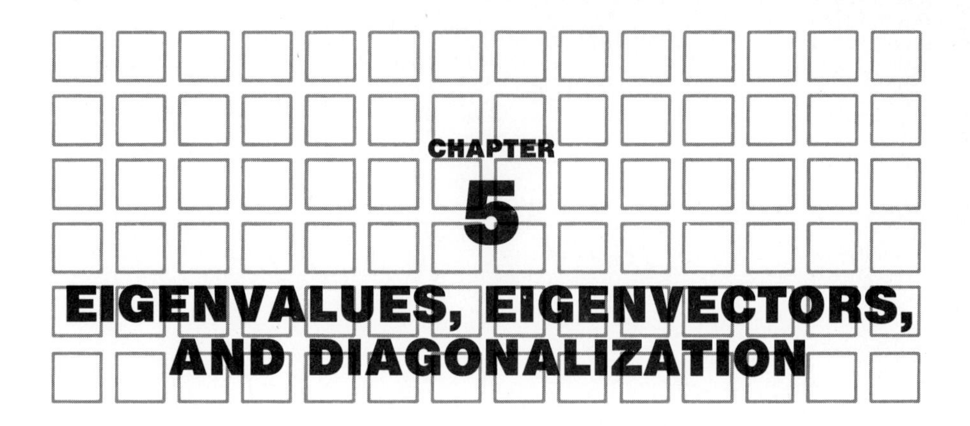

5.1 THE IMPORTANCE OF DIAGONAL SIMILARITY; APPLICATION TO MARKOV CHAINS

Markov chains are a powerful tool for forecasting future events. Effective use of Markov chains involves the calculation of high powers of matrices. Such calculations can be tedious (by hand, at any rate). Extra information about the matrix can make the job easier, as Examples 1 and 2 show.

EXAMPLE 1 Calculate A^6, where

$$A = \begin{pmatrix} 1 & 1 \\ -2 & 4 \end{pmatrix}$$

Solution By definition,

$$\begin{aligned} A^6 &= \underbrace{A \cdot A}\cdot\underbrace{A \cdot A}\cdot\underbrace{A \cdot A} \\ &= \begin{pmatrix} -1 & 5 \\ -10 & 14 \end{pmatrix}\begin{pmatrix} -1 & 5 \\ -10 & 14 \end{pmatrix}\begin{pmatrix} -1 & 5 \\ -10 & 14 \end{pmatrix} \\ &= \begin{pmatrix} -49 & 65 \\ -130 & 146 \end{pmatrix}\begin{pmatrix} -1 & 5 \\ -10 & 14 \end{pmatrix} \\ &= \begin{pmatrix} -601 & 665 \\ -1330 & 1394 \end{pmatrix} \end{aligned}$$

EXAMPLE 2 Calculate A^6, where

$$A = \begin{pmatrix} 1 & 1 \\ -2 & 4 \end{pmatrix}$$

this time given that

$$A = P\begin{pmatrix} 2 & 0 \\ 0 & 3 \end{pmatrix}P^{-1}$$

where

$$P = \begin{pmatrix} 1 & 1 \\ 1 & 2 \end{pmatrix} \quad \text{and} \quad P^{-1} = \begin{pmatrix} 2 & -1 \\ -1 & 1 \end{pmatrix}$$

Solution Let

$$D = \begin{pmatrix} 2 & 0 \\ 0 & 3 \end{pmatrix}.$$

Since

$$A^6 = PDP^{-1}PDP^{-1}PDP^{-1}PDP^{-1}PDP^{-1}PDP^{-1} = PD^6P^{-1}$$

and

$$D^6 = \begin{pmatrix} 2^6 & 0 \\ 0 & 3^6 \end{pmatrix} = \begin{pmatrix} 64 & 0 \\ 0 & 729 \end{pmatrix}$$

we see that

$$A^6 = \begin{pmatrix} 1 & 1 \\ 1 & 2 \end{pmatrix}\begin{pmatrix} 64 & 0 \\ 0 & 729 \end{pmatrix}\begin{pmatrix} 2 & -1 \\ -1 & 1 \end{pmatrix} = \begin{pmatrix} 64 & 729 \\ 64 & 1458 \end{pmatrix}\begin{pmatrix} 2 & -1 \\ -1 & 1 \end{pmatrix}$$
$$= \begin{pmatrix} -601 & 665 \\ -1330 & 1394 \end{pmatrix}$$

Example 2 shows that if a matrix A is similar to a diagonal matrix D, then computing A^n can be done by computing D^n, which is easy.

Two questions must be answered at this point:

1. Given A, can we find P and D so that $A = PDP^{-1}$?

2. How are powers of matrices used in Markov chains?

First, we answer question 2; question 1 is covered in Secs. 5.2 and 5.3. We begin with a simple discussion of Markov chains.

Markov chains are used to analyze *systems* which at a given time can be in only one of a finite number of states. For example, a person may or may not be in debt, the weather may be dry or wet, a mechanical system may be in equilibrium or not. In each case the system may change from state to state. In addition, there is a probability of transition from one state to another between successive observation times. The objective of Markov analysis is to calculate the probability that a system will be in a particular state at some future time and to determine the long-range behavior of the system.

DEFINITION 5.1.1 A *probability*[1] is a number p with $0 \le p \le 1$.

Intuitively, the *empirical probability* of an event E in an experiment is

$$p = \frac{\text{number of occurrences of } E \text{ in a large number } N \text{ of experiments}}{N}$$

Notice that the minimum and maximum values of p are 0 and 1, respectively.

EXAMPLE 3 A coin is tossed 1000 times and allowed to land, and the result, heads or tails, is recorded. The observations are 700 heads, 300 tails. What is the empirical probability of the coin landing heads in a single toss? Is the coin a "fair" coin?

Solution The empirical probability of heads is $\frac{700}{1000} = .7$. The coin is not fair because by experience, the tossing of a fair coin would likely result in approximately 500 heads in 1000 flips.

The idea of the empirical probability makes possible a definition of the concept of a Markov chain.

DEFINITION 5.1.2 Let a system $\mathscr{S}$ have possible states $s_1, s_2, \ldots, s_n$. Suppose that we observe $\mathscr{S}$ at given times $T_1, T_2, \ldots, T_m, \ldots$. A *Markov chain* is a process in which the empirical probability that $\mathscr{S}$ is in a particular state at observation time T_k depends only on which state $\mathscr{S}$ is in at time T_{k-1}.

EXAMPLE 4 Consider the system of a student $\mathscr{S}$ who is in states

$$s_1\text{:}\quad \text{semester grade point average} < 3.0$$

or

$$s_2\text{:}\quad \text{semester grade point average} \ge 3.0$$

As a result of observations from high school, it is found that if $\mathscr{S}$ is in state s_1 in one semester, then she or he will work harder the next semester and achieve state s_2 with probability .8, state s_1 with the lower probability of .2. But if $\mathscr{S}$ is in state s_2 in one semester, then $\mathscr{S}$ relaxes the next semester and falls below 3.0 with probability .3; and $\mathscr{S}$ stays above 3.0 with probability .7. This is an example of a Markov chain with observation times at the end of each semester. We are assuming that $\mathscr{S}$'s achievement in one semester is determined only by

[1] We are using a brief intuitive notion of probability here; for a detailed development, see a text on probability theory.

her or his motivation resulting from the previous semester's grade point average.

The probabilities in Example 4 are called *transition probabilities* and can be arranged in a matrix

$$\begin{array}{c} \text{State at time } T_{k-1} \\ M = \begin{array}{cc} s_1 & s_2 \end{array} \\ \begin{pmatrix} .2 & .3 \\ .8 & .7 \end{pmatrix} \begin{array}{c} s_1 \\ s_2 \end{array} \quad \text{State at time } T_k \end{array}$$

Thus .8 is the transition probability of changing from state s_1 to s_2, and so on. Note that the columns add to 1. This matrix is called the *transition matrix* for the Markov chain.

The importance of the transition matrix is that it can be used to find the probability of being in state s_1 or s_2 at later times. To illustrate, suppose the student achieves a 2.5 grade point average (GPA) in the first semester. We know that the student is in state s_1 and can represent this with a state vector from $\mathcal{M}_{21}$ as

$$S = \begin{pmatrix} 1 \\ 0 \end{pmatrix}$$

The first slot is filled with the probability that $\mathcal{S}$ is in s_1 and the second slot with the probability she or he is in s_2. Multiplying S by M, we have

$$MS = \begin{pmatrix} .2 \\ .8 \end{pmatrix}$$

Recalling our example, we see that

$$\begin{pmatrix} .2 \\ .8 \end{pmatrix}$$

represents the probabilities of being in s_1 and s_2 after a semester with GPA < 3.0. That is, MS is exactly the state vector for the next observation time. In general, this is true:

> If $\mathcal{S}$, with transition matrix M, has state vector S at time T_{k-1}, then MS is the state vector for time T_k.

Note that we require state vectors to have nonnegative entries which sum to 1.

EXAMPLE 5 Show that if $\mathcal{S}$, with transition matrix M, has state vector S at time T_0, then the state vector at time T_4 is M^4S.

Solution Using the statement above, the state vector at time T_1 is MS. To find the state vector at time T_2, we multiply the state vector at time T_1 by M. That is, we compute

$$M(MS) = M^2S$$

Continuing in this way, we have

	TIME				
	T_0	T_1	T_2	T_3	T_4
State vector	S	MS	$M(MS)$	$M(M(MS))$	$M(M(M(MS)))$
Simplified state vector	S	MS	M^2S	M^3S	M^4S

From Example 5 we see the importance of the power of a matrix for Markov chains: It is used to compute future state vectors. The next example is more specific.

EXAMPLE 6 Recall the student from Example 4. If the student achieves a 2.5 GPA in the first semester, what is the probability that $\mathscr{S}$ will have a GPA above 3.0 after the fourth semester?

Solution The initial state vector is

$$\begin{pmatrix} 1 \\ 0 \end{pmatrix}$$

Schematically, we have the following:

	TIME			
	T_0	T_1	T_2	T_3
Semester	1	2	3	4
State vector	$\begin{pmatrix} 1 \\ 0 \end{pmatrix}$	$M\begin{pmatrix} 1 \\ 0 \end{pmatrix}$	$M^2\begin{pmatrix} 1 \\ 0 \end{pmatrix}$	$M^3\begin{pmatrix} 1 \\ 0 \end{pmatrix}$

So we see that

$$M^3\begin{pmatrix} 1 \\ 0 \end{pmatrix}$$

must be calculated. First, we calculate M^3.

$$M = \begin{pmatrix} .2 & .3 \\ .8 & .7 \end{pmatrix}$$

$$M^2 = \begin{pmatrix} .28 & .27 \\ .72 & .73 \end{pmatrix}$$

$$M^3 = \begin{pmatrix} .272 & .273 \\ .728 & .727 \end{pmatrix}$$

Thus the state vector at the end of the fourth semester is

$$M^3S = \begin{pmatrix} .272 \\ .728 \end{pmatrix}$$

The probability of the student having a GPA above 3.0 in the fourth semester is .728. The student is much more likely to have a GPA above 3.0 than below.

EXAMPLE 7 A utility company finds that, in general, if a typical customer pays a bill late one month, that person will pay before the due date on the next billing $\frac{1}{2}$ of the time. But if a customer pays early one month, that person is likely to pay late the following month $\frac{6}{10}$ of the time. For the March billing virtually all 10,000 customers pay their bills on time. About how many customers will pay late for the July billing?

Solution We treat the payment system as a Markov chain. The states are

$$s_1 = \text{payment on time} \qquad s_2 = \text{payment late}$$

the transition matrix is

$$\begin{array}{cc} & \text{This month} \\ & \begin{matrix} s_1 & s_2 \end{matrix} \\ M = & \begin{pmatrix} \frac{4}{10} & \frac{1}{2} \\ \frac{6}{10} & \frac{1}{2} \end{pmatrix} \begin{matrix} s_1 \\ s_2 \end{matrix} \quad \text{Next month} \end{array}$$

To find the state vector for July, we can make a chart

	MONTH				
	MARCH	APRIL	MAY	JUNE	JULY
State vector	$S = \begin{pmatrix} 1 \\ 0 \end{pmatrix}$	MS	M^2S	M^3S	M^4S

Or we could realize that July is month 7, March is month 3, and so we must calculate $M^{7-3} = M^4$. Calculating, we find

$$M^4 = \begin{pmatrix} .4546 & .4545 \\ .5454 & .5455 \end{pmatrix}$$

Therefore the state vector is

$$M^4 \begin{pmatrix} 1 \\ 0 \end{pmatrix} = \begin{pmatrix} .4546 \\ .5454 \end{pmatrix}$$

The utility can expect about (.5454)(10,000) = 5454 people to pay late. The company probably needs to institute greater penalties for late payment.

We discuss more properties of Markov chains in Secs. 5.2 and 5.3.

PROBLEMS 5.1

1. Given that

$$A = \begin{pmatrix} -1 & -3 \\ \frac{3}{2} & \frac{7}{2} \end{pmatrix} = \begin{pmatrix} 2 & -1 \\ -1 & 1 \end{pmatrix} \begin{pmatrix} \frac{1}{2} & 0 \\ 0 & 2 \end{pmatrix} \begin{pmatrix} 1 & 1 \\ 1 & 2 \end{pmatrix}$$

calculate A^{10}.

2. Which of the following matrices cannot be transition matrices for a Markov chain?

(a) $\begin{pmatrix} -.1 & .3 \\ .9 & .7 \end{pmatrix}$ (b) $\begin{pmatrix} \frac{1}{2} & \frac{1}{3} \\ \frac{1}{2} & \frac{2}{3} \end{pmatrix}$

(c) $\begin{pmatrix} \frac{1}{2} & \frac{1}{2} \\ \frac{1}{3} & \frac{2}{3} \end{pmatrix}$ (d) $\begin{pmatrix} .1 & .3 & .1 \\ .2 & .7 & .8 \\ .7 & 0 & .1 \end{pmatrix}$

(e) $\begin{pmatrix} \frac{1}{2} & \frac{1}{3} & \frac{1}{10} \\ \frac{1}{2} & \frac{1}{3} & \frac{5}{10} \\ 0 & \frac{1}{6} & \frac{4}{10} \end{pmatrix}$ (f) $\begin{pmatrix} .1 & .6 & .3 \\ .9 & .4 & .7 \end{pmatrix}$

3. Consider a system of laundry detergent consumers. Consumers buy either liquid or dry detergent each week. During a previous advertising campaign for liquid detergents, a market research firm found that $\frac{4}{10}$ of the people who bought dry detergent one week bought liquid the next. But $\frac{8}{10}$ of the people who bought liquid one week bought dry the next week. Assume similar market dynamics in the new advertising campaign. If at the beginning of the campaign half the people buy liquid and half buy dry detergent, what share of the market will liquid detergent have after 4 weeks? Can this result be used to evaluate the advertising agency?

4. In a certain town it is known from past experience that if the weather is wet one day, there is a $\frac{4}{10}$ probability of wet weather the next day. But if the weather is dry one day, then there is a $\frac{9}{10}$ probability of dry weather the next

day. On Monday the weather is wet, and a company's picnic is scheduled for Saturday. Should the picnic be rescheduled?

5. Regarding Prob. 4, would an appreciable advantage be gained in having the picnic on Sunday instead of Saturday?

5.2 THE EIGENPROBLEM

We saw in Sec. 5.1 that it is important to know when we can write a matrix A as

$$A = PDP^{-1}$$

where D is a diagonal matrix. To see where D comes from, let us suppose that A is 2×2,

$$A = P\begin{pmatrix} 2 & 0 \\ 0 & 3 \end{pmatrix}P^{-1}$$

and that P represents a rotation of θ radians in E^2. Let X in $\mathcal{M}_{21}$ be a vector that P^{-1} rotates to

$$\begin{pmatrix} 1 \\ 0 \end{pmatrix} \qquad \text{or} \qquad P^{-1}X = \begin{pmatrix} 1 \\ 0 \end{pmatrix}$$

Then

$$AX = P\begin{pmatrix} 2 & 0 \\ 0 & 3 \end{pmatrix}\begin{pmatrix} 1 \\ 0 \end{pmatrix} = P\begin{pmatrix} 2 \\ 0 \end{pmatrix} = 2P\begin{pmatrix} 1 \\ 0 \end{pmatrix} = 2X$$

(We know that

$$P\begin{pmatrix} 1 \\ 0 \end{pmatrix} = X$$

since P just undoes the rotation of P^{-1}.) We see then that $AX = 2X$. In this case we say the number 2 is an *eigenvalue* (pronounced "eye-genvalue") of A and X is an *eigenvector* of A. We can also see that 3 is an eigenvalue of A by letting Y be a vector which P rotates to

$$\begin{pmatrix} 0 \\ 1 \end{pmatrix}$$

Then

$$AY = P\begin{pmatrix} 0 \\ 3 \end{pmatrix} = 3P\begin{pmatrix} 0 \\ 1 \end{pmatrix} = 3Y$$

The calculations in the last paragraph showed that for the given matrix, the entries of D satisfied the equations $AX = 2X$ and $AY = 3Y$. This suggests that the following problem, *the fourth basic problem of linear algebra,* is important in finding diagonal matrices similar to a given matrix.

The Matrix Eigenproblem

Given an $n \times n$ matrix A in $\mathscr{C}_{nn}$, find all numbers λ and all *nonzero* vectors X in $\mathscr{C}_{n1}$ such that $AX = \lambda X$. The numbers λ are called *eigenvalues*[2] of A, and the vectors X are called *eigenvectors*[2] of A.

The solution of the matrix eigenproblem reduces to the solution of linear equations (the first basic problem of linear algebra), as Example 1 shows.

EXAMPLE 1 Find the eigenvalues and eigenvectors of

$$A = \begin{pmatrix} 1 & 1 \\ -2 & 4 \end{pmatrix}$$

Solution We must solve $AX = \lambda X$. In expanded form this is

$$\begin{pmatrix} 1 & 1 \\ -2 & 4 \end{pmatrix}\begin{pmatrix} x_1 \\ x_2 \end{pmatrix} = \lambda\begin{pmatrix} x_1 \\ x_2 \end{pmatrix} = \begin{pmatrix} \lambda x_1 \\ \lambda x_2 \end{pmatrix}$$

Reducing this to linear equations, we find

$$(A - \lambda I)X = \begin{pmatrix} 1 & 1 \\ -2 & 4 \end{pmatrix}\begin{pmatrix} x_1 \\ x_2 \end{pmatrix} - \begin{pmatrix} \lambda & 0 \\ 0 & \lambda \end{pmatrix}\begin{pmatrix} x_1 \\ x_2 \end{pmatrix} = \begin{pmatrix} 0 \\ 0 \end{pmatrix}$$

or

$$\begin{aligned} (1 - \lambda)x_1 + x_2 &= 0 \\ -2x_1 + (4 - \lambda)x_2 &= 0 \end{aligned} \tag{5.2.1}$$

Since we have only two equations and three unknowns (λ, x_1, x_2), we expect a third condition to make solution possible. Because *eigenvectors are required to be nonzero,* we realize that the system (5.2.1) must have a nontrivial solution for

$$\begin{pmatrix} x_1 \\ x_2 \end{pmatrix}$$

From our previous work on homogeneous equations, we know that a nontrivial solution exists if and only if

$$\det\begin{pmatrix} 1 - \lambda & 1 \\ -2 & 4 - \lambda \end{pmatrix} = 0$$

---

[2] They are also called *characteristic values* and *characteristic vectors.* In many applications, the matrices, eigenvalues, and eigenvectors are required to be real.

This condition reduces to

$$(1 - \lambda)(4 - \lambda) + 2 = 6 - 5\lambda + \lambda^2 = (\lambda - 3)(\lambda - 2) = 0$$

which means that $\lambda = 2$ and $\lambda = 3$ are solutions to $AX = \lambda X$, with X still to be found.

To find X corresponding to $\lambda = 2$, substitute $\lambda = 2$ into (5.2.1), to obtain

$$\begin{aligned} -x_1 + x_2 &= 0 \\ -2x_1 + 2x_2 &= 0 \end{aligned}$$

which has solution

$$\begin{pmatrix} k \\ k \end{pmatrix}$$

Thus the general *eigenvector corresponding to the eigenvalue* 2 is

$$\begin{pmatrix} k \\ k \end{pmatrix}$$

Specific eigenvectors are obtained by putting k equal to specific numbers.

To calculate X corresponding to $\lambda = 3$, substitute $\lambda = 3$ into (5.2.1) to find

$$\begin{aligned} -2x_1 + x_2 &= 0 \\ -2x_1 + x_2 &= 0 \end{aligned}$$

This has solution

$$\begin{pmatrix} j \\ 2j \end{pmatrix}$$

which is the general *eigenvector of A corresponding to the eigenvalue 3.*

Finally, we write the solution to the eigenproblem for A by listing the *eigenpairs*

$$\left(2, \begin{pmatrix} k \\ k \end{pmatrix}\right), \quad \left(3, \begin{pmatrix} j \\ 2j \end{pmatrix}\right)$$

Solution procedure for the eigenproblem for $A_{n \times n}$

1. Solve $\det(A - \lambda I) = 0$ for eigenvalues $\lambda_1, \lambda_2, \ldots, \lambda_n$.
2. To find the general eigenvector corresponding to λ, solve the homogeneous equations $(A - \lambda I)X = 0$.

Closely related to the matrix eigenproblem is the eigenproblem for linear transformations.

Given $L: V \to V$, a linear transformation, find all numbers λ and nonzero vectors $\mathbf{v}$ in V such that

$$L(\mathbf{v}) = \lambda\mathbf{v}$$

To solve this problem, we need to solve the matrix eigenproblem for a matrix which represents L. We can solve the problem this way because *all matrices representing L have the same eigenvalues.* This follows from Theorems 5.2.1 and 5.2.2.

THEOREM 5.2.1 If A is an $n \times n$ matrix, then

(*a*) Det $(A - \lambda I)$ is a polynomial $p(\lambda)$ of degree n.

(*b*) The eigenvalues of A are the solutions of $p(\lambda) = 0$.

(*c*) If λ_0 is an eigenvalue, any nontrivial solution of $(A - \lambda_0 I)X = 0$ is an eigenvector of A corresponding to λ_0.

Proof (*a*) If A is 2×2, then $\det(A - \lambda I) = (a_{11} - \lambda)(a_{22} - \lambda) - a_{12}a_{21}$, which is a polynomial in λ of degree 2. To proceed by induction, suppose that for a $k \times k$ matrix the determinant of $A - \lambda I$ is a polynomial of degree k. Now if B is a $(k + 1) \times (k + 1)$ matrix, then we calculate $\det(B - \lambda I)$ by using the definition and column 1. In this way the determinant is seen to be a polynomial of degree $k + 1$, and the principle of mathematical induction proves part (*a*).

(*b*) A number λ is an eigenvalue if and only if $AX = \lambda X$ and $X \neq 0$. Thus, $AX - \lambda X = (A - \lambda I)X = 0$. Homogeneous equations have nontrivial solutions if and only if the determinant of the coefficient matrix is zero. Therefore λ is an eigenvalue if and only if $p(\lambda) = \det(A - \lambda I) = 0$.

(*c*) Let X be any nontrivial solution of $(A - \lambda_0 I)X = 0$. Then $AX = \lambda_0 IX = \lambda_0 X$.

The polynomial $p(\lambda)$ mentioned in Theorem 5.2.1 is called the *characteristic polynomial of A*. If A is $n \times n$, its characteristic polynomial has n roots. Therefore an $n \times n$ matrix has n eigenvalues.

THEOREM 5.2.2 If A and B are similar, then the set of eigenvalues of A is equal to the set of eigenvalues of B. If the similarity transform is P (that is, if $A = P^{-1}BP$) and (λ_0, X) is an eigenpair of A, then (λ_0, PX) is an eigenpair of B.

Proof Let A be similar to B, and let λ_0 be an eigenvalue of A. We have, using $A = P^{-1}BP$,

$$\begin{aligned} 0 = \det(A - \lambda_0 I) &= \det(P^{-1}BP - \lambda_0 I) \\ &= \det(P^{-1}BP - \lambda_0 P^{-1}IP) \\ &= \det(P^{-1}BP - P^{-1}(\lambda_0 I)P) \\ &= \det(P^{-1}(B - \lambda_0 I)P) \\ &= \det(P^{-1}) \det(B - \lambda_0 I) \det P \\ &= \det(B - \lambda_0 I) \end{aligned}$$

Thus, λ_0 is an eigenvalue of B. So every eigenvalue of A is also an eigenvalue of B. The argument is virtually the same to show that every eigenvalue of B is an eigenvalue of A.

Now let (λ_0, X) be an eigenpair of A. We have

$$AX = \lambda_0 X$$

so

$$(P^{-1}BP)X = \lambda_0 X$$

Multiplying both sides on the left by P and reassociating the products lead to

$$B(PX) = \lambda_0(PX)$$

Now we can state a solution to the eigenproblem for linear transformations.

Choose a basis $\mathscr{S}$ for V. Let M be the matrix representing L with respect to the basis $\mathscr{S}$. If (λ_0, X) is an eigenpair of M, then letting $\mathbf{v}$ be the vector such that $(\mathbf{v})_{\mathscr{S}} = X$, we have

$$L(\mathbf{v}) = \lambda_0 \mathbf{v}$$

Because of this correspondence we work mainly with the matrix eigenvalue problem. Now we consider some more examples of this problem.

EXAMPLE 2 Show that

$$A = \begin{pmatrix} 0 & 1 \\ -1 & 0 \end{pmatrix}$$

has complex eigenvalues.

Solution We have

$$\det(A - \lambda I) = \det\begin{pmatrix} -\lambda & 1 \\ -1 & -\lambda \end{pmatrix} = \lambda^2 + 1$$

The equation $\det(A - \lambda I) = 0$ is $\lambda^2 + 1 = 0$, which has solutions $\lambda_1 = i$ and $\lambda_2 = -i$.

For the eigenvectors we solve $(A - \lambda I)X = 0$ for X when λ is i or $-i$. The equations are

$\lambda_1 = i$:
$$\left(\begin{array}{cc|c} -i & 1 & 0 \\ -1 & -i & 0 \end{array}\right) \xrightarrow{iR1 + R2} \left(\begin{array}{cc|c} -i & 1 & 0 \\ 0 & 0 & 0 \end{array}\right)$$

$$x_2 = k \qquad x_1 = \frac{1}{i}k = -ik$$

$$X = k\begin{pmatrix} -i \\ 1 \end{pmatrix}$$

$\lambda_2 = -i$:
$$\left(\begin{array}{cc|c} i & 1 & 0 \\ -1 & i & 0 \end{array}\right) \xrightarrow{-iR1 + R2} \left(\begin{array}{cc|c} i & 1 & 0 \\ 0 & 0 & 0 \end{array}\right)$$

$$x_2 = r \qquad x_1 = \frac{-1}{i}r = ir$$

$$X = r\begin{pmatrix} i \\ 1 \end{pmatrix}$$

Therefore the solution to the eigenproblem is

$$\left(i, c\begin{pmatrix} -i \\ 1 \end{pmatrix}\right) \qquad \left(-i, d\begin{pmatrix} i \\ 1 \end{pmatrix}\right)$$

where c and d are arbitrary constants.

In the first two examples the eigenvalues were distinct. However, this need not always be the case.

EXAMPLE 3 Solve the eigenproblem for

$$A = \begin{pmatrix} 3 & 0 \\ 0 & 3 \end{pmatrix}$$

Solution Solving $\det(A - \lambda I) = (3 - \lambda)^2 = 0$, we find $\lambda = 3$ is a solution of multiplicity 2 since $3 - \lambda$ is twice a factor of $\det(A - \lambda I)$. To find the eigenvector corresponding to $\lambda = 3$, we substitute $\lambda = 3$ into $(A - \lambda I)X = 0$ to obtain

$$0x_1 + 0x_2 = 0$$
$$0x_1 + 0x_2 = 0$$

which means that *any* vector X is an eigenvalue corresponding to $\lambda = 3$. That is, the eigenpairs are

$$\left(3, \begin{pmatrix} k \\ j \end{pmatrix}\right) \qquad \left(3, \begin{pmatrix} r \\ s \end{pmatrix}\right)$$

where r, s, k, and j are arbitrary.

Before proceeding to more examples, we summarize what we have done so far.

An *eigenvalue* of a matrix A is a *number* λ which satisfies the equation $\det(A - \lambda I) = 0$. An eigenvector corresponding to λ is any nonzero solution of $(A - \lambda I)X = 0$. The equation $\det(A - \lambda I) = 0$ is called the *characteristic equation* of A. The expression $\det(A - \lambda I)$ is always a polynomial (of degree n if A is $n \times n$) and is called the *characteristic polynomial* of A. The eigenvalues of A are the roots of the characteristic polynomial and the solutions of the characteristic equation. From algebra we know that a polynomial of degree n has n complex roots; therefore an $n \times n$ matrix A has n eigenvalues. In counting eigenvalues, multiplicities must be taken into account. For instance, in Example 3, the 2×2 matrix had two eigenvalues, because the root, 3, of the characteristic polynomial was a repeated root.

If (λ_0, X) and (λ_0, Y) are two eigenpairs of A, then $(\lambda_0, X + Y)$ and (λ_0, cX) are eigenpairs of A, provided $X + Y \neq 0$ and $c \neq 0$. Therefore, if we adjoin the zero vector to the set of all eigenvectors associated with λ_0, we obtain a subspace of $\mathscr{C}_{n1}$. This subspace is denoted $E_{(\lambda_0)}$, and is called the *eigenspace* of λ_0. Thus for matrix A in Example 1,

$$E_{(2)} = \operatorname{span}\left\{\begin{pmatrix}1\\1\end{pmatrix}\right\} \qquad \text{and} \qquad E_{(3)} = \operatorname{span}\left\{\begin{pmatrix}1\\2\end{pmatrix}\right\}$$

EXAMPLE 4 Show that

$$A = \begin{pmatrix} 0 & 1 & 0 \\ -1 & 0 & 0 \\ 0 & 0 & 2 \end{pmatrix}$$

has only one real eigenvalue. Find the eigenspace corresponding to that eigenvalue.

Solution The characteristic polynomial is

$$\det\begin{pmatrix} -\lambda & 1 & 0 \\ -1 & -\lambda & 0 \\ 0 & 0 & 2-\lambda \end{pmatrix} = (\lambda^2 + 1)(2 - \lambda)$$

Since $\lambda^2 + 1$ is never zero for real λ, $\lambda = 2$ is the only real eigenvalue. The eigenvector is found by solving $A - \lambda I = 0$ with $\lambda = 2$:

$$\left(\begin{array}{ccc|c} -2 & 1 & 0 & 0 \\ -1 & -2 & 0 & 0 \\ 0 & 0 & 0 & 0 \end{array}\right) \xrightarrow[\substack{R1 + R2 \\ -\frac{2}{5}R2}]{-\frac{1}{2}R1} \left(\begin{array}{ccc|c} 1 & -\frac{1}{2} & 0 & 0 \\ 0 & 1 & 0 & 0 \\ 0 & 0 & 0 & 0 \end{array}\right)$$

The general solution is

$$\begin{pmatrix} 0 \\ 0 \\ k \end{pmatrix}$$

and the eigenpair is

$$\left(2, \begin{pmatrix} 0 \\ 0 \\ k \end{pmatrix}\right)$$

The eigenspace

$$E_{(2)} = \text{span}\left\{\begin{pmatrix} 0 \\ 0 \\ 1 \end{pmatrix}\right\}$$

EXAMPLE 5 Solve the eigenproblem for

$$A = \begin{pmatrix} -4 & -4 & -8 \\ 4 & 6 & 4 \\ 6 & 4 & 10 \end{pmatrix}$$

Solution The characteristic polynomial $p(\lambda)$ is

$$\det\begin{pmatrix} -4-\lambda & -4 & -8 \\ 4 & 6-\lambda & 4 \\ 6 & 4 & 10-\lambda \end{pmatrix} = -\lambda^3 + 12\lambda^2 - 44\lambda + 48$$

To solve $p(\lambda) = 0$, we first guess at one root of $p(\lambda)$. First we try integer factors of 48: ± 1, ± 2, ± 3, ± 4, ± 6, ± 8, ± 12, ± 16, ± 24, ± 48. Since $p(1) \neq 0$ and $p(-1) \neq 0$, neither 1 nor -1 is a root. However, $p(2) = 0$, so 2 is a root and $\lambda - 2$ is a factor. Dividing, we have

$$\begin{array}{r|l}
 & -\lambda^2 + 10\lambda - 24 \\ \hline
\lambda - 2 & -\lambda^3 + 12\lambda^2 - 44\lambda + 48 \\
 & -\lambda^3 + 2\lambda^2 \\ \hline
 & 10\lambda^2 - 44\lambda \\
 & 10\lambda^2 - 20\lambda \\ \hline
 & -24\lambda + 48 \\
 & -24\lambda + 48 \\ \hline
 & 0
\end{array}$$

so that

$$p(\lambda) = (\lambda - 2)(-\lambda^2 + 10\lambda - 24)$$

Now we can factor the quadratic part of $p(\lambda)$ easily into $(-\lambda + 4)(\lambda - 6)$. Therefore

$$p(\lambda) = (\lambda - 2)(-\lambda + 4)(\lambda - 6)$$

and the eigenvalues are 2, 4, and 6.

To find the eigenvectors, we substitute $\lambda = 2$, $\lambda = 4$, and $\lambda = 6$ into $(A - \lambda I)X = 0$ and solve. The resulting equations are

$$\lambda = 2: \quad \left(\begin{array}{ccc|c} -6 & -4 & -8 & 0 \\ 4 & 4 & 4 & 0 \\ 6 & 4 & 8 & 0 \end{array}\right) \longrightarrow \left(\begin{array}{ccc|c} 3 & 2 & 4 & 0 \\ 0 & 4 & -4 & 0 \\ 0 & 0 & 0 & 0 \end{array}\right) \Rightarrow X = \begin{pmatrix} -2k \\ k \\ k \end{pmatrix}$$

$$\lambda = 4: \quad \left(\begin{array}{ccc|c} -8 & -4 & -8 & 0 \\ 4 & 2 & 4 & 0 \\ 6 & 4 & 6 & 0 \end{array}\right) \longrightarrow \left(\begin{array}{ccc|c} 2 & 1 & 2 & 0 \\ 0 & 4 & 0 & 0 \\ 0 & 0 & 0 & 0 \end{array}\right) \Rightarrow X = \begin{pmatrix} -j \\ 0 \\ j \end{pmatrix}$$

$$\lambda = 6: \quad \left(\begin{array}{ccc|c} -10 & -4 & -8 & 0 \\ 4 & 0 & 4 & 0 \\ 6 & 4 & 4 & 0 \end{array}\right) \longrightarrow \left(\begin{array}{ccc|c} -5 & -2 & -4 & 0 \\ 0 & -2 & 1 & 0 \\ 0 & 0 & 0 & 0 \end{array}\right) \Rightarrow X = \begin{pmatrix} -r \\ \frac{1}{2}r \\ r \end{pmatrix}$$

The eigenpairs are

$$\left(2, \begin{pmatrix} -2k \\ k \\ k \end{pmatrix}\right) \quad \left(4, \begin{pmatrix} -j \\ 0 \\ j \end{pmatrix}\right) \quad \left(6, \begin{pmatrix} -r \\ \frac{1}{2}r \\ r \end{pmatrix}\right)$$

and the eigenspaces are

$$E_{(2)} = \operatorname{span}\left\{\begin{pmatrix} -2 \\ 1 \\ 1 \end{pmatrix}\right\} \qquad E_{(4)} = \operatorname{span}\left\{\begin{pmatrix} -1 \\ 0 \\ 1 \end{pmatrix}\right\} \qquad E_{(6)} = \operatorname{span}\left\{\begin{pmatrix} -2 \\ 1 \\ 2 \end{pmatrix}\right\}$$

EXAMPLE 6 The matrices

$$A = \begin{pmatrix} 3 & -1 & 0 \\ 0 & 3 & 0 \\ 0 & 0 & 5 \end{pmatrix} \quad \text{and} \quad B = \begin{pmatrix} 3 & 0 & 0 \\ 0 & 3 & 0 \\ 0 & 0 & -5 \end{pmatrix}$$

both have eigenvalues 3 and -5, with 3 being an eigenvalue of multiplicity 2. Compare the eigenspaces for these two matrices.

Solution For A the general eigenpairs are

$$\left(3, \begin{pmatrix} k \\ 0 \\ 0 \end{pmatrix}\right) \quad \text{and} \quad \left(-5, \begin{pmatrix} 0 \\ 0 \\ r \end{pmatrix}\right)$$

so the eigenspaces are

$$E_{(3)} = \text{span}\left\{\begin{pmatrix}1\\0\\0\end{pmatrix}\right\} \qquad E_{(-5)} = \text{span}\left\{\begin{pmatrix}0\\0\\1\end{pmatrix}\right\}$$

For B the general eigenpairs are

$$\left(3, \begin{pmatrix}r\\s\\0\end{pmatrix}\right) \quad \text{and} \quad \left(-5, \begin{pmatrix}0\\0\\t\end{pmatrix}\right)$$

so the eigenspaces for B are

$$E_{(3)} = \text{span}\left\{\begin{pmatrix}1\\0\\0\end{pmatrix}, \begin{pmatrix}0\\1\\0\end{pmatrix}\right\} \qquad E_{(-5)} = \text{span}\left\{\begin{pmatrix}0\\0\\1\end{pmatrix}\right\}$$

For A, dim $E_{(3)} <$ (multiplicity of $\lambda = 3$), while for B, dim $E_{(3)} =$ (multiplicity of $\lambda = 3$).

In the preceding examples we have been able to solve the eigenproblem exactly. In general, this is not the case; sometimes numerical methods must be used. These are discussed in Chap. 6.

Fixed Vectors of Matrices If a matrix A has an eigenpair $(1, X)$, (that is, if $AX = X$), then X is called a *fixed vector* (or fixed point) of A. This concept is important for Markov chains. Some Markov chains have the property that as n grows larger and larger, M^n (powers of the transition matrix) begins to look the same. For example, the matrix

$$M = \begin{pmatrix} .2 & .3 \\ .8 & .7 \end{pmatrix}$$

from Example 5 of Sec. 5.1 has powers

$$M^2 = \begin{pmatrix} .28 & .27 \\ .72 & .73 \end{pmatrix}$$

$$M^3 = \begin{pmatrix} .272 & .273 \\ .728 & .727 \end{pmatrix}$$

$$M^4 = \begin{pmatrix} .2728 & .2727 \\ .7272 & .7273 \end{pmatrix}$$

$$M^5 = \begin{pmatrix} .27272 & .27273 \\ .72728 & .72727 \end{pmatrix}$$

$$M^6 = \begin{pmatrix} .272728 & .272727 \\ .727272 & .727273 \end{pmatrix}$$

which appear to be approximating a matrix

$$T = \begin{pmatrix} .27\overline{27}\cdots & .27\overline{27}\cdots \\ .7272\cdots & .7272\cdots \end{pmatrix}$$
$$= \begin{pmatrix} \frac{27}{99} & \frac{27}{99} \\ \frac{72}{99} & \frac{72}{99} \end{pmatrix}$$

(see Prob. 5). In this case we write

$$M^n \to T \qquad \text{as} \qquad n \to +\infty$$

If a Markov chain has this property, it is called *regular*. In a regular Markov chain, for *any* initial state vector S

$$M^n S \to TS$$

This means that regardless of initial state, the Markov chain settles into an *equilibrium state* $E = TS$.

The important fact is this:

If a Markov chain with transition matrix M is regular, then the equilibrium state E is a fixed point of M. That is,

$$ME = E$$

EXAMPLE 7 It can be shown that a Markov chain is *regular* if its transition matrix M, or some power of M, has only positive entries. Consider a Markov chain with transition matrix

$$M = \begin{pmatrix} .3 & 1 \\ .7 & 0 \end{pmatrix}$$

Show that the chain is regular. Find the equilibrium state vector for the chain.

Solution Since M does not have all positive entries, we must look at powers of M to establish regularity. Since

$$M^2 = \begin{pmatrix} .79 & .3 \\ .21 & .7 \end{pmatrix}$$

has all positive entries, the chain is regular. To find the fixed point E, we must solve

$$ME = E$$

with E being a state vector. That is, we must solve

$$ME = E$$
$$E^+ = 1 \qquad E \geq 0$$

where $E \geq 0$ means all the components of E are nonnegative and E^+ means the sum of the components of E.

First, we solve $ME = E$. Letting

$$ME = E = \begin{pmatrix} e_1 \\ e_2 \end{pmatrix}$$

we have

$$\begin{pmatrix} .3 & 1 \\ .7 & 0 \end{pmatrix}\begin{pmatrix} e_1 \\ e_2 \end{pmatrix} = \begin{pmatrix} e_1 \\ e_2 \end{pmatrix}$$

which reduces to

$$\begin{aligned} .-7e_1 + e_2 &= 0 \\ .7e_1 - e_2 &= 0 \end{aligned}$$

or $-.7e_1 + e_2 = 0$. Therefore

$$E = \begin{pmatrix} \frac{10}{7}k \\ k \end{pmatrix}$$

The requirement $E^+ = 1$ forces the condition

$$\begin{aligned} \tfrac{10}{7}k + k &= 1 \\ k &= \tfrac{7}{17} \end{aligned}$$

so that

$$E = \begin{pmatrix} \frac{10}{17} \\ \frac{7}{17} \end{pmatrix}$$

is the equilibrium state for the chain. The interpretation of this result is that if M is the transition matrix for a Markov chain, then in the long run the system is in state number 1 for $\frac{10}{17}$ of the time and in state number 2 for $\frac{7}{17}$ of the time. Note that the condition $E^+ = 1$ forced us to choose only one of the eigenvectors with eigenvalue 1.

EXAMPLE 8 Show that the matrix

$$M = \begin{pmatrix} 0 & 1 \\ 1 & 0 \end{pmatrix}$$

cannot be the transition matrix for a *regular* Markov chain.

Solution We will show that powers of M do not tend to a particular matrix. The powers are

$$M^2 = \begin{pmatrix} 1 & 0 \\ 0 & 1 \end{pmatrix} \qquad M^3 = \begin{pmatrix} 0 & 1 \\ 1 & 0 \end{pmatrix} \qquad M^4 = \begin{pmatrix} 1 & 0 \\ 0 & 1 \end{pmatrix} \qquad \cdots$$

Thus

$$M^{2n} = \begin{pmatrix} 1 & 0 \\ 0 & 1 \end{pmatrix} \qquad M^{2n+1} = \begin{pmatrix} 0 & 1 \\ 1 & 0 \end{pmatrix}$$

and the associated Markov chain is not regular.

EXAMPLE 9 A robotics company is to manufacture a robot arm which attempts to pick parts off one conveyer belt and put them on another. The arm occasionally fails to grasp the part securely and does not transfer the part to the second belt successfully. The robot is designed so that in case of failure it "tries harder" in the sense that secondary circuits are activated. In trials it is found that if the arms fails at one time, it succeeds the next time 97 percent of the time. If the arms succeeds at a given time, the secondary circuits are deactivated and the arm will fail the next time only 2 percent of the time. Will the arm satisfy the customer's requirement that it work successfully 98 percent of the time?

Solution The transition matrix is

State at time T_{k-1}

Success Failure

$$M = \begin{pmatrix} .98 & .97 \\ .02 & .03 \end{pmatrix} \begin{matrix} \text{Success} \\ \text{Failure} \end{matrix} \qquad \text{State at time } T_k$$

To find the fixed point $ME = E$, $E^+ = 1$, $E \geq 0$, we solve

$$-.02x_1 + .97x_2 = 0$$
$$.02x_1 - .97x_2 = 0$$

to find

$$X = \begin{pmatrix} 48.5k \\ k \end{pmatrix}$$

The condition $E^+ = 1$ forces $k = .02\overline{02}$, so that

$$E = \begin{pmatrix} .97\overline{97} \\ .02\overline{02} \end{pmatrix}$$

In the long run the arm will work successfully $97.\overline{97}$ percent of the time. Strictly speaking, the arm is not satisfactory; however, $97.\overline{97}$ percent is very close to 98 percent, and perhaps the customer would accept it.

EXAMPLE 10 A space vehicle has three navigational computers. Each onboard navigation computer constantly performs internal checks on its own circuits. If a circuit has failed, the computer will cease navigation functions and repair itself. The computer is therefore in one of three states

s_1 performing navigation functions

s_2 failed, repair not begun

s_3 failed, repairing itself

In trials the transition matrix is

$$\begin{array}{c} \text{State at time } T_{k-1} \\ \begin{array}{ccc} s_1 & s_2 & s_3 \end{array} \end{array}$$

$$M = \begin{pmatrix} .98 & 0 & .9 \\ .01 & .1 & 0 \\ .01 & .9 & .1 \end{pmatrix} \begin{array}{c} s_1 \\ s_2 \\ s_3 \end{array} \quad \text{State at time } T_k$$

Approximately what percentage of the time is the computer operational?

Solution First, by checking M^2 we find that the chain is regular. Solving $ME = E$, $E^+ = 1$, $E \geq 0$, we obtain

$$E = \begin{pmatrix} .96774\cdots \\ .01075\cdots \\ .021505\cdots \end{pmatrix}$$

The conclusion is that the computer is operational about 96.77 percent of the time. By using laws of probability which we have not discussed, the probability that all three computers would be in failure at the same time is less than 0.000034 percent.

Although Markov chains are not the only application in which fixed vectors of matrices are used, they are one of the most important.

PROBLEMS 5.2

1. Solve the eigenproblem for the following matrices.

(a) $\begin{pmatrix} 0 & 1 \\ -1 & 0 \end{pmatrix}$ (b) $\begin{pmatrix} 1 & 0 \\ 1 & 1 \end{pmatrix}$

(c) $\begin{pmatrix} 0 & -1 \\ -1 & 0 \end{pmatrix}$ (d) $\begin{pmatrix} 1 & 1 & 0 \\ 0 & -1 & 1 \\ 0 & 0 & 2 \end{pmatrix}$

(e) $\begin{pmatrix} 2 & -1 \\ -4 & 2 \end{pmatrix}$ (f) $\begin{pmatrix} 1 & -1 & 4 \\ 3 & 2 & -1 \\ 2 & 1 & -1 \end{pmatrix}$

(g) $\begin{pmatrix} 0 & 1 & 0 & 0 \\ 1 & 0 & 0 & 0 \\ 0 & 0 & 1 & 1 \\ 0 & 0 & -2 & 4 \end{pmatrix}$ (h) $\begin{pmatrix} 0 & i \\ i & 0 \end{pmatrix}$

(*i*) $\begin{pmatrix} i & 1 \\ 0 & i \end{pmatrix}$ (*j*) $\begin{pmatrix} 0 & -i \\ i & 0 \end{pmatrix}$

2. Find fixed points (if they exist) of the following matrices.

(*a*) $\begin{pmatrix} \frac{1}{2} & \frac{1}{2} \\ \frac{1}{2} & \frac{1}{2} \end{pmatrix}$ (*b*) $\begin{pmatrix} 0 & -1 \\ -1 & 0 \end{pmatrix}$

(*c*) $\begin{pmatrix} 0 & 1 \\ 1 & 0 \end{pmatrix}$ (*d*) $\begin{pmatrix} 2 & 3 \\ 0 & -1 \end{pmatrix}$

(*e*) $\begin{pmatrix} -1 & 0 & 1 \\ 0 & 2 & -1 \\ 0 & 0 & 3 \end{pmatrix}$ (*f*) $\begin{pmatrix} 2 & 0 & 0 \\ 1 & 1 & 0 \\ 7 & 3 & 0 \end{pmatrix}$

3. For Prob. 3 of Sec. 5.1 find the equilibrium state for the Markov chain.

4. For Prob. 4 of Sec. 5.1 find the equilibrium state for the Markov chain.

5. Show that $0.2727\cdots = \frac{27}{99}$ by writing

$$0.272727\cdots = 0.27 + 0.0027 + \cdots = \tfrac{27}{100} + \tfrac{27}{10{,}000} + \cdots$$
$$= \tfrac{27}{100} + \tfrac{27}{100}\left(\tfrac{1}{100}\right) + \tfrac{27}{100}\left(\tfrac{1}{100}\right)^2 + \cdots$$

and using the formula for the sum of

$$a + ar + ar^2 + \cdots \qquad |r| < 1$$

which is $a/(1 - r)$. Do the same for $0.7272\cdots = \frac{72}{99}$.

6. Show that if X is an eigenvector of A, then so is kX for any $k \neq 0$. [*Hint*: Look at $A(kX)$.]

7. Show that if λ is an eigenvalue of A, then $k\lambda$ is an eigenvalue of kA ($k \neq 0$).

8. Show that if λ is an eigenvalue of A, then λ^2 in an eigenvalue of A^2. [*Hint*: Look at $A(AX)$.]

9. Show that if $\lambda \neq 0$ is an eigenvalue of A and if A^{-1} exists, then $1/\lambda$ is an eigenvalue of A^{-1}.

10. Show that if $\lambda = 0$ is an eigenvalue of A, then A is singular. [*Hint*: λ must satisfy $\det(A - \lambda I) = 0$.]

11. Show that the eigenvalues of a triangular matrix are just the diagonal entries.

12. Show that if X and Y are eigenvectors belonging to the same eigenvalue λ, then $aX + bY$ is also an eigenvector belonging to λ (provided $aX + bY \neq 0$).

13. How are the eigenvalues of A^T related to the eigenvalues of A? [*Hint*: Use $(B + C)^T = B^T + C^T$ on the characteristic equation.]

14. State conditions on a, b, c, and d that will guarantee real eigenvalues for

$$A = \begin{pmatrix} a & b \\ c & d \end{pmatrix}$$

15. How are the eigenvalues of $\bar{A}$ related to the eigenvalues of A?

16. How are the eigenvalues of A^* related to the eigenvalues of A?

17. Let (λ, X) be an eigenpair of A. Show that (λ^n, X) is an eigenpair of A^n.

18. Show that if (λ, X) is an eigenpair of A, then

$$a_n\lambda^n + a_{n-1}\lambda^{n-1} + \cdots + a_2\lambda^2 + a_1\lambda + a_0$$

is an eigenvalue of

$$a_nA^n + a_{n-1}A^{n-1} + \cdots + a_2A^2 + a_1A + a_0I$$

19. Let A be a matrix which is nilpotent. Show that zero is the only eigenvalue of A.

5.3 DIAGONALIZATION OF MATRICES

Markov chains are a prime example of the importance of being able to write a matrix A as PDP^{-1}, where D is diagonal. When this can be done, we call A diagonalizable.

DEFINITION 5.3.1 A matrix A is *diagonalizable* when there exist a diagonal matrix and an invertible matrix P such that $A = PDP^{-1}$. When D and P are found for a given A, we say that A has been *diagonalized.* Note that $D = P^{-1}AP$.

With the concept of diagonalization defined, we can state another major problem of linear algebra, our fifth.

Diagonalization Problem

Given a matrix $A_{n\times n}$, determine whether A is diagonalizable. If A is diagonalizable, find P and D in the equation

$$A = PDP^{-1}$$

To approach the diagonalization problem, we first ask: If A is diagonalizable, what must be true about P and D? If A is diagonalizable, then $A = PDP^{-1}$ which means that $AP = PD$. Now writing

$$P = \begin{pmatrix} p_{11} & p_{12} & \cdots & p_{1n} \\ p_{21} & p_{22} & \cdots & p_{2n} \\ \cdots & \cdots & \cdots & \cdots \\ p_{n1} & p_{n2} & \cdots & p_{nn} \end{pmatrix}$$

and

$$D = \begin{pmatrix} c_1 & & & \mathbf{0} \\ & c_2 & & \\ & & \ddots & \\ \mathbf{0} & & & c_n \end{pmatrix}$$

we see that

$$AP_1 = c_1P_1 \qquad AP_2 = c_2P_2 \qquad \ldots \qquad AP_n = c_nP_n$$

where P_k is the vector made of the kth column of P. Therefore P is the matrix made up of columns which are eigenvectors of A. The diagonal elements of D are the corresponding eigenvalues. Moreover, since P is invertible, the columns are linearly independent. Therefore we have the following theorem.

THEOREM 5.3.1 If $A_{n\times n}$ is diagonalizable, then A has n linearly independent eigenvectors. Also, in the equation $A = PDP^{-1}$, P is a matrix whose columns are eigenvectors, and the diagonal entries of D are the eigenvalues corresponding column by column to their respective eigenvectors.

This theorem tells us what P and D must look like if A is diagonalizable. We would like this theorem to be reversible. That is, we hope that if we

1. Solve the eigenproblem for $A_{n\times n}$: $AX_1 = \lambda_1 X_1, \ldots, AX_n = \lambda_n X_n$,
2. Find that the eigenvectors can be chosen as linearly independent,
3. Set $P = (X_1 X_2 \cdots X_n)$,

then we would have

$$A = P\begin{pmatrix} \lambda_1 & & & \mathbf{0} \\ & \lambda_2 & & \\ & & \ddots & \\ \mathbf{0} & & & \lambda_n \end{pmatrix}P^{-1}$$

In fact, this is true. Before proving that this procedure works we give an example.

EXAMPLE 1 If possible, diagonalize

$$A = \begin{pmatrix} 1 & 1 \\ -2 & 4 \end{pmatrix}$$

Solution The eigenproblem for A was solved in Sec. 5.2. The general eigenpairs are

$$\left(2, \begin{pmatrix} k \\ k \end{pmatrix}\right) \qquad \left(3, \begin{pmatrix} j \\ 2j \end{pmatrix}\right)$$

and specific choices are

$$\left(2, \begin{pmatrix} 1 \\ 1 \end{pmatrix}\right) \qquad \left(3, \begin{pmatrix} 1 \\ 2 \end{pmatrix}\right)$$

Now

$$\left\{\begin{pmatrix} 1 \\ 1 \end{pmatrix}, \begin{pmatrix} 1 \\ 2 \end{pmatrix}\right\}$$

is a linearly independent set, so we form

$$P = \begin{pmatrix} 1 & 1 \\ 1 & 2 \end{pmatrix}$$

Then

$$P^{-1} = \begin{pmatrix} 2 & -1 \\ -1 & 1 \end{pmatrix}$$

Finally we check to see whether

$$A = PDP^{-1}$$

We have

$$PDP^{-1} = \begin{pmatrix} 1 & 1 \\ 1 & 2 \end{pmatrix}\begin{pmatrix} 2 & 0 \\ 0 & 3 \end{pmatrix}\begin{pmatrix} 2 & -1 \\ -1 & 1 \end{pmatrix}$$

Eigenvector for $\lambda = 2$ in 1st column

Eigenvalue $\lambda = 2$ in 1st column

Eigenvector for $\lambda = 3$ in 2d column

Eigenvalue $\lambda = 3$ in 2d column

$$= \begin{pmatrix} 2 & 3 \\ 2 & 6 \end{pmatrix}\begin{pmatrix} 2 & -1 \\ -1 & 1 \end{pmatrix}$$

$$= \begin{pmatrix} 1 & 1 \\ -2 & 4 \end{pmatrix} = A$$

Thus A has been diagonalized. If we had formed

$$P = \begin{pmatrix} 1 & 1 \\ 2 & 1 \end{pmatrix}$$

then D would be

$$\begin{pmatrix} 3 & 0 \\ 0 & 2 \end{pmatrix}$$

Now we state and prove the theorem which the last example illustrates.

THEOREM 5.3.2 If $A_{n \times n}$ has n linearly independent eigenvectors $X_1, X_2, \ldots, X_n$ with

$$AX_1 = \lambda_1 X_1 \qquad \cdots \qquad AX_n = \lambda_n X_n$$

then A is diagonalizable and $A = PDP^{-1}$, where

$$P = (X_1 X_2 \cdots X_n) \qquad D = \begin{pmatrix} \lambda_1 & & & \mathbf{0} \\ & \lambda_2 & & \\ & & \ddots & \\ \mathbf{0} & & & \lambda_n \end{pmatrix}$$

Proof If $\{X_1, \ldots, X_n\}$ is a linearly independent set, then $P = (X_1 \cdots X_n)$ is invertible. Now

$$AP = (\underbrace{AX_1}_{\text{Column 1}} \quad \underbrace{AX_2}_{\text{Column 2}} \quad \cdots \quad \underbrace{AX_n}_{\text{Column } n}) = (\underbrace{\lambda_1 X_1}_{\text{Column 1}} \quad \underbrace{\lambda_2 X_2}_{\text{Column 2}} \quad \cdots \quad \underbrace{\lambda_n X_n}_{\text{Column } n})$$

$$= (X_1 X_2 \cdots X_n) \begin{pmatrix} \lambda_1 & & & \mathbf{0} \\ & \lambda_2 & & \\ & & \ddots & \\ \mathbf{0} & & & \lambda_n \end{pmatrix}$$

$$= PD$$

Since $AP = PD$, we have $A = PDP^{-1}$ and A has been diagonalized.

Theorems 5.3.1 and 5.3.2 together give us an important result.

THEOREM 5.3.3 Solution to the diagonalization problem

Matrix $A_{n \times n}$ is diagonalizable if and only if A has n linearly independent eigenvectors. In that case, if $X_1, X_2, \ldots, X_n$ are the linearly independent eigenvectors and the eigenpairs are

$$(\lambda_1, X_1), \qquad (\lambda_2, X_2), \qquad \ldots, \qquad (\lambda_n, X_n)$$

then setting $P = (X_1\ X_2 \cdots X_n)$ and

$$D = \begin{pmatrix} \lambda_1 & & & \mathbf{0} \\ & \lambda_2 & & \\ & & \ddots & \\ \mathbf{0} & & & \lambda_n \end{pmatrix}$$

we have

$$A = PDP^{-1} \qquad \text{and} \qquad D = P^{-1}AP$$

The result in Theorem 5.3.3 can be stated in two other equivalent ways.

Matrix $A_{n \times n}$ is diagonalizable if and only if there exists a basis of $\mathscr{C}_{n1}$ consisting of eigenvectors of A. In that case, if $\{X_1, \ldots, X_n\}$ is the basis of eigenvectors, and the eigenpairs are $(\lambda_1, X_1), \ldots, (\lambda_n, X_n)$, then the construction of P and D proceeds as in the statement above.

Let $A_{n \times n}$ have eigenvalues $\lambda_1, \ldots, \lambda_m$ with $m \leq n$ (m is strictly less than n if some of the eigenvalues have multiplicity 2 or more). Then A is diagonalizable if and only if

$$\sum_{k=1}^{m} \dim E_{(\lambda_k)} = n$$

Note that to solve the diagonalization problem for A, we first solve the eigenproblem for A.

EXAMPLE 2 Solve the diagonalization problem for

$$A = \begin{pmatrix} 0 & 1 \\ -1 & 0 \end{pmatrix}$$

Solution We found in a previous example that the eigenpairs for A are

$$\left(i, c\begin{pmatrix} -i \\ 1 \end{pmatrix}\right) \quad \left(-i, d\begin{pmatrix} i \\ 1 \end{pmatrix}\right)$$

We have two linearly independent eigenvectors for the 2×2 matrix. Thus A is diagonalizable. Choosing $c = d = 1$, we can select

$$X_1 = \begin{pmatrix} -i \\ 1 \end{pmatrix} \quad X_2 = \begin{pmatrix} i \\ 1 \end{pmatrix}$$

and construct

$$P = \begin{pmatrix} -i & i \\ 1 & 1 \end{pmatrix}$$

Thus

$$P^{-1}\begin{pmatrix} \dfrac{i}{2} & \dfrac{1}{2} \\ \dfrac{-i}{2} & \dfrac{1}{2} \end{pmatrix}$$

and

$$\begin{pmatrix} 0 & 1 \\ -1 & 0 \end{pmatrix} = \begin{pmatrix} -i & i \\ 1 & 1 \end{pmatrix} \begin{pmatrix} i & 0 \\ 0 & -i \end{pmatrix} \begin{pmatrix} \frac{i}{2} & \frac{1}{2} \\ \frac{-i}{2} & \frac{1}{2} \end{pmatrix}$$

EXAMPLE 3 Solve the diagonalization problem for

$$A = \begin{pmatrix} 1 & 1 \\ 0 & 1 \end{pmatrix}$$

Solution The characteristic equation is

$$\det \begin{pmatrix} 1-\lambda & 1 \\ 0 & 1-\lambda \end{pmatrix} = 0$$

or $(1-\lambda)^2 = 0$. The eigenvalues are $\lambda_1 = 1$ and $\lambda_2 = 1$. For the eigenvectors we solve $A - \lambda I = 0$ which reduces to

$$\lambda = 1 \qquad \begin{aligned} 0x_1 + x_2 &= 0 \\ 0x_1 + 0x_2 &= 0 \end{aligned}$$

Therefore,

$$X_1 = \begin{pmatrix} k \\ 0 \end{pmatrix} \qquad X_2 = \begin{pmatrix} j \\ 0 \end{pmatrix}$$

and we cannot obtain two linearly independent eigenvectors for the 2×2 matrix. Thus A is *not* diagonalizable. Another way to say this is that the eigenvectors of A do not form a basis for $\mathscr{C}_{21}$.

EXAMPLE 4 Solve the diagonalization problem for

$$A = \begin{pmatrix} -4 & -4 & -8 \\ 4 & 6 & 4 \\ 6 & 4 & 10 \end{pmatrix}$$

Solution In Example 5 of Sec. 5.2 we solved the eigenproblem for A. Referring to that example, we see that specific eigenpairs are

$$\left(2, \begin{pmatrix} -2 \\ 1 \\ 1 \end{pmatrix}\right) \quad \left(4, \begin{pmatrix} -1 \\ 0 \\ 1 \end{pmatrix}\right) \quad \left(6, \begin{pmatrix} -2 \\ 1 \\ 2 \end{pmatrix}\right)$$

Since

$$\left\{ \begin{pmatrix} -2 \\ 1 \\ 1 \end{pmatrix} \begin{pmatrix} -1 \\ 0 \\ 1 \end{pmatrix} \begin{pmatrix} -2 \\ 1 \\ 2 \end{pmatrix} \right\}$$

is linearly independent, the 3×3 matrix possesses three linearly independent eigenvectors and is diagonalizable. A choice for P is

$$P = \begin{pmatrix} -2 & -1 & -2 \\ 1 & 0 & 1 \\ 1 & 1 & 2 \end{pmatrix}$$

which gives

$$P^{-1} = \begin{pmatrix} -1 & 0 & -1 \\ -1 & -2 & 0 \\ 1 & 1 & 1 \end{pmatrix}$$

and

$$A = P \begin{pmatrix} 2 & 0 & 0 \\ 0 & 4 & 0 \\ 0 & 0 & 6 \end{pmatrix} P^{-1}$$

If we wanted D to have the eigenvalues of A in descending order of magnitude, we would choose

$$P = \begin{pmatrix} -2 & -1 & -2 \\ 1 & 0 & 1 \\ 2 & 1 & 1 \end{pmatrix}$$

In that case

$$D = \begin{pmatrix} 6 & 0 & 0 \\ 0 & 4 & 0 \\ 0 & 0 & 2 \end{pmatrix} \qquad \text{and} \qquad P^{-1} = \begin{pmatrix} 1 & 1 & 1 \\ -1 & -2 & 0 \\ -1 & 0 & -1 \end{pmatrix}$$

EXAMPLE 5 Show that the matrices

$$A = \begin{pmatrix} 3 & -1 & 0 \\ 0 & 3 & 0 \\ 0 & 0 & -5 \end{pmatrix} \qquad \text{and} \qquad B = \begin{pmatrix} 0 & -3 & -3 \\ -3 & 0 & -3 \\ -3 & -3 & 0 \end{pmatrix}$$

both have 3 as an eigenvalue of multiplicity 2. Show that A is not diagonalizable, but B is diagonalizable.

Solution The characteristic polynomial for B factors as $-(\lambda + 6)(\lambda - 3)^2$, so that 3 is an eigenvalue of multiplicity 2. We already know that 3 is an eigenvalue of multiplicity 2 for A from Example 6 of Sec. 5.2. The eigenpairs for B are

$$\left(3, \begin{pmatrix} -r - s \\ r \\ s \end{pmatrix}\right) \qquad \left(\left(-6, \begin{pmatrix} t \\ t \\ t \end{pmatrix}\right)\right)$$

and we can choose

$$\mathscr{S} = \left\{ \begin{pmatrix} -1 \\ 1 \\ 0 \end{pmatrix}, \begin{pmatrix} -1 \\ 0 \\ 1 \end{pmatrix}, \begin{pmatrix} 1 \\ 1 \\ 1 \end{pmatrix} \right\}$$

$\uparrow$ $r = 1$, $s = 0$ $\qquad$ $\uparrow$ $r = 0$, $s = 1$

as a basis of eigenvectors for $\mathscr{C}_{31}$. Thus B is diagonalizable. But the eigenpairs for A do not generate a basis for $\mathscr{C}_{31}$, and A is not diagonalizable. The key observation here is that for A, dim $E_{(3)}$ is strictly less than the multiplicity of the eigenvalue 3; for B, dim $E_{(3)}$ equals the multiplicity of the eigenvalue 3.

EXAMPLE 6 Determine whether the shear linear transformation as defined in previous examples is diagonalizable.

Solution To solve this problem, we use a matrix which represents shear. The reason this can be done is that if A and B are similar matrices and one is similar to a diagonal matrix D, then the other is also similar to the same diagonal matrix (Prob. 14 in Sec. 4.4). Thus *diagonalizability is invariant under similarity*, and we say a linear transformation is diagonalizable if some representing matrix of the transformation is diagonalizable. It is sufficient to use the matrix with respect to the standard basis

$$\begin{pmatrix} 1 & k \\ 0 & 1 \end{pmatrix}$$

to represent shear. The general eigenpair for this matrix is

$$\left(1, \begin{pmatrix} r \\ 0 \end{pmatrix}\right)$$

A basis for $\mathscr{C}_{21}$ cannot be constructed from the eigenvectors of the representing matrix. Therefore, the shear transformation is not diagonalizable.

We now know that an $n \times n$ matrix A is diagonalizable if and only if A has n linearly independent eigenvectors. If n is large, checking for linear independence can be tedious. There is a simple sufficient condition for diagonalizability.

THEOREM 5.3.4 If $A_{n \times n}$ has n distinct eigenvalues $\lambda_1, \lambda_2, \ldots, \lambda_n$, then the eigenvectors $X_1, X_2, \ldots, X_n$ in the eigenpairs

$$(\lambda_1, X_1) \qquad (\lambda_2, X_2) \qquad \ldots \qquad (\lambda_n, X_n)$$

form a linearly independent set, and A is therefore diagonalizable.

EXAMPLE 7 Show that

$$A = \begin{pmatrix} 1 & 7 & 6 \\ 0 & -1 & 3 \\ 0 & 0 & 2 \end{pmatrix}$$

is diagonalizable.

Solution The characteristic equation is $\det(A - \lambda I) = 0$ which is $(1-\lambda)(-1-\lambda)(2-\lambda) = 0$. Thus the eigenvalues of A are 1, -1, and 2. Since A is 3×3 and A has three distinct eigenvalues, Theorem 5.3.4 implies that A is diagonalizable.

Proof of Theorem 5.3.4 Suppose $A_{n \times n}$ has distinct eigenvalues $\lambda_1, \ldots, \lambda_n$, with associated eigenvectors $X_1, X_2, \ldots, X_n$. If we show that $\mathscr{S} = \{X_1, \ldots, X_n\}$ is a linearly independent set, then A is diagonalizable. We will suppose that $\mathscr{S}$ is linearly dependent and derive a contradiction.

Suppose that $\mathscr{S}$ is linearly dependent (LD). We can find a set $\mathscr{S}_k = \{X_1, \ldots, X_k\}$, $k < n$, which is linearly independent (LI) by the following process. Since eigenvectors are nonzero, $\{X_1\}$ is an LI set. If $\{X_1, X_2\}$ is LD, we stop with $\mathscr{S}_1$ as our LI set. If $\mathscr{S}_2$ is LI but $\mathscr{S}_3 = \{X_1, X_2, X_3\}$ is LD, we stop with $\mathscr{S}_2$, and so on. Suppose we have completed this process and have $\mathscr{S}_k$ LI but $\mathscr{S}_{k+1}$ LD ($k + 1$ could equal n).

So there exist constants $c_1, \ldots, c_{k+1}$, not all zero, with

$$c_1X_1 + c_2X_2 + \cdots + c_kX_k + c_{k+1}X_{k+1} = 0 \tag{5.3.1}$$

Also $c_{k+1} \neq 0$, because if c_k is zero, then $c_1 = c_2 = \cdots = c_k = 0$ (by the linear independence of $\mathscr{S}_k$), and $\mathscr{S}_{k+1}$ would be LI, a contradiction.

Now we multiply both sides of Eq. (5.3.1) and use matrix algebra to obtain

$$c_1AX_1 + c_2AX_2 + \cdots + c_kAX_k + c_{k+1}AX_{k+1} = 0$$

or

$$c_1\lambda_1X_1 + c_2\lambda_2X_2 + \cdots + c_k\lambda_kX_k + c_{k+1}\lambda_{k+1}X_{k+1} = 0 \tag{5.3.2}$$

Multiplying (5.3.1) by λ_{k+1} and subtracting from Eq. (5.3.2), we have

$$c_1(\lambda_1 - \lambda_{k+1})X_1 + \cdots + c_k(\lambda_k - \lambda_{k+1})X_k = 0$$

Since the eigenvalues are distinct and $\mathscr{S}_k$ is LI, we must have $c_1 = c_2 = \cdots = c_k = 0$. Substituting these values in (5.3.1), we have

$$c_{k+1}X_{k+1} = 0$$

Now $c_{k+1} \neq 0$ so X_{k+1} must be the zero vector; this contradicts the hypothesis that X_{k+1} is an eigenvector. Therefore it is impossible for $\mathscr{S}$ to be LD, and $\mathscr{S}$ must be LI.

Checking Diagonalizations of Matrices Once a matrix has been diagonalized, we may want some easy checks on the diagonal form D. Since

$$A = PDP^{-1}$$

means that A is similar to D, then by previous results on similar matrices we must have

$$\operatorname{tr} A = \operatorname{tr} D$$
$$\det A = \det D$$
$$\operatorname{rank} A = \operatorname{rank} D$$

EXAMPLE 8 Use the three checks above to check the diagonalization in Example 4.

Solution In Example 4 we found that

$$A = \begin{pmatrix} -4 & -4 & -8 \\ 4 & 6 & 4 \\ 6 & 4 & 10 \end{pmatrix} \qquad D = \begin{pmatrix} 2 & 0 & 0 \\ 0 & 4 & 0 \\ 0 & 0 & 6 \end{pmatrix}$$

In this case $\operatorname{tr} A = \operatorname{tr} D = 12$, $\det A = \det B = 48$, and $\operatorname{rank} A = \operatorname{rank} D = 3$. This does not *guarantee* that the diagonalization is correct but gives more confidence in the answer.

EXAMPLE 9 A person claims that the matrix

$$A = \begin{pmatrix} 1 & 2 & 3 \\ 4 & 5 & 6 \\ 7 & 8 & 9 \end{pmatrix}$$

is similar to

$$D = \begin{pmatrix} 15 & 0 & 0 \\ 0 & 0 & 0 \\ 0 & 0 & 0 \end{pmatrix}$$

Is this correct?

Solution We have

$$\operatorname{tr} A = 15 = \operatorname{tr} D$$
$$\det A = 0 = \det D$$
$$\operatorname{rank} A = 2 \neq 1 = \operatorname{rank} D$$

Since the ranks are unequal, the person has incorrectly diagonalized A.

Note that these checks are not sufficient to prove a diagonalization correct; they can only help you find an incorrect diagonalization. For example,

$$\begin{pmatrix} 1 & 1 \\ 0 & 1 \end{pmatrix} \quad \text{and} \quad \begin{pmatrix} 1 & 0 \\ 0 & 1 \end{pmatrix}$$

have the same trace, determinant, and rank, but

$$\begin{pmatrix} 1 & 0 \\ 0 & 1 \end{pmatrix}$$

is not the diagonalization of

$$\begin{pmatrix} 1 & 1 \\ 0 & 1 \end{pmatrix}$$

because

$$\begin{pmatrix} 1 & 1 \\ 0 & 1 \end{pmatrix}$$

is not even diagonalizable. See Example 3.

Geometric Meaning of Diagonalization In the real vector space case, the diagonalization of $A_{n\times n}$ gives information about the geometric action of the transformation $T_A: \mathscr{M}_{n1} \to \mathscr{M}_{n1}$ generated by A. For example, since

$$\begin{pmatrix} 1 & 1 \\ -2 & 4 \end{pmatrix} = \begin{pmatrix} 1 & 1 \\ 1 & 2 \end{pmatrix}\begin{pmatrix} 2 & 0 \\ 0 & 3 \end{pmatrix}\begin{pmatrix} 2 & -1 \\ -1 & 1 \end{pmatrix} \qquad \begin{pmatrix} 1 & 1 \\ -2 & 4 \end{pmatrix}\begin{pmatrix} 1 \\ 1 \end{pmatrix} = 2\begin{pmatrix} 1 \\ 1 \end{pmatrix}$$

$$\begin{pmatrix} 1 & 1 \\ -2 & 4 \end{pmatrix}\begin{pmatrix} 1 \\ 2 \end{pmatrix} = 3\begin{pmatrix} 1 \\ 1 \end{pmatrix}$$

we know that T_A simply stretches the vectors

$$\begin{pmatrix} 1 \\ 1 \end{pmatrix} \quad \text{and} \quad \begin{pmatrix} 1 \\ 2 \end{pmatrix}$$

(See Fig. 5.3.1.) Since

$$\left\{\begin{pmatrix} 1 \\ 1 \end{pmatrix}, \begin{pmatrix} 1 \\ 2 \end{pmatrix}\right\}$$

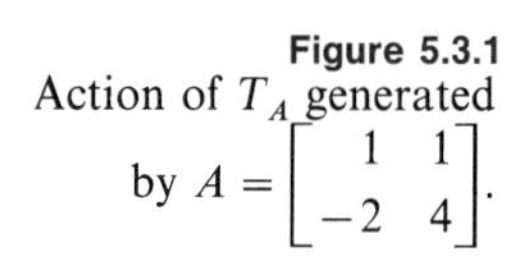

Figure 5.3.1
Action of T_A generated by $A = \begin{bmatrix} 1 & 1 \\ -2 & 4 \end{bmatrix}$.

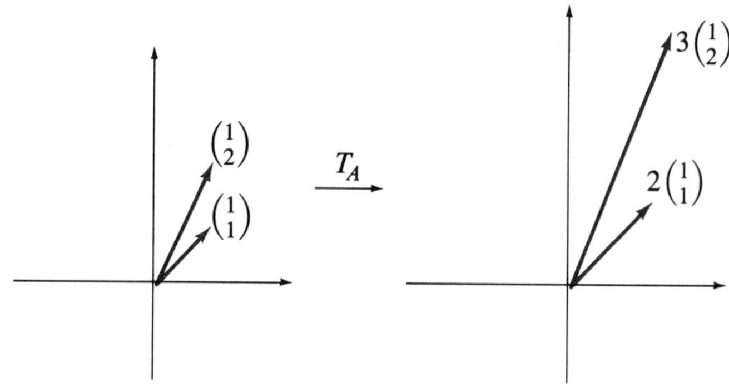

is a basis for E^2, if a vector X has coordinate vector

$$\begin{pmatrix} a \\ b \end{pmatrix}$$

then the coordinate vector of

$$T_A\left(\begin{pmatrix} a \\ b \end{pmatrix}\right) \quad \text{is} \quad \begin{pmatrix} 2a \\ 3b \end{pmatrix}$$

This illustrates the use of the words *characteristic values and characteristic vectors* in that the vectors

$$\begin{pmatrix} 1 \\ 1 \end{pmatrix} \quad \text{and} \quad \begin{pmatrix} 1 \\ 2 \end{pmatrix}$$

are "characteristic" of the action of T. The vectors define directions which can be called *principal directions* of T.

EXAMPLE 10 Show that if the coordinate vector of X with respect to

$$\mathscr{S} = \left\{\begin{pmatrix} 1 \\ 1 \end{pmatrix}, \begin{pmatrix} 1 \\ 2 \end{pmatrix}\right\} \quad \text{is} \quad \begin{pmatrix} a \\ b \end{pmatrix}$$

then

$$(T_A(X))_{\mathscr{S}} = \begin{pmatrix} 2a \\ 3b \end{pmatrix}$$

Solution For a given vector X, the coordinates are given by PX, where P is the transition matrix from $\mathscr{M}_{21}$ with the standard basis to $\mathscr{M}_{21}$ with basis $\mathscr{S}$. In this case the matrix is

$$P = \begin{pmatrix} 2 & -1 \\ -1 & 1 \end{pmatrix}$$

Now

$$AX = PDP^{-1}X = PDP^{-1}X = PD\begin{pmatrix} a \\ b \end{pmatrix} = P\begin{pmatrix} 2 & 0 \\ 0 & 3 \end{pmatrix}\begin{pmatrix} a \\ b \end{pmatrix} = P\begin{pmatrix} 2a \\ 3b \end{pmatrix}$$

Since

$$P\begin{pmatrix} 2a \\ 3b \end{pmatrix}$$

gives the vector AX, and P is the inverse of the transition matrix P^{-1},

$$\begin{pmatrix} 2a \\ 3b \end{pmatrix}$$

is the coordinate vector with respect to $\mathscr{S}$ for AX.

The last example illustrates the following fact.

If $A = A_{n\times n} = PDP^{-1}$, then the action of the transformation T_A on $\mathscr{M}_{n1}$ can be thought of in terms of the action of D on $\mathscr{M}_{n1}$ with the basis of eigenvectors of A.

We will see in the next section that if A is a real symmetric matrix, then the basis of eigenvectors can always be chosen as orthonormal.

EXAMPLE 11 Analyze $T:\mathscr{M}_{21} \to \mathscr{M}_{21}$ defined by

$$T(X) = \begin{pmatrix} \frac{1}{2} & \frac{1}{2} \\ \frac{1}{2} & \frac{1}{2} \end{pmatrix} X \qquad X \in \mathscr{M}_{21}$$

by diagonalizing the matrix.

Solution Let

$$A = \begin{pmatrix} \frac{1}{2} & \frac{1}{2} \\ \frac{1}{2} & \frac{1}{2} \end{pmatrix}$$

and calculate det $(A - \lambda I)$. The characteristic equation $\lambda^2 - \lambda = 0$ yields eigenvalues $\lambda_1 = 1$ and $\lambda_2 = 0$. The eigenpairs are

$$\left(1, \begin{pmatrix} k \\ k \end{pmatrix}\right) \qquad \left(0, \begin{pmatrix} -j \\ j \end{pmatrix}\right)$$

Choosing k and j so that the eigenvectors have length 1, we have

$$\left(1, \begin{pmatrix} 1/\sqrt{2} \\ 1/\sqrt{2} \end{pmatrix}\right) \qquad \left(0, \begin{pmatrix} -1/\sqrt{2} \\ 1/\sqrt{2} \end{pmatrix}\right)$$

Therefore

$$\begin{pmatrix} \frac{1}{2} & \frac{1}{2} \\ \frac{1}{2} & \frac{1}{2} \end{pmatrix} = \begin{pmatrix} \frac{1}{\sqrt{2}} & \frac{-1}{\sqrt{2}} \\ \frac{1}{\sqrt{2}} & \frac{1}{\sqrt{2}} \end{pmatrix} \begin{pmatrix} 1 & 0 \\ 0 & 0 \end{pmatrix} \begin{pmatrix} \frac{1}{\sqrt{2}} & \frac{1}{\sqrt{2}} \\ \frac{-1}{\sqrt{2}} & \frac{1}{\sqrt{2}} \end{pmatrix}$$

This means that the action of T on a standard coordinate matrix for a vector is as follows:

First:	Rotation 45° clockwise
Second:	Projection on x axis
Third:	Rotation 45° counterclockwise

As a final application of these remarks, we note that since the shear transformation is not diagonalizable (Example 6), the shear transformation does not "stretch" objects in two independent directions. This reflects our intuitive feelings about shear, which result from a "sideways" deformation of the cube illustrated in previous encounters with this example.

PROBLEMS 5.3

1. Solve the diagonalization problem for the following matrices. If the matrix is diagonalizable, write P, P^{-1}, and D in $A = PDP^{-1}$.

(*a*) $\begin{pmatrix} 0 & 1 & 0 \\ -1 & 0 & 0 \\ 0 & 0 & 2 \end{pmatrix}$ (*b*) $\begin{pmatrix} \frac{1}{2} & \frac{1}{3} \\ \frac{1}{2} & \frac{2}{3} \end{pmatrix}$ (*c*) $\begin{pmatrix} -2 & -2 & -4 \\ 2 & 3 & 2 \\ 3 & 2 & 5 \end{pmatrix}$

(*d*) $\begin{pmatrix} 1 & 4 & 5 \\ 0 & 2 & 6 \\ 0 & 0 & 3 \end{pmatrix}$ (*e*) $\begin{pmatrix} 1 & 1 & 0 & 0 \\ -2 & 4 & 0 & 0 \\ 0 & 0 & 2 & 1 \\ 0 & 0 & 1 & 2 \end{pmatrix}$ (*f*) $\begin{pmatrix} 1 & -1 & 4 \\ 3 & 2 & -1 \\ 2 & 1 & -1 \end{pmatrix}$

(*g*) $\begin{pmatrix} 1 & 0 & 0 \\ 1 & 1 & 0 \\ 0 & 1 & 1 \end{pmatrix}$ (*h*) $\begin{pmatrix} 1 & 2 \\ -2 & 2 \end{pmatrix}$ (*i*) $\begin{pmatrix} 0 & i \\ i & 0 \end{pmatrix}$

(*j*) $\begin{pmatrix} i & 1 \\ 0 & i \end{pmatrix}$ (*k*) $\begin{pmatrix} 1 & i \\ -i & 1 \end{pmatrix}$

2. Show that if a diagonalization of A is D, then D^2 is a diagonalization of A^2. What is a diagonalization of A^m?

3. Diagonalize

$$A = \begin{pmatrix} 1 & 1 \\ 0 & \frac{1}{2} \end{pmatrix}$$

and use the diagonalization to calculate A^{12}.

4. If A is diagonalizable, is A^T diagonalizable?

5. If A is nonsingular and diagonalizable, is A^{-1} diagonalizable?

6. (*a*) If A is diagonalizable and trace $A > 0$, must A have a positive eigenvalue?

   (*b*) What if A is required to have real entries and all real eigenvalues?

7. If A is diagonalizable, has all positive entries, and has all real eigenvalues, must A have a positive eigenvalue?

8. Let A be a nilpotent matrix, with $A^m = 0$. If A is diagonalizable, what are its eigenvalues?

9. Let A be an involutory matrix ($A^2 = \mathrm{I}$). If A is diagonalizable, what are its eigenvalues?

10. Let A be an orthogonal matrix ($AA^T = I$). If A is diagonalizable, what can you say about its eigenvalues?

11. If A and B are diagonalizable with

$$A = P^{-1}D_1P \qquad B = Q^{-1}D_2Q$$

is D_1D_2 necessarily a diagonalization of AB? Is $D_1 + D_2$ necessarily a diagonalization of $A + B$?

12. Analyze the action of $T:\mathcal{M}_{21} \to \mathcal{M}_{21}$ defined by

$$T(X) = \begin{pmatrix} 1 & 1 \\ 1 & 1 \end{pmatrix} X \qquad X \in \mathcal{M}_{21}$$

by diagonalizing the matrix.

5.4 DIAGONALIZATION: SYMMETRIC AND HERMITIAN MATRICES

Symmetric and hermitian matrices, which arise in many applications, enjoy the property of always being diagonalizable. Also the set of eigenvectors of such matrices can always be chosen as orthonormal. The diagonalization procedure is essentially the same as outlined in Sec. 5.3, as we will see in our examples.

EXAMPLE 1 The horizontal motion of the system of masses and springs

Wall [spring–mass–spring–mass–spring] Wall (No gravity)

where all the masses are the same and the springs are the same, can be analyzed by diagonalizing the symmetric matrix

$$A = \begin{pmatrix} 2 & -1 \\ -1 & 2 \end{pmatrix}$$

Diagonalize A.

Solution We have

$$\det(A - \lambda I) = \det\begin{pmatrix} 2-\lambda & -1 \\ -1 & 2-\lambda \end{pmatrix} = (2-\lambda)^2 - 1 = \lambda^2 - 4\lambda + 3$$

so that the eigenvalues are $\lambda_1 = 3$ and $\lambda_2 = 1$. Eigenvectors are found by solving $(A - \lambda I)X = 0$; these equations and solutions are

$$\lambda_1 = 3: \qquad \begin{matrix} -1x_1 - 1x_2 = 0 \\ -1x_1 - 1x_2 = 0 \end{matrix} \Rightarrow X_1 = \begin{pmatrix} k \\ -k \end{pmatrix}$$

$$\lambda_2 = 1: \qquad \begin{aligned} 1x_1 - 1x_2 &= 0 \\ -1x_1 + 1x_2 &= 0 \end{aligned} \Rightarrow X_2 = \begin{pmatrix} j \\ j \end{pmatrix}$$

Now X_1 and X_2 are orthogonal since $X_1 \cdot X_2 = kj - kj = 0$. If we normalize X_1 and X_2, we have the choices for eigenvectors

$$\mathcal{O}_1 = \begin{pmatrix} 1/\sqrt{2} \\ -1/\sqrt{2} \end{pmatrix} \qquad \mathcal{O}_2 = \begin{pmatrix} 1/\sqrt{2} \\ 1/\sqrt{2} \end{pmatrix}$$

and $\mathcal{S} = \{\mathcal{O}_1, \mathcal{O}_2\}$ forms an orthonormal set. Finally, with

$$P = \begin{pmatrix} 1/\sqrt{2} & 1/\sqrt{2} \\ -1/\sqrt{2} & 1/\sqrt{2} \end{pmatrix}$$

we can write

$$\begin{pmatrix} 2 & -1 \\ -1 & 2 \end{pmatrix} = \begin{pmatrix} 1/\sqrt{2} & 1/\sqrt{2} \\ -1/\sqrt{2} & 1/\sqrt{2} \end{pmatrix} \begin{pmatrix} 3 & 0 \\ 0 & 1 \end{pmatrix} \begin{pmatrix} 1/\sqrt{2} & -1/\sqrt{2} \\ 1/\sqrt{2} & 1/\sqrt{2} \end{pmatrix}$$

Since P is an orthogonal matrix, $P^{-1} = P^T$. Also P represents a rotation of $\pi/4$ radians clockwise.

EXAMPLE 2 Diagonalize

$$A = \begin{pmatrix} 0 & 1 & 0 \\ 1 & 0 & 0 \\ 0 & 0 & 1 \end{pmatrix}$$

Choose P as an orthogonal matrix.

Solution The characteristic equation is $(1 - \lambda)(\lambda^2 - 1) = 0$, which has solutions 1 and -1. The eigenvalue 1 has multiplicity 2, and the eigenvalue -1 is simple (multiplicity 1). Now to determine the eigenvectors, we solve equations.

$$\lambda_1 = 1: \qquad \begin{aligned} -x_1 + x_2 + 0x_3 &= 0 \\ x_1 - x_2 + 0x_3 &= 0 \\ 0x_1 + 0x_2 + 0x_3 &= 0 \end{aligned} \Rightarrow X_1 = \begin{pmatrix} j \\ j \\ k \end{pmatrix}$$

$$\lambda_2 = -1: \qquad \begin{aligned} 1x_1 + 1x_2 + 0x_3 &= 0 \\ 1x_1 + 1x_2 + 1x_3 &= 0 \\ 0x_1 + 0x_2 + 2x_3 &= 0 \end{aligned} \Rightarrow X_2 = \begin{pmatrix} r \\ -r \\ 0 \end{pmatrix}$$

We see that corresponding to $\lambda_1 = 1$ the eigenspace is two-dimensional. Therefore, we can choose a basis for $E_{(1)}$ (the *basis problem* has returned!) of orthonormal vectors. In fact, choosing first $j = 1$, $k = 0$ and second $k = 1$, $j = 0$, we have

$$V_1 = \begin{pmatrix} 1 \\ 1 \\ 0 \end{pmatrix} \quad \text{and} \quad V_2 = \begin{pmatrix} 0 \\ 0 \\ 1 \end{pmatrix}$$

which are orthogonal. Normalizing V_1 (V_2 is already normalized) yields

$$\mathcal{O}_1 = \begin{pmatrix} 1/\sqrt{2} \\ 1/\sqrt{2} \\ 0 \end{pmatrix} \qquad \mathcal{O}_2 = \begin{pmatrix} 0 \\ 0 \\ 1 \end{pmatrix}$$

Not as much work is required for $\lambda_2 = -1$ since the eigenspace for this eigenvalue is one-dimensional. Normalizing X_2, we have

$$\mathcal{O}_3 = \begin{pmatrix} 1/\sqrt{2} \\ -1/\sqrt{2} \\ 0 \end{pmatrix}$$

Therefore

$$P = \begin{pmatrix} 1/\sqrt{2} & 0 & 1/\sqrt{2} \\ 1/\sqrt{2} & 0 & -1/\sqrt{2} \\ 0 & 1 & 0 \end{pmatrix} \quad \text{and} \quad P^{-1} = \begin{pmatrix} 1/\sqrt{2} & 1/\sqrt{2} & 0 \\ 0 & 0 & 1 \\ 1/\sqrt{2} & -1/\sqrt{2} & 0 \end{pmatrix}$$

$$= P^T \qquad (!!)$$

Diagonalize A:

$$A = P \begin{pmatrix} 1 & 0 & 0 \\ 0 & 1 & 0 \\ 0 & 0 & -1 \end{pmatrix} P^T$$

The last two examples illustrate the basic results for diagonalization of symmetric matrices.

THEOREM 5.4.1 If $A_{n \times n}$ is symmetric with real entries, then

(*a*) The eigenvalues are real.

(*b*) Eigenvectors corresponding to distinct eigenvalues are orthogonal.

(*c*) The eigenspaces of each eigenvalue have orthogonal bases. The dimension of an eigenspace corresponds to the multiplicity of the eigenvalue.

(*d*) Matrix A is orthogonally diagonalizable; that is, there exists an orthogonal matrix P such that

$$A = PDP^T \qquad (\text{and so} \qquad D = P^T AP)$$

Proof We prove only parts (*a*) and (*b*). Parts (*c*) and (*d*) are proved in more advanced texts.

(*a*) Suppose that $\lambda = a + bi$ and that $X + iY$ is the corresponding eigenvector. Therefore,

$$[A - (a + bi)I](X + iY) = 0 + 0i$$

Carrying out the multiplications and setting real and imaginary parts equal, we find

$$aIX - AX - bIY = 0$$
$$aIY - AY + bIX = 0$$

Now take the dot product of the first equation with Y and the second with X to find

$$aY \cdot IX - Y \cdot AX - bY \cdot IY = Y \cdot 0 = 0$$
$$aX \cdot IY - X \cdot AY + bX \cdot IX = X \cdot 0 = 0$$

or

$$aY^TIX - Y^TAX - bY^TIY = 0$$
$$aX^TIY - X^TAY + bX^TIX = 0$$

Because A and I are symmetric $I = I^T$ and $A = A^T$. Thus

$$aY^TI^TX^{TT} - Y^TA^TX^{TT} - bY^TIY = 0$$
$$aX^T\,IY - X^T\,AY + bX^T\,IX = 0$$

and since the transpose of a product is the product of the transposes in reverse order, we have

$$a(X^T\,IY)^T - (X^T\,AY)^T - bY^T\,IY = 0$$
$$aX^TIY - X^TAY + bX^TIX = 0$$

Finally, taking the transpose of both sides of the second equation results in

$$a(X^TIY)^T - (X^TAY)^T - bY^TIY = 0$$
$$a(X^TIY)^T - (X^TAY)^T + bX^TIX = 0$$

Subtracting the equations yields

$$b(Y^TIY + X^TIX) = 0$$

or

$$b(|Y|^2 + |X|^2) = 0$$

Because both $|X|$ and $|Y|$ cannot be zero (for if so, $X + iY = 0$ and could not be an eigenvector), we must have $b = 0$. Therefore λ is real.

(*b*) Let X_1 and X_2 be eigenvectors corresponding to λ_1 and λ_2, $\lambda_1 \neq \lambda_2$, $\lambda_1 \neq 0$. We want to show that $X_1 \cdot X_2 = 0$. Now

$$X_1 \cdot X_2 = \frac{1}{\lambda_1}\lambda_1 X_1 \cdot X_2 = \frac{1}{\lambda_1} AX_1 \cdot X_2$$
$$= \frac{1}{\lambda_1}(AX_1)^T X_2$$
$$= \frac{1}{\lambda_1}(X_1{}^T A^T)X_2$$

$$\begin{aligned}
\text{(By symmetry)} \qquad &= \frac{1}{\lambda_1}(X_1{}^T A)X_2 \\
&= \frac{1}{\lambda_1} X_1{}^T(AX_2) \\
&= \frac{1}{\lambda_1} X_1{}^T \lambda_2 X_2 \\
&= \frac{\lambda_2}{\lambda_1} X_1{}^T X_2 \\
&= \frac{\lambda_2}{\lambda_1}(X_1 \cdot X_2)
\end{aligned}$$

If $\lambda_2 = 0$, then $X_1 \cdot X_2 = 0$. If $\lambda_2 \neq 0$, then $\lambda_2/\lambda_1 \neq 1$ and we must still have $X_1 \cdot X_2 = 0$.

Theorem 5.4.1*c* tells us that we can find an orthogonal basis for each eigenspace. This may require the Gram-Schmidt process, as the next example shows.

EXAMPLE 3 *Orthogonally diagonalize*

$$A = \begin{pmatrix} 0 & 1 & 1 \\ 1 & 0 & 1 \\ 1 & 1 & 0 \end{pmatrix}$$

That is, diagonalize A with an orthogonal matrix P.

Solution The characteristic polynomial is $-\lambda^3 + 3\lambda + 2$ which has roots -1 (multiplicity 2) and 2 (simple). To determine eigenvectors, we solve $(A - \lambda I)X = 0$:

$$\lambda_1 = 2: \qquad \begin{aligned} -2x_1 + x_2 + x_3 &= 0 \\ x_1 - 2x_2 + x_3 &= 0 \\ x_1 + x_2 - 2x_3 &= 0 \end{aligned} \Rightarrow X_1 = \begin{pmatrix} k \\ k \\ k \end{pmatrix}$$

$$\lambda_2 = -1: \qquad \begin{aligned} x_1 + x_2 + x_3 &= 0 \\ x_1 + x_2 + x_3 &= 0 \\ x_1 + x_2 + x_3 &= 0 \end{aligned} \Rightarrow X_2 = \begin{pmatrix} -k-j \\ k \\ j \end{pmatrix}$$

Since rank $(A - \lambda_2 I) = 1$, the dimension of $E_{(\lambda_2)}$ is 2.

Looking at X_2 and putting $k = 1, j = 0$, we have

$$V_1 = \begin{pmatrix} -1 \\ 1 \\ 0 \end{pmatrix}$$

in the eigenspace. Setting $k = 0, j = 1$, we find

$$V_2 = \begin{pmatrix} -1 \\ 0 \\ 1 \end{pmatrix}$$

in the eigenspace. Now V_1 and V_2 are not orthogonal to each other, but they are linearly independent and span the eigenspace. Using the Gram-Schmidt process on $\{V_1, V_2\}$, we find

$$\mathcal{O}_1 = \begin{pmatrix} -1/\sqrt{2} \\ 1/\sqrt{2} \\ 0 \end{pmatrix} \quad \text{and} \quad \mathcal{O}_2 = \begin{pmatrix} -1/\sqrt{6} \\ -1/\sqrt{6} \\ 2/\sqrt{6} \end{pmatrix}$$

as orthonormal basis vectors for the eigenspace of $\lambda_2 = -1$. Letting $\mathcal{O}_3 = X_1/|X_1|$, we obtain an orthonormal basis (for E^3) of eigenvectors of A. Choosing P as

$$\begin{pmatrix} -1/\sqrt{2} & -1/\sqrt{6} & 1/\sqrt{3} \\ 1/\sqrt{2} & -1/\sqrt{6} & 1/\sqrt{3} \\ 0 & 2/\sqrt{6} & 1/\sqrt{3} \end{pmatrix}$$

we have

$$A = P\begin{pmatrix} -1 & 0 & 0 \\ 0 & -1 & 0 \\ 0 & 0 & 2 \end{pmatrix}P^T$$

Now that we can orthogonally diagonalize symmetric matrices, we can consider an application to analytic geometry.

Quadratic Forms and Conic Sections A classical problem of analytic geometry is the following:

For the conic section centered at the origin of the xy plane, described by

$$ax^2 + bxy + cy^2 = d \tag{5.4.1}$$

determine whether the conic section is an ellipse, a hyperbola, or a parabola. Graph the conic section.

A symmetric matrix can be used to describe the left-hand side of Eq. 5.4.1. In particular,

$$(x \quad y)\begin{pmatrix} a & b/2 \\ b/2 & c \end{pmatrix}\begin{pmatrix} x \\ y \end{pmatrix} = ax^2 + bxy + cy^2 = d$$

Let us call

$$A = \begin{pmatrix} a & b/2 \\ b/2 & c \end{pmatrix}$$

the *matrix of the conic section.*[3] Making a change of basis with the orthogonal matrix P which diagonalizes A, we write

$$\begin{pmatrix} x' \\ y' \end{pmatrix} = P^T \begin{pmatrix} x \\ y \end{pmatrix} \qquad \text{or} \qquad P\begin{pmatrix} x' \\ y' \end{pmatrix} = \begin{pmatrix} x \\ y \end{pmatrix}$$

Substituting and denoting

$$\begin{pmatrix} x' \\ y' \end{pmatrix}$$

by X', we have

$$X^T AX = (PX')^T A(PX') = (X'^T P^T)A(PX') = X'^T(P^T AP)X' = X'^T DX'$$

which reduces to

$$\begin{pmatrix} x' \\ y' \end{pmatrix}^T \begin{pmatrix} \lambda_1 & 0 \\ 0 & \lambda_2 \end{pmatrix} \begin{pmatrix} x' \\ y' \end{pmatrix} = d$$

or

$$\lambda_1 x'^2 + \lambda_2 y'^2 = d$$

This last equation is easy to classify and graph in the $x'y'$ plane since it has no "mixed" term $x'y'$.

EXAMPLE 4 Classify $xy = 1$ and graph it.

Solution (We already know the graph since the equation can be rewritten as $y = 1/x$. This will make it easy to check our answer.) The matrix of the conic section is

$$A = \begin{pmatrix} 0 & \frac{1}{2} \\ \frac{1}{2} & 0 \end{pmatrix}$$

and A has eigenpairs

$$\left(\frac{1}{2}, \begin{pmatrix} 1/\sqrt{2} \\ 1/\sqrt{2} \end{pmatrix}\right) \qquad \left(-\frac{1}{2}, \begin{pmatrix} -1/\sqrt{2} \\ 1/\sqrt{2} \end{pmatrix}\right)$$

so that

---

[3] Several matrices are possible:

$$\begin{pmatrix} a & b \\ 0 & c \end{pmatrix} \quad \text{and} \quad \begin{pmatrix} a & 0 \\ b & c \end{pmatrix}$$

are two. We used the symmetric matrix so we can be assured of diagonalizability.

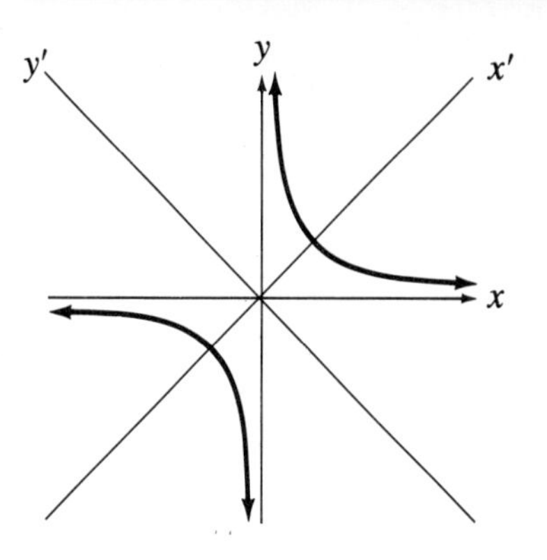

Figure 5.4.1 Graph of $xy = 1$.

$$P = \begin{pmatrix} 1/\sqrt{2} & -1/\sqrt{2} \\ 1/\sqrt{2} & 1/\sqrt{2} \end{pmatrix} \qquad P^T = \begin{pmatrix} 1/\sqrt{2} & 1/\sqrt{2} \\ -1/\sqrt{2} & 1/\sqrt{2} \end{pmatrix}$$

and

$$A = P\begin{pmatrix} \frac{1}{2} & 0 \\ 0 & -\frac{1}{2} \end{pmatrix}P^T$$

Therefore, with the change of variables (basis)

$$\begin{pmatrix} x' \\ y' \end{pmatrix} = P^T\begin{pmatrix} x \\ y \end{pmatrix} = \begin{pmatrix} 1/\sqrt{2} & 1/\sqrt{2} \\ -1/\sqrt{2} & 1/\sqrt{2} \end{pmatrix}\begin{pmatrix} x \\ y \end{pmatrix}$$

which is the same as

$$x' = \frac{1}{\sqrt{2}}x + \frac{1}{\sqrt{2}}y \qquad y' = -\frac{1}{\sqrt{2}}x + \frac{1}{\sqrt{2}}y$$

we have

$$\tfrac{1}{2}x'^2 - \tfrac{1}{2}y'^2 = 1$$

or

$$\frac{x'^2}{(\sqrt{2})^2} - \frac{y'^2}{(\sqrt{2})^2} = 1$$

Therefore, the conic section is a hyperbola. To sketch the graph, we must determine the $x'y'$ axes. Since the point $(x, y) = (1, 1)$ gives $(x', y') = (2/\sqrt{2}, 0)$ and $(x, y) = (-1, 1)$ gives $(x', y') = (0, 2/\sqrt{2})$, the x' and y' axes are as shown in Fig. 5.4.1 The graph of the hyperbola is also shown in Fig. 5.4.1. Note that the new axes contain the eigenvectors.

EXAMPLE 5 Classify the conic section

$$2x^2 + 2xy + 2y^2 = 27$$

and graph it.

Solution The matrix of the conic section is

$$A = \begin{pmatrix} 2 & 1 \\ 1 & 2 \end{pmatrix}$$

Eigenpairs of A are

$$\left(3, \begin{pmatrix} 1/\sqrt{2} \\ 1/\sqrt{2} \end{pmatrix}\right) \quad \text{and} \quad \left(1, \begin{pmatrix} -1/\sqrt{2} \\ 1/\sqrt{2} \end{pmatrix}\right)$$

Therefore in x', y' coordinates defined as in Example 4 (the eigenvectors are the same) we have

$$3x'^2 + y'^2 = 27$$

or

$$\frac{x'^2}{3^2} + \frac{y'^2}{(3\sqrt{3})^2} = 1$$

which describes an ellipse. The graph of the ellipse is shown in Fig. 5.4.2. Note that the new axes contain the eigenvectors of the matrix. Also note that $x'y'$ axes are obtained by a 45° counterclockwise rotation, which is the action of P. Moreover, x' is defined by the first eigenvector, and y' is defined by the second eigenvector.

Those who have solved these types of conic section problems in calculus realize that this linear algebra method of removing the xy term is much simpler. Of course, a lot of powerful machinery had to be developed to get to this point.

In general, the problem of removing the xy term in $ax^2 + bxy + cy^2$ is known as the problem of *diagonalizing a quadratic form.* This problem arises in many

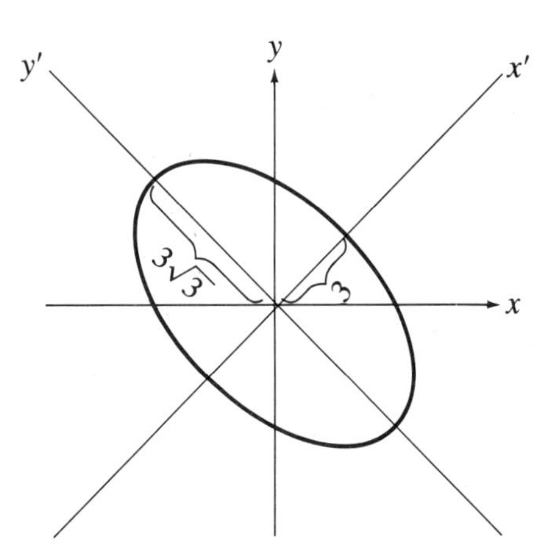

Figure 5.4.2
Graph of $2x^2 + 2xy + 2y^2 = 27$.

areas; statistics and physics are two. A real quadratic form in the variables x_1, $x_2, \ldots, x_n$ is a function $Q: E^n \to \mathbb{R}$ given by

$$Q\left(\begin{pmatrix} x_1 \\ x_2 \\ \vdots \\ x_n \end{pmatrix}\right) = (x_1 \quad x_2 \quad \cdots \quad x_n) A \begin{pmatrix} x_1 \\ \vdots \\ x_n \end{pmatrix} \tag{5.4.2}$$

where A is in $\mathscr{M}_{nn}$. Written out, a real quadratic form in x_1, x_2, x_3 looks like

$$a{x_1}^2 + b{x_2}^2 + c{x_3}^2 + dx_1x_2 + ex_1x_3 + fx_2x_3$$

where a through f are real numbers. Note that each term has degree 2—hence the name *quadratic form*. Our basic theorem about diagonalization of symmetric matrices means that any real quadratic form can be diagonalized. So there are new variables $x'_1, \ldots, x'_n$ such that in the new variables $X' = P^T X$

$$Q\begin{pmatrix} x'_1 \\ \vdots \\ x'_n \end{pmatrix} = \lambda_1 x_1'^2 + \lambda_2 x_2'^2 + \cdots + \lambda_n x'^n$$

This follows from the fact that the matrix A in Eq. (5.4.2) can always be chosen as symmetric, and symmetric matrices are orthogonally diagonalizable.

Diagonalization in the Hermitian Case Theorem 5.4.1 with a slight change of wording holds true for hermitian matrices.

If $A_{n \times n}$ is hermitian, then

1. The eigenvalues are real.
2. Eigenvectors corresponding to distinct eigenvalues are orthogonal.
3. The eigenspaces of each eigenvalue have orthogonal bases. The dimension of an eigenspace corresponds to the multiplicity of the eigenvalue.
4. Matrix A is *unitarily diagonalizable*. That is, there exists a unitary matrix U ($U^{-1} = U^*$) such that

$$A = UDU^* \qquad (\text{thus } D = U^*AU)$$

The proofs of 1 and 2 are almost the same as in Theorem 5.4.1*a* and *b*. The difference is that A^* is used instead of A^T and in $\mathscr{C}_{n1}$, $X \cdot Y = X^*Y$.

EXAMPLE 6 Can

$$A = \begin{pmatrix} 1 & 1-i \\ 1+i & 0 \end{pmatrix}$$

be unitarily diagonalized? If so, perform the diagonalization.

Solution

$$A^* = \bar{A}^T = \begin{pmatrix} 1 & 1+i \\ 1-i & 0 \end{pmatrix}^T = \begin{pmatrix} 1 & 1-i \\ 1+i & 0 \end{pmatrix} = A$$

Because A is hermitian, it can be unitarily diagonalized. Now to find the eigenpairs,

$$\det(A - \lambda I) = (1-\lambda)(-\lambda) - (1-i)(1+i) = \lambda^2 - \lambda - 2$$

So we have $\lambda_1 = 2$ and $\lambda_2 = -1$. Eigenpairs are

$$\left(2, \begin{pmatrix} 1-i \\ 1 \end{pmatrix}\right) \qquad \left(-1, \begin{pmatrix} 1-i \\ -2 \end{pmatrix}\right)$$

To find U, we normalize the eigenvectors and use them for the columns of U. The normalized eigenvectors are found by calculating

$$\left|\begin{pmatrix} 1-i \\ 1 \end{pmatrix}\right| = \sqrt{(1+i, 1)\begin{pmatrix} 1-i \\ 1 \end{pmatrix}} = \sqrt{3}$$

and

$$\left|\begin{pmatrix} 1-i \\ -2 \end{pmatrix}\right| = \sqrt{(1+i, -2)\begin{pmatrix} 1-i \\ -2 \end{pmatrix}} = \sqrt{6}$$

(Remember that $|X| = \sqrt{\langle X, X\rangle} = \sqrt{X^*X}$.) Thus we have

$$U = \begin{pmatrix} \dfrac{1-i}{\sqrt{3}} & \dfrac{1-i}{\sqrt{6}} \\ \dfrac{1}{\sqrt{3}} & \dfrac{-2}{\sqrt{6}} \end{pmatrix} \qquad U^* = \begin{pmatrix} \dfrac{1+i}{\sqrt{3}} & \dfrac{1}{\sqrt{3}} \\ \dfrac{1+i}{\sqrt{6}} & \dfrac{-2}{\sqrt{6}} \end{pmatrix}$$

and

$$\begin{pmatrix} 1 & 1-i \\ 1+i & 0 \end{pmatrix} = U \begin{pmatrix} 2 & 0 \\ 0 & -1 \end{pmatrix} U^*$$

When a hermitian matrix A is diagonalized, the set of orthonormal eigenvectors of A is called the set of *principal axes* of A and the associated matrix U is called a *principal axis transformation.* For a real hermitian matrix, the principal axis transformation allows us to analyze A geometrically.

EXAMPLE 7 Consider $T_A: \mathcal{M}_{21} \to \mathcal{M}_{21}$ defined by

$$T_A\left(\begin{pmatrix} a \\ b \end{pmatrix}\right) = \begin{pmatrix} 1 & 2 \\ 2 & 1 \end{pmatrix}\begin{pmatrix} a \\ b \end{pmatrix}$$

This can be diagonalized with

$$U = \begin{pmatrix} 1/\sqrt{2} & -1/\sqrt{2} \\ 1/\sqrt{2} & 1/\sqrt{2} \end{pmatrix}$$

so that

Figure 5.4.3 Action of D on the unit circle.

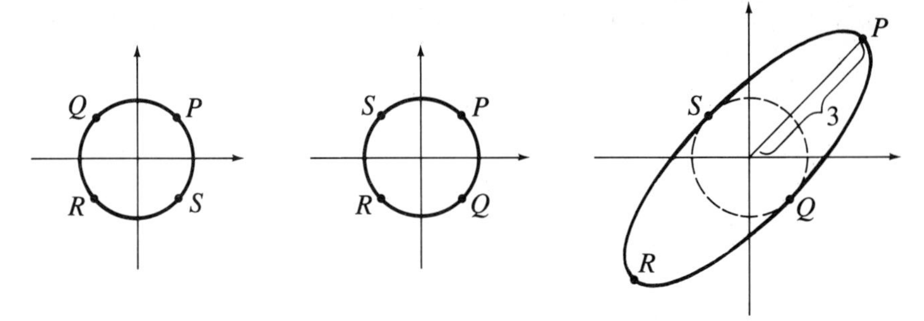

$$\begin{pmatrix} 1 & 2 \\ 2 & 1 \end{pmatrix} = \begin{pmatrix} 1/\sqrt{2} & -1/\sqrt{2} \\ 1/\sqrt{2} & 1/\sqrt{2} \end{pmatrix} \begin{pmatrix} 3 & 0 \\ 0 & -1 \end{pmatrix} \begin{pmatrix} 1/\sqrt{2} & 1/\sqrt{2} \\ -1/\sqrt{2} & 1/\sqrt{2} \end{pmatrix}$$

Now U represents rotation of 45° counterclockwise while U^T represents rotation of 45° clockwise. If we want to see what T_A does to

$$\begin{pmatrix} a \\ b \end{pmatrix}$$

we can look at

$$\begin{pmatrix} 1/\sqrt{2} & -1/\sqrt{2} \\ 1/\sqrt{2} & 1/\sqrt{2} \end{pmatrix} \begin{pmatrix} 3 & 0 \\ 0 & -1 \end{pmatrix} \begin{pmatrix} 1/\sqrt{2} & 1/\sqrt{2} \\ -1/\sqrt{2} & 1/\sqrt{2} \end{pmatrix} \begin{pmatrix} a \\ b \end{pmatrix}$$

and we see that T_A is a

1. Rotation of 45° clockwise

2. Stretch of 3 in the first component and a reversal in the second component

3. Rotation of 45° counterclockwise

We can say even more by determining what T_A does to the unit circle. In the new coordinates, the unit circle is unchanged because U and U^T represent rotations. However, in the new coordinates we have the action of D as changing the unit circle by reflecting about the line which is defined by

$$\text{span}\left\{\begin{pmatrix} 1/\sqrt{2} \\ 1/\sqrt{2} \end{pmatrix}\right\}$$

and then transforming the reflected circle to an ellipse, as shown in Fig. 5.4.3.

Finally, we note that in diagonalizing a quadratic form for a conic section, the new axes obtained from the rotation are exactly the principal axes of the matrix for the quadratic form.

PROBLEMS 5.4 Diagonalize the matrices in Probs. 1 to 7. If possible, orthogonally diagonalize.

1. $\begin{pmatrix} -1 & 2 & 0 \\ 1 & 2 & 1 \\ 0 & 2 & -1 \end{pmatrix}$ **2.** $\begin{pmatrix} 2 & -1 & 0 \\ -1 & 2 & -1 \\ 0 & -1 & 2 \end{pmatrix}$ **3.** $\begin{pmatrix} 1 & 1 \\ 1 & 1 \end{pmatrix}$

4. $\begin{pmatrix} 0 & 0 & -2 \\ 0 & -4 & 0 \\ -2 & 0 & 3 \end{pmatrix}$ **5.** $\begin{pmatrix} 2 & 2 & 0 & 0 \\ 2 & 2 & 0 & 0 \\ 0 & 0 & 0 & -1 \\ 0 & 0 & -1 & 0 \end{pmatrix}$

6. $\begin{pmatrix} 0 & 1 & 0 & 0 \\ 1 & 0 & 0 & 0 \\ 0 & 0 & 0 & 1 \\ 0 & 0 & 1 & 0 \end{pmatrix}$ **7.** $\begin{pmatrix} 10 & 8 & 4 \\ 8 & 10 & 4 \\ 4 & 4 & 4 \end{pmatrix}$

8. Classify and graph the following conic sections.

(*a*) $4x^2 + 4xy + 4y^2 = 18$ (*b*) $3x^2 + 10xy + 3y^2 = 8$

9. Show that the rotation matrix

$$R_\theta = \begin{pmatrix} \cos\theta & -\sin\theta \\ \sin\theta & \cos\theta \end{pmatrix} \qquad \theta \neq 0, \pi$$

is not diagonalizable if complex eigenvalues are not allowed. What if no restriction is placed on eigenvalues?

10. If an antisymmetric matrix is diagonalizable, must one of the eigenvalues be zero?

11. Unitarily diagonalize

$$\begin{pmatrix} 2 & 3+3i \\ 3-3i & 5 \end{pmatrix}$$

12. Unitarily diagonalize

$$\begin{pmatrix} 1 & i \\ -i & 0 \end{pmatrix}$$

13. Unitarily diagonalize

$$\begin{pmatrix} 1 & i & 0 \\ -i & 0 & 0 \\ 0 & 0 & -1 \end{pmatrix}$$

14. Unitarily diagonalize

$$\begin{pmatrix} 2 & 0 & 1+i \\ 0 & 2 & i \\ 1-i & -i & 0 \end{pmatrix}$$

15. A matrix is called a *Hadamard matrix* if its entries are only 1 or -1 and its rows are mutually orthogonal. Show that

$$\begin{pmatrix} 1 & 1 \\ -1 & 1 \end{pmatrix} \quad \text{and} \quad \begin{pmatrix} 1 & 1 & 1 & 1 \\ -1 & 1 & -1 & 1 \\ 1 & 1 & -1 & -1 \\ -1 & 1 & 1 & -1 \end{pmatrix}$$

are Hadamard matrices. (Hadamard matrices are used in coding theory.)

16. If $H_{n \times n}$ is a Hadamard matrix, must the $2n \times 2n$ matrix

$$\begin{pmatrix} H & H \\ H & -H \end{pmatrix}$$

be a Hadamard matrix?

17. Let A be real symmetric. Show that $\det A$ equals the product of the eigenvalues of A.

18. Let A be hermitian. Show that $\det A$ is real.

19. Let A be symmetric with all eigenvalues positive. Show that there is a matrix S such that $A = S^2$. Matrix S is called the *square root* of A. [*Hint*: Use $A = PDP^T$ and ask yourself, Does D have a square root?]

20. Use the diagonalization of

$$A = \begin{pmatrix} 2 & 1 \\ 1 & 2 \end{pmatrix}$$

(see Example 5) to analyze the transformation of $\mathbb{R}^2$ defined by A.

5.5 POSTCALCULUS: SYSTEMS OF LINEAR FIRST-ORDER DIFFERENTIAL EQUATIONS, FUNDAMENTAL FREQUENCIES, AND STABILITY

A linear homogeneous first-order system of differential equations, with constant coefficients, in the three functions $y_1(t)$, $y_2(t)$, $y_3(t)$ is a set of equations of the form (the prime denotes differentiation)

$$\begin{pmatrix} y_1'(t) \\ y_2'(t) \\ y_3'(t) \end{pmatrix} = \begin{pmatrix} a_{11} & a_{12} & a_{13} \\ a_{21} & a_{22} & a_{23} \\ a_{31} & a_{32} & a_{33} \end{pmatrix} \begin{pmatrix} y_1(t) \\ y_2(t) \\ y_3(t) \end{pmatrix}$$

Denoting

$$\begin{pmatrix} y_1 \\ y_2 \\ y_3 \end{pmatrix} \quad \text{by} \quad Y$$

and

$$\begin{pmatrix} y_1' \\ y_2' \\ y_3' \end{pmatrix} \quad \text{by} \quad Y'$$

the system can be written

$$Y' = AY \tag{5.5.1}$$

where A is the 3×3 matrix in the first equation. This system is called a 3×3 system. An $n \times n$ system can be written as Eq. (5.5.1) if we have

$$Y = \begin{pmatrix} y_1 \\ y_2 \\ \vdots \\ y_n \end{pmatrix}$$

and A is an $n \times n$ matrix.

Such systems arise in simple diffusion problems in bioengineering. Consider two cells (Fig. 5.5.1) which contain equal volumes V of a salt solution. The boundary between the cells is a semipermeable membrane. Let $y_1(t)$ and $y_2(t)$ represent the amount of salt dissolved in cells 1 and cells 2, respectively. A reasonable assumption is that the rate of change of the amount of salt in a cell due to passage through the membrane is proportional to the difference in the concentrations in the cells. This leads to the equations

$$\underbrace{\frac{y_1'(t)}{V}}_{\text{Rate of change of concentration}} = \underbrace{k}_{\text{Proportionality constant } (>0)} \underbrace{\left(\frac{y_2(t)}{V} - \frac{y_1(t)}{V}\right)}_{\text{Difference in concentrations}}$$

$$\frac{y_2'(t)}{V} = k\left(\frac{y_1(t)}{V} - \frac{y_2(t)}{V}\right)$$

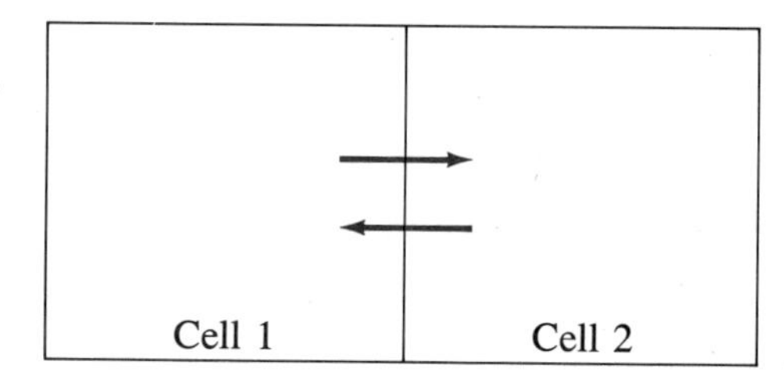

Figure 5.5.1
Cells of salt solution.

and the system

$$Y' = \begin{pmatrix} -k & +k \\ +k & -k \end{pmatrix} Y$$

(we canceled the V's). The matrix of the system is symmetric. This happens often in applications.

A first-order system such as Eq. (5.5.1), once derived as a model for a physical problem, must be solved. That is, functions $y_1(t), y_2(t), \ldots, y_n(t)$ must be found which, when substituted into (5.5.1), give a true mathematical equation.

EXAMPLE 1 Show that

$$Y = \begin{pmatrix} ce^{-2t} \\ -ce^{-2t} \end{pmatrix}$$

is a solution of

$$Y' = \begin{pmatrix} -1 & +1 \\ +1 & -1 \end{pmatrix} Y$$

Show that

$$V = \begin{pmatrix} c \\ c \end{pmatrix}$$

is also a solution.

Solution We have

$$Y' = \begin{pmatrix} -2ce^{-2t} \\ +2ce^{-2t} \end{pmatrix} \quad \text{and} \quad \begin{pmatrix} -1 & +1 \\ +1 & -1 \end{pmatrix} Y = \begin{pmatrix} -2ce^{-2t} \\ +2ce^{-2t} \end{pmatrix}$$

Therefore,

$$Y = \begin{pmatrix} ce^{-2t} \\ -ce^{-2t} \end{pmatrix}$$

is a solution.

For

$$V = \begin{pmatrix} c \\ c \end{pmatrix}$$

we obtain

$$V' = \begin{pmatrix} 0 \\ 0 \end{pmatrix} \quad \text{and} \quad \begin{pmatrix} -1 & 1 \\ 1 & -1 \end{pmatrix} \begin{pmatrix} c \\ c \end{pmatrix} = \begin{pmatrix} 0 \\ 0 \end{pmatrix}$$

Therefore,

$$V = \begin{pmatrix} c \\ c \end{pmatrix}$$

is a solution also.

The solutions given in Example 1 are closely related to the eigenvalues and eigenvectors of

$$\begin{pmatrix} -1 & 1 \\ 1 & -1 \end{pmatrix}$$

EXAMPLE 2 Find the eigenvalues and eigenvectors of

$$A = \begin{pmatrix} -1 & 1 \\ 1 & -1 \end{pmatrix}$$

Relate them to the solutions of $Y' = AY$ given in Example 1.

Solution The characteristic polynomial of A is $p(\lambda) = \lambda^2 + 2\lambda$ which has roots $\lambda_1 = -2$ and $\lambda_2 = 0$. The eigenpairs are

$$\left(-2, \begin{pmatrix} k \\ -k \end{pmatrix}\right) \quad \text{and} \quad \left(0, \begin{pmatrix} j \\ j \end{pmatrix}\right)$$

The first eigenvalue is the exponent of the exponential function e^{-2t}. If we multiply the eigenvector for -2 by e^{-2t}, we obtain the first solution in Example 1. Using the second eigenpair, if we multiply

$$e^{0t}\begin{pmatrix} j \\ j \end{pmatrix} = \begin{pmatrix} j \\ j \end{pmatrix}$$

we obtain the second solution in Example 1.

The results of Example 2 indicate that if (λ, X) is an eigenpair of A, then

$$e^{\lambda t}X$$

is a solution of $Y' = AY$. This is basically true.

THEOREM 5.5.1 Let A be a matrix with real entries. Let X be an eigenvector corresponding to to the real eigenvalue λ of A. The vector $Y = e^{\lambda t}X$ is a solution of $Y' = AY$ for all t.

Proof Let $Y = e^{\lambda t}X$. Then $Y' = \lambda e^{\lambda t}X$, since X is a constant vector. However, $AY = A(e^{\lambda t}X) = e^{\lambda t}(AX) = e^{\lambda t}\lambda X = \lambda e^{\lambda t}X$. Thus $Y' = AY$.

THEOREM 5.5.2 If A is diagonalizable, then $Y' = AY$ has n linearly independent solutions of the form in Theorem 5.5.1.

THEOREM 5.5.3 If Y_1 and Y_2 are two solutions of $Y' = AY$, then $c_1 Y_1 + c_2 Y_2$ is a solution also.

Proof If $Y' = AY$ and $Y' = AY$, then

$$A(c_1 Y_1 + c_2 Y_2) = c_1 A Y_1 + c_2 A Y_2 = c_1 Y_1' + c_2 Y_2'$$
$$= (c_1 Y_1 + c_2 Y_2)'$$

Therefore, $c_1 Y_1 + c_2 Y_2$ is a solution.

These theorems tell us that if we solve the eigenproblem for a diagonalizable matrix A, then we have solved

$$Y' = AY$$

EXAMPLE 3 Solve the diffusion problem from the beginning of this section:

$$Y' = \begin{pmatrix} -k & k \\ k & -k \end{pmatrix} Y$$

Solution The characteristic polynomial of

$$A = \begin{pmatrix} -k & k \\ k & -k \end{pmatrix}$$

is $p(\lambda) = \lambda^2 + 2k\lambda$, so the eigenvalues are 0 and $-2k$. Since

$$\left(0, \begin{pmatrix} 1 \\ 1 \end{pmatrix}\right) \quad \text{and} \quad \left(-2k, \begin{pmatrix} 1 \\ -1 \end{pmatrix}\right)$$

are eigenpairs, we have

$$\begin{pmatrix} 1 \\ 1 \end{pmatrix} \quad \text{and} \quad \begin{pmatrix} e^{-2kt} \\ -e^{-2kt} \end{pmatrix}$$

as independent solutions. By Theorem 5.5.3 a general solution of the system is

$$c_1 \begin{pmatrix} 1 \\ 1 \end{pmatrix} + c_2 \begin{pmatrix} e^{-2kt} \\ -e^{-2kt} \end{pmatrix} = \begin{pmatrix} c_1 + c_2 e^{-2kt} \\ c_1 - c_2 e^{-2kt} \end{pmatrix}$$

Since the solution of the diffusion problem is (and it is the most general solution, a fact we will not prove)

$$\begin{pmatrix} y_1(t) \\ y_2(t) \end{pmatrix} = \begin{pmatrix} c_1 + c_2 e^{-2kt} \\ c_1 - c_2 e^{-2kt} \end{pmatrix} \tag{5.5.2}$$

we note that as $t \to +\infty$,

$$\begin{pmatrix} y_1 \\ y_2 \end{pmatrix} \longrightarrow \begin{pmatrix} c_1 \\ c_1 \end{pmatrix}$$

That is, as $t \to \infty$, the diffusion system achieves the *equilibrium state* of equal concentrations of salt in each compartment. This makes sense and gives us confidence that the mathematical model of our differential equation may be an accurate model for the diffusion system.

Diffusion problems ordinarily include an *initial condition*: the amount of salt in each compartment at the starting time ($t = 0$). For example, if we began the system with $y_1(0) = 2$ and $y_2(0) = 1$, the solution would have to satisfy

$$\begin{pmatrix} y_1(0) \\ y_2(0) \end{pmatrix} = \begin{pmatrix} 2 \\ 1 \end{pmatrix}$$

This means, by putting $t = 0$ into Eq. (5.5.2), that

$$\begin{pmatrix} c_1 + c_2 \\ c_1 - c_2 \end{pmatrix} = \begin{pmatrix} 2 \\ 1 \end{pmatrix}$$

or $c_1 = \frac{3}{2}$ and $c_2 = \frac{1}{2}$. Therefore, the solution of the diffusion problem would be

$$\begin{pmatrix} y_1(t) \\ y_2(t) \end{pmatrix} = \begin{pmatrix} \frac{3}{2} + \frac{1}{2}e^{-2kt} \\ \frac{3}{2} - \frac{1}{2}e^{-2kt} \end{pmatrix}$$

This solution reflects some physical principles:

1. As $t \to \infty$, the salt will balance between the two compartments. As $t \to \infty$,

$$\begin{pmatrix} y_1(t) \\ y_2(t) \end{pmatrix} \longrightarrow \begin{pmatrix} \frac{3}{2} \\ \frac{3}{2} \end{pmatrix}$$

   Since the volume of each cell is the same, the concentrations are equal.

2. The total amount of salt is conserved. At the beginning we have $2 + 1 = 3$ units of salt. For all t,

$$\begin{aligned} y_1(t) + y_2(t) &= \tfrac{3}{2} + \tfrac{1}{2}e^{-2kt} + \tfrac{3}{2} - \tfrac{1}{2}e^{-2kt} \\ &= 3 \text{ units of salt} \end{aligned}$$

3. The first compartment will always have more salt than the second. In our solution $y_1(t) > y_2(t)$ for any finite time t since $\frac{1}{2}e^{-2kt} > -\frac{1}{2}e^{-2kt}$.

4. The amount of salt in compartment 1 (the saltier one at the beginning) will always be decreasing. Since $y_1(t) = \frac{3}{2} + \frac{1}{2}e^{-2kt}$, we have the derivative

$$y_1'(t) = -ke^{-2kt} < 0 \qquad \text{for all } t$$

   Thus $y_1(t)$ is decreasing.

5. The amount of salt in compartment 2 will always be increasing. This is shown in the same way as principle 4 was shown.

6. As the permeability increases (k gets larger), the balancing of the salt occurs more quickly. If $k_1 > k_2$, then

$$e^{-2k_1t} < e^{-2k_2t} \qquad \text{and} \qquad \tfrac{3}{2} + \tfrac{1}{2}e^{-2k_1t} < \tfrac{3}{2} + \tfrac{1}{2}e^{-2k_2t}$$

   (See Fig. 5.5.2.)

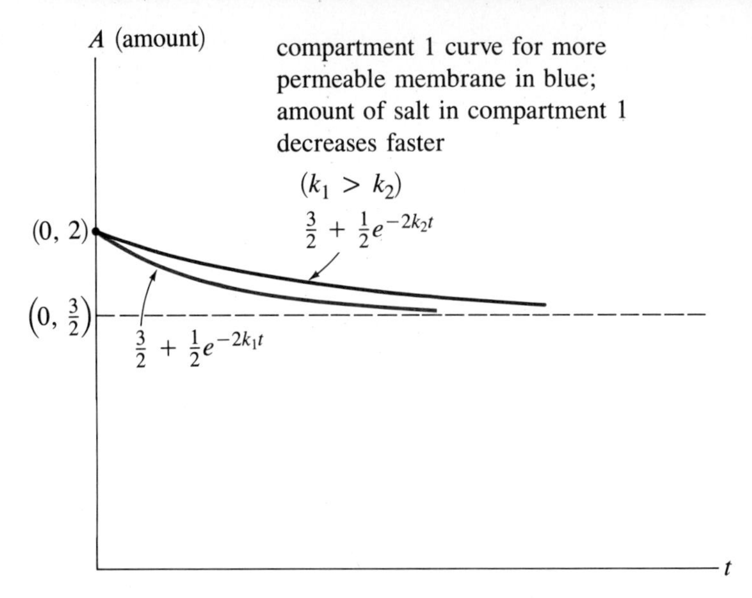

Figure 5.5.2 Amounts of salt.

Since the solution to the system of differential equations adheres to these physical intuitions we have about the system, we have confidence in the mathematical model. Of course, conducting an experiment and making observations are the ultimate test for the model. It turns out that the model is good for dilute solutions; moreover, solutes other than salt can be modeled.

For large systems, direct numerical methods are often used instead of attempting diagonalization of A. Diagonalization of A has importance in finding *canonical cooordinates* for a system of differential equations.

To summarize: The general solution to

$$Y' = AY$$

where A is diagonalizable with distinct eigenvalues, can be written as

$$Y = c_1 e^{\lambda_1 t} X_1 + c_2 e^{\lambda_2 t} X_2 + \cdots + c_n e^{\lambda_n t} X_n$$

where $c_1, \ldots, c_n$ are arbitrary constants, $\lambda_1, \ldots, \lambda_n$ are the eigenvalues of A, and $X_1, \ldots, X_n$ are the independent eigenvectors of A. The constants $c_1, \ldots, c_n$ are determined exactly when an initial condition

$$Y(0) = \begin{pmatrix} y_1(0) \\ \vdots \\ y_n(0) \end{pmatrix} = \begin{pmatrix} a_1 \\ \vdots \\ a_n \end{pmatrix}$$

is given.

Finally, if we allow complex solutions of $\det(A - \lambda I) = 0$ to be used for eigenvalues, we can find solutions which are sine and cosine functions. For example,

$$\begin{pmatrix} y_1' \\ y_2' \end{pmatrix} = \begin{pmatrix} 0 & -1 \\ 1 & 0 \end{pmatrix} \begin{pmatrix} y_1 \\ y_2 \end{pmatrix}$$

has solution

$$Y = \begin{pmatrix} \cos t \\ \sin t \end{pmatrix}$$

Note that the eigenvalues of

$$\begin{pmatrix} 0 & -1 \\ 1 & 0 \end{pmatrix}$$

are i and $-i$. The connection to the solution is that

$$e^{it} = \cos t + i \sin t$$

a fundamental identity for the analysis of differential equations—especially as models of vibrating systems in mechanics.

Vibrations The system of trolleys in Fig. 5.5.3 will vibrate if they are pulled and released. However, only certain vibrations can be sustained without outside forces. The frequencies of these vibrations are called the *fundamental frequencies* of the system. These frequencies can be found by solving the eigenvalue problem for a certain matrix. The matrix is obtained as follows: With $y_1(t)$ and $y_2(t)$ as the displacements of trolley 1 and trolley 2, respectively, from equilibrium (see Fig. 5.5.4), the equations of motion (with no external forces) are

$$y_1'' = -2\frac{k}{m} y_1 + \frac{k}{m} y_2$$

$$y_2'' = \frac{k}{m} y_1 - 2\frac{k}{m} y_2$$

Assuming solutions of the form $y_1 = A \cos \omega t + B \sin \omega t$ and $y_2 = C \cos \omega t + D \sin \omega t$, where ω is the frequency, and substituting we find

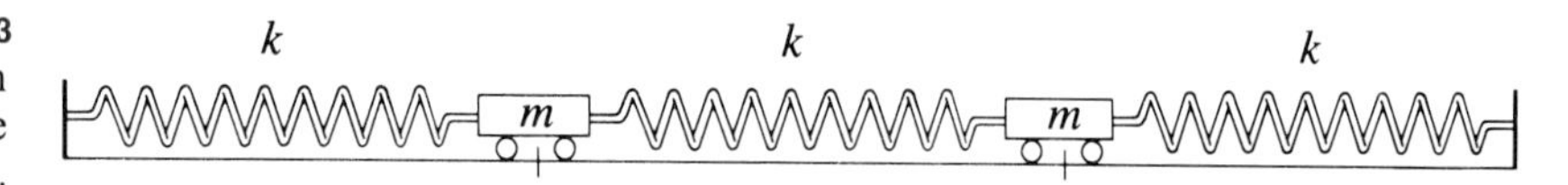

Figure 5.5.3 Trolleys of mass m in equilibrium; k is the spring constant.

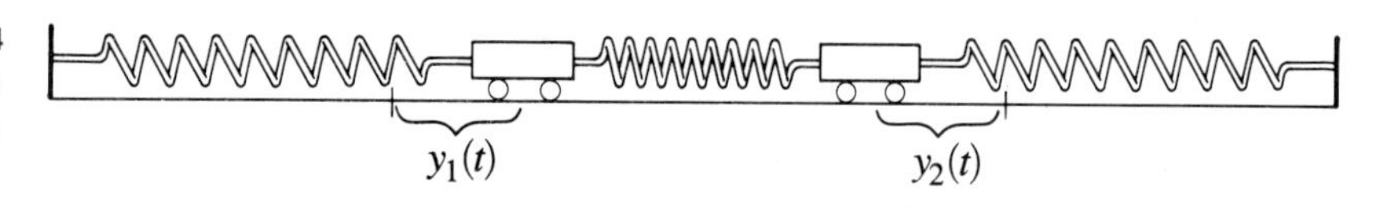

Figure 5.5.4 Trolleys not in equilibrium.

$$-\omega^2 y_1 = -2\frac{k}{m}y_1 + \frac{k}{m}y_2$$

$$-\omega^2 y_2 = \frac{k}{m}y_1 - \frac{2k}{m}y_2$$

or

$$\begin{pmatrix} \dfrac{-2k}{m} & \dfrac{k}{m} \\ \dfrac{k}{m} & \dfrac{-2k}{m} \end{pmatrix}\begin{pmatrix} y_1 \\ y_2 \end{pmatrix} = -\omega^2\begin{pmatrix} y_1 \\ y_2 \end{pmatrix} \tag{5.5.3}$$

This last equation means that solutions of the form specified can exist only if the matrix (call it A) in Eq. (5.5.3) has negative eigenvalues; for if λ is an eigenvalue, $\lambda = -\omega^2$ and ω is real.

The matrix A has eigenpairs

$$\left(-3\frac{k}{m}, \begin{pmatrix} 1 \\ -1 \end{pmatrix}\right) \quad \left(-\frac{k}{m}, \begin{pmatrix} 1 \\ 1 \end{pmatrix}\right)$$

The interpretation is, since $-\omega^2$ is $-3k/m$ or $-k/m$, that the characteristic frequencies are $\sqrt{3k/m}$ and $\sqrt{k/m}$. Also the opposite signs of the components in the first eigenvector indicate that $y_1 = -y_2$ (trolleys moving in opposite directions), and the same signs of the components in the second eigenvector indicate that $y_1 = y_2$ (trolleys moving in the same direction).

Stability For vibrating systems as well as diffusion systems in equilibrium, stability of equilibrium is an important consideration. Roughly speaking, an equilibrium state for a system described by systems of differential equations is a state for which the velocity of the components is zero. For example, a pendulum has two equilibrium states, as shown in Fig. 5.5.5. However, one equilibrium

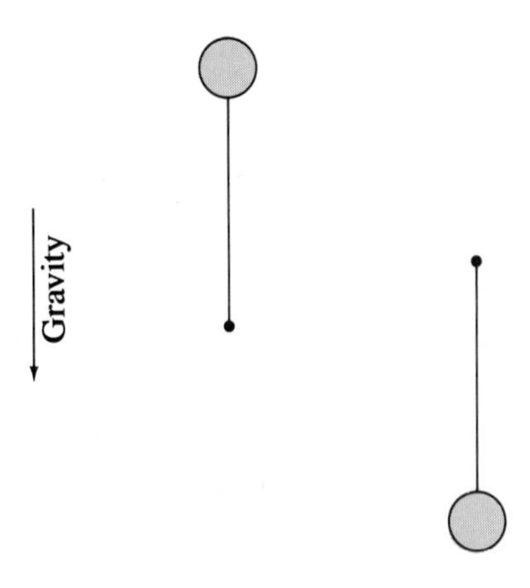

Figure 5.5.5 Two equilibrium positions for a pendulum.

state is fundamentally different from the other. The second is *stable*: After a slight push the pendulum will return to its original position. In the first position a slight push will cause the pendulum to leave that equilibrium state, never to return.

Stability enters into the design of structures, chemical reactors, and machines. Often, several equilibrium configurations are theoretically possible in chemical reactors, but only the stable equilibrium can be maintained in practice.

Roughly speaking, for the types of systems we have briefly discussed, stability is guaranteed if the eigenvalues of matrix A of the differential equation $Y'' = AY$ have negative real parts. This means that in both of our examples the system was stable. Rigorous discussions of stability can be found in many texts on ordinary differential equations.

PROBLEMS 5.5

1–7. Find the general solution to the system $Y' = AY$ for the matrices A given in Probs. 1 to 7 of the last section, provided the given matrix has distinct eigenvalues.

Use the general solutions in Probs. 1 to 7 to solve the differential equations in Probs. 8 to 10 with initial conditions as given:

8. For 3×3 matrices $\qquad Y(0) = \begin{pmatrix} 1 \\ 1 \\ 0 \end{pmatrix}$

9. For 2×2 matrices $\qquad Y(0) = \begin{pmatrix} -1 \\ 1 \end{pmatrix}$

10. For 4×4 matrices $\qquad Y(0) = \begin{pmatrix} 1 \\ -1 \\ 0 \\ 1 \end{pmatrix}$

11. Consider a three-cell diffusion system with cells of equal volume separated by semipermeable membranes, as shown.

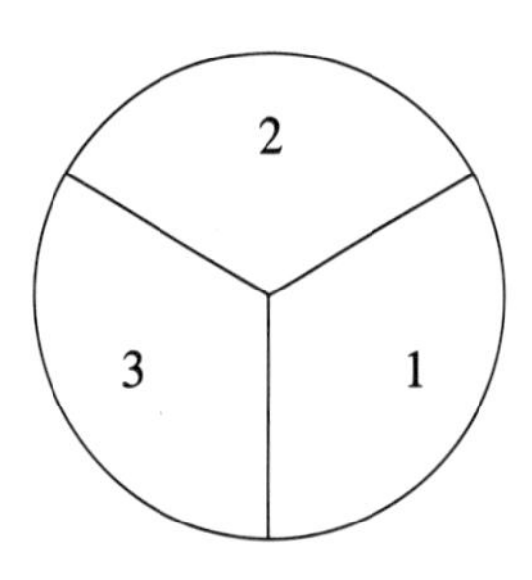

Suppose the diffusion coefficients are all equal to k (>0). Show that the first-order system modeling the diffusion is

$$\begin{aligned} y_1'(t) &= k[y_2(t) - y_1(t)] + k[y_3(t) - y_1(t)] \\ y_2'(t) &= k[y_1(t) - y_2(t)] + k[y_3(t) - y_2(t)] \\ y_3'(t) &= k[y(t) - y_3(t)] + k[y_2(t) - y_3(t)] \end{aligned}$$

where y_1, y_2, and y_3 are the amounts of salt in compartments 1, 2, and 3, respectively. Show that the matrix of the system has an eigenvalue of multiplicity 2.

12. The system below has equations of motion (gravity neglected).

$$\begin{pmatrix} y_1'' \\ y_2'' \end{pmatrix} = \begin{pmatrix} \dfrac{-(k_1 + k_2)}{m_1} & \dfrac{k_2}{m_2} \\ \dfrac{k_2}{m_2} & \dfrac{-k_2}{m_2} \end{pmatrix} \begin{pmatrix} y_1 \\ y_2 \end{pmatrix}$$

where $y_1(t)$ and $y_2(t)$ are the displacements of masses m_1 and m_2, respectively, from equilibrium.

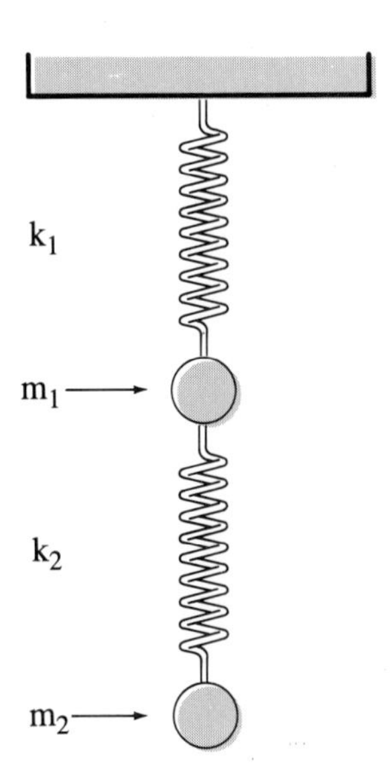

(*a*) For $k_1 = k_2 = k$ and $m_1 = m_2 = m$, find the characteristic frequencies.

(*b*) For $k_1 = k_2 = k$ and $m_1 = m_2 = m$, show that the eigenvalues of A are negative to show that the system is stable.

(*c*) If $k_1 \neq k_2$ and $m_1 \neq m_2$, is the system stable?

SUMMARY

Representation of a linear transformation by a diagonal matrix is advantageous but not possible for all transformations. The problem reduces to asking when a *matrix* can be diagonalized. *An $n \times n$ matrix can be diagonalized if and only if it possesses a set of n linearly independent eigenvectors.* Thus the solution of the *diagonalization problem* depends on finding the *eigenvalues* and *eigenvectors* of the matrix. In applications, the eigenvalues are related to frequencies of vibration and stability of mechanical systems.

Real symmetric matrices are always orthogonally diagonalizable, and *hermitian matrices are always unitarily diagonalizable.* To carry out these diagonalizations may require the Gram-Schmidt procedure to construct an orthonormal basis of eigenvectors of the matrix.

As a result of the efforts in the first five chapters, we have presented solutions of the five basic problems of linear algebra:

1. *Solution of linear equations.* Gaussian elimination was the primary method of solution, and attention was given to numerical considerations.
2. *The basis problem.* Using the concepts of linear independence and dependence, we showed how to solve this problem in a constructive, "basis-building" way.
3. *The matrix representation problem.* Linear transformations can be represented by matrices; the solution involved the representation of images under the transformation of basis elements.
4. *The eigenvalue-eigenvector problem.* This problem was solved by using the methods for solving linear equations and determinants. The solution to this problem is critical to success in solving the last basic problem.
5. *The diagonalization problem.* Using the techniques of the solutions of the first four basic problems, we solved this problem which has many applications in science and engineering.

Certainly, a good knowledge of the solution methods for these five problems equips a person to go on to more advanced study of linear algebra. Meanwhile it gives one the tools to handle basic linear algebraic problems and methods in science, engineering, and numerical analysis. As the reader has noticed, the problems and examples of this chapter were constructed so that eigenvalues and eigenvectors were reasonably calculable by hand. In general, this is not the case so numerical methods have been devised to handle these problems. Two of these methods are discussed in the upcoming chapter.

ADDITIONAL PROBLEMS

1. Is the set of all diagonalizable $n \times n$ matrices a subspace of $\mathscr{M}_{nn}$?
2. Let A be an $n \times n$ diagonalizable matrix. Discuss the solvability of $AX_{n \times 1} = B$.

3. Find the characteristic polynomials of

$$\begin{pmatrix} 0 & 1 \\ -a_1 & -a_2 \end{pmatrix} \quad \text{and} \quad \begin{pmatrix} 0 & 1 & 0 \\ 0 & 0 & 1 \\ -a_1 & -a_2 & -a_3 \end{pmatrix}$$

4. Find the characteristic polynomial of

$$\begin{pmatrix} 0 & 1 & 0 & 0 & \cdots & 0 \\ 0 & 0 & 1 & 0 & \cdots & 0 \\ 0 & 0 & 0 & 1 & \cdots & 0 \\ \vdots & \vdots & \vdots & \vdots & \ddots & \vdots \\ 0 & 0 & 0 & 0 & \cdots & 1 \\ -a_1 & -a_2 & -a_3 & -a_4 & \cdots & -a_n \end{pmatrix}$$

5. Let $A_{n \times n}$ be defined by $a_{ij} = 1$ for all i and j. Show that the characteristic equation is $(-\lambda)^{n-1}(n - \lambda) = 0$.

6. Suppose A and B are $n \times n$ matrices and $A = PDP^{-1}$ and $B = PEP^{-1}$, where D and E are diagonal matrices. Compare AB and BA.

7. If a diagonalizable matrix is nilpotent what can you say about its trace?

8. Let

$$A = \begin{pmatrix} -(b + a) & a & b \\ a & -(a + c) & c \\ b & c & -(b + c) \end{pmatrix}$$

Show that A has an eigenvalue of zero, with corresponding eigenvector

$$\begin{pmatrix} 1 \\ 1 \\ 1 \end{pmatrix}$$

What can you say about the other two eigenvalues?

9. A vector $X_{n \times 1}$ is called a *root vector* of order k for the matrix $A_{n \times n}$ if there exists a number λ such that

$$(A - \lambda I)^{k-1} X \neq 0 \qquad \text{and} \qquad (A - \lambda I)^k X = 0$$

Root vectors are useful in studying systems of differential equations. Show that

$$\begin{pmatrix} 2 \\ 0 \\ 1 \end{pmatrix}$$

is a root vector of order 3 for

$$A = \begin{pmatrix} 0 & 1 & 0 \\ 0 & 0 & 1 \\ 0 & 0 & 0 \end{pmatrix}$$

with $\lambda = 0$.

10. Show that

$$\begin{pmatrix} 3 & 1 \\ 2 & 2 \end{pmatrix}$$

has no root vector of order 2.

11. If A is real and diagonalizable with all positive eigenvalues, then does A have a square root? That is, does there exist a real matrix B with $A = B^2$?

12. In mathematical physics, lagrangian equations of motion are sometimes used to study vibrating systems. These equations generally involve two quadratic forms, and so we have the problem of diagonalizing the matrices of both quadratic forms at the same time. If A and B are real symmetric matrices and one of A and B is positive definite, then the matrices can be simultaneously diagonalized. That is, there exists a matrix P such that $P^T AP$ and $P^T BP$ are diagonal. Show by computation of $P^T AP$ and $P^T BP$ that

$$P = \begin{pmatrix} \frac{1}{2\sqrt{2}} & \frac{1}{2\sqrt{2}} \\ \frac{1}{2} & -\frac{1}{2} \end{pmatrix}$$

simultaneously diagonalizes

$$\begin{pmatrix} a & \frac{a}{2} \\ \frac{a}{2} & \frac{a}{2} \end{pmatrix} \quad \text{and} \quad \begin{pmatrix} b & 0 \\ 0 & \frac{b}{2} \end{pmatrix}$$

13. Consider the vibrating system with damper (like a shock absorber) in Fig. AP5.13. The system of differential equations for this system has matrix

Figure AP5.13

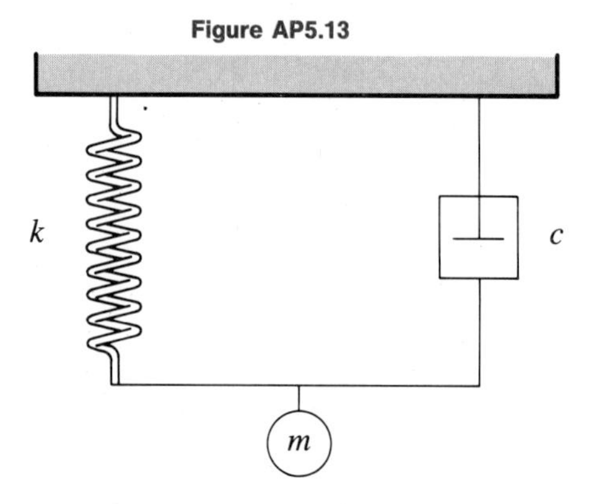

$$\begin{pmatrix} 0 & 1 \\ \dfrac{-c}{m} & \dfrac{-k}{m} \end{pmatrix}$$

What conditions on k, c, and m guarantee complex eigenvalues? This corresponds to the case of underdamping or light damping—the mass oscillates, with amplitude going to zero as time goes on. The imaginary part of the eigenvalue gives the frequency of oscillation.

14. In Prob. 12 of the Additional Problems in Chap. 4 we presented a matrix associated with age distributions in a population. An eigenvector associated with the largest positive eigenvalue (if such an eigenvalue exists) is called a *stable age distribution.* The components of the eigenvector give the relative proportions of the age groups. For example, the matrix

$$A = \begin{pmatrix} 0 & 0 & 6 \\ \frac{1}{2} & 0 & 0 \\ 0 & \frac{1}{3} & 0 \end{pmatrix}$$

represents a species with life span of 3 years (hence the 3×3 matrix) which produces six offspring in the third year (the entry a_{13}) but produces none when younger. Show that A has eigenvector

$$\begin{pmatrix} 6 \\ 3 \\ 1 \end{pmatrix}$$

corresponding to its largest positive eigenvalue. This means that in the stable age distribution, the ratio of the age groups 0 to 1, 1 to 2, and 2 to 3 years is 6:3:1.

15. The circuit in Fig. AP5.15 leads to consideration of a system of differential equations involving the matrix

$$A = \begin{pmatrix} 0 & \dfrac{-R}{L} \\ \dfrac{1}{RC} & \dfrac{-1}{RC} \end{pmatrix}$$

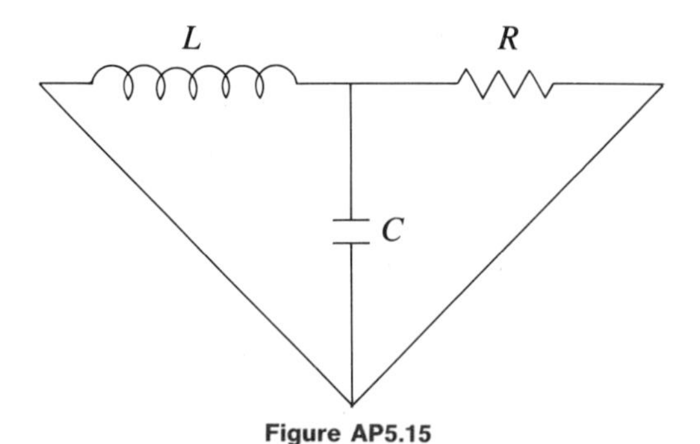

Figure AP5.15

Under what conditions on L, R, and C does A have complex eigenvalues with nonzero imaginary part? In that case, show that the real parts of the eigenvalues are negative. This means that the currents involved all die to zero in an oscillatory fashion.

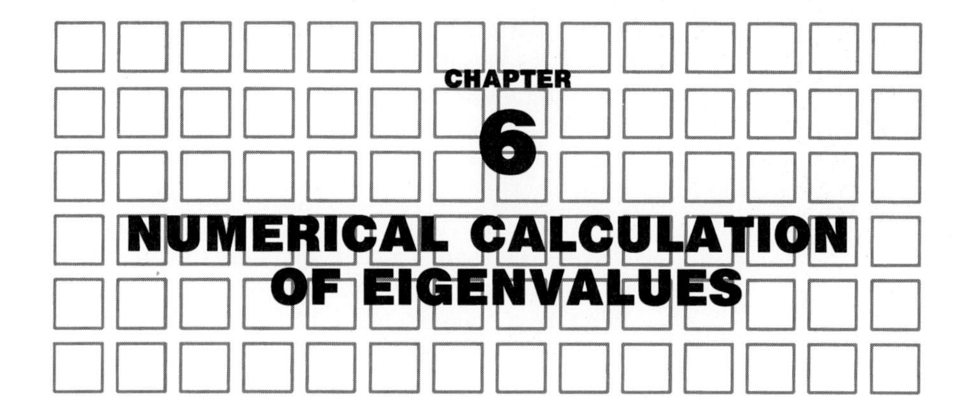

6.1 STABILITY OF THE NUMERICAL EIGENVALUE PROBLEM

In this chapter, this problem is considered.

> Given $A_{n \times n}$ with real entries, find numerical approximations to the eigenvalues and eigenvectors of A.

We outline two methods of solution for this problem in Secs. 6.2 and 6.3. However, there is a warning to be heeded:

> If A and B are almost equal, the eigenvalues of A and B need *not* be almost equal.

Another way of saying this is as follows:

> Small changes in A do *not* necessarily lead to small changes in the eigenvalues of A.

These are important points, since the computer stored form of A may not be exactly equal to A.

To illustrate, consider the matrices

$$A = \begin{pmatrix} 1 & 1000 \\ 0 & 1 \end{pmatrix} \qquad \text{and} \qquad B = \begin{pmatrix} 1 & 1000 \\ 0.001 & 1 \end{pmatrix}$$

Matrix A has eigenvalues 1 and 1, and B has eigenvalues 0 and 2. Even though A differs from B by only 0.001 in the row 2, column 1 entry, the eigenvalues differ by 1. That is, a change of only 0.001 in one entry of A led to a 100 percent change in the eigenvalue.

For a more extreme example, consider matrices

$$Z = \begin{pmatrix} 1 & 1000 \\ 0 & 1 \end{pmatrix} \quad \text{and} \quad C = \begin{pmatrix} 1 & 1000 \\ -0.001 & 1 \end{pmatrix}$$

Matrix C differs from A by only 0.001 in the second row and first column. The eigenvalues of A are 1 and 1, but C has no real eigenvalues at all since the characteristic polynomial is $2 - 2\lambda + \lambda^2$.

What these examples tell us is that when error enters into the calculation of the entries of a matrix A and we later try to calculate the eigenvalues of A, we must view the results carefully. But if A is symmetric, small changes in A will generally not lead to large changes in the eigenvalues. So in applications involving symmetric matrices, numerical methods are generally quite successful in computing the needed eigenvalues.[1]

To make a concrete statement, we define the *Frobenius norm* of a matrix A as

$$\|A\|_F = \sqrt{\sum_{1 \le i, j \le n} |a_{ij}|^2}$$

STABILITY THEOREM Let A be an $n \times n$ matrix, and let E be an $n \times n$ "error matrix." Suppose that A and E are real and symmetric, and set $\hat{A} = A + E$ (that is, $\hat{A}$ is the "error version" of A). Let $\lambda_1, \ldots, \lambda_n$ be the eigenvalues of A and $\hat{\lambda}_1, \ldots, \hat{\lambda}_n$ be the eigenvalues of $\hat{A}$. Then

$$(\lambda_1 - \hat{\lambda}_1)^2 + (\lambda_2 - \hat{\lambda}_2)^2 + \cdots + (\lambda_n - \hat{\lambda}_n)^2 \le \|E\|_F^2$$

STABILITY COROLLARY With the same hypotheses of the stability theorem,

$$|\lambda_k - \hat{\lambda}_k| \le \|E\|_F$$

for $k = 1, 2, \ldots, n$. This means that the process of finding $\lambda_1, \ldots, \lambda_n$ is *stable*: Small errors in A ($\|E\|_F$ small) lead to small errors in determination of $\lambda_1, \ldots, \lambda_n$ ($|\lambda_k - \hat{\lambda}_k|$ small).

We cannot prove the theorem, but the corollary is obtained by noting that by the theorem

$$(\lambda_k - \hat{\lambda}_k)^2 \le (\lambda_1 - \hat{\lambda}_1)^2 + \cdots + (\lambda_k - \hat{\lambda}_k)^2 + \cdots + (\lambda_n - \hat{\lambda}_n)^2 \le \|E\|_F^2$$

Thus

---

[1] There exist certain matrices which can be used to test numerical methods. If a method performs well on these matrices, confidence in the method increases. See, for example, *A Collection of Matrices for Testing Computational Algorithms* by R. Gregory and D. L. Karney. Wiley, NY, 1969.

$$\sqrt{(\lambda_k - \hat{\lambda}_k)^2} \leq \|E\|_F$$

This corollary tells us that if $\|E\|_F$ is less than or equal to a small number ε then all the computed eigenvalues $\hat{\lambda}_k$ of $A + E$ will be within ε of the true eigenvalues λ_k of A. That is, small error in real symmetric A leads to small absolute error in a calculation of the eigenvalues. Our initial example in this section does not involve a symmetric matrix.

EXAMPLE 1 Suppose we need to calculate the eigenvalues of

$$A = \begin{pmatrix} 1 & -1 & 2 \\ -1 & 2 & 7 \\ 2 & 7 & 5 \end{pmatrix}$$

but because of errors in calculation of the entries of A, the computer works with

$$\hat{A} = \begin{pmatrix} 1.01 & -1.05 & 2.1 \\ -1.05 & 1.97 & 7.1 \\ 2.1 & 7.1 & 4.9 \end{pmatrix}$$

What will be the maximum difference in the calculated eigenvalues and the eigenvalues of A?

Solution We have $\hat{A} = A + E$, so $E = \hat{A} - A$. Therefore

$$E = \begin{pmatrix} 0.01 & -0.05 & 0.1 \\ -0.05 & -0.03 & 0.1 \\ 0.1 & 0.1 & -0.1 \end{pmatrix}$$

Now

$$\begin{aligned} \|E\|_F &= \sqrt{(0.01)^2 + 2(-0.05)^2 + 4(0.1)^2 + (0.03)^2 + (-0.1)^2} \\ &= 0.23664 \cdots \end{aligned}$$

so we are guaranteed that

$$|\lambda_k - \hat{\lambda}_k| \leq 0.23664$$

The maximum error is 0.23664.

EXAMPLE 2 Suppose that we want to calculate the eigenvalues of

$$A = \begin{pmatrix} 2 & -1 & 0 \\ -1 & 2 & -1 \\ 0 & -1 & 2 \end{pmatrix}$$

in a computer which keeps only six significant digits. What is a bound for the error in calculating the eigenvalues of A, assuming an error of 0.000001 in each entry of A?

Solution Suppose that

$$E = \begin{pmatrix} \varepsilon & \varepsilon & \varepsilon \\ \varepsilon & \varepsilon & \varepsilon \\ \varepsilon & \varepsilon & \varepsilon \end{pmatrix}$$

where $\varepsilon = 0.000001$. Then by the stability corollary

$$|\lambda - \hat{\lambda}_k| \le \sqrt{9 \cdot (0.000001)^2} = 0.000003$$

Of course, this estimate neglects errors introduced by calculations used in a particular method of computing eigenvalues.

The main point has been made: In general, small errors in A need *not* lead to small errors in computation of $\lambda_1, \ldots, \lambda_n$; however, if A is symmetric, with symmetrically distributed error, then small errors in A *do* lead to small absolute errors in determination of $\lambda_1, \ldots, \lambda_n$.

Finally, even if the absolute error is small

$$|\lambda_k - \hat{\lambda}_k| \le \varepsilon$$

where ε is small, the *relative* error could be large. For instance, if $\lambda_k \doteq \varepsilon/2$, then

$$\frac{|\lambda_k - \hat{\lambda}_k|}{|\lambda_k|} \doteq 200\%$$

PROBLEMS 6.1

1. Consider the matrices

$$H_{10\times 10} = \begin{pmatrix} 1 & & & & \mathbf{0} \\ 1 & 1 & & & \\ 0 & 1 & 1 & & \\ \vdots & \ddots & \ddots & \ddots & \\ 0 & \cdots & 0 & 1 & 1 \end{pmatrix} \qquad \text{and} \qquad H + E$$

where

$$E_{10\times 10} = \begin{pmatrix} 0 & 0 & \cdots & 0 & \frac{1}{2^{10}} \\ 0 & 0 & \cdots & 0 & 0 \\ \cdots & \cdots & \cdots & \cdots & \cdots \\ 0 & 0 & \cdots & 0 & 0 \end{pmatrix}$$

(*a*) What are the eigenvalues of H?

(*b*) Show that $\lambda = \frac{1}{2}$ is an eigenvalue of $H + E$.

(*c*) Show that $\|E\|_F = 1/2^{10}$.

(*d*) Why does the stability corollary not apply to H and $H + E$?

2. Calculate the Frobenius norm for the following matrices.

(*a*)
$$D = \begin{pmatrix} d_1 & & & \mathbf{0} \\ & d_2 & & \\ & & \ddots & \\ \mathbf{0} & & & d_n \end{pmatrix}$$
(*b*) I (*c*) 0 (zero matrix) (*d*) $D + I$

3. Regarding Prob. 2, which is larger?

$$\|D + I\|_F \qquad \text{or} \qquad \|D\|_F + \|I\|_F$$

4. Let A be an $n \times n$ matrix. If $\|A\|_F = 0$, must A be the zero matrix?

5. Let $A = (a_{ij})_{n \times n}$. Define the 1 norm of A by

$$\|A\|_1 = \sum_{1 \le i, j \le n} |a_{ij}|$$

Let

$$A = \begin{pmatrix} 1 & -2 \\ 2 & 0 \end{pmatrix}$$

Calculate $\|A\|_1$ and $\|A\|_F$. Which norm is larger?

6. Suppose the eigenvalues of an $n \times n$ symmetric matrix A are to be computed. Because of a data entry error, every entry of A has 0.0001 added to it. What is the error bound for $|\lambda_k - \hat{\lambda}_k|$, as given in the stability corollary? How does the error bound change as n increases? What can you say about the stability of the eigenvalue problem for large versus small matrices?

6.2 POWER METHOD

The problem we are considering is this: Given an $n \times n$ real matrix A, find numerical approximations to the eigenvalues and eigenvectors of A. This *numerical eigenproblem* is difficult to solve in general. In many applications, A may be symmetric, or tridiagonal or have some other special form or property. Consequently, most numerical methods are designed for special matrices.

The *power method*, the subject of this section, can be used when

1. $A_{n \times n}$ has n linearly independent eigenvectors (6.2.1)

2. The eigenvalues can be ordered in magnitude as

$$|\lambda_1| > |\lambda_2| \ge |\lambda_3| \ge \cdots \ge |\lambda_n| \tag{6.2.2}$$

↑ Note the strict inequality

When this ordering can be done, λ_1 is called the *dominant* eigenvalue of A.

EXAMPLE 1 Let A have eigenvalues 2, 5, 0, -7, and -2. Does A have a dominant eigenvalue? If so, which is dominant?

Solution Since

$$|-7| > |5| > |2| \geq |-2| > |0|$$

A has a dominant eigenvalue of $\lambda_1 = -7$.

Now suppose A satisfies conditions (6.2.1) and (6.2.2). What must be true about A? If X_0 is any vector, it can be written as

$$X_0 = c_1V_1 + c_2V_2 + \cdots + c_nV_n$$

where $\{V_1, \ldots, V_n\}$ is the set of n linearly independent eigenvectors. Thus

$$AX_0 = c_1\lambda_1V_1 + c_2\lambda_2V_2 + \cdots + c_n\lambda_nV_n$$
$$A^2X_0 = c_1\lambda_1{}^2V_1 + c_2\lambda_2{}^2V_2 + \cdots + c_n\lambda_n{}^2V_n$$
$$\cdots\cdots\cdots\cdots\cdots\cdots\cdots\cdots\cdots\cdots\cdots\cdots$$
$$A^mx_0 = c_1\lambda_1{}^mV_1 + c_2\lambda_2{}^mV_2 + \cdots + c_n\lambda_n{}^mV_n$$

And if we divide the last equation by $\lambda_1{}^m$, we have

$$\frac{1}{\lambda_1{}^m}A^mX_0 = c_1V_1 + c_2\left(\frac{\lambda_2}{\lambda_1}\right)^m V_2 + \cdots + \left(\frac{\lambda_n}{\lambda_1}\right)^m c_nV_n \tag{6.2.3}$$

As m gets larger and larger, the terms $(\lambda_2/\lambda_1)^m, \ldots, (\lambda_n/\lambda_1)^m$ all get closer and closer to zero (remember: $|\lambda_2/\lambda_1| < 1, \ldots, |\lambda_n/\lambda_1| < 1$). Therefore, for large m

$$\frac{1}{\lambda_1{}^m}A^mX_0 \doteq c_1V_1 \tag{6.2.4}$$

As long as $c_1 \neq 0$, this last equation can give us an approximation to λ_1. (To guarantee that $c_1 \neq 0$, X_0 should not be orthogonal to V_1.) To obtain the approximation, note that in addition to (6.2.4) we would have, since $m + 1 > m$, that

$$\frac{1}{\lambda_1{}^{m+1}}A^{m+1}X_0 \doteq c_1V_1 \tag{6.2.5}$$

Now taking the dot product of both sides of (6.2.4) and (6.2.5) with any Y which is not orthogonal to V_1, we have

$$\frac{1}{\lambda_1{}^m}(A^mX_0 \cdot Y) \doteq c_1V_1 \cdot Y$$

$$\frac{1}{\lambda_1^{m+1}}(A^{m+1}X_0 \cdot Y) \doteq c_1V_1 \cdot Y$$

and so

$$\frac{1}{\lambda_1^{m+1}}(A^{m+1}X_0 \cdot Y) \doteq \frac{1}{\lambda_1{}^m}(A^m X_0 \cdot Y) \neq 0$$

Finally, by dividing we have

$$\frac{A^{m+1}X_0 \cdot Y}{A^m X_0 \cdot Y} \doteq \frac{\lambda_1^{m+1}}{\lambda_1{}^m} = \lambda_1$$

The powers of A give the power method its name.

Power Method

Suppose that A has n linearly independent eigenvectors and has a dominant eigenvalue λ_1 with eigenvector V_1. Let Y be any vector not orthogonal to V_1. Then for large m, if X_0 is not orthogonal to V_1,

$$\frac{A^{m+1}X_0 \cdot Y}{A^m X_0 \cdot Y} \doteq \lambda_1 \qquad (6.2.6)$$

Clearly the concepts of linear independence and basis were essential to showing that the power method works. Once again, the theory of linear algebra supplies the foundation for a numerical method.

EXAMPLE 2 Use the power method to estimate the largest eigenvalue of

$$A = \begin{pmatrix} 1 & 3 \\ 2 & 2 \end{pmatrix}$$

Solution Let

$$X_0 = \begin{pmatrix} 1 \\ 0 \end{pmatrix}$$

We compute AX_0, A^2X_0, and so on.[2]

$$AX_0 = \begin{pmatrix} 1 \\ 3 \end{pmatrix} \qquad A^4X_0 = A\begin{pmatrix} 34 \\ 36 \end{pmatrix} = \begin{pmatrix} 142 \\ 140 \end{pmatrix}$$

$$A^2X_0 = A\begin{pmatrix} 1 \\ 3 \end{pmatrix} = \begin{pmatrix} 10 \\ 8 \end{pmatrix} \qquad A^5X_0 = A\begin{pmatrix} 142 \\ 140 \end{pmatrix} = \begin{pmatrix} 562 \\ 564 \end{pmatrix}$$

$$A^3X_0 = A\begin{pmatrix} 10 \\ 8 \end{pmatrix} = \begin{pmatrix} 34 \\ 36 \end{pmatrix} \qquad A^6X_0 = A\begin{pmatrix} 562 \\ 564 \end{pmatrix} = \begin{pmatrix} 2254 \\ 2252 \end{pmatrix}$$

[2] Note that the components of the computed vectors are getting larger. To overcome this, one can modify the power method to the power method with scaling, which is outlined in Prob. 6.

Now let

$$Y = \begin{pmatrix} 1 \\ 0 \end{pmatrix}$$

In Eq. (6.2.6) with $m = 5$:

$$\lambda_1 \doteq \frac{A^6 X_0 \cdot Y}{A^5 X_0 \cdot Y} = \frac{\begin{pmatrix} 2254 \\ 2252 \end{pmatrix} \cdot \begin{pmatrix} 1 \\ 0 \end{pmatrix}}{\begin{pmatrix} 562 \\ 564 \end{pmatrix} \cdot \begin{pmatrix} 1 \\ 0 \end{pmatrix}} = \frac{2254}{562} = 4.0106 \cdots$$

Checking by the methods of Chap. 5, we find

$$\left(4, \begin{pmatrix} 1 \\ 1 \end{pmatrix}\right)$$

is the dominant eigenpair; the approximation to λ_1 has absolute error $=$ $0.0106 \cdots$, which is less than 0.3 percent relative error.

In Example 2 the vectors $A^m X_0$ appear to be almost parallel to the eigenvector which corresponds to λ_1. In fact, this will always be the case. To see this, first we note that for any eigenpair (λ, V) of A we have

$$AV = \lambda V$$

So if we take the dot product of both sides with V, we see that

$$\frac{AV \cdot V}{V \cdot V} = \lambda \tag{6.2.7}$$

We can choose Y in Eq. (6.2.6) as

$$Y = A^m X_0$$

to find

$$\lambda_1 \doteq \frac{A^{m+1} X_0 \cdot A^m X_0}{A^m X_0 \cdot A^m X_0} \doteq \frac{A(A^m X_0) \cdot A^m X_0}{A^m X_0 \cdot A^m X_0}$$

The last expression matches Eq. (6.2.7) with $V = A^m X_0$. Therefore, it is reasonable to believe that *$A^m X_0$ is (approximately) an eigenvector corresponding to λ_1.*

EXAMPLE 3 Use the power method to calculate an approximation to the dominant eigenpair for

$$A = \begin{pmatrix} -7 & 2 \\ 8 & -1 \end{pmatrix}$$

Solution Let

$$X_0 = \begin{pmatrix} 1 \\ 0 \end{pmatrix}$$

Then

$$A(X_0) = \begin{pmatrix} -7 \\ 8 \end{pmatrix}$$

$$A^2(X_0) = \begin{pmatrix} 65 \\ -64 \end{pmatrix}$$

$$A^3(X_0) = \begin{pmatrix} -583 \\ 584 \end{pmatrix}$$

We stop here because the vectors $A^m X_0$ already appear to be approaching a multiple of

$$\begin{pmatrix} -1 \\ 1 \end{pmatrix}$$

Now

$$\lambda_1 \doteq \frac{A^3 X_0 \cdot A^2 X_0}{A^2 X_0 \cdot A^2 X_0} = \frac{-583 \cdot 65 - 584 \cdot 64}{65 \cdot 65 + 64 \cdot 64} = -9.05$$

So a dominant eigenpair is (approximately)

$$\left(-9.05, \frac{1}{583}\begin{pmatrix} -583 \\ 584 \end{pmatrix}\right) = \left(-9.05, \begin{pmatrix} -1 \\ 1.002 \end{pmatrix}\right)$$

Since the actual dominant eigenpair is

$$\left(-9, \begin{pmatrix} -1 \\ 1 \end{pmatrix}\right)$$

the method has worked well in this example.

Although the power method has worked well in these examples, we must say something about cases in which the power method may fail. There are basically three such cases:

1. Using the power method when A is not diagonalizable. Recall that A has n linearly independent eigenvectors *if and only if* A is diagonalizable. Of course, it is not easy to tell by just looking at A whether it is diagonalizable.

2. Using the power method when A does not have a dominant eigenvalue, or when the dominant eigenvalue is such that

$$|\lambda_1| > |\lambda_2| \qquad \text{but} \qquad |\lambda_1| \doteq |\lambda_2|$$

(Then $|\lambda_1/\lambda_2|$ is barely less than 1, and high powers of $|\lambda_1/\lambda_2|$ do not tend to zero quickly.) Again, it is not easy to determine whether A has this defect by just looking at A.

3. If the entries of A contain significant error. Powers A^m of A will have significant roundoff error in their entries.

Here is a rule of thumb for using the power method:

1. Try it, and if the numbers

$$\frac{A^{m+1}X_0 \cdot A^m X_0}{A^m X_0 \cdot A^m X_0}$$

approach a single number λ_1, then stop and go to step 2.

2. Check whether $(\lambda_1, A^m X_0)$ is an eigenpair by checking whether

$$A(A^m X_0) \doteq \lambda_1(A^m X_0)$$

3. If step 2 checks, accept

$$(\lambda_1, A^m X_0)$$

as a dominant eigenpair.

EXAMPLE 4 Check the answer to Example 3.

Solution The proposed eigenpair is

$$\left(-9.05, \begin{pmatrix} -1 \\ 1.005 \end{pmatrix}\right)$$

To check, we calculate

$$A\begin{pmatrix} -1 \\ 1.005 \end{pmatrix} \quad \text{and} \quad -9.05\begin{pmatrix} -1 \\ 1.005 \end{pmatrix}$$

We have

$$A\begin{pmatrix} -1 \\ 1.005 \end{pmatrix} = \begin{pmatrix} -7 & 2 \\ 8 & -1 \end{pmatrix}\begin{pmatrix} -1 \\ 1.005 \end{pmatrix} = \begin{pmatrix} 9.01 \\ -9.005 \end{pmatrix}$$

and

$$-9.05\begin{pmatrix} -1 \\ 1.005 \end{pmatrix} = \begin{pmatrix} 9.05 \\ -9.09525 \end{pmatrix}$$

Thus

$$A\begin{pmatrix} -1 \\ 1.005 \end{pmatrix} - (-9.05)\begin{pmatrix} -1 \\ 1.005 \end{pmatrix} = \begin{pmatrix} -0.04 \\ 0.09025 \end{pmatrix} \doteq \begin{pmatrix} 0 \\ 0 \end{pmatrix}$$

so

$$A\begin{pmatrix}-1\\1.005\end{pmatrix} \doteq -9.05\begin{pmatrix}-1\\1.005\end{pmatrix}$$

and the answer checks.

To illustrate possible failure of the power method, we show an example.

EXAMPLE 5 Try the power method on

$$A = \begin{pmatrix}1 & 1\\0 & -1\end{pmatrix}$$

with

$$X_0 = \begin{pmatrix}1\\1\end{pmatrix} \quad \text{and} \quad X_0 = \begin{pmatrix}1\\-1\end{pmatrix}$$

Explain the results.

Solution Let

$$X_0 = \begin{pmatrix}1\\1\end{pmatrix}$$

Then we have

$$AX_0 = \begin{pmatrix}2\\-1\end{pmatrix}$$

$$A^2X_0 = A\begin{pmatrix}2\\-1\end{pmatrix} = \begin{pmatrix}1\\1\end{pmatrix}$$

$$A^3X_0 = A\begin{pmatrix}1\\1\end{pmatrix} = \begin{pmatrix}2\\-1\end{pmatrix}$$

$$A^4X_0 = A\begin{pmatrix}2\\-1\end{pmatrix} = \begin{pmatrix}1\\1\end{pmatrix}$$

$$\vdots$$

We see that

$$A^{2n}X_0 = \begin{pmatrix}1\\1\end{pmatrix} \quad \text{and} \quad A^{2n+1}X_0 = \begin{pmatrix}2\\-1\end{pmatrix}$$

so that A^mX_0 is not becoming parallel to any vector. Also

$$\frac{A^{2n+1}X_0 \cdot A^{2n}X_0}{A^{2n}X_0 \cdot A^{2n}X_0} = \frac{\begin{pmatrix}2\\-1\end{pmatrix}\cdot\begin{pmatrix}1\\1\end{pmatrix}}{\begin{pmatrix}1\\1\end{pmatrix}\cdot\begin{pmatrix}1\\1\end{pmatrix}} = \frac{1}{2}$$

and

$$\frac{A^{2n}X_0 \cdot A^{2n-1}X_0}{A^{2n-1}X_0 \cdot A^{2n-1}X_0} = \frac{\begin{pmatrix}1\\1\end{pmatrix} \cdot \begin{pmatrix}2\\-1\end{pmatrix}}{\begin{pmatrix}2\\-1\end{pmatrix} \cdot \begin{pmatrix}2\\-1\end{pmatrix}} = \frac{1}{5}$$

So we have no approximation to the eigenvalues. The power method has failed when

$$X_0 = \begin{pmatrix}1\\1\end{pmatrix}$$

When

$$X_0 = \begin{pmatrix}1\\-1\end{pmatrix}$$

We find

$$AX_0 = \begin{pmatrix}0\\1\end{pmatrix}$$

$$A^2X_0 = A\begin{pmatrix}0\\1\end{pmatrix} = \begin{pmatrix}1\\-1\end{pmatrix}$$

$$A^3X_0 = A\begin{pmatrix}1\\-1\end{pmatrix} = \begin{pmatrix}0\\1\end{pmatrix}$$

$$A^4X_0 = A\begin{pmatrix}0\\1\end{pmatrix} = \begin{pmatrix}1\\-1\end{pmatrix}$$

$$\vdots$$

and for the same reasons as before, the power method fails. Also in this case the approximations to λ_1 oscillate between $-\frac{1}{2}$ and -1.

An explanation for the failure is that A has no dominant eigenvalue. In fact, the eigenvalues are 1 and -1, which have the same magnitude.

When to Stop in the Power Method We would like to have a rule about when to stop in using the power method. Usually we would stop when $|\lambda_1^{\text{calc}} - \lambda_1^{\text{actual}}|$ is small. However, in most realistic problems we do not know $\lambda_1^{\text{actual}}$, so we can only estimate $|\lambda_1^{\text{calc}} - \lambda_1^{\text{actual}}|$. This can be done for symmetric A.

THEOREM 6.2.1 Let A be real symmetric with dominant eigenvalue λ_1. Then if $\lambda_1^{\text{calc}} = (AX \cdot X)/(X \cdot X)$, where $X = A^m X_0$ as in the power method, we have

$$|\lambda_1^{\text{calc}} - \lambda_1^{\text{actual}}| \leq \sqrt{\frac{AX \cdot AX}{X \cdot X} - (\lambda_1^{\text{calc}})^2}$$

EXAMPLE 6 Apply the power method to the symmetric matrix

$$A = \begin{pmatrix} 5 & -2 \\ -2 & 8 \end{pmatrix} \quad \text{with} \quad X_0 = \begin{pmatrix} 1 \\ 1 \end{pmatrix}$$

Estimate the error in using

$$\frac{A^6X_0 \cdot A^5X_0}{A^5X_0 \cdot A^5X_0}$$

for λ. (Eigenvalues are 4 and 9.)

Solution With

$$X_0 = \begin{pmatrix} 1 \\ 1 \end{pmatrix}$$

we find

$$AX_0 = \begin{pmatrix} 3 \\ 6 \end{pmatrix}$$

$$A^2X_0 = A\begin{pmatrix} 3 \\ 6 \end{pmatrix} = \begin{pmatrix} 3 \\ 42 \end{pmatrix}$$

$$A^3X_0 = A\begin{pmatrix} 3 \\ 42 \end{pmatrix} = \begin{pmatrix} -69 \\ 330 \end{pmatrix}$$

$$A^4X_0 = A\begin{pmatrix} -69 \\ 330 \end{pmatrix} = \begin{pmatrix} -1005 \\ 2778 \end{pmatrix}$$

$$A^5X_0 = A\begin{pmatrix} -1005 \\ 2778 \end{pmatrix} = \begin{pmatrix} -10{,}581 \\ 24{,}234 \end{pmatrix}$$

$$A^6X_0 = A\begin{pmatrix} -10{,}581 \\ 24{,}234 \end{pmatrix} = \begin{pmatrix} -101{,}373 \\ 215{,}034 \end{pmatrix} = \frac{1}{101{,}373}\begin{pmatrix} -1 \\ 2.1212157 \end{pmatrix}$$

Now using A^6X_0, we have

$$\lambda_1 \doteq \frac{A^6X_0 \cdot A^5X_0}{A^5X_0 \cdot A^5X_0} \doteq 8.9865$$

The error estimate is

$$\begin{aligned} |8.9865 - \lambda_1^{\text{actual}}| &\leq \sqrt{\frac{A^6X_0 \cdot A^6X_0}{A^5X_0 \cdot A^5X_0} - (8.9865)^2} \\ &\doteq \sqrt{\frac{565.16106 \times 10^8}{6.9924435 \times 10^8} - (8.9865)^2} \\ &= \sqrt{80.824544 - 80.757182} \\ &= \sqrt{0.067362} \\ &\doteq 0.26 \end{aligned}$$

From the error estimate we know our error is at most 0.26. In practice, we would not stop here because the percentage error could be

$$\frac{\text{Maximum absolute error}}{\text{Calculated value}} = \frac{0.26}{8.9865} \doteq 2.9\%$$

which is fairly large. Ordinarily, we would continue calculations until the error estimate were smaller than some preset tolerance.

We should make two observations concerning Example 6. First, the error estimate is *pessimistic*. The estimate is 0.26, but 8.9865 is only 0.0135 from the actual dominant eigenvalue of 9. That is, the approximation is much closer to the actual eigenvalue than the error estimate predicts. Because of this, another check for accuracy (although not totally reliable) is to calculate the relative error

$$E_{n+1} = \frac{\left|\lambda_1^{\text{calc at step } n} - \lambda_1^{\text{calc at step } n+1}\right|}{\left|\lambda_1^{\text{calc at step } n+1}\right|} \tag{6.2.8}$$

and stop if E is small. In fact, this second check must be used for nonsymmetric A since the estimate in Theorem 6.2.1 is for symmetric matrices only.

Second, the components of the approximate eigenvectors get quite large. To overcome this problem, we can scale the approximate eigenvectors at each stage by multiplying the approximate eigenvector by the reciprocal of the largest (in absolute value) component of the approximate eigenvector, then use the scaled vector in the next step (see Prob. 6). If we repeat Example 6 with this scaling process, we obtain

$$X_0 = \begin{pmatrix} 1 \\ 1 \end{pmatrix}$$

Step 1: $AX_0 = \begin{pmatrix} 3 \\ 6 \end{pmatrix} \xrightarrow{\text{Scale}} \begin{pmatrix} \frac{1}{2} \\ 1 \end{pmatrix} = W_1$

Step 2: $AW_1 = \begin{pmatrix} 0.5 \\ 7 \end{pmatrix} \xrightarrow{\text{Scale}} \begin{pmatrix} \frac{1}{14} \\ 1 \end{pmatrix} = W_2$

Step 3: $AW_2 = \begin{pmatrix} -1.64\cdots \\ 7.85\cdots \end{pmatrix} \xrightarrow{\text{Scale}} \begin{pmatrix} -0.209\cdots \\ 1 \end{pmatrix} = W_3$

$\vdots$

Step 10: $\xrightarrow{\text{Scale}} \begin{pmatrix} -0.4988\cdots \\ 1 \end{pmatrix} = W_{10}$

Step 11: $AW_{10} = \begin{pmatrix} -4.49\cdots \\ 8.997\cdots \end{pmatrix} \quad \begin{pmatrix} -0.4994\cdots \\ 1 \end{pmatrix}$

The approximate eigenvalue is (see Prob. 6)

$$\frac{AW_{10} \cdot W_{10}}{W_{10} \cdot W_{10}} \doteq 9.002$$

The approximate eigenvector as calculated above is close to the actual (any multiple of)

$$\begin{pmatrix} -0.5 \\ 1 \end{pmatrix}$$

EXAMPLE 7 Calculate E_{n+1} [from Eq. (6.2.8)] in Example 2, for $n = 3, 4, 5$.

Solution We have

$$\lambda_1^{\text{step }3} = 3.829268\cdots \qquad \lambda_1^{\text{step }5} = 3.9926566\cdots$$
$$\lambda_1^{\text{step }4} = 4.0244698\cdots \qquad \lambda_1^{\text{step }6} = 4.0017604\cdots$$

Therefore,

$$E_4 = \frac{|\lambda_1^{\text{step }3} - \lambda_1^{\text{step }4}|}{|\lambda_1^{\text{step }4}|} \doteq 0.04858037 \doteq 4.9\%$$

$$E_5 = \frac{|\lambda_1^{\text{step }4} - \lambda_1^{\text{step }5}|}{|\lambda_1^{\text{step }5}|} \doteq 0.0079679 \doteq 0.8\%$$

$$E_6 = \frac{|\lambda_1^{\text{step }5} - \lambda_1^{\text{step }6}|}{|\lambda_1^{\text{step }6}|} \doteq 0.0022749 \doteq 0.23\%$$

If we had decided to accept $\lambda_1^{\text{step }n}$ as the approximate eigenvalue when $E_{n+1} < 0.005$, then we would accept 4.0017604 as λ_1. But if we had initially decided to accept $\lambda_1^{\text{step }n}$ when $E_{n+1} < 0.01$, then we would accept 3.9926566.

Finding Nondominant Eigenvalues Once the dominant eigenpair (λ_1, V_1) of A is computed, we may wish to compute λ_2. (Recall that $|\lambda_1| > |\lambda_2| \geq |\lambda_3| \geq \cdots \geq |\lambda_n|$.) If A is symmetric, it can be proved that if $U_1 = V_1/|V_1|$, then

$$\mathscr{A} = A - \lambda_1 U_1 U_1^T$$

has eigenvalues $0, \lambda_2, \lambda_3, \ldots, \lambda_n$ and the eigenvectors of $\mathscr{A}$ are eigenvectors of A. Therefore, to find λ_2, we could apply the power method to $\mathscr{A}$. However, a *warning* is in order. Since λ_1 is not exact, some error will be introduced in the power method applied to $\mathscr{A}$. The application of the power method to $\mathscr{A}$ to find λ_2 is called the *method of deflation.*

EXAMPLE 8 Apply the method of deflation to the matrix of Example 6 to find λ_2. Assume that $\lambda_1 = 9$ and

$$V_1 = \begin{pmatrix} 1 \\ -2 \end{pmatrix}$$

Solution We form $\mathscr{A}$:

$$\mathscr{A} = \begin{pmatrix} 5 & -2 \\ -2 & 8 \end{pmatrix} - 9 \begin{pmatrix} \frac{1}{\sqrt{5}} \\ \frac{-2}{\sqrt{5}} \end{pmatrix} \begin{pmatrix} \frac{1}{\sqrt{5}} & \frac{-2}{\sqrt{5}} \end{pmatrix}$$

$$= \begin{pmatrix} \frac{16}{5} & \frac{8}{5} \\ \frac{8}{5} & \frac{4}{5} \end{pmatrix} = \frac{4}{5} \begin{pmatrix} 4 & 2 \\ 2 & 1 \end{pmatrix}$$

Now, applying the power method to $\mathscr{A}$ with

$$X_0 = \begin{pmatrix} 1 \\ 1 \end{pmatrix}$$

we have

$$\mathscr{A}X_0 = \frac{4}{5}\begin{pmatrix} 6 \\ 3 \end{pmatrix}$$

$$\mathscr{A}^2X_0 = \frac{4}{5}\mathscr{A}\begin{pmatrix} 6 \\ 3 \end{pmatrix} = \frac{16}{25}\begin{pmatrix} 30 \\ 15 \end{pmatrix}$$

$$\mathscr{A}^3X_0 = \frac{16}{25}\mathscr{A}\begin{pmatrix} 30 \\ 15 \end{pmatrix} = \frac{64}{125}\begin{pmatrix} 150 \\ 75 \end{pmatrix}$$

Clearly, the vectors generated by the power method are all multiples of

$$\begin{pmatrix} 2 \\ 1 \end{pmatrix}$$

Also

$$\lambda_2^{\text{step 2}} = \frac{\frac{16}{25}\begin{pmatrix} 30 \\ 15 \end{pmatrix} \cdot \frac{4}{5}\begin{pmatrix} 6 \\ 3 \end{pmatrix}}{\frac{4}{5}\begin{pmatrix} 6 \\ 3 \end{pmatrix} \cdot \frac{4}{5}\begin{pmatrix} 6 \\ 3 \end{pmatrix}} = 4$$

$$\lambda_2^{\text{step 3}} = \frac{\frac{64}{125}\begin{pmatrix} 150 \\ 75 \end{pmatrix} \cdot \frac{16}{25}\begin{pmatrix} 30 \\ 15 \end{pmatrix}}{\frac{16}{25}\begin{pmatrix} 30 \\ 15 \end{pmatrix} \cdot \frac{16}{25}\begin{pmatrix} 30 \\ 15 \end{pmatrix}} = 4$$

Therefore, the first (and only, since A is 2×2) nondominant eigenvalue for A is 4, with eigenvector

$$\begin{pmatrix} 2 \\ 1 \end{pmatrix}$$

In fact, this is exactly correct.

EXAMPLE 9 Rework Example 8 with λ_1 and V_1 *as calculated in Example 6.* Compare the results with Example 8.

Solution We have $\lambda_1 = 8.987$ and

$$V_1 = \begin{pmatrix} -1 \\ 2.12 \end{pmatrix}$$

Now

$$U_1 = \begin{pmatrix} -1 \\ 2.12 \end{pmatrix} \Big/ \left|\begin{pmatrix} -1 \\ 2.12 \end{pmatrix}\right| = \frac{1}{2.344}\begin{pmatrix} -1 \\ 2.12 \end{pmatrix} = \begin{pmatrix} -0.427 \\ 0.904 \end{pmatrix}$$

so

$$\begin{aligned} \mathscr{A} &= \begin{pmatrix} 5 & -2 \\ -2 & 8 \end{pmatrix} - 8.987\begin{pmatrix} -0.427 \\ 0.904 \end{pmatrix}(-0.427 \quad 0.904) \\ &= \begin{pmatrix} 5 & -2 \\ -2 & 8 \end{pmatrix} - 8.987\begin{pmatrix} 0.182 & -0.386 \\ -0.386 & 0.817 \end{pmatrix} \\ &= \begin{pmatrix} 5 & -2 \\ -2 & 8 \end{pmatrix} - \begin{pmatrix} 1.636 & -3.469 \\ -3.469 & 7.342 \end{pmatrix} = \begin{pmatrix} 3.364 & 1.469 \\ 1.469 & 0.658 \end{pmatrix} \end{aligned}$$

Now we apply the power method with

$$X_0 = \begin{pmatrix} 1 \\ 1 \end{pmatrix}$$

to find

$$\begin{aligned} \mathscr{A}X_0 &= \begin{pmatrix} 4.833 \\ 2.127 \end{pmatrix} \\ \mathscr{A}^2X_0 &= \mathscr{A}\begin{pmatrix} 4.833 \\ 2.127 \end{pmatrix} = \begin{pmatrix} 19.383 \\ 8.5 \end{pmatrix} = \frac{1}{8.5}\begin{pmatrix} 2.28 \\ 1 \end{pmatrix} \\ \mathscr{A}^3X_0 &= \mathscr{A}\begin{pmatrix} 19.383 \\ 8.5 \end{pmatrix} = \begin{pmatrix} 77.691 \\ 34.066 \end{pmatrix} = \frac{1}{34.066}\begin{pmatrix} 2.28 \\ 1 \end{pmatrix} \\ \mathscr{A}^4X_0 &= \mathscr{A}\begin{pmatrix} 77.691 \\ 34.066 \end{pmatrix} = \begin{pmatrix} 311.4 \\ 136.54 \end{pmatrix} = \frac{1}{136.54}\begin{pmatrix} 2.28 \\ 1 \end{pmatrix} \end{aligned}$$

Since the vectors generated seem to be fixed at multiples of

$$\begin{pmatrix} 2.28 \\ 1 \end{pmatrix}$$

we know that

$$\lambda_2^{\text{step 3}} = \frac{\mathscr{A}^3X_0 \cdot \mathscr{A}^2X_0}{\mathscr{A}^2X_0 \cdot \mathscr{A}^2X_0} = 4.00813$$

$$\lambda_2^{\text{step 4}} = \frac{\mathscr{A}^4 X_0 \cdot \mathscr{A}^3 X_0}{\mathscr{A}^3 X_0 \cdot \mathscr{A}^3 X_0} = 4.00817$$

The relative error in the eigenvalue is very small, and we will accept

$$\left(4.008, \begin{pmatrix} 2.28 \\ 1 \end{pmatrix}\right)$$

as the first nondominant eigenpair of A. However, the eigenvector is inaccurate because

$$\begin{pmatrix} 2 \\ 1 \end{pmatrix}$$

is correct. The results are less accurate than in Example 8 because we started with inaccurate λ_1 and V_1.

EXAMPLE 10 Rework Example 9, using $\lambda_1 = 9.002$ and

$$V_1 = \begin{pmatrix} -0.4994 \\ 1 \end{pmatrix}$$

as found in the calculations following Example 6.

Solution This time,

$$U_1 = \begin{pmatrix} -0.4467 \\ 0.8946 \end{pmatrix} \quad \text{and} \quad \mathscr{A} = \begin{pmatrix} 3.203 & 1.597 \\ 1.597 & 0.795 \end{pmatrix}$$

After four steps, beginning with

$$\begin{pmatrix} 1 \\ 1 \end{pmatrix}$$

we find $\lambda_2 \doteq 3.997$ and

$$V_2 \doteq \begin{pmatrix} 1 \\ 0.498 \end{pmatrix}$$

The calculated approximations are more accurate this time because we started with more accurate values for λ_1 and V_1.

For an $n \times n$ matrix (symmetric), to find λ_3 by deflation, we proceed as follows. After finding (λ_2, V_2), we form

$$\mathscr{B} = \mathscr{A} - \lambda_2 U_2 U_2^T$$

where $U_2 = V_2/|V_2|$. Then $\mathscr{B}$ will have eigenvalues $0, 0, \lambda_3, \lambda_4, \ldots, \lambda_n$, and the eigenvalues of $\mathscr{B}$ will be eigenvectors of A. The power method applied to $\mathscr{B}$ will yield λ_3 as the dominant eigenvalue of $\mathscr{B}$. To find $\lambda_4, \lambda_5, \ldots$, continue the procedure.

In general, the deflation method becomes more inaccurate as we calculate more eigenvalues, because error is introduced in each eigenvalue and eigenvector and this error accumulates as the process continues. Luckily, in many applications only the dominant eigenvalue of A is needed.

PROBLEMS 6.2 In Probs. 1 to 5, use the power method to calculate approximations to the dominant eigenpair (if a dominant eigenpair exists). If the method does not work, give a reason.

1. $\begin{pmatrix} 1 & 5 \\ 5 & 6 \end{pmatrix}$ **2.** $\begin{pmatrix} 3 & 4 & 0 \\ 1 & 3 & 0 \\ 0 & 0 & 2 \end{pmatrix}$ **3.** $\begin{pmatrix} 2 & 3 \\ -2 & 1 \end{pmatrix}$ **4.** $\begin{pmatrix} 3 & 3 \\ 3 & 5 \end{pmatrix}$ **5.** $\begin{pmatrix} 3 & 3 & 0 \\ 3 & 5 & 0 \\ 0 & 0 & 6 \end{pmatrix}$

6. *The power method with scaling.* From the examples in this chapter we saw vectors with large components generated by the power method. To avoid this problem, we can at each step multiply the vector

$$X = \begin{pmatrix} x_1 \\ \vdots \\ x_n \end{pmatrix} \quad \text{by} \quad \frac{1}{\max\{|x_1|, |x_2|, \ldots, |x_n|\}}.$$

This is called the scaling of X. For example, the scaling of

$$\begin{pmatrix} 7 \\ 5 \end{pmatrix} \quad \text{is} \quad \frac{1}{7}\begin{pmatrix} 7 \\ 5 \end{pmatrix} = \begin{pmatrix} 1 \\ \frac{5}{7} \end{pmatrix}$$

and the scaling of

$$\begin{pmatrix} 3 \\ -6 \end{pmatrix} \quad \text{is} \quad \frac{1}{6}\begin{pmatrix} 3 \\ -6 \end{pmatrix} = \begin{pmatrix} \frac{1}{2} \\ -1 \end{pmatrix}$$

The power method with scaling proceeds as follows: Choose X_0.

Step 1. Calculate AX_0. Let V_1 = scaled version of AX_0.
Step 2. Calculate AV_1. Let V_2 = scaled version of AX_0.
Step 3. Calculate AV_2. Let V_3 = scaled version of AX_0.

Continue in this way. We then have at step m

$$\lambda_1 \doteq \frac{AV_{m-1} \cdot V_{m-1}}{V_{m-1} \cdot V_{m-1}}$$

and V_m is an approximate eigenvector.

Use the power method with scaling on Probs. 1, 2, and 5.

7. If a matrix has complex eigenvalues, can the power method as described in this section work?

8. Use the relative error E_{n+1} from Eq. (6.2.8) to estimate the error in the computed dominant eigenvalue in Probs. 1, 2, and 5.

9. Use deflation to fluid the nondominant eigenvalues for Probs. 1, 2, and 5.

6.3 QR METHOD

The basis of the QR method for calculating the eigenvalues of A is the fact that an $n \times n$ real matrix can be written as

$$A = QR \qquad (QR \text{ factorization of } A)$$

where Q is orthogonal and R is upper triangular. The method is efficient for the calculation of all the eigenvalues of a matrix.

The construction of Q and R proceeds as follows. Matrices $P_1, P_2, \ldots, P_{n-1}$ are constructed so that $P_{n-1}P_{n-2} \cdots P_2P_1A = R$ is upper triangular. These matrices can be chosen as orthogonal matrices and are called *householder matrices.* Since the P's are orthogonal, the stability of the eigenvalue problem will not be worsened (this is proved in numerical analysis texts). If we let

$$Q^T = P_{n-1}P_{n-2} \cdots P_2P_1$$

then we have $Q^TA = R$ and

$$QQ^TA = QR$$
$$IA = QR$$
$$A = QR$$

We discuss the construction of the P's presently. First, we state how the QR factorization of A is used to find eigenvalues of A. We define sequences of matrices $A_1, A_2, \ldots, A_m, \ldots, Q_1, Q_2, \ldots, Q_m, \ldots,$ and $R_1, R_2, \ldots, R_m, \ldots$ by this process:

Step 1. Set $A_1 = A$, $Q_1 = Q$, and $R_1 = R$.

Step 2. First set $A_2 = R_1Q_1$; then factor A_2 as $A_2 = Q_2R_2$ (QR factorization of A_2).

Step 3. First, set $A_3 = R_2Q_2$; then factor A_3 as $A_3 = Q_3R_3$ (QR factorization of A_3).

Step *m*. Set $A_m = R_{m-1}Q_{m-1}$; then factor A_m as $A_m = Q_mR_m$ (QR factorization of A_m).

At the kth step, a matrix A_k is found, first, by using Q_{k-1} and R_{k-1} from the previous step; second, A_k is factored into Q_kR_k. Thus a QR factorization takes place at each step. Matrix A_m will tend toward a triangular or nearly triangular form. Thus the eigenvalues of A_m will be easy to calculate. The importance is that if the eigenvalues can be ordered as $|\lambda_1| > |\lambda_2| > \cdots > |\lambda_n| > 0$, then the following is true:

> As m increases, the eigenvalues of A_m approach the eigenvalues of A.

The proof of this fact is well beyond the scope of this book. Before applying the QR algorithm to some examples, we discuss the QR factorization of a matrix A.

The idea in QR factorization is to first find P_1 which, when multiplied on the left of A, will produce zeros below a_{11}. That is, we want

$$P_1 \begin{pmatrix} a_{11} & a_{12} & \cdots & a_{1n} \\ a_{21} & a_{22} & \cdots & a_{2n} \\ \cdots & \cdots & \cdots & \cdots \\ a_{n1} & a_{n2} & \cdots & a_{nn} \end{pmatrix} = \begin{pmatrix} \tilde{a}_{11} & \tilde{a}_{12} & \cdots & \tilde{a}_{1n} \\ 0 & \tilde{a}_{22} & \cdots & \tilde{a}_{2n} \\ \cdots & \cdots & \cdots & \cdots \\ 0 & \tilde{a}_{n2} & \cdots & \tilde{a}_{nn} \end{pmatrix}$$

After this is done, we find P_2 which will produce

$$P_2P_1A = P_2 \begin{pmatrix} \tilde{a}_{11} & \tilde{a}_{12} & \cdots & \tilde{a}_{1n} \\ \mathbf{0} & \tilde{a}_{22} & \cdots & \tilde{a}_{2n} \\ \cdots & \cdots & \cdots & \cdots \\ \mathbf{0} & \tilde{a}_{n2} & \cdots & \tilde{a}_{nn} \end{pmatrix} = \begin{pmatrix} \hat{a}_{11} & \hat{a}_{12} & \hat{a}_{13} & \cdots & \hat{a}_{1n} \\ \mathbf{0} & \hat{a}_{22} & \hat{a}_{23} & \cdots & \hat{a}_{2n} \\ \mathbf{0} & \mathbf{0} & \hat{a}_{33} & \cdots & \hat{a}_{3n} \\ \cdots & \cdots & \cdots & \cdots & \cdots \\ \mathbf{0} & \mathbf{0} & \hat{a}_{n3} & \cdots & \hat{a}_{nn} \end{pmatrix}$$

The process is continued until we have

$$P_{n-1}P_{n-2}\cdots P_2P_1A = R = \begin{pmatrix} \diagdown \\ \mathbf{0\text{'s}} \end{pmatrix}$$

The problem is to find the P matrices. It turns out that the matrices P_k can be chosen as orthogonal matrices. In fact, the construction proceeds as follows.

To construct P_k:

1. Pull column k out of the matrix $P_{k-1}P_{k-2}\cdots P_2P_1A$ (just A if $k = 0$):

$$\begin{pmatrix} a_{1k} \\ a_{2k} \\ \vdots \\ a_{kk} \\ a_{k+1,k} \\ \vdots \\ a_{nk} \end{pmatrix}$$

2. Normalize this column vector, and call the new vector

$$\begin{pmatrix} d_1 \\ d_2 \\ \vdots \\ d_{k-1} \\ d_k \\ \vdots \\ d_n \end{pmatrix}$$

3. Set $D = \pm\sqrt{d_k{}^2 + \cdots + d_n{}^2}$ (choose $+$ if $d_k \le 0$).

4. Set $v_1 = v_2 = \cdots = v_{k-1} = 0$. Also set

$$v_k = \sqrt{\frac{1}{2}\left(1 - \frac{d_k}{D}\right)} \qquad p = -Dv_k$$

and

$$v_j = \frac{d_j}{2p} \qquad \text{for} \qquad j = k+1, k+2, \ldots, n$$

5. Write

$$V = \begin{pmatrix} 0 \\ 0 \\ \vdots \\ 0 \\ v_k \\ v_{k+1} \\ \vdots \\ v_n \end{pmatrix}$$

Note that $|V| = 1$.

6. Form the matrix

$$P_k \equiv I - 2VV^T$$

The matrix P_k will work for finding a QR factorization of A. These matrices, because of their form, are called *householder matrices.* It can be shown that householder matrices are orthogonal.

DEFINITION 6.3.1 A householder matrix is any matrix of the form $H = I - 2VV^T$, where $|V| = 1$.

THEOREM 6.3.1 Householder matrices are orthogonal.

Proof We show that $H^TH = I$. By definition, then, H would be orthogonal. First we note that H is symmetric:

$$H^T = (I - 2VV^T)^T = I^T - 2V^{T^T}V^T = I - 2VV^T = H$$

Now

$$\begin{aligned} H^TH = H^2 = (I - 2VV^T)^2 &= I - 4VV^T + 4(VV^T)(VV^T) \\ &= I - 4VV^T + 4V(V^TV)V^T = I - 4VV^T + 4V|V|^2V^T \\ &= I - 4VV^T + 4VV^T = I \end{aligned}$$

EXAMPLE 1 Find a QR factorization of

$$A = \begin{pmatrix} 2 & 1 & 1 \\ 1 & 2 & 1 \\ 1 & 1 & 2 \end{pmatrix}$$

keeping four digits to the right of the decimal point.

Solution The first column normalized is

$$\begin{pmatrix} 2/\sqrt{6} \\ 1/\sqrt{6} \\ 1/\sqrt{6} \end{pmatrix} = \begin{pmatrix} 0.8165 \\ 0.4082 \\ 0.4082 \end{pmatrix} = \begin{pmatrix} d_1 \\ d_2 \\ d_3 \end{pmatrix}$$

The "diagonal" element is 0.8165. We want zeros below it. First we calculate D:

$$D = -\sqrt{(0.8165)^2 + (0.4082)^2 + (0.4082)^2} = -1$$

The minus sign was chosen since $0.8165 > 0$. Now we set

$$v_1 = \sqrt{\frac{1}{2}\left(1 - \frac{0.8165}{-1}\right)} = 0.9530 \qquad P = -Dv_1 = 0.9530$$

$$v_2 = \frac{d_2}{2p} = \frac{0.4082}{2(0.9530)} = 0.214 \qquad v_3 = \frac{d_3}{2p} = 0.214$$

Thus

$$P_1 = I - 2\begin{pmatrix} v_1 \\ v_2 \\ v_3 \end{pmatrix}(v_1 v_2 v_3) = I - 2\begin{pmatrix} 0.9082 & 0.204 & 0.204 \\ 0.204 & 0.0458 & 0.0458 \\ 0.204 & 0.0458 & 0.0458 \end{pmatrix}$$

$$= \begin{pmatrix} -0.8164 & -0.408 & -0.408 \\ -0.408 & 0.9084 & -0.0916 \\ -0.408 & -0.0916 & 0.9084 \end{pmatrix}$$

Multiply A by P_1; we obtain

$$P_1 A = \begin{pmatrix} -2.4488 & -2.04 & -2.04 \\ 0^* & 1.3172 & 0.3172 \\ 0^* & 0.3172 & 1.3172 \end{pmatrix}$$

To construct P_2, we look at column 2 of $P_1 A$:

$$\begin{pmatrix} -2.04 \\ 1.3172 \\ 0.3172 \end{pmatrix}$$

Normalized, this is

* Actually we obtained 0.0008. Because of rounding, we called these zero.

$$\begin{pmatrix} d_1 \\ d_2 \\ d_3 \end{pmatrix} = \begin{pmatrix} -.833 \\ 0.5379 \\ 0.1295 \end{pmatrix}$$

In this case $D = \sqrt{(0.5379^2 + (0.1295)^2} = -0.5533$. The minus sign is chosen since $0.5379 > 0$. We set

$$v_1 = 0 \qquad \text{and} \qquad v_2 = \sqrt{\frac{1}{2}\left(1 - \frac{d_2}{D}\right)} = \sqrt{\frac{1}{2}\left[1 - \frac{0.5379}{(-0.5533)}\right]} = 0.993$$

So

$$p = 0.5533v_2 = 0.5494$$

Therefore,

$$v_3 = \frac{d_3}{2p} = \frac{0.1295}{2(0.5494)} = 0.1179$$

and we have

$$\begin{pmatrix} v_1 \\ v_2 \\ v_3 \end{pmatrix} = \begin{pmatrix} 0.0 \\ 0.993 \\ 0.1179 \end{pmatrix}$$

$$P_2 = I - 2\begin{pmatrix} v_1 \\ v_2 \\ v_3 \end{pmatrix}(v_1 v_2 v_3) = I - 2\begin{pmatrix} 0 & 0 & 0 \\ 0 & 0.986 & 0.1171 \\ 0 & 0.1171 & 0.0139 \end{pmatrix}$$

$$= \begin{pmatrix} 1 & 0 & 0 \\ 0 & -0.972 & -0.2342 \\ 0 & 0.2342 & 0.972 \end{pmatrix}$$

Finally,

$$P_2 P_1 A = P_2 \begin{pmatrix} -2.4488 & -2.04 & -2.04 \\ 0 & 1.3172 & 0.3172 \\ 0 & -0.3172 & 1.3172 \end{pmatrix}$$

$$\begin{pmatrix} -2.4488 & -2.04 & -2.04 \\ 0 & -1.355 & -0.6168 \\ 0 & 0^* & 1.206 \end{pmatrix} = R$$

Therefore,

$$A = \begin{pmatrix} 2 & 1 & 1 \\ 1 & 2 & 1 \\ 1 & 1 & 2 \end{pmatrix} = QR = (P_2 P_1)^T R$$

---

* Actually 0.0001. Because of rounding we call this zero.

From the example just calculated, we see that finding the QR factorization for a 3×3 matrix is tedious by hand. A computer, of course, is necessary to find QR factorizations and, therefore, to use the QR method for finding eigenvalues.

For a 2×2 matrix, only one householder matrix must be found, so we consider the QR factorization for a general 2×2 matrix

$$A = \begin{pmatrix} a & b \\ c & d \end{pmatrix}$$

The first column, normalized, is

$$\begin{pmatrix} \dfrac{a}{\sqrt{a^2 + c^2}} \\ \dfrac{c}{\sqrt{a^2 + c^2}} \end{pmatrix} = \begin{pmatrix} d_1 \\ d_2 \end{pmatrix}$$

Now $D = \pm 1$, so we can write $D = -\text{sign } a$, where sign $a = 1$ if $a \geq 0$ and sign $a = -1$ if $a < 0$.

$$v_1 = -\text{sign } a \sqrt{\frac{1}{2}\left(1 + \frac{a}{\sqrt{a^2 + c^2}}\right)}$$

where sign $a = 1$ if $a \geq 0$ and sign $a = -1$ if $a < 0$. Since $p = -Dv_1$, we have $p = (\text{sign } a)v_1$. For v_2 we have

$$v_2 = \frac{d_2}{2v_1} = \frac{-(\text{sign } a)c}{\sqrt{2}\sqrt{a^2 + c^2}\sqrt{1 + a/\sqrt{a^2 + c^2}}}$$

Therefore,

$$P_1 = I - 2\begin{pmatrix} v_1 \\ v_2 \end{pmatrix}(v_1 v_2) = \begin{pmatrix} 1 - 2v_1{}^2 & -2v_1 v_2 \\ -2v_1 v_2 & 1 - 2v_2{}^2 \end{pmatrix}$$
$$= \frac{1}{\sqrt{a^2 + c^2}}\begin{pmatrix} -a & -c \\ -c & a \end{pmatrix}$$

and

$$P_1 A = \frac{1}{\sqrt{a^2 + c^2}}\begin{pmatrix} -a & -c \\ -c & a \end{pmatrix}\begin{pmatrix} a & b \\ c & d \end{pmatrix} = \frac{1}{\sqrt{a^2 + c^2}}\begin{pmatrix} -(a^2 + c^2) & -ab - cd \\ 0 & ad - bc \end{pmatrix}$$

Since P_1 is symmetric, $Q = P_1$.

EXAMPLE 2 Find a QR factorization of

$$\begin{pmatrix} 3 & 7 \\ 4 & 4 \end{pmatrix}$$

Solution Using the formulas as above, we find that

$$\begin{pmatrix} 3 & 7 \\ 4 & 4 \end{pmatrix} = \begin{pmatrix} -\frac{3}{5} & -\frac{4}{5} \\ -\frac{4}{5} & \frac{3}{5} \end{pmatrix} \begin{pmatrix} -5 & -\frac{37}{5} \\ 0 & -\frac{16}{5} \end{pmatrix}$$

Using the formulas for the 2×2 case, we now calculate the eigenvalues of

$$\begin{pmatrix} 5 & -2 \\ -2 & 8 \end{pmatrix}$$

by the QR method.

EXAMPLE 3 Use the QR method to calculate the eigenvalues of

$$A = \begin{pmatrix} 5 & -2 \\ -2 & 8 \end{pmatrix}$$

(The true eigenvalues are 4 and 9.)

Solution We use the formulas for the 2×2 case each time we need a QR factorization. The calculated matrices are listed below (rounded); after step 3, only A_m is listed.

Step 1: $A_1 = A = \begin{pmatrix} 5 & -2 \\ -2 & 8 \end{pmatrix} \qquad Q_1 = Q = \begin{pmatrix} 0.928 & 0.371 \\ 0.371 & 0.928 \end{pmatrix}$

$R_1 = R = \begin{pmatrix} -5.385 & 4.828 \\ 0 & 6.685 \end{pmatrix}$

Step 2: $A_2 = R_1Q_1 = \begin{pmatrix} 6.793 & -2.482 \\ -2.482 & 6.207 \end{pmatrix} \qquad Q_2 = \begin{pmatrix} -0.939 & -0.343 \\ -0.343 & 0.939 \end{pmatrix}$

$R_2 = \begin{pmatrix} -7.233 & -4.462 \\ 0 & 4.977 \end{pmatrix}$

Step 3: $A_3 = R_2Q_2 = \begin{pmatrix} 8.324 & -1.708 \\ -1.708 & 4.675 \end{pmatrix} \qquad Q_3 = \begin{pmatrix} -0.979 & 0.201 \\ 0.201 & 0.979 \end{pmatrix}$

Step 4: $A_4 = \begin{pmatrix} 8.850 & 0.852 \\ 0.852 & 4.149 \end{pmatrix}$

Step 5: $A_5 = \begin{pmatrix} 8.969 & -0.387 \\ -0.387 & 4.030 \end{pmatrix}$

Step 6: $A_6 = \begin{pmatrix} 8.993 & 0.173 \\ 0.173 & 4.006 \end{pmatrix}$

$\vdots$

Step 12: $A_{12} = \begin{pmatrix} 8.9999996 & 0.00134 \\ 0.00134 & 4.000018 \end{pmatrix}$

Approximate eigenvalues are on the diagonal.

In Example 3, A_m appeared to be converging to a diagonal matrix; of course, the diagonal elements are the approximate eigenvalues. This illustrates the following important result.

THEOREM 6.3.2 Let A be a real $n \times n$ matrix with eigenvalues satisfying

$$|\lambda_1| > |\lambda_2| > \cdots > |\lambda_n| > 0$$

Then matrices A_m in the QR method will converge to an upper triangular matrix with diagonal entries $\{\lambda_k\}$, $k = 1, 2, \ldots, n$. If A is symmetric, matrices A_m converge to a diagonal matrix with the eigenvalues on the diagonal.

If the hypotheses of Theorem 6.3.2 are not satisfied by A, the QR method may fail. If the difference in the magnitudes of the eigenvalues is small, convergence of the QR method can be slow.

EXAMPLE 4 Apply the QR method to attempt calculation of eigenvalues of

$$\begin{pmatrix} 3 & 3 \\ 0.33333 & 5 \end{pmatrix}$$

Note that the true eigenvalues are 4.001 and 3.999.

Solution After computation we find

$$A_{10} = \begin{pmatrix} 4.7058 & -3.1764 \\ 0.1568 & 3.2941 \end{pmatrix}$$

$$A_{20} = \begin{pmatrix} 4.2582 & -3.3132 \\ 0.0201 & 3.7418 \end{pmatrix}$$

$$A_{30} = \begin{pmatrix} 4.1571 & -3.3259 \\ 0.0074 & 3.8428 \end{pmatrix}$$

We see that as the theorem guarantees, A_m is converging to an upper triangular matrix and the diagonal elements are heading in the right direction; however, the convergence is very slow. The slowness of convergence is due to the fact that $|\lambda_1| \doteq |\lambda_2|$.

As we have seen, the convergence of the QR method can be slow: This costs money because of the computer time used. There exist methods for accelerating the convergence of the QR method; these are covered in advanced numerical analysis texts.

Finally, after we find the eigenvalues of A, the corresponding eigenvectors can be found by solving $(\lambda I - A)X = 0$, subject to some side condition such as $|X| = 1$.

PROBLEMS 6.3

In Probs. 1 to 5, find QR factorizations for the given matrices.

1. $\begin{pmatrix} 1 & 5 \\ 5 & 6 \end{pmatrix}$ **2.** $\begin{pmatrix} 3 & 4 & 0 \\ 1 & 3 & 0 \\ 0 & 0 & 2 \end{pmatrix}$ **3.** $\begin{pmatrix} 2 & 3 \\ -2 & 1 \end{pmatrix}$ **4.** $\begin{pmatrix} 3 & 3 \\ 3 & 5 \end{pmatrix}$ **5.** $\begin{pmatrix} 3 & 3 & 0 \\ 3 & 5 & 0 \\ 0 & 0 & 6 \end{pmatrix}$

6. Use the QR method to find approximations to the eigenvalues of the matrices in Probs. 1, 3, and 4.

7. The matrix

$$\begin{pmatrix} 1 & 1 \\ -3 & 1 \end{pmatrix}$$

has complex eigenvalues. Use the QR method to attempt to calculate the eigenvalues. What happens?

8. Do as in Prob. 7 with

$$A = \begin{pmatrix} 1 & 4 \\ -4 & 1 \end{pmatrix}$$

SUMMARY

The *power method* and *QR algorithm* are two methods for numerical calculation of eigenvalues of real matrices. The *stability* of a numerical eigenvalue problem depends on the matrix under consideration. If the matrix is symmetric with symmetrically distributed error, then the calculated eigenvalues will approximate the actual eigenvalues, provided the eigenvalues are all simple. Otherwise, the numerical methods may fail to find all eigenvalues.

Space has not permitted the presentation of all the numerical methods of linear algebra. To cover all these methods requires a fairly intensive course in numerical analysis. But the issues we have briefly addressed in the sections on numerical methods are important issues: error propagation, approximation, iteration, roundoff error, ill-conditioning, and theoretical underpinnings of numerical methods.

The next chapter is similar in flavor. To understand the methods discussed requires a course in optimization. Not having the luxury of that much time, we are content to present the bare bones of the simplex algorithm for linear programming problems and hope that it whets the reader's appetite for a more in-depth experience.

ADDITIONAL PROBLEMS

Shown in Probs. 1 through 5 are physical systems and a matrix associated with the differential equations for the systems. Compute the largest eigenvalue.

1. Coupled pendula.

$$\begin{pmatrix} \dfrac{g}{L} + \dfrac{k}{m} & \dfrac{-k}{m} \\ \dfrac{-k}{m} & \dfrac{g}{L} + \dfrac{k}{m} \end{pmatrix}$$

Use $g = -32$, $k = 0.1$, $L = 2$, and $m = 1$.

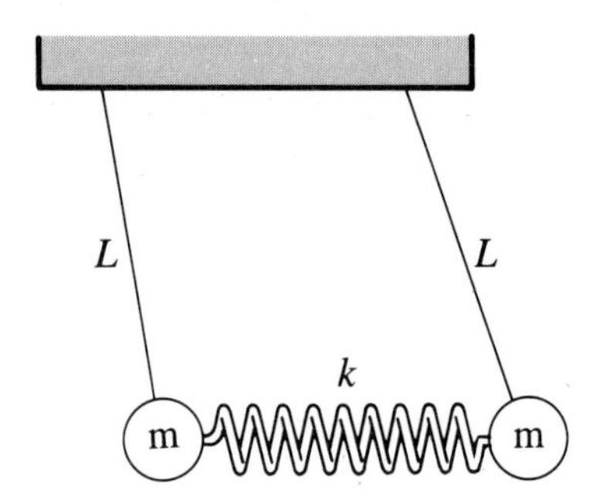

2. Masses in a groove.

$$\left(\frac{k}{m}\right)\begin{pmatrix} 2 & -1 & -1 \\ -1 & 2 & -1 \\ -1 & -1 & 2 \end{pmatrix}$$

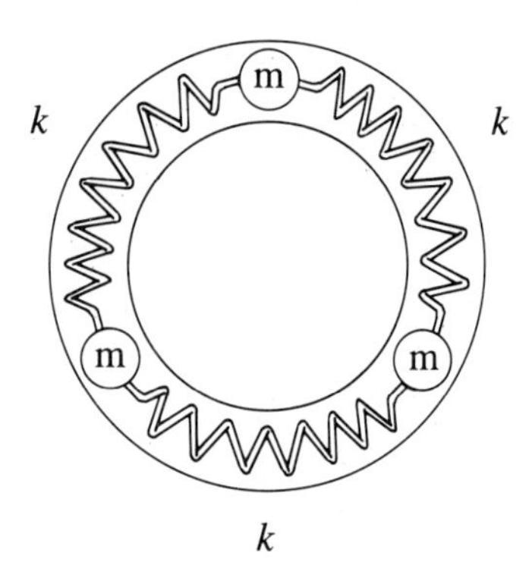

3. An electric circuit.

$$\begin{pmatrix} 0 & \dfrac{-R}{L} \\ \dfrac{1}{RC} & \dfrac{-1}{RC} \end{pmatrix}$$

Use $R = 0.25$, $L = 1$, and $C = 3$.

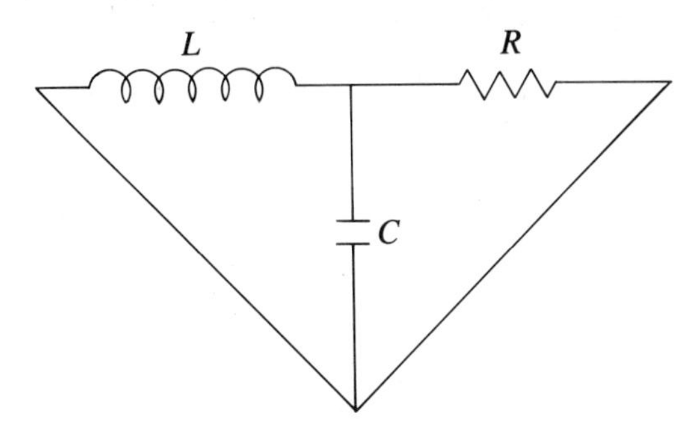

4. Trolleys connected with springs.

$$\begin{pmatrix} \dfrac{k_1 + k_2}{m_1} & \dfrac{-k_2}{m_1} & 0 \\ \dfrac{-k_2}{m_2} & \dfrac{k_2 + k_3}{m_2} & \dfrac{-k_3}{m_2} \\ 0 & \dfrac{-k_3}{m_3} & \dfrac{-k_3}{m_3} \end{pmatrix}$$

Use $k_i = 1$ and $m_i = 2$.

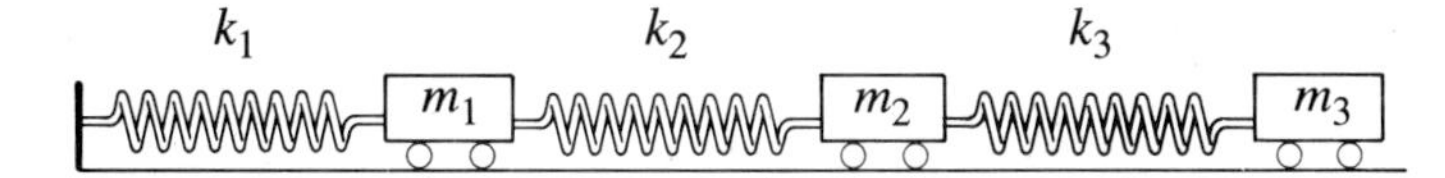

5. Solute diffusion.

$$\begin{pmatrix} -(k_1 + k_3) & k_1 & k_3 \\ k_1 & -(k_1 + k_2) & k_2 \\ k_3 & k_2 & -(k_2 + k_3) \end{pmatrix}$$

Use $k_1 = 2k_2$ and $k_1 = 3k_3$.

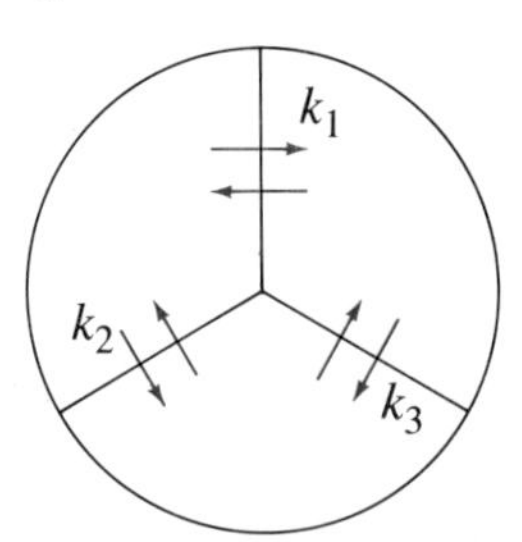

6. Gershgorin's theorem states for an $n \times n$ matrix A that any eigenvalue must lie in one of the circles in the complex plane

$$\left\{ z \text{ such that } |z - a_{kk}| \le \sum_{\substack{j=1,\\ j\neq k}}^{n} |a_{kj}| \right\}$$

Thus for the matrix

$$A = \begin{pmatrix} i & 0.1 & 0.5 \\ 0.3 & 2 & 0.1 \\ i/2 & -1 & -3 \end{pmatrix}$$

we have the circles as shown in Fig. AP6.6. Gershgorin's theorem gives us an estimate as to where to look for the eigenvalues, real or complex. Apply Gershgorin's theorem to the matrices in Probs. 1 through 5.

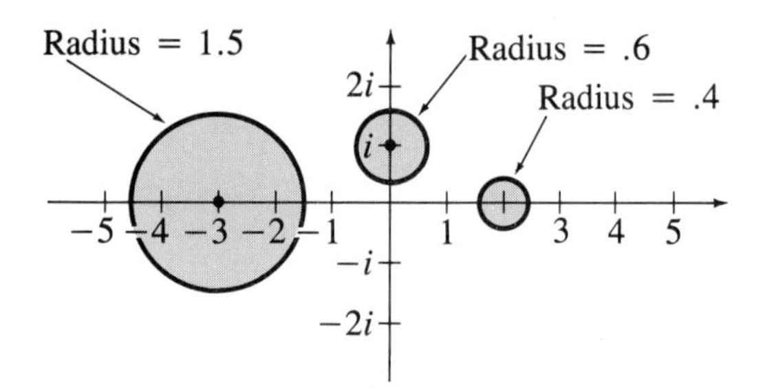

Figure AP6.6

7. Use Gershgorin's theorem to show that

$$A = \begin{pmatrix} i & 0.4 & 0.1i \\ 0.4 & 1+i & 0 \\ 0.1i & 0 & -3i \end{pmatrix}$$

has no real eigenvalues. Note that A is symmetric, but not real symmetric.

8. Use Gershgorin's theorem to show that

$$A = \begin{pmatrix} -4 & 1 & 1 \\ 0 & -6 & 2 \\ 1 & 0 & -2 \end{pmatrix}$$

has no eigenvalues with positive real part.

CHAPTER

7

INTRODUCTION TO LINEAR PROGRAMMING

7.1 SIMPLE EXAMPLES

A basic problem of applied science is optimization, for example, maximization of output of a chemical process. Often the problem is to maximize the range values of a linear transformation $T: E^n \to \mathbb{R}$.

EXAMPLE 1 A chemical company produces two components, Chem-A and Chem-B. Chem-A mix sells for \$3 per pound and Chem-B for \$2 per pound. If x_1 lb of Chem-A and x_2 lb of Chem-B are produced per day, write the transformation which has range values as revenue.

Solution Total revenue equals $3x_1 + 2x_2$. Therefore, we can represent revenue by $T((x_1, x_2)) = 3x_1 + 2x_2$, which is a linear transformation on E^2 with range values of revenue in E^1. We can think of this as

$$T: \text{(production space)} \longrightarrow \text{(revenue space)}$$

If the chemical company in Example 1 wants to maximize daily revenue, the value of T should be maximized. Unless further information is given, a decision by a chemical engineer on a production schedule (values of x_1 and x_2) cannot be made. Only a finite number of pounds of each chemical compound can be produced each day. Also, as often happens, production of one commodity often limits production of the other: machinery, labor, or resources must be shared.

EXAMPLE 2 Referring to Example 1, suppose that the maximum total production per day is 3000 lb and that no more than 2000 lb of Chem-A can be made. Express these *constraints* on production as inequalities.

Solution The constraints are

$$x_1 + x_2 \leq 3000$$
$$x_1 \leq 2000$$

In addition to the constraints listed above, we have $x_1 \geq 0$ and $x_2 \geq 0$. Therefore, to maxmize daily revenue from Chem-A and Chem-B, the chemical company should consider the problem

$$\begin{array}{lll} \text{Maximize} & T((x_1, x_2)) = 3x_1 + 2x_2 & \\ \text{Subject to} & x_1 + x_2 \leq 3000 & x_1 \geq 0 \\ & x_1 \leq 2000 & x_2 \geq 0 \end{array}$$

The problem of maximizing a *linear* function of the variables $x_1, \ldots, x_n$ (that is, maximizing the range values of a linear transformation $T: E^n \to \mathbb{R}$) subject to *linear* inequality constraints is called a *linear program* (LP). The expression for T is called the *objective function*, and the region defined by the inequality constraints is called the *feasible region*. The feasible region is just the subset of E^n containing possible production schedules.

EXAMPLE 3 Graph the feasible region for the problem in Examples 1 and 2. State several feasible production schedules.

Figure 7.1.1

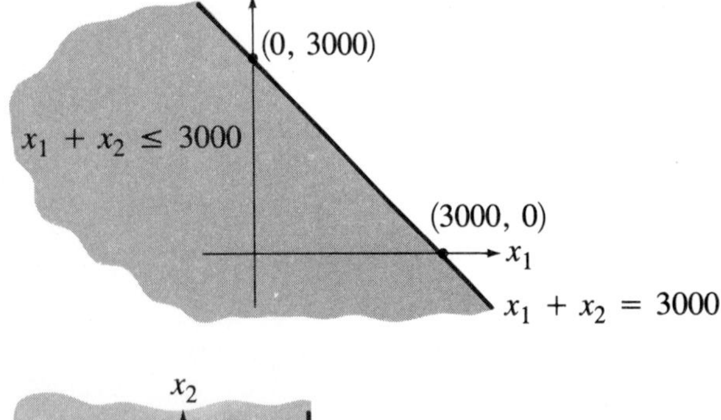

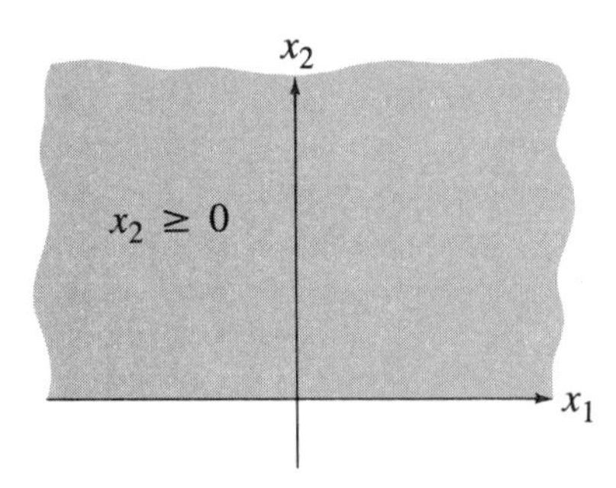

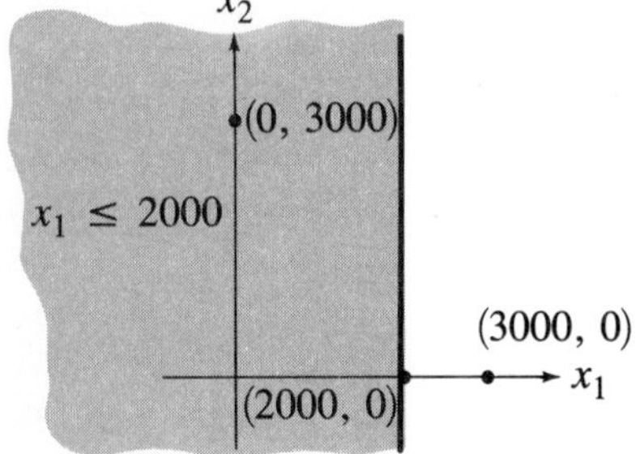

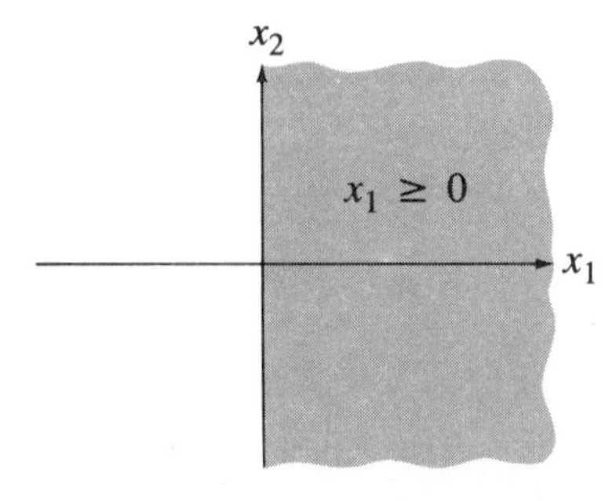

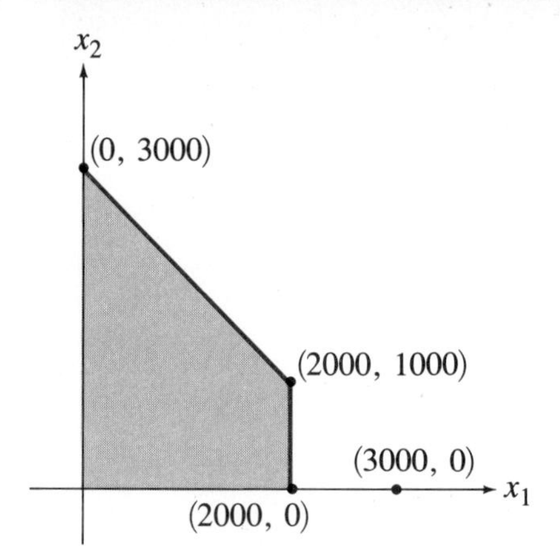

Figure 7.1.2 Feasible region.

Solution We graph each inequality separately and then find the intersection of the graphs (see Fig. 7.1.1). The intersection of the regions in Fig. 7.1.1 is shown in Fig. 7.1.2. Any points in the region are feasible production schedules. Some examples are (0, 0), (2000, 0), (0, 3000), (2000, 1000), (1000, 2000), (1000, 0), (2000, 500), (0, 2500), (1000, 1000), (500, 2000), (1500, 750), and (700, 300).

The first four points listed in Example 3 are vertex points, the second four are face points, and the last four are interior points. We will see that the vertex points are the most important. For LP problems in two variables, the feasible region always has a polygonal boundary.

A key observation for the solution of an LP problem in two dimensions (such as our present example) is that T takes on constant values on straight lines. In particular, for a number k, the equation $T((x_1, x_2)) = k$ is $3x_1 + 2x_2 = k$, which represents a straight line in the x_1x_2 plane. Such lines are called *level lines* for T. Along the level lines, revenue is the constant value k; some level lines are drawn in Fig. 7.1.3. Note that the lines are parallel and that the revenue increases as the lines move out into the first quadrant.

Now we superimpose on Fig. 7.1.3 the feasible region. (See Fig. 7.1.4.) From Fig. 7.1.4 we see that the maximum feasible revenue is $8000, achieved with a production schedule of 2000 lb of Chem-A and 1000 lb of Chem-B. We say that this is the optimal production schedule. We have found the line of maximum constant revenue which also intersects the feasible region.

In the general case of n products, the same method works. Instead of *lines* of constant revenue, *planes* or *hyperplanes* of constant revenue will arise. The feasible region will be a *polyhedron* in n-space, instead of a polygon in two-space. The optimal production schedule will still occur on a *vertex* (corner point) of the polyhedron. This is ensured by the next theorem, a fundamental theorem of line programming.

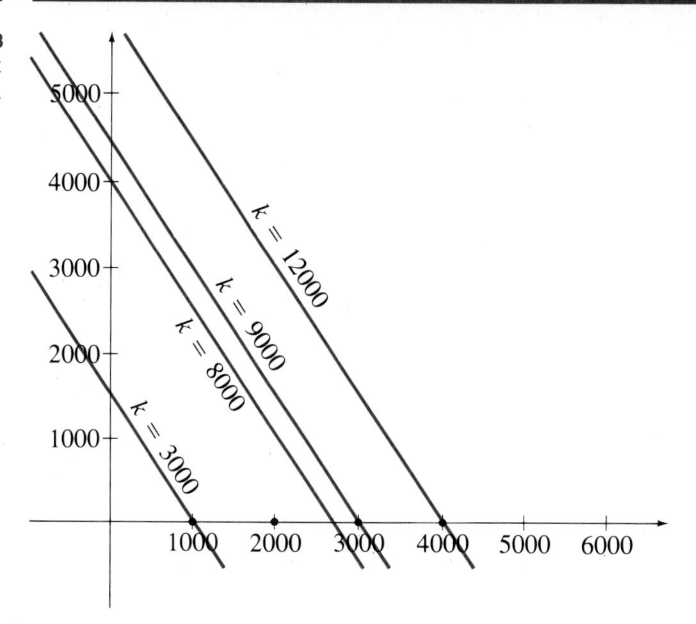

Figure 7.1.3 Lines of constant revenue $3x_1 + 2x_2 = k$.

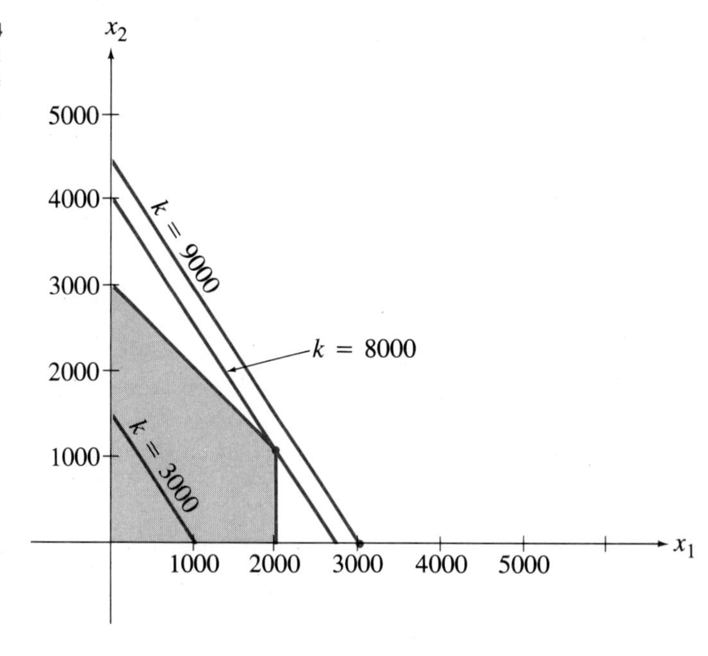

Figure 7.1.4 Lines of constant revenue and the feasible region.

THEOREM 7.1.1 Let $T: E^n \to \mathbb{R}$ be linear. On the restricted domain of a polyhedron P the maximum and minimum values of T are achieved at vertices of P.

We cannot prove this theorem in general, but a proof in the two-dimensional case could be generated by looking at level lines of T. Some of the several

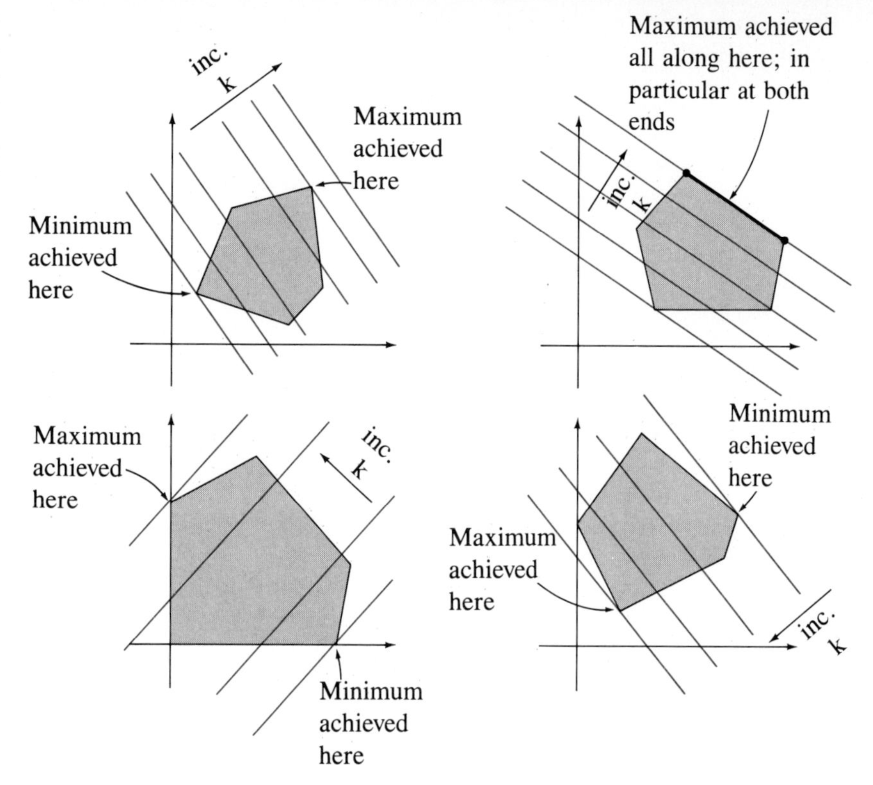

Figure 7.1.5 Some possibilities in the two-dimensional LP problem. The lines are $ax_1 + bx_2 = k$. Direction of increasing k is indicated by the arrow.

possible cases for maximum and minimum values are shown in Fig. 7.1.5.

The theorem says that in order to maximize or minimize T on a bounded feasible region, we need only find all the vertices $v_1, v_2, v_3, v_4, \ldots, v_m$; calculate $T(v_1), \ldots, T(v_m)$ and pick the largest and smallest values from the list.

EXAMPLE 4 Maximize and minimize $T((x_1, x_2)) = 3x_1 - 7x_2$ subject to $x_1 \geq 0$, $x_2 \geq 0$, $4x_1 + 5x_2 \leq 20$, and $x_1 + 2x_2 \leq 10$.

Solution We graph the feasible region. Note that the sides of the region are the lines $x_1 = 0, x_2 = 0, 4x_1 + 5x_2 = 20$, and $5x_1 + 2x_2 = 10$. That is, we find the borders by changing all the inequality constraints to equations. We have the region shown in Fig. 7.1.6. It has vertices (0, 0), (0, 4), (2, 0), $(\frac{10}{17}, 3\frac{9}{17})$. Now we calculate

$$T((0, 0)) = 0 \qquad T((2, 0)) = 6 \qquad T((0, 4)) = -28 \qquad T((\tfrac{10}{17}, 3\tfrac{9}{17})) = -22$$

to find that

$$\text{Maximum of } T \text{ is } 6 \text{ achieved at } x_1 = 2, x_2 = 0$$

$$\text{Minimum of } T \text{ is } -28 \text{ achieved at } x_1 = 0, x_2 = 4$$

Figure 7.1.6 Feasible region of Example 4.

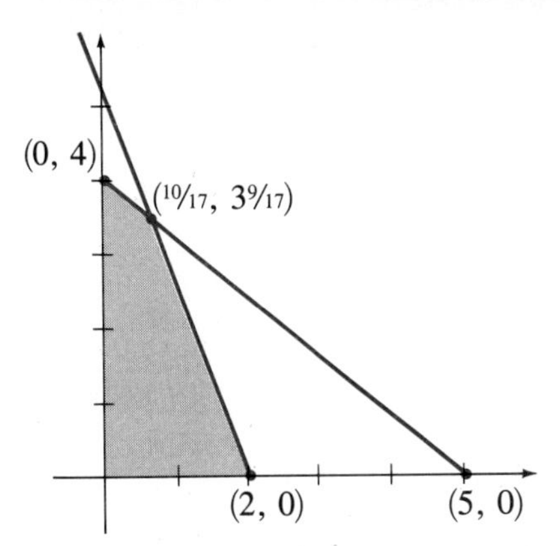

Production schedules involving several products obviously exist. Since we lose the ability to graph feasible regions for four products or more, an *analytical* method for solving the LP program is necessary. We might think that the way to do this would be to find all the vertices of the feasible region and calculate T at the vertices. However, if n is large, this is a difficult problem in itself. A widely used method for solving n-dimensional LP problems is the *simplex algorithm.* In this method we essentially keep track of values of T as we move from one vertex to a neighboring vertex where T has a larger or equal value. Once we find a vertex where T takes the same or lesser values at all neighboring vertices, the optimal vertex has been found. Note that the simplex algorithm can fail (owing to time limitations or an unbounded feasible region) in some cases, but works for almost all LP problems which arise in practice.

PROBLEMS 7.1

In Probs. 1 to 4, a linear objective function and some constraints are given. Find the maximum and minimum values of the objective function.

1. $T((x, y)) = 3x - 2y$
$$0 \le x \le 5$$
$$0 \le y \le 8$$
$$0 \le x + y \le 4$$

2. $T((x, y)) = 4x + 7y$
$$0 \le x \le 6$$
$$0 \le y \le 10$$
$$0 \le 2x + y \le 8$$

3. $T((x, y)) = -x - 5y$
$$0 \le x \le 5$$
$$0 \le y \le 7$$
$$0 \le x + y \le 10$$

4. $T((x, y)) = 2x + 3y$
$$0 \le x \le 10$$
$$0 \le y \le 20$$
$$0 \le x + y \le 15$$
$$0 \le 2x + 3y \le 36$$

5. In optimization, if the objective function is nonlinear, then the maximum may not occur at a vertex of the feasible region. Show that the maximum of $T((x, y)) = xy$ over the region

$$0 \le x \qquad 0 \le y \qquad 0 \le x + y \le 8$$

occurs at $x = 4$, $y = 4$, which is not a vertex. [*Hint*: What do the curves $xy = k$ look like?]

6. In optimization, if the objective function is linear but the constraints are nonlinear, the maximum may not occur at a corner point. Show that the maximum of $T((x, y)) = x + y$ over the region

$$0 \le x \qquad 0 \le y \qquad 0 \le x^2 + y^2 \le 16$$

occurs at the point $x = 2^{3/2}$, $y = 2^{3/2}$. The "corner points" of the feasible region are (0, 0), (0, 4), and (4, 0). However, $(2^{3/2}, 2^{3/2})$ is what is called an *extreme point* of the region.

7. Develop an example for Prob. 5 with the word *maximum* replaced by *minimum.*

8. Develop an example for Prob. 6 with the word *maximum* replaced by *minimum.*

9. A lawn mower company makes two types of lawn mowers: a basic model and a self-propelled model. The company has only one assembly line. At most 150 basic model lawn mowers will be made in a day. At most 120 self-propelled lawn mowers will be made in a day. These constraints are to prevent oversupply and a resulting decrease in prices. Because of time limitations at most 200 lawn mowers can be made per day. The lawn mowers sell for \$180 (basic model) and \$250 (self-propelled model). What production schedule will maximize revenue?

10. Regarding Prob. 9, suppose that to introduce a new self-propelled model, it is sold to retail outlets for \$170. What production schedule maximizes revenue?

11. The boundedness of the feasible region can be important. Consider

$$\begin{array}{ll} \text{Maximize} & T((x, y)) = x + y \\ \text{Subject to} & x \ge 0 \qquad y \ge 0 \qquad 2x - y - 1 \ge 0 \end{array}$$

Show that T has no maximum on this feasible region.

7.2 REWRITING THE LP PROBLEM

The simplex algorithm is a method for solving LP problems. To prepare for the use of the algorithm, the LP problem is rewritten in a certain form. We consider in detail the problem

$$\begin{array}{ll} \text{Maximize} & T = c_1x_1 + c_2x_2 + \cdots + c_nx_n \\ \text{Subject to} & a_{11}x_1 + a_{12}x_2 + \cdots + a_{1n}x_n \le b_1 \\ & a_{21}x_1 + a_{22}x_2 + \cdots + a_{2n}x_n \le b_2 \\ & \cdots\cdots\cdots\cdots\cdots\cdots\cdots\cdots\cdots \\ & a_{m1}x_1 + a_{m2}x_2 + \cdots + a_{mn}x_n \le b_m \\ & x_1 \ge 0 \qquad x_2 \ge 0 \qquad \cdots \qquad x_n \ge 0 \\ & \text{where} \quad b_1 > 0 \qquad b_2 > 0 \qquad \cdots \qquad b_m > 0 \end{array} \tag{7.2.1}$$

The problem is to find vectors

$$X = \begin{pmatrix} x_1 \\ \vdots \\ x_2 \end{pmatrix}$$

whose components maximize T and satisfy the given inequalities. In shorter notation, the problem may be written

$$\begin{array}{ll} \text{Maximize} & C^T X \\ \text{Subject to} & AX \leq B \qquad B > 0 \qquad X \geq 0 \end{array} \tag{7.2.2}$$

Problem (7.2.2) is, of course, difficult. The simplex algorithm approach is based first on *changing the difficult problem to one which we have a chance of solving*, namely, linear equations. "Slack variables" $z_1, z_2, \ldots, z_m$ are introduced in the inequalities $AX \leq B$ by the rule

$$AX + Z = B$$

Thus the inequality constraints are changed to equality constraints. If we rename variables $y_1 = x_1$, $y_2 = x_2, \ldots, y_n = x_n$, $y_{n+1} = z_1$, $y_{n+2} = z_2, \ldots, y_{n+m} = z_m$, then we have the constraints in short form

$$\mathscr{A} Y = B$$

where $\mathscr{A}$ and Y are partitioned:

$$\mathscr{A} = \left(A \;\middle|\; \underbrace{\begin{matrix} 1 & & \mathbf{0} \\ & \ddots & \\ \mathbf{0} & & 1 \end{matrix}}_{I_{m \times m}} \right) \qquad \text{and} \qquad Y = \begin{pmatrix} X \\ \hline z_1 \\ z_2 \\ \vdots \\ z_m \end{pmatrix}$$

Of course, $z_1 \geq 0$, $z_2 \geq 0, \ldots, z_m \geq 0$.

EXAMPLE 1 Find matrices A, $\mathscr{A}$, Y, and B for Example 1 of Sec. 7.1.

Solution The problem was

$$\begin{array}{lll} \text{Maximize} & 3x_1 + 2x_2 & \\ \text{Subject to} & x_1 + x_2 \leq 3000 & x_1 \geq 0 \\ & x_1 \leq 2000 & x_2 \geq 0 \end{array}$$

Therefore,

$$C = \begin{pmatrix} 3 \\ 2 \end{pmatrix} \qquad X = \begin{pmatrix} x_1 \\ x_2 \end{pmatrix} \qquad B = \begin{pmatrix} 3000 \\ 2000 \end{pmatrix}$$

Introducing slack variables (they "take up the slack" in the inequalities to make equations), we have

$$\begin{array}{lllll} \text{Maximize} & 3x_1 + 2x_2 + 0z_1 + 0z_2 & & & \\ \text{Subject to} & x_1 + x_2 + z_1 = 3000 & x_1 \geq 0 & z_1 \geq 0 \\ & x_1 + z_2 = 2000 & y_1 \geq 0 & z_2 \geq 0 \end{array}$$

so that

$$A = \begin{pmatrix} 1 & 1 \\ 1 & 0 \end{pmatrix} \qquad \mathscr{A} = \begin{pmatrix} 1 & 1 & 1 & 0 \\ 1 & 0 & 0 & 1 \end{pmatrix} \qquad Y = \begin{pmatrix} x_1 \\ x_2 \\ z_1 \\ z_2 \end{pmatrix}$$

So far, all we have done is rewrite the LP problem. It is in the form

$$\text{Maximize} \qquad P^T Y \qquad \text{where } P = \begin{pmatrix} C \\ \hline 0 \\ \vdots \\ 0 \end{pmatrix}$$

$$\text{Subject to} \qquad \mathscr{A} Y = B \qquad Y \geq 0 \tag{7.2.3}$$

Now let

$$A_j = \begin{pmatrix} a_{1j} \\ a_{2j} \\ \vdots \\ a_{mj} \end{pmatrix} \qquad j = 1, 2, \ldots, m + n = N$$

(that is, A_j is the jth column of $\mathscr{A}$) and

$$A_0 = B$$

With this notation we can write (7.2.3) as (find Y which will)

$$\begin{array}{l} \text{Maximize} \qquad P^T Y \\ \text{With} \qquad Y \geq 0 \\ A_0 = \sum_{j=1}^{N} y_j A_j \end{array} \tag{7.2.4}$$

EXAMPLE 2 Write the problem from Example 1 in the form (7.2.4).

Solution

$$A_0 = \begin{pmatrix} 3000 \\ 2000 \end{pmatrix} \qquad A_1 = \begin{pmatrix} 1 \\ 1 \end{pmatrix} \qquad A_2 = \begin{pmatrix} 1 \\ 0 \end{pmatrix} \qquad A_3 = \begin{pmatrix} 1 \\ 0 \end{pmatrix} \qquad A_4 = \begin{pmatrix} 0 \\ 1 \end{pmatrix}$$

Also $y_1 = x_1$, $y_2 = x_2$, $y_3 = z_1$, $y_4 = z_2$,

$$P = \begin{pmatrix} c_1 \\ c_2 \\ 0 \\ 0 \end{pmatrix} = \begin{pmatrix} 3 \\ 2 \\ 0 \\ 0 \end{pmatrix} \qquad \text{and} \qquad N = n + m = 4$$

Writing out (7.2.4) yields

$$A_0 = \sum_1^4 y_j A_j \leftrightarrow \begin{pmatrix} 3000 \\ 2000 \end{pmatrix} = x_1 \begin{pmatrix} 1 \\ 1 \end{pmatrix} + x_2 \begin{pmatrix} 1 \\ 0 \end{pmatrix} + z_1 \begin{pmatrix} 1 \\ 0 \end{pmatrix} + z_2 \begin{pmatrix} 0 \\ 1 \end{pmatrix}$$

$$= \begin{pmatrix} x_1 + x_2 + z_1 & \\ x_1 & + z_2 \end{pmatrix}$$

$$\text{Maximize } P^T Y \leftrightarrow \text{maximize } (3 \quad 2 \quad 0 \quad 0) \begin{pmatrix} x_1 \\ x_2 \\ x_3 \\ x_4 \end{pmatrix} = 3x_1 + 2x_2$$

$$Y \geq 0 \leftrightarrow x_1 \geq 0, x_2 \geq 0, z_1 \geq 0, z_2 \geq 0$$

In the form (7.2.4) the LP problem becomes this:

> What should the coefficients in the expansion of A_0 in terms of $A_1, A_2, \ldots, A_n$ be, in order for $P^T Y$ to be a maximum with $Y \geq 0$?

Thus the LP problem has been reformulated as a problem of picking a particular linear combination of vectors to represent a given vector.

PROBLEMS 7.2

1. Rewrite Prob. 1 of Sec. 7.1 in form (7.2.4).
2. Rewrite Prob. 2 of Sec. 7.1 in form (7.2.4).
3. Rewrite Prob. 3 of Sec. 7.1 in form (7.2.4).
4. Rewrite Prob. 4 of Sec. 7.1 in form (7.2.4).
5. Rewrite

$$\begin{array}{ll} \text{Maximize} & T((x, y, z)) = 2x + y + 3z \\ \text{Subject to} & 3x + 2y \leq 450 \qquad 3x + 2y + 2z \leq 1200 \\ & \qquad\qquad\qquad\qquad\qquad y + 2z \leq 900 \\ & x \geq 0 \qquad y \geq 0 \qquad z \geq 0 \end{array}$$

in form (7.2.4).

7.3 LINEAR INDEPENDENCE, FEASIBLE SOLUTIONS OF LP, AND THE SIMPLEX ALGORITHM

Vectors Y which satisfy the restraints

$$A_0 = \sum_{j=1}^{N} y_j A_j \qquad Y \geq 0$$

from (7.2.4) are called *feasible solutions* of (7.2.4). Any feasible solution which maximizes $P^T Y$ is called an *optimal feasible solution.* Optimal feasible solutions are solutions to the original LP problem.

Feasible solutions have an important property. To discuss it, call the values of y_j in

$$\sum_{j=1}^{N} y_j A_j$$

the *weights* of the A_j's. That is, y_1 is the weight of A_1, and so on. Also Y is called the *weight vector.* The fundamental property of an optimal feasible solution is the following:

> If Y is an optimal feasible solution of (7.2.4), then the vectors A_j with positive weights form a linearly independent set in $\mathcal{M}_{N1}$.

This theorem can be proved with some machinery beyond the scope of this course: convex sets, extreme points, and linear functionals.

The theorem tells us that a *feasible solution cannot be optimal if the vectors with positive weight form a linearly dependent set.*

EXAMPLE 1 The problem from Example 1 in Sec. 7.1 has (7.2.4) form

$$\begin{pmatrix} 3000 \\ 2000 \end{pmatrix} = x_1 \begin{pmatrix} 1 \\ 1 \end{pmatrix} + x_2 \begin{pmatrix} 1 \\ 0 \end{pmatrix} + z_1 \begin{pmatrix} 1 \\ 0 \end{pmatrix} + z_2 \begin{pmatrix} 1 \\ 1 \end{pmatrix}$$

$$= y_1 \begin{pmatrix} 1 \\ 1 \end{pmatrix} + y_2 \begin{pmatrix} 1 \\ 0 \end{pmatrix} + y_3 \begin{pmatrix} 1 \\ 0 \end{pmatrix} + y_4 \begin{pmatrix} 0 \\ 1 \end{pmatrix}$$

To have an optimal feasible solution, at most two of the weights can be nonzero because the vectors A_j, $j = 1, 2, 3, 4$, are all from $\mathcal{M}_{21}$ which has dimension 2. The solution found earlier—$x_1 = 2000$ and $x_2 = 1000$—forces $y_3 = z_1 = 0$ and $y_4 = z_2 = 0$. Thus the set of vectors A with positive weights is

$$\left\{ \begin{pmatrix} 1 \\ 1 \end{pmatrix}, \begin{pmatrix} 1 \\ 0 \end{pmatrix} \right\}$$

which is linearly independent.

EXAMPLE 2 Regarding Example 1, we could arbitrarily choose $x_1 = 0$, $x_2 = 0$, $z_1 = 3000$, and $z_2 = 2000$ and have

$$A_0 = \sum_{j=1}^{4} y_j A_j$$

satisfied, with the set of vectors with positive weights

$$\left\{\begin{pmatrix}1\\0\end{pmatrix}, \begin{pmatrix}0\\1\end{pmatrix}\right\}$$

being linearly independent. However, $P^T Y = 0$ and is clearly not maximized. This example emphasizes the fact that *linear independence* of the set of vectors with positive weights *is not sufficient* to yield an optimal feasible solution Y.

The question now has to be, How do we compute Y which solves (7.2.4)? Roughly speaking, we can do this.

1. Start with a feasible solution

$$Y^{(1)} = \begin{pmatrix}0\\0\\\vdots\\b_1\\\vdots\\b_m\end{pmatrix} = \begin{pmatrix}y_1\\\vdots\\y_n\\y_{n+1}\\\vdots\\y_{n+m}\end{pmatrix}$$

That is, put $x_1 = x_2 = \cdots = x_n = 0$ and $z_1 = b_1, \ldots, z_m = b_m$. The restraints are satisfied. The vectors A_j, $j = n + 1$ to $j = n + m(= N)$, simply form the standard basis for $\mathcal{M}_{m1}$. Call this basis $\mathscr{B}_1$.

2. Change the basis $\mathscr{B}_1$ by deleting a vector from it and adding one of the vectors from $A_1, A_2, \ldots, A_n$ to get a new basis $\mathscr{B}_2$ in such a way that the new weight vector, call it $Y^{(2)}$, makes

$$P^T Y^{(2)} \geq P^T Y^{(1)}$$

(The basis problem returns again!)

3. Repeat the procedure until at step S

$$P^T Y^{(S)}$$

can be increased no more by changing bases.

It can be proved that this procedure will work; we do not prove it, but we do apply it to our example.

We begin with

$$Y^{(1)} = \begin{pmatrix} 0 \\ 0 \\ 3000 \\ 2000 \end{pmatrix}$$

the matrix

$$\mathscr{A} = \begin{pmatrix} 1 & 1 & 1 & 0 \\ 1 & 0 & 0 & 1 \end{pmatrix}$$

and the basis

$$\mathscr{B}_1 = \left\{ \begin{pmatrix} 1 \\ 0 \end{pmatrix} \begin{pmatrix} 0 \\ 1 \end{pmatrix} \right\} = \{A_3, A_4\}$$

the last two vectors in $\mathscr{A}$. Note that $P^T Y^{(1)} = 0$. To build $\mathscr{B}_2$, we could do the following:

1. Use

$$\{A_2, A_4\} = \left\{ \begin{pmatrix} 1 \\ 0 \end{pmatrix}, \begin{pmatrix} 0 \\ 1 \end{pmatrix} \right\}$$

in which case the restraint gives $x_2 = 3000$, $z_2 = 2000$, and $P^T Y = 6000$.

2. Use

$$\{A_1, A_4\} = \left\{ \begin{pmatrix} 1 \\ 1 \end{pmatrix}, \begin{pmatrix} 0 \\ 1 \end{pmatrix} \right\}$$

in which case the restraint yields

$$\begin{aligned} x_1 \qquad &= 3000 \\ x_1 + z_2 &= 2000 \end{aligned}$$

which gives $x_1 = 3000$, $z_2 = -1000$, which violates the constraint $Y \geq 0$.

3. Use

$$\{A_1, A_3\} = \left\{ \begin{pmatrix} 1 \\ 1 \end{pmatrix} \begin{pmatrix} 1 \\ 0 \end{pmatrix} \right\}$$

in which case the restraint gives

$$\begin{aligned} x_1 + z_1 &= 3000 \\ x_1 \qquad &= 2000 \end{aligned}$$

and we have $x_1 = 2000$, $z_1 = 1000$. In this case $P^T Y = 6000$.

Because $P^T Y = 6000$ in both case 1 and case 3, we can use either basis for $\mathscr{B}_2$. We choose

$$\mathscr{B}_2 = \left\{\begin{pmatrix}1\\1\end{pmatrix}, \begin{pmatrix}1\\0\end{pmatrix}\right\} = \{A_1, A_3\}$$

Now we want to build $\mathscr{B}_3$. Our choices for the A_j's are

$\{A_1, A_2\}$

$\{A_2, A_3\}$ but this is not a basis

$\{A_2, A_4\}$, $\{A_1, A_4\}$ $\Big\{$ not used because already used in building $\mathscr{B}_1$

For

$$\mathscr{B}_3 = \{A_1, A_2\} = \left\{\begin{pmatrix}1\\1\end{pmatrix}\begin{pmatrix}1\\0\end{pmatrix}\right\}$$

we find from the restraint that

$$\begin{aligned} x_1 + x_2 &= 3000 \\ x_1 \qquad &= 2000 \\ x_1 \qquad &= 2000 \\ x_2 &= 1000 \end{aligned} \qquad \left\{\begin{array}{l} z_1 = 0 \text{ and } z_2 = 0 \\ \text{because they are weights of } A_3 \\ \text{and } A_4 \text{ which are not in } \mathscr{B}_3 \end{array}\right.$$

and obtain

$$P^T Y = 8000$$

Now if we try to find $\mathscr{B}_4$, we obtain only bases which have already been constructed and yielded smaller values of $P^T Y$. Thus the solution to the LP problem is

$$x_1 = 2000 \qquad x_2 = 1000 \qquad \text{and} \qquad \max P^T Y = 8000$$

just as we found before by geometric methods.

The procedure just used would be quite cumbersome for large problems. An algorithm for simplifying the computations is the simplex algorithm. Its foundation is basis searching, as we have done in the example. We now state the method and use it in some examples.

Simplex Algorithm Consider this maximum problem:

$$\begin{array}{ll} \text{Maximize} & T = c_1x_1 + c_2x_2 + \cdots + c_nx_n \\ \text{Subject to} & a_{11}x_1 + \cdots + a_{1n}x_n \le b_1 \\ & \cdots\cdots\cdots\cdots\cdots\cdots \\ & a_{m1}x_1 + \cdots + a_{mn}x_n \le b_m \\ & x_1 \ge 0 \quad \cdots \quad x_n \ge 0 \end{array}$$

1. Write the *initial simplex table* [which is an $(m + 1) \times (m + n + 2)$ matrix]

$$\left(\begin{array}{cccccc|cccc|c} 1 & -c_1 & -c_2 & \cdots & c_n & & 0 & 0 & \cdots & 0 & 0 \\ \hline 0 & a_{11} & a_{12} & \cdots & a_{1n} & & 1 & 0 & \cdots & 0 & b_1 \\ & & & & & & 0 & 1 & & \vdots & \\ \vdots & \vdots & \vdots & & \vdots & & \vdots & & \ddots & 0 & \vdots \\ 0 & a_{m1} & a_{m2} & \cdots & a_{mn} & & 0 & \cdots & 0 & 1 & b_m \end{array}\right)$$

2. Select a negative entry in the first row, and call the column in which it is found the *pivot column.*
3. Select a *pivot row* by listing all the quotients obtained by dividing each positive entry of the pivot column into the corresponding entries in the last column and choosing any row which leads to the smallest quotient.
4. The entry common to the pivot row and pivot column is called the *pivot.* Using row operations, make the pivot element equal to 1 and obtain 0s above and below the pivot.
5. If there are no negative entries in row 1, stop and go to step 6. Otherwise repeat steps 2, 3, and 4.
6. Call the resulting table the *final simplex table.* The maximum value of T in the LP problem is the first entry in the last column of the final simplex table.

One more time we will solve our example problem—this time with the simplex algorithm. The problem is

$$\begin{array}{lll} \text{Maximize} & T = 3x_1 + 2x_2 & \\ \text{Subject to} & x_1 + x_2 \leq 3000 & x_1 \geq 0 \\ & x_1 \leq 2000 & x_2 \geq 0 \end{array}$$

The initial simplex table is

$$\left(\begin{array}{ccc|cc|c} 1 & -3 & -2 & 0 & 0 & 0 \\ \hline 0 & 1 & 1 & 1 & 0 & 3000 \\ 0 & 1 & 0 & 0 & 1 & 2000 \end{array}\right)$$

Step 2 requires choice of a negative entry from row 1. One choice is -3, making column 2 the pivot column. The quotients from step 3 are $\frac{3000}{1}$ and $\frac{2000}{1}$; thus the pivot row is row 3. The row operations required in step 4 are $3R3 + R1$ and $-R3 + R2$. This yields the table

$$\left(\begin{array}{ccc|cc|c} 1 & 0 & -2 & 0 & 3 & 6000 \\ \hline 0 & 0 & 1 & 1 & -1 & 1000 \\ 0 & 1 & 0 & 0 & 1 & 2000 \end{array}\right)$$

Because row 1 still has a negative entry, we repeat steps 2, 3, and 4. The pivot element is in the second row and third column. Using $2R2 + R1$, we have the final simplex table

$$\left(\begin{array}{ccc|cc|c} 1 & 0 & 0 & 0 & 1 & 8000 \\ \hline 0 & 0 & \textcircled{1} & 1 & -1 & \textcircled{1000} \\ 0 & \textcircled{1} & 0 & 0 & 1 & \textcircled{2000} \end{array}\right) \quad x_2 = 1000$$

$$x_1 = 2000$$

and the maximum value of T in the LP problem is 8000, and $x_2 = 1000$, $x_1 = 2000$.

For another example, consider a chemical company which uses three chemicals, A, B, and C, to make three compounds D, E, and F. The maximum amounts of A, B, and C available each day are 75, 200, and 300 lb, respectively. Each pound of D requires $\frac{1}{2}$ lb of A and $\frac{1}{2}$ lb of B; each pound of E requires $\frac{1}{3}$ lb each of A, B, and C; and each pound of F requires $\frac{1}{3}$ lb of B and $\frac{2}{3}$ lb of C. The final products sell as follows: D, \$2 per pound; E, \$1 per pound; and F, \$3 per pound. What production schedule maximizes daily revenue?

Let x_D, x_E, and x_F represent the pounds of D, E, and F, respectively, produced daily. Then revenue $R = 2x_D + x_E + 3x_F$. The constraints on X must be derived now. Because of the limitations on A, B, and C we have

$$\begin{aligned} \text{Total A, lb} &= \tfrac{1}{2}x_D + \tfrac{1}{3}x_E \qquad\qquad \leq 75 \\ \text{Total B, lb} &= \tfrac{1}{2}x_D + \tfrac{1}{3}x_E + \tfrac{1}{3}x_F \leq 200 \\ \text{Total C, lb} &= \qquad\quad\;\; \tfrac{1}{3}x_E + \tfrac{2}{3}x_F \leq 300 \end{aligned}$$

The problem of maximizing R is an LP maximum problem. The simplex table is

$$\left(\begin{array}{cccc|ccc|c} 1 & -2 & -1 & -3 & 0 & 0 & 0 & 0 \\ \hline 0 & \frac{1}{2} & \frac{1}{3} & 0 & 1 & 0 & 0 & 75 \\ 0 & \frac{1}{2} & \frac{1}{3} & \frac{1}{3} & 0 & 1 & 0 & 200 \\ 0 & 0 & \frac{1}{3} & \frac{2}{3} & 0 & 0 & 1 & 300 \end{array}\right)$$

By using -3 to identify the pivot column, the pivot row is row 3; after row operations we have

$$\left(\begin{array}{cccc|ccc|c} 1 & -2 & \frac{1}{2} & 0 & 0 & 0 & \frac{9}{2} & 1350 \\ \hline 0 & \frac{1}{2} & \frac{1}{6} & 0 & 1 & 0 & 0 & 75 \\ 0 & \frac{1}{2} & \frac{1}{6} & 0 & 0 & 1 & -\frac{1}{2} & 50 \\ 0 & 0 & \frac{1}{2} & 1 & 0 & 0 & \frac{3}{2} & 450 \end{array}\right)$$

At the next step, the pivot column is column 2, and the pivot row is row 2. After row operations we obtain the final simplex table.

$$\left(\begin{array}{cccc|ccc|c} 1 & 0 & \frac{7}{6} & 0 & 0 & 4 & \frac{5}{2} & 1550 \\ \hline 0 & 0 & \frac{1}{6} & 0 & 1 & -1 & \frac{1}{12} & 25 \\ 0 & 1 & \frac{1}{3} & 0 & 0 & 2 & -1 & 100 \\ 0 & 0 & \frac{1}{2} & 1 & 0 & 0 & \frac{3}{2} & 450 \end{array}\right)$$

The revenue maximum is \$1550. Using

$$\left(\begin{array}{ccc|c} 0 & \frac{1}{6} & 0 & 25 \\ 1 & \frac{1}{3} & 0 & 100 \\ 0 & \frac{1}{2} & 1 & 450 \end{array}\right)$$

we find $x_D = 50$, $x_E = 150$, and $x_F = 375$ for the optimal production schedule.

The proof that the simplex algorithm works for maximum problems can be found in most advanced linear programming texts. Also the treatment of minimum problems, the phenomenon of cycling, and other subtleties of the simplex method can be found in those texts.

Epilog The simplex algorithm was devised by the mathematician George Dantzig of Stanford University in 1947, and the algorithm has been the basic tool for LP programs since then. Summer 1984 brought the announcement by Narendra Karmarkar, a mathematician at AT&T Bell Laboratories, of a new algorithm, claimed to be significantly faster than the simplex algorithm. As late as December 1985 (*IEEE Spectrum*, December 1985, pp. 54, 55), controversy existed in the mathematical community as to whether Karmarkar's algorithm works as well as claimed. Speed is crucial in some applications of LP to large problems, such as rescheduling airlines when severe weather disrupts regular schedules. Widespread successful application of the new algorithm to problems inaccessible to the simplex method will determine the acceptance of the new algorithm as a working tool for applied mathematics. (See *Business Week*, Sept. 21, 1987, p. 68.)

Regardless of the outcome of the controversy, the excitement caused by Karmarkar's algorithm shows that mathematics and linear algebra are far from cut-and-dried disciplines; they are growing, living fields of study and application.

PROBLEMS 7.3

1. Solve Prob. 1 of Sec. 7.1, (*a*) by basis building and (*b*) by using the simplex algorithm.

2. Solve Prob. 2 of Sec. 7.1, (*a*) by basis building and (*b*) by using the simplex algorithm.

3. Solve Prob. 3 of Sec. 7.1, (*a*) by basis building and (*b*) by using the simplex algorithm.

4. Solve Prob. 4 of Sec. 7.1, (*a*) by basis building and (*b*) by using the simplex algorithm.

5. Use the simplex algorithm to solve

$$\begin{array}{ll} \text{Maximize} & T((x_1, x_2, x_3)) = 2x_1 + 4x_2 + x_3 \\ \text{Subject to} & x_1 + 2x_2 + x_3 \le 400 \\ & \quad x_2 + x_3 \le 100 \\ & \quad x_1 + 3x_3 \le 200 \\ & x_1 \ge 0 \qquad x_2 \ge 0 \qquad x_3 \ge 0 \end{array}$$

SUMMARY

The basis problem is a fundamental tool in the development of the *simplex algorithm* for solving a type of optimization problem known as a linear programming problem. The literature on LP problems is vast; we have barely touched the surface of the subject. Recently (1984) a new algorithm—Karmarkar's algorithm—has been developed for LP problems. This new algorithm holds the promise of faster solution of large-scale LP problems.

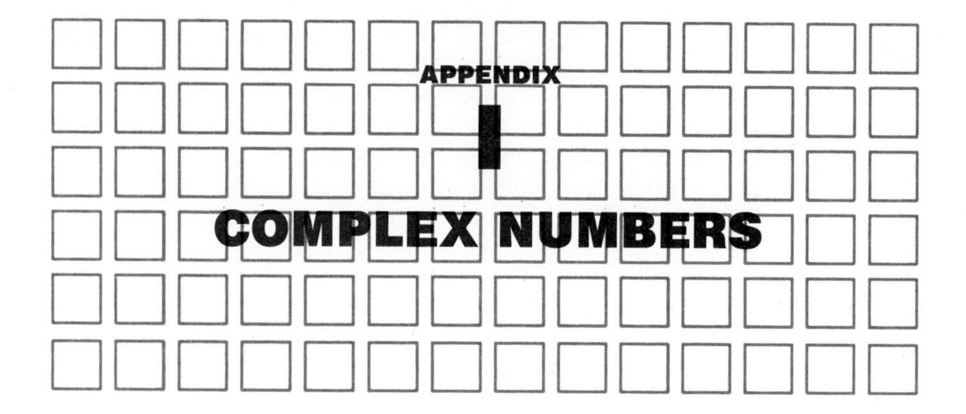

A complex number z is a number of the form $a + bi$, where a and b are real and i is the *imaginary unit* which has the property $i^2 = -1$. The set of complex numbers is denoted $\mathbb{C}$. We write $z = a + bi$, and we call a the real part of z and b the imaginary part of z. For $z = 3 - 5i$, the real part of z is 3, and the imaginary part of z is -5. This is written

$$\operatorname{Re} z = 3 \qquad \operatorname{Im} z = -5$$

Two complex numbers are equal if and only if their real and imaginary parts are equal. The zero for $\mathbb{C}$ is $0 + 0i$.

The rules for addition, subtraction, multiplication, and division of complex numbers are as follows:

$$(a + bi) + (c + di) = (a + c) + (b + d)i$$

$$(a + bi) - (c + di) = (a - c) + (b - d)i$$

$$(a + bi)(c + di) = (ac - bd) + (bc + ad)i \qquad \text{(formally use distributive laws)}$$

$$\frac{a + bi}{c + di} = \frac{(a + bi)}{(c + di)}\frac{(c - di)}{(c - di)} = \frac{ac + bd}{c^2 + d^2} + \frac{bc - ad}{c^2 + d^2}i \qquad c + di \neq 0$$

Thus if

then

$$z_1 = 2 - i \qquad z_2 = -3 + 4i$$

$$z_1 + z_2 = -1 + 3i$$

$$z_1 - z_2 = 5 - 5i$$

$$z_1 z_2 = -6 + 8i + 3i - 4i^2 = -6 + 4 + 11i = -2 + 11i$$

$$\frac{z_1}{z_2} = \frac{2 - i}{-3 + 4i} \cdot \frac{-3 - 4i}{-3 - 4i} = \frac{-6 - 8i + 3i - 4}{25} = \frac{-10 - 5i}{25} = -\frac{2}{5} - \frac{1}{5}i$$

The *conjugate* of a complex number $z = a + bi$ is written $\bar{z}$ and defined

$$\bar{z} = a - bi$$

Note that $\bar{z} = z$ if and only if $b = 0$—that is, if and only if z is real. Important properties of the conjugate are as follows:

If z_1 and z_2 are complex numbers, then

$$\overline{z_1 + z_2} = \bar{z}_1 + \bar{z}_2$$

$$\overline{z_1 z_2} = \bar{z}_1 \bar{z}_2$$

$$\overline{\left(\frac{z_1}{z_2}\right)} = \frac{\bar{z}_1}{\bar{z}_2}$$

$$\bar{z}_1 = z_1$$

For example, consider the numbers used above $z_1 = 2 - i$ and $z_2 = -3 + 4i$. We have $\bar{z}_1 = 2 + i$, $\bar{z}_2 = -3 - 4i$, and $\bar{z}_1 + \bar{z}_2 = -1 - 3i = \overline{-1 + 3i} = \overline{z_1 + z_2}$.

The *magnitude* of $z = a + bi$ is denoted $|z|$ and defined by

$$|z| = \sqrt{z\bar{z}} = \sqrt{a^2 + b^2}$$

Therefore, $|-3 + 4i| = \sqrt{3^2 + 4^2} = 5$. If we associate $a + bi$ with the point (a, b) on the axes in Fig. I.1, we see that $|z|$ is the distance of the complex number from zero.

The complex numbers along with the operations defined are a *number field*. This means that $\mathbb{C}$ enjoys the same properties as $\mathbb{R}$:

$z_1 + z_2 = z_2 + z_1$ $z_1 z_2 = z_2 z_1$	Commutativity of addition and multiplication
$(z_1 + z_2) + z_3 = z_1 + (z_2 + z_3)$ $(z_1 z_2) z_3 = z_1(z_2 z_3)$	Associativity of addition and multiplication

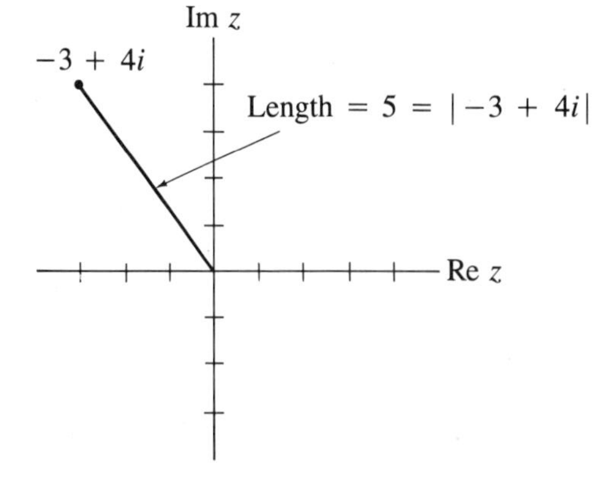

Figure I.1

$z_1 + 0 = z_1$	Additive identity
$z_1 + (-z_1) = 0$	Additive inverse
$z_1 1 = z_1$	Multiplicative identity
$z_1 \cdot \dfrac{1}{z_1} = 1 \qquad (z_1 \neq 0)$	Multiplicative inverse
$z_1(z_2 + z_3) = z_1 z_2 + z_1 z_3$	Distributive law

Because these properties hold, computing with complex numbers differs from computing with real numbers only in the operations themselves.

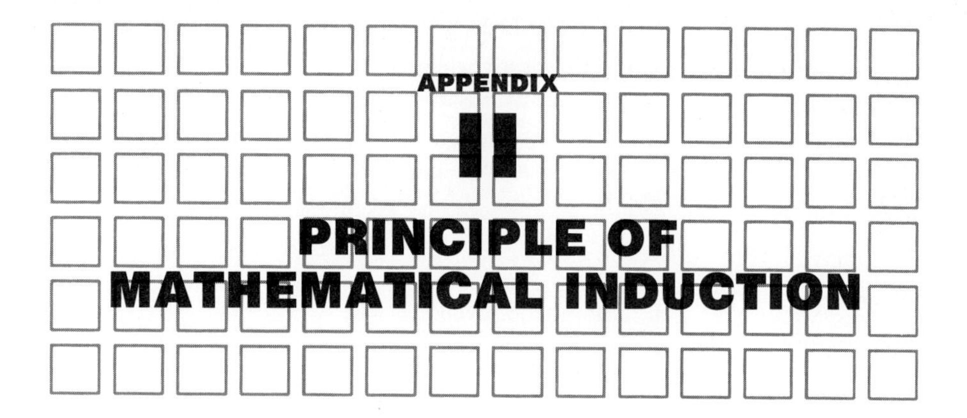

To prove a statement about the positive integers, we can use the principle of mathematical induction (PMI). Let $P(n)$ denote a proposition about the positive integers. For example,

$$1 + 2 + \cdots + n = \frac{n(n+1)}{2} \tag{II.1}$$

and

$$1 + x + x^2 + \cdots + x^n = \frac{1 - x^{n+1}}{1 - x} \tag{II.2}$$

and

$$n! > 2^n \qquad n \geq 4 \tag{II.3}$$

are all examples of such propositions.

To prove that $P(n)$ is true, the PMI tells us that it is sufficient to do two things.

1. Prove that $P(n)$ is true for the first admissible value for n. Usually $n = 1$, but this need not be so: For (II.3) the first value of n is 4.

   Note: This is the *induction hypothesis.*

2. Prove that if $P(k)$ is true, it must follow that $P(k+1)$ is true.

It may seem strange that we need prove only two statements to show the truth of a proposition for an infinite set of values of n. Consider, however, the following instructions for stringing an infinite necklace of beads:

1. If the last bead was red, string another red bead.

2. String a red bead first.

Only two instructions suffice to complete the infinite necklace.

As an example, consider (II.3) [(II.1) was considered in the text]. Then $P(n)$ is

$$n! > 2^4 \qquad n \geq 4$$

1. Is $P(4)$ true? Well

$$4! = 24 \qquad 2^4 = 16$$

so

$$4! > 2^4$$

and $P(4)$ is true.

2. Suppose that $P(k)$ is true. That is, suppose $k! > 2^k$. We need to show that $P(k + 1)$ is true—that is, we need to show that $(k + 1)! > 2^{k+1}$. Starting with $k! > 2^k$, we multiply both sides by $k + 1$ to find

Because $k! > 2^k$ (induction hypothesis)

$$(k + 1)k! > (k + 1)2^k > 2 \cdot 2^k = 2^{k+1}$$

$k \geq 4$, so $k + 1 > 2$

Therefore

$$(k + 1)! > 2^{k+1}$$

By the PMI it is true that $n! > 2^n$, $n \geq 4$.

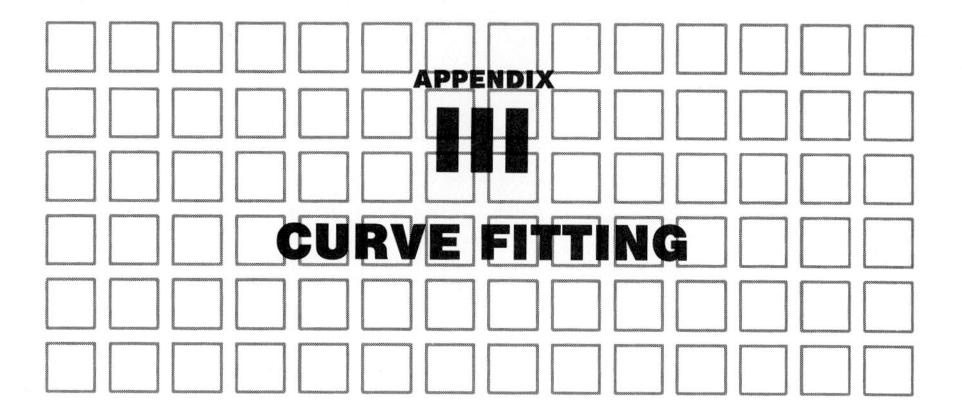

APPENDIX III CURVE FITTING

An application of linear equations is the fitting of curves to data which can be represented by points in the plane. For example, if a person records at different temperatures the pressure of a gas, the information could be represented as in Fig. III.1. One reason for "fitting" a curve to data points of this type is to estimate pressures at higher temperatures. One way to fit a curve to data is to use a curve which passes through all the data points—this process is called *collocation.* Another way is to find a straight line about which the data clusters—this can be done with a process known as *linear least-squares regression.* In Fig. III.2 we show these two possibilities.

Figure III.1

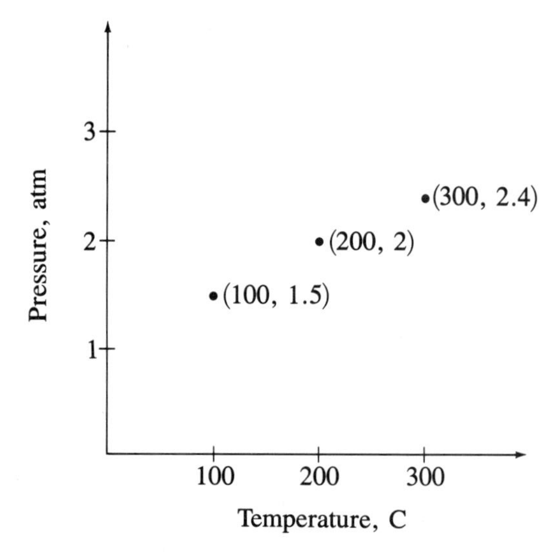

Figure III.2

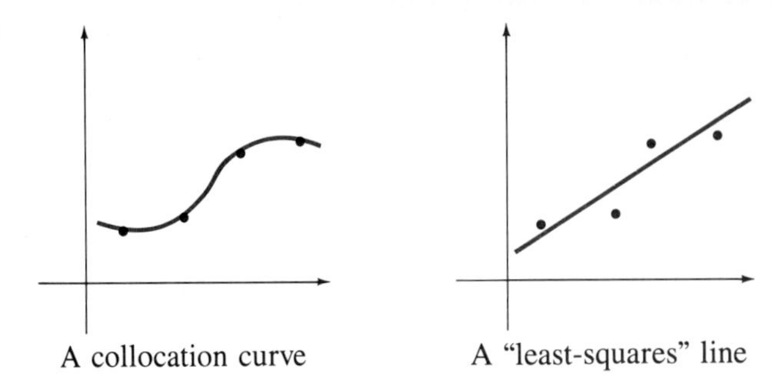

A collocation curve

A "least-squares" line

EXAMPLE 1 Find a quadratic function $p(x) = ax^2 + bx + c$ which passes through the points $(-1, 6)$, $(1, 4)$, and $(2, 12)$ in the xy plane.

Solution We must find the coefficients a, b, and c in $p(x)$. Since the graph of $y = p(x)$ must pass through all the points, we must have

$$\begin{aligned} 6 &= p(-1) = a - b + c \\ 4 &= p(1) = a + b + c \\ 12 &= p(2) = 4a + 2b + c \end{aligned}$$

Thus the equations for a, b, and c are represented by the augmented matrix

$$\left(\begin{array}{ccc|c} 1 & -1 & 1 & 6 \\ 1 & 1 & 1 & 4 \\ 4 & 2 & 1 & 12 \end{array}\right)$$

This row-reduces to

$$\left(\begin{array}{ccc|c} 1 & -1 & 1 & 6 \\ 0 & 2 & 0 & -2 \\ 0 & 0 & -3 & -6 \end{array}\right)$$

and $c = 2$, $b = -1$, $a = 3$. The quadratic function is $p(x) = 3x^2 - x + 2$ (see Fig. III.3).

For our second example we need to know the so-called normal equations for the least squares. These equations are derived in courses in calculus; we simply state them here.

Given N points $(x_1, y_1), (x_2, y_2), \ldots, (x_N, y_N)$ in the xy plane, the equation of the line of best least-squares fit is $y = Ax + B$, where A and B are the solutions to the normal equations

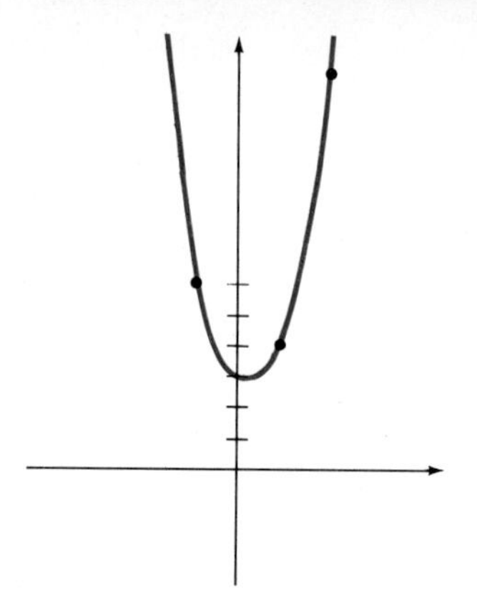

Figure III.3

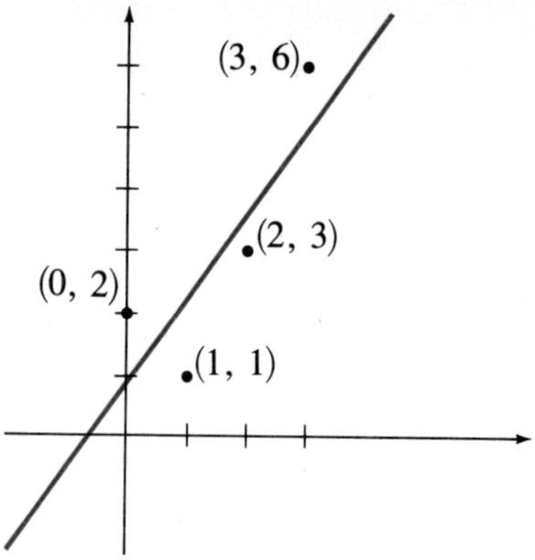

(3, 6)
(2, 3)
(0, 2)
(1, 1)

Figure III.4

$$(x_1^2 + \cdots + x_N^2)A + (x_1 + \cdots + x_N)B = x_1y_1 + \cdots + x_Ny_N$$
$$(x_1 + \cdots + x_N)A + NB = y_1 + \cdots + y_N$$

Note that the matrix of coefficients is symmetric.

From *Cramer's rule* this system always has a unique solution if the x values are distinct.

EXAMPLE 2 Find the line of best least-squares fit for the set of points (0, 2), (1, 1), (2, 3), (3, 6). Sketch the data and the line.

Solution Identifying $(x_1, y_1) = (0, 2)$, $(x_2, y_2) = (1, 1)$, and so on, we have

$$x_1^2 + x_2^2 + x_3^2 + x_4^2 = 14$$
$$x_1 + x_2 + x_3 + x_4 = 6$$
$$x_1y_1 + x_2y_2 + x_3y_3 + x_4y_4 = 25$$
$$y_1 + y_2 + y_3 + y_4 = 12$$

and the normal equations are

$$14A + 6B = 25$$
$$6A + 4B = 12$$

and so $A = \frac{7}{5}$, $B = \frac{9}{10}$. The line of best least-squares fit is $y = \frac{7}{5}x + \frac{9}{10}$, which, as can be seen from Fig. III.4, follows the trend of the data points.

Whether to use collocation or least-squares methods in a given problem depends on the specific application. Most of the time least-squares methods are a good choice. There exist least-squares methods in which nonlinear functions are used to fit data. Generally speaking, computer programs exist for carrying out most data-fitting methods.

Higher Dimensions Consider a set of points in three-space $\{(x_1, y_1, z_1), (x_2, y_2, z_2), \ldots, (x_n, y_n, z_n)\}$. In this case for *linear* description of the data, we may try to "fit" a plane to the points. The least-squares problem becomes, by letting $z = Ax = By + C$,

$$\text{Minimize (over all choices of } A, B, C) \sum_{k=1}^{n} [z_k - (Ax_k + By_k + C)]^2$$

Setting the partial derivatives equal to zero to locate possible extremes, we obtain three equations in three unknowns for the normal equations. These can be solved to determine the plane of best least-squares fit. As the number of dimensions exceeds 3, we must talk about "hyperplanes of best least-squares fit," and the normal equations have an $n \times n$ coefficient matrix if the data are from E^n.

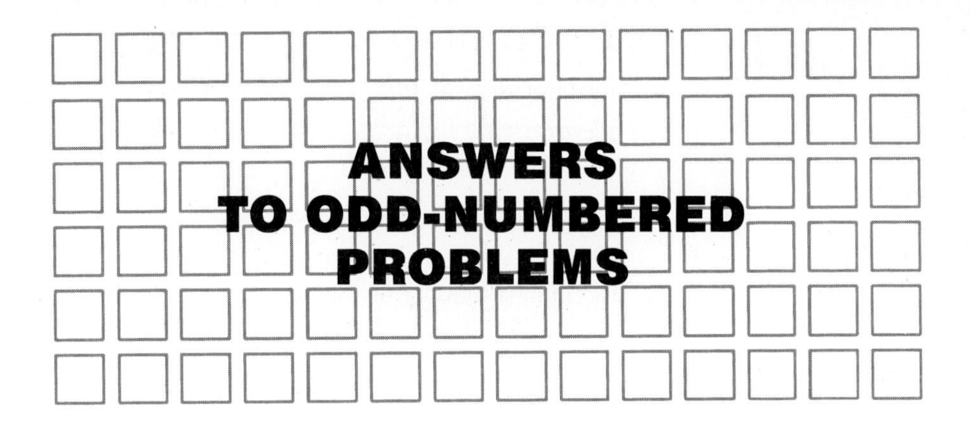

PROBLEMS 1.1

1. (*a*) Nonlinear. (*b*) Nonlinear. (*c*) Linear. (*d*) Linear. (*e*) Nonlinear. (*f*) Nonlinear. (*g*) Nonlinear. (*h*) Linear.

3. Let L be the number of batches of lean, E be the number of batches of extra lean. The equations are

$$\text{Total fat} = 10 = 1.5L + E$$
$$\text{Total red} = 80 = 8.5L + 9E$$

The solution is $L = 2$, $E = 7$.

5. $x = 2.1$, $y = -.9$ 7. $x = 76/13$, $y = -5/13$

9. $B = 400C + 0.35T + 120H$

PROBLEMS 1.2

1. (*a*) $x_1 = 1$, $x_2 = -1$, $x_3 = 4$

(*b*) $x_1 = (14 + 2s)/5$, $x_2 = (1 + 3s)/5$, $x_3 = s$, s arbitrary

(*c*) No solution. (*d*) $x_1 = 11/3$, $x_2 = -1/3$.

(*e*) $x_1 = (30 + s)/8$, $x_2 = (11s - 6)/16$, $x_3 = s$, s arbitrary

(*f*) No solution.

3. (*a*), (*b*), (*d*), (*g*), and (*h*) are in row echelon form.

(*c*) Fails because $a_{13} \neq 0$. (*e*) Fails because $a_{13} \neq 0$.

(*f*) Fails because $a_{44} \neq 1$ or 0.

5. In each case, eliminate variables.

(a) Only one solution regardless of the choice of a and b.

(b) If $-2a + b \neq 0$, then no solution; if $-2a + b = 0$, then an infinite number of solutions.

(c) Only one solution regardless of the choice of a and b.

(d) If $-a - 3b + c \neq 0$, then no solution; if $-a - 3b + c = 0$, then an infinite number of solutions.

7. Row reduction leads to

$$\left(\begin{array}{cc|c} a & b & r \\ c & d & s \end{array}\right) \longrightarrow \left(\begin{array}{cc|c} 1 & b/a & r/a \\ c & d & s \end{array}\right) \longrightarrow \left(\begin{array}{cc|c} 1 & b/a & r/a \\ 0 & d - cb/a & s - cr/a \end{array}\right)$$

Therefore if $d - cb/a \neq 0$ we have a unique solution. Equivalently, the condition is $ad - bc \neq 0$.

9. Row reduction as in Prob. 7 leads to

$$\left(\begin{array}{cccc|c} 1 & b/a & 0 & 0 & r/a \\ 0 & ad - bc & 0 & 0 & as - cr \\ 0 & 0 & 1 & f/e & t/e \\ 0 & 0 & 0 & eh - fg & eu - gt \end{array}\right)$$

If $ad - bc \neq 0$ and $eh - fg \neq 0$, then we have a unique solution.

11. Suppose first that $a \neq 0$. Then we can row reduce

$$\begin{pmatrix} a & b \\ c & d \end{pmatrix} \quad \begin{pmatrix} 1 & b/a \\ c & d \end{pmatrix} \longrightarrow \begin{pmatrix} 1 & b/a \\ 0 & d - bc/a \end{pmatrix} \longrightarrow \begin{pmatrix} 1 & b/a \\ 0 & ad - bc \end{pmatrix} = \begin{pmatrix} 1 & b/a \\ 0 & 0 \end{pmatrix}$$

This is the reduced row echelon form. Suppose on the other hand that $a = 0$. Then we have $ad - bc = -bc = 0$, so either $b = 0$ or $c = 0$.

If $b = 0$ *and* $c = 0$, we have the matrix

$$\begin{pmatrix} 0 & 0 \\ 0 & d \end{pmatrix}$$

Which row reduces to $\begin{pmatrix} 0 & d \\ 0 & 0 \end{pmatrix}$. If $d \neq 0$ we have reduced row echelon form of

$$\begin{pmatrix} 0 & 1 \\ 0 & 0 \end{pmatrix}$$

If $d = 0$ we have

$$\begin{pmatrix} 0 & 0 \\ 0 & 0 \end{pmatrix}$$

If $b = 0$ and $c \neq 0$, we have the matrix

$$\begin{pmatrix} 0 & 0 \\ c & d \end{pmatrix}$$

The reduced row echelon form is

$$\begin{pmatrix} 1 & d/c \\ 0 & 0 \end{pmatrix}$$

If $b \neq 0$ and $c = 0$, we have the matrix

$$\begin{pmatrix} 0 & b \\ 0 & d \end{pmatrix}$$

The reduced row echelon form is $\begin{pmatrix} 0 & 1 \\ 0 & 0 \end{pmatrix}$.

13. Substitute:

$$a(u + r) + b(v + s) = au + ar + bv + bs = (au + bv) + (ar + bs) = m + 0 = m$$
$$c(u + r) + d(v + s) = cu + cr + dv + ds = (cu + dv) + (cr + ds) = n + 0 = n$$

15. $x_1 = t$, $x_2 = \sqrt{2}t$, $x_3 = t$, t arbitrary.

17. (*a*) See the solution to Prob. 11.

(*b*) Simply consider all possible pairings. For instance compare

$$\begin{pmatrix} 0 & 1 \\ 0 & 0 \end{pmatrix} \quad \text{and} \quad \begin{pmatrix} 1 & r \\ 0 & 0 \end{pmatrix}$$

No row operations can introduce a nonzero entry in the first row and first column of the first matrix. Thus these two matrices cannot be row equivalent.

(*c*) Suppose a 2×2 matrix had two distinct reduced row echelon forms. These forms would have to come from the list in *b*. Let A be the original matrix and B and C be the two reduced row echelon forms. Then by Prob. 16, since C is row equivalent to A and A is row equivalent to B, then C has to be row equivalent to B, a contradiction.

19. The last row is all zeros. Perform the row operations $-R1 + R2$, $-R1 + R3, \ldots, -R1 + Rn$. Then perform

$$\frac{1}{n} R2, \frac{1}{2n} R3, \ldots, \frac{1}{(n-1)n} Rn$$

to obtain

$$\begin{pmatrix} 1 & 2 & 3 & \cdots & n \\ 1 & 1 & 1 & \cdots & 1 \\ 1 & 1 & 1 & \cdots & 1 \\ \cdots & \cdots & \cdots & \cdots & \cdots \\ 1 & 1 & 1 & \cdots & 1 \end{pmatrix}$$

Now it is easy to obtain all zeros in rows 3 through n.

21. $x = \overline{2 - is} = 2 + is$, $y = s$, s arbitrary.

23. For part 1 suppose $s_1, \ldots, s_n$ is a solution to the system. Therefore

$$\begin{aligned} a_{11}s_1 + a_{12}s_2 + \cdots + a_{1n}s_n &= b_1 \\ \cdots\cdots\cdots\cdots&\cdots\cdots \\ a_{i1}s_1 + a_{i2}s_2 + \cdots + a_{in}s_n &= b_i \\ \cdots\cdots\cdots\cdots&\cdots\cdots \\ a_{m1}s_1 + a_{m2}s_2 + \cdots + a_{mn}s_n &= b_m \end{aligned} \tag{1}$$

Now multiply equation i by r to obtain for the ith equation

$$ra_{i1}x_1 + ra_{i2}x_2 + \cdots + ra_{in}x_n = rb_i$$

Now factor:

$$r(a_{i1}x_1 + a_{i2}x_2 + \cdots + a_{in}x_n) = rb_i$$

From (1), $a_{i1}s_1 + \cdots + a_{in}s_n = b_i$; therefore,

$$rb_i = rb_i$$

All the other equations being unchanged have $(s_1, \ldots, s_n)$ as a solution also. Therefore $(s_1, \ldots, s_n)$ is a solution of the new system. This argument is reversible, by multiplying by $1/r$.

For part 2, note that interchanging two equations simply reorganizes the system; no algebraic operation has occurred.

PROBLEMS 1.3

1. (*a*) .0008. (*b*) $.5480 \times 10^7$. (*c*) .7930. (*d*) $.5000 \times 10^4$

3. $\dfrac{a/b}{c/d} = \dfrac{-1}{(.0003)/2} = \dfrac{-1}{.0002} = -5000 \qquad \dfrac{ad}{bc} = \dfrac{-6}{.0009} = -6667$

5. (*a*) $x_1 = 10$, $x_2 = 1$ (*b*) $x_1 = -2.571$, $x_2 = .9997$

(*c*) $x_1 = -2.582$, $x_2 = 1.001$

The actual solution is $x_1 = -2.580979\ldots$, $x_2 = 1.000396\ldots$. Sample calculations for *a*

$$\left(\begin{array}{cc|c} .0001 & 3.172 & 3.173 \\ .6721 & 4.227 & 2.494 \end{array}\right) \xrightarrow{\frac{1}{.0001}R1} \left(\begin{array}{cc|c} 1 & 31720 & 31730 \\ .6721 & 4.227 & 2.494 \end{array}\right)$$

Now using $-.6721R1 + R2$ we have

$$\begin{aligned} (-.6721)(31720) + 4.227 &= (-.6721 \times 10^0)(.3172 \times 10^5) + (.4227 \times 10^1) \\ &= -(.2131901\cdots \times 10^5) + (.4227 \times 10^1) \\ &\overset{\text{Round}}{=} -(.2132 \times 10^5) + .4227 \times 10^1 \\ &= -.2132 \times 10^5 + .00004 \times 10^5 \\ &\overset{\text{Round}}{=} -.2132 \times 10^5 + 0 = -21320 \end{aligned}$$

So we have

$$\left(\begin{array}{cc|c} 1 & 31720 & 31730 \\ 0 & -21320 & -21330 \end{array}\right)$$

Thus after rounding, $x_2 = 1$ and $x_1 = 10$.

7. (*a*) $x_1 = 99.9$, $x_2 = .999$, $x_3 = .1$ (*b*) $x_1 = 100$, $x_2 = 1$, $x_3 = .1$

(*c*) $x_1 = 100$, $x_2 = 1$, $x_3 = .1$

Note: in *b* after partial pivoting and back substitution the equation for x_1 is

$$x_1 = 99.1 + x_2 - x_3 = 99.1 + 1 - .1$$

If we calculate $(99.1 + 1) - .1$, we get 99.9. If we calculate $99.1 + (1 - .1)$, we get 100. In the first instance the subtraction of a small number from a large one leads to inaccuracy. The exact solution is $x_1 = 100$, $x_2 = 1$, $x_3 = .1$.

9. (*a*) $(x_1, x_2, x_3) = (1, 1, 1)$. (*b*) $(x_1, x_2, x_3) = (1.09, .49, 1.5)$. Pivoting gives no advantage.

PROBLEMS 1.4

1. (*a*) Row one, column two. (*b*) Row three, column two.

(*c*) Row two, column one. (*d*) Row four, column one.

(*e*) Row two, column three. (*f*) Row six, column one.

3. (*a*) $\left(\begin{array}{ccc|c} 1 & -1 & 1 & 6 \\ 1 & 1 & 2 & 8 \\ 2 & -3 & -1 & 1 \end{array}\right)$, 3×4 (*b*) $\left(\begin{array}{ccc|c} 2 & -3 & 1 & 5 \\ 1 & 1 & -1 & 3 \\ 4 & -1 & -1 & 11 \end{array}\right)$, 3×4

(*c*) $\left(\begin{array}{ccc|c} 1 & -3 & 1 & 6 \\ 2 & 1 & -3 & -2 \\ 1 & 4 & -4 & 0 \end{array}\right)$, 3×4 (*d*) $\left(\begin{array}{cc|c} 1 & -1 & 4 \\ 2 & 1 & 7 \\ 5 & -2 & 19 \end{array}\right)$, 3×3

(*e*) $\left(\begin{array}{ccc|c} 3 & -2 & 1 & 12 \\ 2 & -6 & 4 & 6 \end{array}\right)$, 2×4 (*f*) $\left(\begin{array}{cc|c} 1 & 2 & 6 \\ 3 & 6 & 8 \\ 5 & 10 & 12 \end{array}\right)$, 3×3

5. For example $A = \begin{pmatrix} 1 & 0 & 0 \\ 0 & 1 & 0 \\ 0 & 0 & 1 \end{pmatrix}$ and any $B_{3 \times 3}$. **7.** $AB = \begin{pmatrix} 2 & 2 \\ 2 & 2 \end{pmatrix} = BC$.

9. No. $\begin{pmatrix} i & i \\ i & i \end{pmatrix}\begin{pmatrix} i & 2i \\ 2i & i \end{pmatrix} = \begin{pmatrix} -3 & -3 \\ -3 & -3 \end{pmatrix}$; other examples are possible.

11. Set $a = GK$ and $b = L$. **13.** All the entries of B are zero.

15. No. Let

$$A = \begin{pmatrix} 1 & 1 \\ 2 & 2 \end{pmatrix} \qquad B = \begin{pmatrix} 1 & 0 \\ 0 & 1 \end{pmatrix} \qquad C = \begin{pmatrix} 1 & 2 \\ 1 & 2 \end{pmatrix}$$

Other examples are possible.

17. (*a*) Not possible. (*b*) $\begin{pmatrix} 3 & 0 & 9 \\ -2 & 6 & 0 \end{pmatrix}$ (*c*) $\begin{pmatrix} 15 & 60 \\ -4 & 49 \end{pmatrix}$ (*d*) $\begin{pmatrix} -5 \\ 0 \\ 12 \\ 3 \end{pmatrix}$

PROBLEMS 1.5

1. (*a*) $A + B = \begin{pmatrix} 1 & 6 \\ -1 & 7 \end{pmatrix} \qquad B + A = \begin{pmatrix} 1 & 6 \\ -1 & 7 \end{pmatrix} \qquad A + B = B + A$

(*b*) $A + (B + C) = \begin{pmatrix} 1 & 2 \\ -3 & 2 \end{pmatrix} + \begin{pmatrix} 1 & 4 \\ 7 & 7 \end{pmatrix} = \begin{pmatrix} 2 & 6 \\ 4 & 9 \end{pmatrix}$

$(A + B) + C = \begin{pmatrix} 1 & 6 \\ -1 & 7 \end{pmatrix} + \begin{pmatrix} 1 & 0 \\ 5 & 2 \end{pmatrix} = \begin{pmatrix} 2 & 6 \\ 4 & 9 \end{pmatrix}$

$A + (B + C) = (A + B) + C$

(*c*) $A(B + C) = \begin{pmatrix} 1 & 2 \\ -3 & 2 \end{pmatrix}\begin{pmatrix} 1 & 4 \\ 7 & 7 \end{pmatrix} = \begin{pmatrix} 15 & 18 \\ 11 & 2 \end{pmatrix}$

$AB + BC = \begin{pmatrix} 4 & 14 \\ 4 & -2 \end{pmatrix} + \begin{pmatrix} 11 & 4 \\ 7 & 4 \end{pmatrix} = \begin{pmatrix} 15 & 18 \\ 11 & 2 \end{pmatrix}$

$A(B + C) = AB + AC$

(*d*) $(B + C)A = \begin{pmatrix} 1 & 4 \\ 7 & 7 \end{pmatrix}\begin{pmatrix} 1 & 2 \\ -3 & 2 \end{pmatrix} = \begin{pmatrix} -11 & 10 \\ -14 & 28 \end{pmatrix}$

$BA + CA = \begin{pmatrix} -12 & 8 \\ -13 & 14 \end{pmatrix} + \begin{pmatrix} 1 & 2 \\ -1 & 14 \end{pmatrix} = \begin{pmatrix} -11 & 10 \\ -14 & 28 \end{pmatrix}$

$(B + C)A = BA + CA$

(*e*) $(AB)C = \begin{pmatrix} 4 & 14 \\ 4 & -2 \end{pmatrix}\begin{pmatrix} 1 & 0 \\ 5 & 2 \end{pmatrix} = \begin{pmatrix} 74 & 28 \\ -6 & -4 \end{pmatrix}$

$A(BC) = \begin{pmatrix} 1 & 2 \\ -3 & 2 \end{pmatrix}\begin{pmatrix} 20 & 8 \\ 27 & 10 \end{pmatrix} = \begin{pmatrix} 74 & 28 \\ -6 & -4 \end{pmatrix} \qquad (AB)C = A(BC)$

(*f*) $r(A + B) = 4\begin{pmatrix} 1 & 6 \\ -1 & 7 \end{pmatrix} = \begin{pmatrix} 4 & 24 \\ -4 & 28 \end{pmatrix}$

$rA + rB = \begin{pmatrix} 4 & 8 \\ -12 & 8 \end{pmatrix} + \begin{pmatrix} 0 & 16 \\ 8 & 20 \end{pmatrix} = \begin{pmatrix} 4 & 24 \\ -4 & 28 \end{pmatrix}$

$r(A + B) = rA + rB$

(g) $(r+s)A = -3\begin{pmatrix} 1 & 2 \\ -3 & 2 \end{pmatrix} = \begin{pmatrix} -3 & -6 \\ 9 & -6 \end{pmatrix}$

$rA + sA = \begin{pmatrix} 4 & 8 \\ -12 & 8 \end{pmatrix} + \begin{pmatrix} -7 & -14 \\ 21 & -14 \end{pmatrix} = \begin{pmatrix} -3 & -6 \\ 9 & -6 \end{pmatrix}$

$(r+s)A = rA + sA$

(h) $(rs)A = -28\begin{pmatrix} 1 & 2 \\ -3 & 2 \end{pmatrix} = \begin{pmatrix} -28 & -56 \\ 84 & -56 \end{pmatrix}$

$r(sA) = 4\begin{pmatrix} -7 & -14 \\ 21 & -14 \end{pmatrix} = \begin{pmatrix} -28 & -56 \\ 84 & -56 \end{pmatrix}$

$s(rA) = -7\begin{pmatrix} 4 & 8 \\ -12 & 8 \end{pmatrix} = \begin{pmatrix} -28 & -56 \\ 84 & -56 \end{pmatrix}$

$(sr)A = -28\begin{pmatrix} 1 & 2 \\ -3 & 2 \end{pmatrix} = \begin{pmatrix} -28 & -56 \\ 84 & -56 \end{pmatrix}$

All are equal.

3. Yes. Let $A = (a_{ij})_{n \times n}$, $B = (b_{ij})_{n \times n}$ with $a_{ij} = 0$, $b_{ij} = 0$, whenever $i \neq j$ (that is, off the diagonal). $A + B = (a_{ij} + b_{ij})$. Now whenever $i \neq j$, $a_{ij} + b_{ij} = 0 + 0 = 0$. Thus $A + B$ is a diagonal matrix.

5. Yes. Let $A = (a_{ij})_{n \times n}$ and $B = (b_{ij})_{n \times n}$ be symmetric. That is, suppose $a_{ij} = a_{ji}$, $b_{ij} = b_{ji}$ for all $i, j = 1, 2, \ldots, n$. $A + B = (a_{ij} + b_{ij}) = (c_{ij})$. Now $c_{ij} = a_{ij} + b_{ij} = a_{ji} + b_{ji} = c_{ji}$. Thus $A + B$ is symmetric.

7. (a) The product of diagonal matrices is diagonal. Let A and B be diagonal and set $C = AB$. Then

$$c_{ij} = \sum_{k=1}^{n} a_{ik}b_{kj}$$

Now if $i \neq j$ the sum expands as

$$\sum_{k=1}^{n} a_{ik}b_{kj} = a_{i1}b_{1j} + \cdots + a_{ii}b_{ij} + \cdots + a_{ij}b_{jj} + \cdots + a_{ij}b_{nj}$$

and it is seen that every summand is zero because at least one factor of each summand has unequal subscripts. Thus $c_{ij} = 0$ when $i \neq j$ and AB is diagonal.

(b) No. Consider $\begin{pmatrix} 1 & 1 & 0 \\ 1 & 1 & 1 \\ 0 & 1 & 1 \end{pmatrix}\begin{pmatrix} 1 & 1 & 0 \\ 1 & 1 & 1 \\ 0 & 1 & 1 \end{pmatrix} = \begin{pmatrix} 2 & 2 & 1 \\ 2 & 3 & 2 \\ 1 & 2 & 2 \end{pmatrix}$.

(c) No. Consider $\begin{pmatrix} 1 & 2 \\ 2 & 1 \end{pmatrix}\begin{pmatrix} 2 & -3 \\ -3 & 4 \end{pmatrix} = \begin{pmatrix} -4 & 5 \\ 1 & -2 \end{pmatrix}$.

(*d*) Yes. Let $A_{n\times n}$ and $B_{n\times n}$ be upper triangular. That is $a_{ij}=0$, $b_{ij}=0$, whenever $i>j$. Let $AB=C$. Consider

$$c_{ij}=\sum_{k=1}^{n} a_{ik}b_{kj} \qquad \text{for } i>j$$

In the expansion of the sum, when $i>k$, $a_{ik}=0$. So

$$c_{ij}=\sum_{k=i}^{n} a_{ik}b_{kj}$$

Now when $k>j$, $b_{kj}=0$; in the sum $k\geq i$ and $i>j$, so $k>j$ and

$$c_{ij}=\sum_{k=1}^{n} a_{ik}(0)=0$$

Thus when $i>j$, $c_{ij}=0$.

9. (*a*) $(A+B)^2=\begin{pmatrix}-5 & 48\\ -8 & 43\end{pmatrix}$ (*b*) $A^2+2AB+B^2=\begin{pmatrix}11 & 54\\ 9 & 27\end{pmatrix}$

11. (*a*) $\begin{pmatrix}0 & -b\\ b & 0\end{pmatrix}$

(*b*) Yes. Let A and B be skew-symmetric, so that $A^T=-A$, $B^T=-B$. Now

$$(A+B)^T=A^T+B^T=-A+(-B)=-(A+B)$$

so $A+B$ is skew-symmetric.

(*c*) No. Let A be skew-symmetric, so that $A^T=-A$. $(A^2)^T=(AA)^T=A^TA^T=(-A)(-A)=A^2$. Therefore A^2 is actually symmetric. (Note: $A=0$ is both symmetric and skew-symmetric.)

(*d*) Yes. Let A be skew-symmetric so that $A^T=-A$. Now $(A^3)^T=(A^T)^3=(-A)^3=-A^3$, so A^3 is skew-symmetric.

(*e*) Not much unless $AB=BA$. Let A be symmetric, B be skew-symmetric so that $A^T=A$, $B^T=-B$. Now $(AB)^T=B^TA^T=-BA$. If $AB=BA$, then $(AB)^T=-BA=-AB$ and AB is skew-symmetric.

(*f*) The diagonal must consist of all zeros. After all, if $a_{ij}=-a_{ji}$ then when $i=j$ we have $a_{ii}=-a_{ii}$ which can be satisfied if and only if $a_{ii}=0$.

13. If and only if $AB=BA$. Calculating, we have

$$(A+B)(A-B)=A(A-B)+B(A-B)=A^2-AB+BA-B^2$$

15. Calculate.

17. (*a*) No. Consider $D = \begin{pmatrix} 1 & 0 \\ 0 & 2 \end{pmatrix}$, $A = \begin{pmatrix} 3 & 4 \\ 5 & 6 \end{pmatrix}$.

$$AD = \begin{pmatrix} 3 & 8 \\ 5 & 12 \end{pmatrix} \neq \begin{pmatrix} 3 & 4 \\ 10 & 12 \end{pmatrix} = DA.$$

(*b*) If D is a scalar matrix $AD = DA$ by Theorem 1.5.2 parts g and h together.

19. Let $A = (a_{ij})$, $a_{ij} \in \mathbb{R}$. Because a_{ij} is real $\overline{a_{ij}} = a_{ij}$. Thus $\bar{A} = (\overline{a_{ij}}) = (a_{ij}) = A$.

21. $$C = \begin{pmatrix} 3+i & 1+i \\ 1 & -1 \end{pmatrix} = \begin{pmatrix} 3 & 1 \\ 1 & -1 \end{pmatrix} + \begin{pmatrix} i & i \\ 0 & 0 \end{pmatrix} = \underbrace{\begin{pmatrix} 3 & 1 \\ 1 & -1 \end{pmatrix}}_{A} + i\underbrace{\begin{pmatrix} 1 & 1 \\ 0 & 0 \end{pmatrix}}_{B}$$

$$D = \begin{pmatrix} 4+4i \\ 2i \end{pmatrix} = \underbrace{\begin{pmatrix} 4 \\ 0 \end{pmatrix}}_{R} + i\underbrace{\begin{pmatrix} 4 \\ 2 \end{pmatrix}}_{S}.$$

Using Prob. 20, the system is

$$\begin{pmatrix} 3 & 1 & -1 & -1 \\ 1 & -1 & 0 & 0 \\ 1 & 1 & 3 & 1 \\ 0 & 0 & 1 & -1 \end{pmatrix} \begin{pmatrix} x_1 \\ x_2 \\ y_1 \\ y_2 \end{pmatrix} = \begin{pmatrix} 4 \\ 0 \\ 4 \\ 2 \end{pmatrix}$$

23. (*a*) $A^{**} = (A^*)^* = (\overline{a_{ji}})^* = (\overline{\overline{a_{ij}}}) = (a_{ij}) = A$.

(*b*) $(A + B)^* = (\overline{A + B})^T = (\bar{A} + \bar{B})^T + \bar{A}^T + \bar{B}^T = A^* + B^*$.

(*c*) $(AB)^* = (\overline{AB})^T = (\bar{A}\bar{B})^T = \bar{B}^T A^T = B^* A^*$.

25. $(A^*A)^* = A^*A^{**} = A^*A$.

27. $(A + A^T)^T = A^T + A^{TT} = A^T + A = A + A^T$.

29. $A = \underbrace{\tfrac{1}{2}(A + A^T)}_{\text{Symmetric}} + \underbrace{\tfrac{1}{2}(A - A^T)}_{\text{Skew-symmetric}}$

31. (*a*) and (*b*) simply calculate.

(*c*) A looks like $\begin{pmatrix} 0 & 1 & 0 & 0 & \cdots & 0 \\ 0 & 0 & 1 & 0 & \cdots & 0 \\ 0 & 0 & 0 & 1 & \cdots & 0 \\ \vdots & \vdots & \vdots & & \ddots & \vdots \\ & & & & & 1 \\ 0 & 0 & 0 & & \cdots & 0 \end{pmatrix}$. As powers of A are computed the diagonal of 1's moves toward the upper right. $A^n = 0$.

33. Let $V = \begin{pmatrix} a & b \\ c & d \end{pmatrix}$. Then $V^T = \begin{pmatrix} a & c \\ b & d \end{pmatrix}$ and

$$V - V^T = \begin{pmatrix} 0 & b-c \\ -(b-c) & 0 \end{pmatrix}$$

(*a*) Solve

$$\begin{pmatrix} 0 & b-c \\ -(b-c) & 0 \end{pmatrix} = \begin{pmatrix} 0 & 1 \\ 1 & 0 \end{pmatrix} \Rightarrow \begin{matrix} b-c = 1 \\ -(b-c) = 1 \end{matrix} \Rightarrow \text{No solution.}$$

(*b*) Solve

$$\begin{pmatrix} 0 & b-c \\ -(b-c) & 0 \end{pmatrix} = \begin{pmatrix} 0 & -1 \\ 1 & 0 \end{pmatrix} \Rightarrow \begin{matrix} b-c = -1 \\ -(b-c) = 1 \end{matrix} \Rightarrow b = c - 1.\ a \text{ and}$$

d are unrestricted.

$$V = \begin{pmatrix} a & c-1 \\ c & d \end{pmatrix}$$

(*c*) V can be any symmetric matrix.

35. Because there are no zero rows, and the number of columns equals the number of rows, the reduced row echelon form at least looks like

$$\begin{pmatrix} 1 & & & & & \\ 0 & 1 & & & ? & \\ 0 & 0 & 1 & & & \\ 0 & 0 & 0 & \ddots & & \\ \vdots & \vdots & \vdots & & \ddots & \\ 0 & 0 & 0 & \cdots & 0 & 1 \end{pmatrix}$$

Now by definition of reduced row echelon form only zeros may lie above each leading 1 in a row. Therefore the form is I_n.

37. $A = \begin{pmatrix} 0 & 1 \\ 1 & 0 \end{pmatrix}$, $B = \begin{pmatrix} 0 & -i \\ i & 0 \end{pmatrix}$, $C = \begin{pmatrix} 1 & 0 \\ 0 & -1 \end{pmatrix}$

A and C are symmetric
B is skew-symmetric
B is Hermitian.

39. Use induction. For $n = 1$, $(AB)^1 = AB = A^1B^1$. On the other hand, by commutativity, $AB = BA = B^1A^1$, so $A^1B^1 = B^1A^1$. The induction hypothesis is that (for $n = k - 1$) $(AB)^{k-1} = A^{k-1}B^{k-1} = B^{k-1}A^{k-1}$. Now to prove $(AB)^k = A^kB^k = B^kA^k$ we have $(AB)^k = (AB)^{k-1}(AB) = A^{k-1}B^{k-1}AB = A^{k-1}BB \cdots BBAB$. Using $BA = AB$ $k - 1$ times the last term is equal to $A^{k-1}A\ BB \cdots BBB = A^kB^k$. The argument for $(BA)^k$ works the same way.

PROBLEMS 1.6

1. (a) -7. (b) 3. (c) -543. **3.** (a) 0. (b) 56. (c) 0.

5. (a) $x = \frac{16}{11} + \frac{7i}{11}$, $y = \frac{24}{11} - \frac{28}{11}i$. (b) $(x, y, z) = \left(-\frac{5}{3}, -\frac{8}{3}, -\frac{2}{3}\right)$.

7. $\det(AB) = \det A \det B$; $\det(BA) = \det B \det A$; Now $\det A$, $\det B$ are just complex numbers. So $\det A \det B = \det B \det A$. Therefore $\det(AB) = \det(BA)$.

9. $\det(AB) = \det\begin{pmatrix} 16 & 19 \\ 9 & 51 \end{pmatrix} = 645 \qquad \det A \det B = (15)(43) = 645.$

11. Use induction. $P(n)$ is: If A is $n \times n$, then $\det A^T = \det A$.

1. $P(2)$ is true. Let

$$A = \begin{pmatrix} a & b \\ c & d \end{pmatrix} \det A = ad - bc$$

$$A^T = \begin{pmatrix} a & c \\ b & d \end{pmatrix} \det A = ad - bc$$

Therefore $\det A = \det A^T$.

2. Suppose $P(k-1)$ is true and prove $P(k)$ is true. Consider

$$A = \begin{pmatrix} a_{11} & \cdots & a_{1n} \\ \vdots & & \vdots \\ a_{k1} & \cdots & a_{kk} \end{pmatrix} \qquad A^T = \begin{pmatrix} a_{11} & \cdots & a_{k1} \\ \vdots & & \vdots \\ a_{1n} & \cdots & a_{kk} \end{pmatrix}$$

Let $\mathscr{A}_{ij}$ = submatrix of A generated by deleting row i and column j.

Let $\mathscr{B}_{ij}$ = submatrix of A^T generated by deleting row i and column j.

Note that $\mathscr{A}_{ij} = \mathscr{B}_{ji}{}^T$ and since $\mathscr{A}_{ij}$ and $\mathscr{B}_{ji}$ are $k-1 \times k-1$, we have by the induction hypothesis that $\det \mathscr{B}_{ji} = \det \mathscr{B}_{ji}{}^T = \det \mathscr{A}_{ij}$. Now using the last row for A

$$\det A = (1)^{k+1} a_{k1} \det \mathscr{A}_{k1} + \cdots + (-1)^{2k} a_{kk} \det \mathscr{A}_{kk}$$

and using the last column for B,

$$\begin{aligned} \det A^T &= (-1)^{k+1} a_{k1} \det \mathscr{B}_{1k} + \cdots + (-1)^{2k} a_{kk} \det \mathscr{B}_{kk} \\ &= (-1)^{k+1} a_{k1} \det \mathscr{A}_{k1} + \cdots + (-1)^{2k} a_{kk} \det \mathscr{A}_{kk} \\ &= \det A \end{aligned}$$

13. $\det A^* = \det(\bar{A})^T = \det \bar{A} \underset{\uparrow}{=} \overline{\det A}$

Prob. 12

15. cI_n is certainly upper triangular, so its determinant is the product of the diagonal elements. So $\det cI_n = c \cdot c \cdots c = c^n$.

17. Because $A^2 = A$, $\det(A^2) = \det A$. Thus

$$\begin{aligned} \det(AA) &= \det A \\ \det A \det A &= \det A \\ (\det A)^2 - \det A &= 0 \\ \det A(\det A - 1) &= 0 \\ \det A &= 0 \quad \text{or} \quad 1 \end{aligned}$$

19. The determinant must be 0. Because

$$A^n = 0, \qquad \det(A^n) = \det 0$$

So

$$\begin{aligned} \det(AA\cdots A) &= 0 \\ (\det A)(\det A)\cdots(\det A) &= 0 \\ (\det A)^n &= 0 \\ \det A &= 0 \end{aligned}$$

21. No. For example

$$A = \begin{pmatrix} 1 & 1 \\ 0 & 0 \end{pmatrix} \quad \text{and} \quad B = \begin{pmatrix} 0 & 0 \\ 1 & 1 \end{pmatrix}$$

both have determinant zero, but are not equal.

23. The determinant is zero *if n is odd*. We have $A^T = -A$ so

$$\det A^T = \det(-A) = (-1)^n \det A$$

On the other hand, $\det A^T = \det A$ (always) so

$$\det A = (-1)^n \det A$$

If n is even there is no information. However if n is odd we have $\det A = -\det A$ and therefore $\det A = 0$.

25. $\det H_2 = \dfrac{1}{12} \qquad \det H_3 = \dfrac{1}{2160} \qquad \det H_4 = \dfrac{1}{6048000}$

PROBLEMS 1.7

1. (*a*) $A^{-1} = \begin{pmatrix} -2 & 1 \\ \frac{3}{2} & -\frac{1}{2} \end{pmatrix}$ (*b*) $A^{-1} = \begin{pmatrix} \frac{2}{5} - \frac{1}{5}i & \frac{4}{5} - \frac{2}{5}i \\ 0 & -\frac{1}{3} \end{pmatrix}$

(*c*) $A^{-1} = \begin{pmatrix} 0 & 1 \\ 1 & 0 \end{pmatrix} = A$ (*d*) $A^{-1} = \dfrac{1}{111}\begin{pmatrix} 14 & -1 & -9 \\ 11 & 23 & -15 \\ 40 & 13 & 6 \end{pmatrix}$

(*e*) $A^{-1} = \begin{pmatrix} \frac{2}{3} & \frac{1}{6} & 0 & 0 \\ -\frac{1}{3} & \frac{1}{6} & 0 & 0 \\ 0 & 0 & -1 & -1 \\ 0 & 0 & 3 & \frac{5}{2} \end{pmatrix}$

3. $\frac{1}{3}\begin{pmatrix} 1 & 2 \\ -3 & 6 \end{pmatrix}$ **5.** Reader's choice!

7. $\det(AB) = \det A \det B = (\det A)0 = 0$. Because $\det(AB) = 0$, AB is not invertible.

9. A is nonsingular so A^{-1} exists. Multiply $AB = AC$ on both sides by A^{-1}.

$$\begin{aligned} A^{-1}(AB) &= A^{-1}(AC) \\ (A^{-1}A)B &= (A^{-1}A)C \\ IB &= IC \\ B &= C \end{aligned}$$

Note that the multiplication was from the left—this is called *premultiplying*.

11. $$AA^T = \begin{pmatrix} \cos\theta & \sin\theta \\ -\sin\theta & \cos\theta \end{pmatrix}\begin{pmatrix} \cos\theta & -\sin\theta \\ \sin\theta & \cos\theta \end{pmatrix}$$
$$= \begin{pmatrix} \cos^2\theta + \sin^2\theta & -\cos\theta\sin\theta + \cos\theta\sin\theta \\ -\sin\theta\cos\theta + \sin\theta\cos\theta & \sin^2\theta + \cos^2\theta \end{pmatrix} = \begin{pmatrix} 1 & 0 \\ 0 & 1 \end{pmatrix} = I$$

Because the inverse of a matrix is unique, $A^T = A^{-1}$.

13. Because $A^{-1} = A^*$, $\det A^{-1} = \det A^* = \overline{\det A}$, so $1/\det A = \overline{\det A}$ and $1 = \det A\, \overline{\det A} = |\det A|^2$. $|\det A|$ being positive then must be equal to 1.

15. No. Let A be 3×3 and skew-symmetric, $A^T = -A$. Now $\det A^T = \det(-A) = (-1)^3 \det A = -\det A$. Also $\det A^T = \det A$, so $\det A = -\det A$ and we must have $\det A = 0$. Generalization: If n is odd, $A_{n\times n}$ is skew-symmetric, then $\det A = 0$ and A is not invertible (see Prob. 23 in Prob. Set 1.6).

17. If A is nilpotent of exponent k, $A^k = 0$. Now $0 = \det 0 = \det(A^k) = (\det A)^k$, so $\det A = 0$ and A is singular.

19. If AB is invertible $\det(AB) \neq 0$. Thus $(\det A)(\det B) \neq 0$ and neither $\det A$ nor $\det B$ can be zero. Hence both A and B are invertible.

21. We have $A^{-1} = B^{-1}$. Multiply both sides by B (from the right):

$$\begin{aligned} A^{-1}B &= B^{-1}B \\ A^{-1}B &= I \end{aligned}$$

Now multiply both sides by A (from the left):

$$\begin{aligned} A(A^{-1}B) &= AI \\ (AA^{-1})B &= A \\ IB &= A \\ B &= A \end{aligned}$$

23. In the last section each determinant was found to be nonzero so H_2, H_3, and H_4 are invertible. Using the values from the last section

$$\det H_2^{-1} = \frac{1}{\det H_2} = 12$$

$$\det H_3^{-1} = \frac{1}{\det H_3} = 2160$$

$$\det H_4^{-1} = \frac{1}{\det H_4} = 6{,}048{,}000$$

PROBLEMS 1.8

1. (*a*) $L = \begin{pmatrix} 1 & 0 \\ 6721 & 1 \end{pmatrix} \qquad U = \begin{pmatrix} 0.0001 & 3.172 \\ 0 & -21314.785 \end{pmatrix}$

(*b*) $L = \begin{pmatrix} 1 & 0 & 0 \\ -\frac{1}{3} & 1 & 0 \\ \frac{1}{6} & -\frac{10}{17} & 1 \end{pmatrix} \qquad U = \begin{pmatrix} 3 & -2 & 4 \\ 0 & -\frac{17}{30} & \frac{1}{3} \\ 0 & 0 & -\frac{75}{51} \end{pmatrix}$

(*c*) $L = \begin{pmatrix} 1 & 0 & 0 \\ -100 & 1 & 0 \\ 0 & \frac{1}{99} & 1 \end{pmatrix} \qquad U = \begin{pmatrix} 0.01 & -1 & 0 \\ 0 & -99 & -1 \\ 0 & 0 & \frac{991}{99} \end{pmatrix}$

3. (*a*) Is not. For the second row $2 \not> |-1| + |-1|$. (*b*) Is.

(*c*) Is not. For a skew-symmetric matrix, the diagonal elements are all zero.

(*d*) Is.

5. Not necessarily. Consider

$$A = \begin{pmatrix} -2 & 1 \\ 0 & 1 \end{pmatrix} \quad \text{and} \quad B = \begin{pmatrix} 1 & 0 \\ 2 & 3 \end{pmatrix}$$

$$AB = \begin{pmatrix} 0 & 3 \\ 2 & 3 \end{pmatrix}$$

7. $L = \begin{pmatrix} 1 & 0 & 0 \\ 2 & 1 & 0 \\ 3 & 2 & 1 \end{pmatrix} \qquad U = \begin{pmatrix} 1 & 1 & -1 \\ 0 & 1 & -3 \\ 0 & 0 & 0 \end{pmatrix}$

9. (*a*) $\begin{pmatrix} 1 & 0 \\ 0 & -1 \end{pmatrix} \qquad \begin{pmatrix} 1 & 0 & 0 \\ 0 & -1 & 0 \\ 0 & 0 & 1 \end{pmatrix}$

These are not positive definition because they are symmetric with negative determinant.

(*b*) $\begin{pmatrix} 1 & 2 \\ 2 & -8 \end{pmatrix} \qquad \begin{pmatrix} 1 & 2 & 0 \\ 2 & -8 & 0 \\ 0 & 0 & 1 \end{pmatrix}$

These are symmetric and the required "subdeterminants" are positive, so they are positive definite, but the first row of each precludes *SRD*.

11. $L = I, \quad U = I. \quad L = rI \quad U = \dfrac{1}{r}I, \quad r \neq 0.$

ADDITIONAL PROBLEMS (CHAPTER 1)

1. Yes, adjoin an equation inconsistent with a previous equation. For example

$$\begin{aligned} x + y &= 2 \\ x - y &= 0 \end{aligned}$$

has a unique solution. Adjoin $x + y = 3$.

3. If A is invertible the answer is yes, because $X = A^{-1}B$ and both A^{-1} and B have only real entries. If A is not invertible, then complex solutions are possible unless for some reason they have been removed from the discussion. For instance

$$\begin{pmatrix} 1 & -1 \\ 0 & 0 \end{pmatrix}\begin{pmatrix} x_1 \\ x_2 \end{pmatrix} = \begin{pmatrix} 0 \\ 0 \end{pmatrix}$$

has solution

$$\begin{pmatrix} x_1 \\ x_2 \end{pmatrix} = \begin{pmatrix} s \\ s \end{pmatrix}$$

where s is completely arbitrary and could be assigned any complex value.

5. Not necessarily. Consider

$$A = \begin{pmatrix} 1 & 0 \\ 0 & 1 \end{pmatrix} \quad \text{and} \quad B = \begin{pmatrix} i & 0 \\ 0 & i \end{pmatrix}$$

$$A + iB = \begin{pmatrix} 0 & 0 \\ 0 & 0 \end{pmatrix}$$

If A and B are real $A + iB$ still need not be invertible. Consider

$$A = \begin{pmatrix} 1 & 2 \\ 1 & 1 \end{pmatrix} \quad \text{and} \quad B = \begin{pmatrix} 1 & 0 \\ 0 & -1 \end{pmatrix}$$

$$A + iB = \begin{pmatrix} 1 + i & 2 \\ 1 & 1 - i \end{pmatrix} \quad \text{and} \quad \det(A + iB) = 0$$

7. Not necessarily.

$$A = \begin{pmatrix} 1 & 0 \\ 0 & -1 \end{pmatrix} \geq \begin{pmatrix} -1 & 0 \\ 0 & -1 \end{pmatrix} = B$$

but $\det A = -1 < 1 = \det B$.

9. $$\begin{aligned}(A+I)(A-I) &= A(A-I)+I(A-I)\\ &= A^2-A+A-I\\ &= A^2-I\\ (A-I)(A+I) &= (A-I)A+(A-I)I\\ &= A^2-IA+AI-I^2\\ &= A^2-A+A-I\\ &= A^2-I.\end{aligned}$$

11. The dominance matrix is

$$A=\begin{pmatrix}0&1&1&0\\0&0&0&1\\0&1&0&0\\1&0&1&0\end{pmatrix}\qquad A^2=\begin{pmatrix}0&1&0&1\\1&0&1&0\\0&0&0&1\\0&2&1&0\end{pmatrix}\qquad A^3=\begin{pmatrix}1&0&1&1\\0&2&1&0\\1&0&1&0\\0&1&0&2\end{pmatrix}$$

$$A+A^2+A^3=\begin{pmatrix}1&2&2&2\\1&2&2&1\\1&1&1&1\\1&3&2&2\end{pmatrix}$$

The powers for competitors 1 through 4 are 7, 6, 4, and 8, respectively. Predict competitor number 4 to win. (But there are various possibilities—what if competitors 1 and 2 decide to work together until competitor 4 is eliminated? Then competitor 2 leaks the plan to competitor 3 and makes the deal that once competitor 4 is eliminated, competitors 2 and 3 work to eliminate competitor 1.)

13. $BA=\begin{pmatrix}1 & R_1\\ 1/R_2 & 1+R_1/R_2\end{pmatrix}\qquad AB\neq BA$

The combined circuits are not equivalent.

15. $A+B=\begin{pmatrix}2 & R_1\\ 1/R_2 & 2\end{pmatrix}$

In parallel circuits, the placement above and below of the boxes is immaterial.

PROBLEMS 2.1

3. (*a*) $(2, 6)$; (*b*) $(-6, -9)$; (*c*) $(3, -2)$; (*d*) $(6, -9)$; (*e*) $(-1, -1, -8)$; (*f*) $(4, 1, 1)$.

5. (*a*) $\sqrt{5}$; (*b*) $\sqrt{10}$; (*c*) $\sqrt{5}$; (*d*) $\sqrt{\pi^2+1}$. **7.** $k=\pm\sqrt{26}/26$

11. In $\mathbb{R}^2$, let $A = (a, b)$. So

$$|A| = \sqrt{a^2 + b^2} \qquad \text{and} \qquad \frac{1}{|A|} A = \left(\frac{a}{\sqrt{a^2 + b^2}}, \frac{b}{\sqrt{a^2 + b^2}} \right)$$

The length of $(1/|A|)\, A$ is

$$\sqrt{\frac{a^2}{a^2 + b^2} + \frac{b^2}{a^2 + b^2}} = \sqrt{1} = 1$$

13. Let $\mathbf{v} = (a, b)$. Then $r\mathbf{v} = (ra, rb)$.

$$(r\mathbf{v})_s = \begin{pmatrix} ra \\ rb \end{pmatrix}$$

$$r\mathbf{v}_s = r \begin{pmatrix} a \\ b \end{pmatrix} = \begin{pmatrix} ra \\ rb \end{pmatrix}$$

15. Let $\mathbf{v} = (v_1, v_2, v_3)$ and $A = (a_{ij})$. The entries of $A\mathbf{v}_s$ are

$$c_i = \sum_{k=1}^{3} a_{ik} v_k$$

The entries of $(\mathbf{v}A^T)_s$ are

$$d_i = \sum_{k=1}^{3} v_k a_{ik} = \sum_{k=1}^{3} a_{ik} v_k$$

Thus $c_i = d_i$, $i = 1, 2, 3$, and $A\mathbf{v}_s = (\mathbf{v}A^T)_s$.

PROBLEMS 2.2

1. (*a*) 17; (*b*) -3; (*c*) 4; (*d*) -1.

3. (*a*) $17\sqrt{290}/290$; (*b*) $-\sqrt{145}/145$; (*c*) $2\sqrt{410}/205$; (*d*) $-\sqrt{2}/22$.

5. (*a*) $\frac{17}{5}(1, 2)$; (*b*) $\frac{-3}{29}(-5, 2)$; (*c*) $\frac{4}{41}(5, 4, 0)$; (*d*) $-\frac{1}{2}(0, 1, -1)$.

7. By direct calculation $\mathbf{i} \cdot \mathbf{j} = (1, 0, 0) \cdot (0, 1, 0) = 0$. Similarly $\mathbf{i} \cdot \mathbf{k} = \mathbf{j} \cdot \mathbf{k} = 0$.

9. Place the cube so that one corner is at the origin and the three edges coming from the corner lie in the x, y, and z axes. If the length of each side is s, then the diagonal forms a vector (s, s, s). The angle between the edge in the x axis and the diagonal is the angle between $(s, 0, 0)$ and (s, s, s).

$$\cos\theta = \frac{(s, 0, 0) \cdot (s, s, s)}{|(s, 0, 0)|\,|(s, s, s)|} = \frac{s^2}{\sqrt{s^2}\sqrt{3s^2}} = \frac{\sqrt{3}}{3}$$

$$\theta = \cos^{-1}\frac{\sqrt{3}}{3} \doteq .955 \text{ radians} \doteq 54.7^\circ$$

11. $|A + B|^2 = |A|^2 + |B|^2 - 2|A|\,|B| \cos(\pi - \theta)$

$$= |A|^2 + |B|^2 + 2|A|\,|B| \cos\theta \qquad (1)$$

$$|A - B|^2 = |A|^2 + |B|^2 - 2|A|\,|B| \cos\theta \qquad (2)$$

(*a*) Take (1) − (2) and use $A \cdot B = |A|\,|B| \cos\theta$. (*b*) Take (1) + (2).

13. $B_{\text{proj } A} = \dfrac{A \cdot B}{|A|^2} A$ so

$$[B - B_{\text{proj } A}] \cdot A = \left(B - \frac{A \cdot B}{|A|^2} A\right) \cdot A = B \cdot A - \frac{(A \cdot B)(A \cdot A)}{|A|^2}$$

$$= B \cdot A - A \cdot B = 0$$

PROBLEMS 2.3

1. (*a*) Dilation, constant 3; (*b*) contraction, constant $\frac{1}{2}$; (*c*) dilation, constant 2; rotation, π radians.

$$R_\pi D_2 = \begin{pmatrix} -1 & 0 \\ 0 & -1 \end{pmatrix}\begin{pmatrix} 2 & 0 \\ 0 & 2 \end{pmatrix} = \begin{pmatrix} -2 & 0 \\ 0 & -2 \end{pmatrix}$$

The order can be reversed.

3. (*a*) One possibility is rotation of $3\pi/2$ radians and then projection onto x_1 axis:

$$P_{x_1} R_{3\pi/2} = \begin{pmatrix} 1 & 0 \\ 0 & 0 \end{pmatrix}\begin{pmatrix} 0 & 1 \\ -1 & 0 \end{pmatrix} = \begin{pmatrix} 0 & 1 \\ 0 & 0 \end{pmatrix}$$

Note that $R_{3\pi/2} P_{x_1}$ does not work. Another possibility is to project on the x_2 axis first and then rotate $3\pi/2$ radians:

$$R_{3\pi/2} P_{x_2} = \begin{pmatrix} 0 & 1 \\ -1 & 0 \end{pmatrix}\begin{pmatrix} 0 & 0 \\ 0 & 1 \end{pmatrix} = \begin{pmatrix} 0 & 1 \\ 0 & 0 \end{pmatrix}$$

(*b*) Rotation of $\pi/2$ radians clockwise.

5. (*a*) $D_k C_{1/k} = \begin{pmatrix} k & 0 \\ 0 & k \end{pmatrix}\begin{pmatrix} 1/k & 0 \\ 0 & 1/k \end{pmatrix} = \begin{pmatrix} 1 & 0 \\ 0 & 1 \end{pmatrix}$

(*b*) $\begin{pmatrix} \cos\theta & -\sin\theta \\ \sin\theta & \cos\theta \end{pmatrix}\begin{pmatrix} \cos(-\theta) & -\sin(-\theta) \\ \sin(-\theta) & \cos(-\theta) \end{pmatrix}$

$$= \begin{pmatrix} \cos\theta & -\sin\theta \\ \sin\theta & \cos\theta \end{pmatrix}\begin{pmatrix} \cos\theta & \sin\theta \\ -\sin\theta & \cos\theta \end{pmatrix}$$

$$= \begin{pmatrix} \cos^2\theta + \sin^2\theta & 0 \\ 0 & \sin^2\theta + \cos^2\theta \end{pmatrix} = \begin{pmatrix} 1 & 0 \\ 0 & 1 \end{pmatrix}$$

(*c*) $P_x = \begin{pmatrix} 1 & 0 \\ 0 & 0 \end{pmatrix}$; det $P_x = 0$. So P_x is not invertible.

7. If $x_1 > 0, x_2 > 0$ then

$$f\begin{pmatrix} x_1 \\ x_2 \end{pmatrix} = \begin{pmatrix} x_1 \\ x_2 \end{pmatrix}$$

and f is the identity. If $x_1 < 0$ and $x_2 < 0$,

$$f\left(\begin{pmatrix} x_1 \\ x_2 \end{pmatrix}\right) = \begin{pmatrix} -x_1 \\ -x_2 \end{pmatrix} = \begin{pmatrix} -1 & 0 \\ 0 & -1 \end{pmatrix}\begin{pmatrix} x_1 \\ x_2 \end{pmatrix} = R_\pi \begin{pmatrix} x_1 \\ x_2 \end{pmatrix}$$

Similar statements can be made for the other cases: $x_1 < 0, x_2 > 0$ (reflection) and $x_1 > 0, x_2 < 0$ (reflection). That is, in each quadrant, f is represented by matrix multiplication; f is "linear on each quadrant."

PROBLEMS 2.4

1. (*a*) $x + 2y + 3z = 0;$ (*b*) $3x - y = 0;$ (*c*) $x + 2 = 0;$

(*d*) $x + y + z = 1.$

3. (*a*) $(-11, 4, 1) \cdot (x - 2, y - 1, z - 3) = 0;\ -11x + 4y + z = -15;$

(*b*) $(1, 1, 1) \cdot (x + 1, y, z) = 0;\ x + y + z = -1;$

(*c*) $(1, 0, 0) \cdot (x - 2, y - 4, z - 6) = 0;\ x = 2.$

5. (*a*) $x = t + 1, y = -t, z = 0;$ (*b*) $x = t, y = t, z = t;$

(*c*) $x = t, y = t, z = 1 - t;$ (*d*) $x = 1 + t, y = -t, z = 1.$

7. $z = 0, y = 0, x = 0.$

9. (*a*) $N = (3, -3, -1)$ is normal to the plane and $V = (2, 1, 3)$ is parallel to the line. $N \cdot V = 0.$

(*b*) Substitution of the expressions for x, y, and z into $3x - 3y - z = 6$ leads to $6t - 18 - 3t - 12 - 3t + 6 = 6$ or $-24 = 6.$

ADDITIONAL PROBLEMS (CHAPTER 2)

1. $B = (0, \frac{3}{2}), A = (\sqrt{3}, 1)$. Let $C = (c_1, c_2)$. $A + B + C = (\sqrt{3}, \frac{5}{2}) + (c_1, c_2) = 0.$ $c_1 = -\sqrt{3}, c_2 = -\frac{5}{2}.$

3. The homogeneous equations are

$$\begin{aligned} 7w_1 - 6w_2 - 11w_3 &= 0 \\ 2w_2 - 3w_3 &= 0 \\ 7w_1 - 8w_2 - 8w_3 &= 0 \end{aligned}$$

The reduced equations are

$$\left(\begin{array}{ccc|c} 7 & -6 & -11 & 0 \\ 0 & 2 & -3 & 0 \end{array}\right)$$

The solution is $(\frac{20}{7}t, \frac{3}{2}t, t).$

5. The system of equations can be written

$$\begin{pmatrix} 1 & 1 & -1 \\ -2 & k & 0 \\ 7 & 5-k & -5 \end{pmatrix}\begin{pmatrix} w_1 \\ w_2 \\ w_3 \end{pmatrix} = \begin{pmatrix} 0 \\ 0 \\ 0 \end{pmatrix}$$

The system has a nontrivial solution if the determinant of the coefficient matrix is zero. The determinant is zero regardless of the value of k. Thus the system can be balanced for any k.

7. If A and B are component forces and

$$R_\theta = \begin{pmatrix} \cos\theta & -\sin\theta \\ \sin\theta & \cos\theta \end{pmatrix} \quad \text{then} \quad R_\theta A + R_\theta B = R_\theta(A + B)$$

The right-hand side represents the rotation of the resultant vector.

9. Let A and B be the component forces and let C_k be the contraction matrix. Then $C_kA + C_kB = C_k(A + B)$. The right-hand side represents the contraction of the resultant vector.

PROBLEMS 3.1

1. $B = E$, $A = C$.

3. Choose for example

$$\mathbf{x} = (1, 1),\ \mathbf{y} = (1, 2),\ \mathbf{z} = (2, 1)$$

$$(\mathbf{x} - \mathbf{y}) - \mathbf{z} = \begin{pmatrix} -2 \\ -2 \end{pmatrix}$$

$$\mathbf{x} - (\mathbf{y} - \mathbf{z}) = \begin{pmatrix} 2 \\ 0 \end{pmatrix}$$

5. $(3, -2, 1) = 3(1, 0, 0) - 2(0, 1, 0) + 1(0, 0, 1)$.

7. $\det\begin{pmatrix} 1 & -2 & 3 \\ 3 & 1 & 0 \\ 4 & -1 & 3 \end{pmatrix} = 0$ **9.** $\det\begin{pmatrix} 1 & -2 & 3 \\ 3 & 1 & 0 \\ a+3b & -2a+b & 3a \end{pmatrix} = 0$

11. The equation is equivalent to

$$\begin{aligned} 2c_1 - c_2 + c_3 &= 1 \\ 3c_1 + 3c_3 &= -1 \\ 5c_1 + 6c_2 + 11c_3 &= 4 \end{aligned}$$

These reduce to

$$\left(\begin{array}{ccc|c} 2 & -1 & 1 & 1 \\ 0 & 1 & 1 & -\frac{5}{3} \\ 0 & 0 & 0 & 1 \end{array}\right)$$

13. Let $\mathbf{x} = (x_1, x_2, \ldots, x_n)$.

$$\begin{aligned}\mathbf{x} + \boldsymbol{\theta} &= (x_1, x_2, \ldots, x_n) + (0, 0, \ldots, 0) = (x_1 + 0, x_2 + 0, \ldots, x_n + 0)\\ &= (x_1, x_2, \ldots, x_n) = \mathbf{x}\end{aligned}$$

Example from E^4: Let $\mathbf{x} = (1, -6, 4, 2)$.

$$(1, -6, 4, 2) + (0, 0, 0, 0) = (1 + 0, -6 + 0, 4 + 0, 2 + 0) = (1, -6, 4, 2)$$

15. Let $\mathbf{x} = (x_1, x_2, \ldots, x_n)$

$$\begin{aligned}(r + s)\mathbf{x} &= (r + s)(x_1, x_2, \ldots, x_n) = ((r + s)x_1, (r + s)x_2, \ldots, (r + s)x_n)\\ &= (rx_1 + sx_1, rx_2 + sx_2, \ldots, rx_n + sx_n)\\ &= (rx_1, rx_2, \ldots, rx_n) + (sx_1, sx_2, \ldots, sx_n)\\ &= r(x_1, x_2, \ldots, x_n) + s(x_1, x_2, \ldots, x_n) = r\mathbf{x} + s\mathbf{x}\end{aligned}$$

Example in E^5:

$$\begin{aligned}(7 + 4)(1, -2, 0, 6, -3) &= 11(1, -2, 0, 6, -3)\\ &= (11, -22, 0, 66, -33)\\ 7(1, -2, 0, 6, -3) + 4(1, -2, 0, 6, -3) &= (7, -14, 0, 42, -21)\\ &\quad + (4, -8, 0, 24, -12)\\ &= (11, -22, 0, 66, -33)\end{aligned}$$

PROBLEMS 3.2

1. V is a vector space. **3.** V is a vector space.

5. V is a vector space. **7.** V is a vector space.

9. V is a vector space. **11.** V is a vector space.

13. V is not a vector space. Closure for addition fails to hold. For example I and $-I$ are in V, but $I + -I = 0$ is not in V.

15. V is not a vector space. Closure for addition fails to hold. For example consider in the case $n = 2$:

$$A = \begin{pmatrix} 0 & 1 \\ 0 & 0 \end{pmatrix} \quad B = \begin{pmatrix} 0 & 0 \\ 1 & 0 \end{pmatrix}$$

Now $A^2 = B^2 = 0$ so A and B are nilpotent. However, $(A + B)^2 = I \neq 0$ and no power of $A + B$ is 0.

17. V is not a vector space. Closure for scalar multiplication fails to hold. Let A be in V so that $A^2 = I$. Consider $2A$: $(2A)^2 = 4A^2 = 4I \neq I$.

19. V is a vector space. The verification is virtually the same as in example 7 of this section.

21. V is a vector space.

23.
$$\begin{aligned} -\mathbf{x} &= (-\mathbf{x}) + \boldsymbol{\theta} = (-\mathbf{x}) + 0\mathbf{x} = (-\mathbf{x}) + (1 + (-1))\mathbf{x} \\ &= (-\mathbf{x}) + 1\mathbf{x} + (-1)\mathbf{x} = (-\mathbf{x}) + \mathbf{x} + (-1)\mathbf{x} \\ &= \mathbf{x} + (-\mathbf{x}) + (-1)\mathbf{x} = \boldsymbol{\theta} + (-1)\mathbf{x} = (-1)\mathbf{x} \end{aligned}$$

25. V is a vector space.

27. No. The existence of vector $\boldsymbol{\theta}$ in 4 is given in 3.

PROBLEMS 3.3

1. W is not subspace because W is not closed under scalar multiplication. For example, if A is in W, $(-1)A$ is not in W.

3. W is a subspace.

5. W is not a subspace because W is not closed under vector addition. For example

$$\begin{pmatrix} 1 & 1 \\ 0 & 0 \end{pmatrix} \quad \text{and} \quad \begin{pmatrix} 0 & 0 \\ 2 & 3 \end{pmatrix}$$

are in W but the sum is not.

7. W is a subspace.

9. W is not a subspace. Note that scalars can have nonzero imaginary part. So, for example, if A is in W, iA is not in W and W fails to be closed under scalar multiplication.

11. W is a subspace. **13.** W is a subspace.

15. Since the set of solutions is a subset W of $\mathbb{C}^n$, the usual operations are those of $\mathbb{C}^n$ (remember a sub*space* is a sub*set* which is a vector space). Now if X and Y are any solutions, then

$$A(X + Y) = AX + AY = 0 + 0 = 0 \qquad \text{and} \qquad A(cX) = c(AX) = c0 = 0$$

by laws of matrix algebra. Thus W is a subspace.

17. $\operatorname{tr} A = \sum_{i=1}^{n} a_{ii} \qquad \operatorname{tr} B = \sum_{i=1}^{n} b_{ii}$

$$\operatorname{tr}(A + B) = \sum_{i=1}^{n} (a_{ii} + b_{ii}) = \sum_{i=1}^{n} a_{ii} + \sum_{i=1}^{n} b_{ii} = \operatorname{tr} A + \operatorname{tr} B$$

Also

$$r \operatorname{tr}(A) = r \sum_{i=1}^{n} a_{ii} = \sum_{i=1}^{n} r a_{ii} = \operatorname{tr}(rA)$$

For the last part, let $AB = C$, $BA = D$, so that

$$c_{ij} = \sum_{k=1}^{n} a_{ik}b_{kj} \qquad d_{ij} = \sum_{q=1}^{n} b_{iq}a_{qj}$$

Now

$$\begin{aligned} \operatorname{tr}(AB) = \operatorname{tr} C = \sum_{i=1}^{n} c_{ii} &= \sum_{i=1}^{n}\left(\sum_{k=1}^{n} a_{ik}b_{ki}\right) \qquad \text{and} \qquad \operatorname{tr}(BA) \\ = \operatorname{tr} D = \sum_{p=1}^{n} d_{pp} &= \sum_{p=1}^{n}\left(\sum_{q=1}^{n} b_{pq}a_{qp}\right) \end{aligned}$$

Working with the last series we interchange the order of summation and use commutativity in $\mathbb{R}$ to obtain

$$\operatorname{tr}(BA) = \sum_{q=1}^{n}\left(\sum_{p=1}^{n} a_{qp}b_{pq}\right)$$

Now the subscripts are only indices which can be renamed without changing the value of the sum. Put $q = i$, $p = k$ to find

$$\operatorname{tr}(BA) = \sum_{i=1}^{n} a_{ik}b_{ki} = \operatorname{tr}(AB)$$

19. Let (x_1, x_2) and (y_1, y_2) be in W. Then $x_2 = mx_1$, $y_2 = my_1$. Now $c(x_1, x_2) = (cx_1, cx_2)$ and $(cx_2) = cmx_1 = m(cx_1)$ and (cx_1, cx_2) is in W. Also $(x_1, x_2) + (y_1, y_2) = (x_1 + y_1, x_2 + y_2)$ and $(x_2 + y_2) = mx_1 + my_1 = m(x_1 + y_1)$ and $(x_1 + y_1, x_2 + y_2)$ is in W. Now in the second case if $x_2 = mx_1 + b$, $b \neq 0$, then $2(x_1, x_2)$ is not in W because $2x_2 = 2(mx_1 + b) = m(2x_1) + 2b$ and $2b \neq b$.

21. If W is a subspace, then $n(0, 0, 0)$ is in W and the equations in the definition must hold for some t. If the equations hold for some $t = T$ we have $k_1 = -aT$, $k_2 = -bT$ and $k_3 = -cT$. Therefore

$$\begin{aligned} W &= \{(x_1, x_2, x_3) \mid (x_1, x_2, x_3) = (a(t - T), b(t - T), c(t - T)) \qquad t \text{ in } \mathbb{R}\} \\ &= \{(x_1, x_2, x_3) \mid (x_1, x_2, x_3) = (a\tau, b\tau, c\tau) \qquad \tau \text{ in } \mathbb{R}\} \end{aligned}$$

Now the closure is not hard to show. If $\mathbf{x} = (x_1, x_2, x_3)$ and $\mathbf{y} = (y_1, y_2, y_3)$ are in W then

$$\begin{aligned} x_1 + y_1 &= a\tau + a\sigma \\ x_2 + y_2 &= b\tau + b\sigma \qquad \sigma, \tau \in \mathbb{R} \\ x_3 + y_3 &= c\tau + c\sigma \end{aligned}$$

and $(x_1 + y_1, x_2 + y_2, x_3 + y_3) = (a(\tau + \sigma), b(\tau + \sigma), c(\tau + \sigma))$, $\tau + \sigma \in \mathbb{R}$. Also $(rx_1, rx_2, rx_3) = (a(r\tau), b(r\tau), c(r\tau))$, $r\tau \in \mathbb{R}$.

PROBLEMS 3.4

1. (*a*) (a, b, c) is in span (S) if and only if $c = \frac{2}{3}b - \frac{4}{3}a$;

(*b*) S is linearly independent.

3. (*a*) (a, b, c) is in span (S) if and only if $c = a + ib$;

(*b*) S is linearly dependent; (*c*) $(i, i - 1, -1) = i(1, i, 0) + i(0, 1, i)$.

5. (*a*) Span(S) is E^2; (*b*) S is linearly dependent;

(*c*) $(7, 12) = \frac{9}{2}(1, 1) + \frac{5}{4}(2, 6)$.

7. (*a*) (a, b, c) is in span (S) if and only if (a, b, c) is a scalar multiple of $(1, -1, 2)$;

(*b*) S is linearly dependent; (*c*) $(0, 0, 0) = 0(1, -1, 2)$.

9. (*a*) Span (S) is all 2×2 real matrices with trace zero;

(*b*) S is linearly independent.

11. Let $\mathbf{x} = (x_1, x_2, x_3)$ and $\mathbf{y} = (y_1, y_2, y_3)$ be two vectors from E^3. Try to solve $c_1\mathbf{x} + c_2\mathbf{y} = (a, b, c)$. The resulting equations are represented by the augmented matrix

$$\left(\begin{array}{cc|c} x_1 & y_1 & a \\ x_2 & y_2 & b \\ x_3 & y_3 & c \end{array}\right)$$

Row reduction leads to a third equation of $0 =$ (expression in a, b, c) and (a, b, c) cannot be arbitrary.

13. Let $\mathbf{x} = (x_1, x_2)$, $\mathbf{y} = (y_1, y_2)$, $\mathbf{z} = (z_1, z_2)$ and consider $c_1\mathbf{x} + c_2\mathbf{y} + c_3\mathbf{z} = (0, 0)$. The resulting equations are represented by the matrix

$$\left(\begin{array}{ccc|c} x_1 & y_1 & z_1 & 0 \\ x_2 & y_2 & z_2 & 0 \end{array}\right)$$

Because there are fewer equations than unknowns, a nontrivial solution exists.

15. The elements of span (S) are of the form $a + bx^2$ where a and b are arbitrary.

17. The elements of span (S) are real matrices of the form

$$A = \begin{pmatrix} b & a \\ a & -b \end{pmatrix}$$

where a and b are arbitrary. Clearly A is symmetric and has trace zero.

19. Let $S = \{\mathbf{v}_1, \mathbf{v}_2, \ldots, \mathbf{v}_m\}$ be linearly dependent and let $T = \{\mathbf{v}_1, \mathbf{v}_2, \ldots, \mathbf{v}_m, \mathbf{v}_{m+1}, \ldots, \mathbf{v}_n\}$. Consider

$$c_1\mathbf{v}_1 + \cdots + c_m\mathbf{v}_n + c_{m+1}\mathbf{v}_{m+1} + \cdots + c_n\mathbf{v}_n = \boldsymbol{\theta}$$

There exists a nontrivial solution to this equation. In fact, put $c_{m+1} = 0$, $c_{m+2} = 0, \ldots, c_n = 0$ and we have

$$c_1\mathbf{v}_1 + \cdots + c_m\mathbf{v}_m = \boldsymbol{\theta}$$

which has a nontrivial solution by the linear dependence of S.

21. $[A, \mathrm{I}] = 0 \qquad [A, B] = \begin{pmatrix} -1 & 15 \\ -12 & 1 \end{pmatrix} \qquad [B, A] = \begin{pmatrix} 1 & -15 \\ 12 & -1 \end{pmatrix}$

$[A, A] = 0$

23. If $AB = BA$, then $AB - BA = 0$ and $[A, B] = 0$; If $[A, B] = 0$, then $AB - BA = 0$ and $AB = BA$.

PROBLEMS 3.5

1. (*a*) Is a basis; (*b*) is not a basis; linearly dependent set;

(*c*) is not a basis; does not span E^2;

(*d*) is not a basis; linearly dependent set;

(*e*) is not a basis; linearly dependent set.

3. $\{(1, -1, 0), (0, 1, -1), (0, 0, 1)\}$ is a basis containing S. There are other possible solutions.

5. $\left\{\begin{pmatrix} 1 & 1 \\ 0 & 0 \end{pmatrix}, \begin{pmatrix} 0 & 0 \\ 1 & 1 \end{pmatrix}, \begin{pmatrix} 1 & 0 \\ 0 & 1 \end{pmatrix}, \begin{pmatrix} 0 & 1 \\ 0 & 0 \end{pmatrix}\right\}$ is a basis containing S. There are other possible solutions.

7. $\{(i, 1, 0), (0, i, 1), (0, 0, 1)\}$ is a basis containing S. There are other possible solutions.

9. S is linearly dependent and any vector in S can be written as a nontrivial linear combination of the others. Deletion of any vector leads to a basis for span (S).

11. Delete $x - x^2$. If $x + x^2$ is deleted the remaining set is linearly dependent.

13. The set

$$\{(1, 0, 0, 0, 0), (0, 1, 0, 0, 0), (0, 0, 1, 0, 0), (0, 0, 0, 1, 0), (0, 0, 0, 0, 1)\}$$

is a basis for E^5; dim $E^5 = 5$.

15. $S = \left\{\begin{pmatrix} 1 & 0 & 0 & 0 \\ 0 & 0 & 0 & 0 \end{pmatrix}, \begin{pmatrix} 0 & 1 & 0 & 0 \\ 0 & 0 & 0 & 0 \end{pmatrix}, \begin{pmatrix} 0 & 0 & 1 & 0 \\ 0 & 0 & 0 & 0 \end{pmatrix}, \begin{pmatrix} 0 & 0 & 0 & 1 \\ 0 & 0 & 0 & 0 \end{pmatrix}, \right.$
$\left. \begin{pmatrix} 0 & 0 & 0 & 0 \\ 1 & 0 & 0 & 0 \end{pmatrix}, \begin{pmatrix} 0 & 0 & 0 & 0 \\ 0 & 1 & 0 & 0 \end{pmatrix}, \begin{pmatrix} 0 & 0 & 0 & 0 \\ 0 & 0 & 1 & 0 \end{pmatrix}, \begin{pmatrix} 0 & 0 & 0 & 0 \\ 0 & 0 & 0 & 1 \end{pmatrix}\right\}$

is a basis for $\mathscr{M}_{24}$; dim $\mathscr{M}_{24} = 8$.

17. $S = \left\{\begin{pmatrix} i & 0 \\ 0 & 0 \\ 0 & 0 \end{pmatrix}, \begin{pmatrix} 0 & 1 \\ 0 & 0 \\ 0 & 0 \end{pmatrix}, \begin{pmatrix} 0 & 0 \\ -i & 0 \\ 0 & 0 \end{pmatrix}, \begin{pmatrix} 0 & 0 \\ 0 & 1 \\ 0 & 0 \end{pmatrix}, \begin{pmatrix} 0 & 0 \\ 0 & 0 \\ 1 & 0 \end{pmatrix}, \begin{pmatrix} 0 & 0 \\ 0 & 0 \\ 0 & i \end{pmatrix}\right\}$

is a basis for $\mathscr{C}_{32}$; dim $\mathscr{C}_{32} = 6$. The standard basis for $\mathscr{M}_{32}$ is also a basis for $\mathscr{C}_{32}$.

19. Form the basis $S = \{M^{ij}, i, j = 1, 2, \ldots, n$, such that the entry in the ith row and jth column is 1 and all other entries are $0\}$. S has n^2 elements, so dim $\mathscr{M}_{nn} = n^2$. A basis for the symmetric matrices is, using the notation above

$$T = \{M^{11}, M^{22}, \ldots, M^{nn}, M^{ij} + M^{ji}, i > j\}.$$

For example if $n = 3$

$$T = \left\{\begin{pmatrix} 1 & 0 & 0 \\ 0 & 0 & 0 \\ 0 & 0 & 0 \end{pmatrix}, \begin{pmatrix} 0 & 0 & 0 \\ 0 & 1 & 0 \\ 0 & 0 & 0 \end{pmatrix}, \begin{pmatrix} 0 & 0 & 0 \\ 0 & 0 & 0 \\ 0 & 0 & 1 \end{pmatrix}, \begin{pmatrix} 0 & 1 & 0 \\ 1 & 0 & 0 \\ 0 & 0 & 0 \end{pmatrix}, \begin{pmatrix} 0 & 0 & 1 \\ 0 & 0 & 0 \\ 1 & 0 & 0 \end{pmatrix}, \begin{pmatrix} 0 & 0 & 0 \\ 0 & 0 & 1 \\ 0 & 1 & 0 \end{pmatrix}\right\}$$

The number of elements in T is $n + [(n^2 - n)/2] = (n^2 + n)/2$. The dimension of the subspace of symmetric matrices is $(n^2 + n)/2$.

21. The dimension of both is mn. Construct bases using Probs. 15, 17, and 19 as a guide.

23. (*a*) rank $A = 2$; (*b*) rank $A = 4$; (*c*) rank $A = 2$.

25. A basis is $S = \left\{\begin{pmatrix} 1 \\ 0 \\ 0 \\ \vdots \\ 0 \end{pmatrix}, \begin{pmatrix} 0 \\ 1 \\ 0 \\ \vdots \\ 0 \end{pmatrix}, \begin{pmatrix} 0 \\ 0 \\ 1 \\ \vdots \\ 0 \end{pmatrix}, \ldots, \begin{pmatrix} 0 \\ 0 \\ 0 \\ \vdots \\ 1 \end{pmatrix}\right\}$

27. (*a*) $(A|B)$ reduces to

$$\left(\begin{array}{ccc|c} 1 & -1 & 1 & 2 \\ 0 & 3 & -3 & -1 \\ 0 & 0 & 6 & 6 \end{array}\right)$$

so rank $(A) =$ rank $((A|B)) = 3$ and a solution exists.

(*b*) $(A|B)$ reduces to

$$\left(\begin{array}{ccc|c} 1 & 1 & 2 & 3 \\ 0 & -2 & 6 & 5 \\ 0 & 0 & 0 & 0 \end{array}\right)$$

So rank $(A) = \text{rank}((A|B)) = 2$ and a solution exists.

(*c*) $(A|B)$ reduces to

$$\left(\begin{array}{ccc|c} 1 & 1 & 2 & 3 \\ 0 & -2 & 6 & 5 \\ 0 & 0 & 0 & 4 \end{array}\right)$$

So rank $(A) \neq \text{rank}((A|B))$ and no solution exists.

PROBLEMS 3.6

1. (*a*) Is an inner product,

(*b*) Is not an inner product; $\langle \mathbf{x}, \mathbf{x} \rangle = 2x_1$, so that $\langle \mathbf{x}, \mathbf{x} \rangle = 0$ need not mean $\mathbf{x}$ is the zero vector.

(*c*) Is not an inner product; $\langle \mathbf{x}, \mathbf{x} \rangle = x_1^2 - x_2^2$, so that $\langle \mathbf{x}, \mathbf{x} \rangle = 0$ need not mean $\mathbf{x}$ is the zero vector.

(*d*) Is not an inner product; $\langle \mathbf{x} + \mathbf{y}, \mathbf{z} \rangle \neq \langle \mathbf{x}, \mathbf{z} \rangle + \langle \mathbf{y}, \mathbf{z} \rangle$.

(*e*) Is not an inner product; note that $\langle (1, 0), (1, 0) \rangle = 0$, but the vector is not the zero vector.

(*f*) Is not an inner product; note that $\langle (1, 0), (1, 0) \rangle = 0$ but the vector is not the zero vector.

(*g*) Is not an inner product; note that $\langle (1, -1), (1, -1) \rangle = 0$, but the vector is not the zero vector.

(*h*) Is not an inner product; $\langle \mathbf{x} + \mathbf{y}, \mathbf{z} \rangle \neq \langle \mathbf{x}, \mathbf{z} \rangle + \langle \mathbf{y}, \mathbf{z} \rangle$.

3. (1) $\langle \mathbf{y}, \mathbf{x} \rangle = ay_1x_1 + by_2x_2 = ax_1y_1 + bx_2y_2 = \overline{ax_1y_1 + bx_2y_2} = \overline{\langle \mathbf{x}, \mathbf{y} \rangle}$

(2) $\langle \mathbf{x} + \mathbf{z}, \mathbf{y} \rangle = \langle (x_1 + z_1, x_2 + z_2), (y_2, y_2) \rangle$
$= a(x_1 + z_1)y_1 + b(x_2 + z_2)y_2$
$= ax_1y_1 + bx_2y_2 + az_1y_1 + bz_2y_2 = \langle \mathbf{x}, \mathbf{y} \rangle + \langle \mathbf{z}, \mathbf{y} \rangle$

(3) $\langle r\mathbf{x}, \mathbf{y} \rangle = \langle (rx_1, rx_2), (y_1, y_2) \rangle = arx_1y_1 + brx_2y_2$
$= r(ax_1y_1 + bx_2y_2) = r\langle \mathbf{x}, \mathbf{y} \rangle$

(4) $\langle \mathbf{x}, \mathbf{x} \rangle = ax_1^2 + bx_2^2 \geq 0$ because $a > 0$, $b > 0$. $ax_1^2 + bx_2^2 = 0$ if and only if $x_1 = x_2 = 0$. If either a or b is less than zero then property (4) fails to hold and we do not have an inner product.

5. (1) $\langle \mathbf{y}, \mathbf{x} \rangle_k = k\langle \mathbf{y}, \mathbf{x} \rangle = k\overline{\langle \mathbf{x}, \mathbf{y} \rangle} = k\overline{\langle \mathbf{x}, \mathbf{y} \rangle} = \overline{\langle \mathbf{x}, \mathbf{y} \rangle_k}$

(2) $\langle \mathbf{x} + \mathbf{z}, \mathbf{y} \rangle_k = k\langle \mathbf{x} + \mathbf{z}, \mathbf{y} \rangle = k\langle \mathbf{x}, \mathbf{y} \rangle + k\langle \mathbf{z}, \mathbf{y} \rangle = \langle \mathbf{x}, \mathbf{y} \rangle_k + \langle \mathbf{x}, \mathbf{z} \rangle_k$

(3) $r\langle \mathbf{x}, \mathbf{y}\rangle_k = rk\langle \mathbf{x}, \mathbf{y}\rangle = kr\langle \mathbf{x}, \mathbf{y}\rangle = k\langle r\mathbf{x}, \mathbf{y}\rangle = \langle r\mathbf{x}, \mathbf{y}\rangle_k$

(4) $\langle \mathbf{x}, \mathbf{x}\rangle_k = k\langle \mathbf{x}, \mathbf{x}\rangle$. Since $k > 0$, $k\langle \mathbf{x}, \mathbf{x}\rangle > 0$ unless $\mathbf{x} = \boldsymbol{\theta}$ and then $k\langle \mathbf{x}, \mathbf{x}\rangle = 0$. Therefore $\langle \mathbf{x}, \mathbf{x}\rangle_k > 0$ and $\langle \mathbf{x}, \mathbf{x}\rangle_k = 0$ if and only if $\mathbf{x} = \boldsymbol{\theta}$. Note that $\|\mathbf{x}\|_k = \sqrt{k\langle \mathbf{x}, \mathbf{x}\rangle} = \sqrt{k}\,\|\mathbf{x}\|$. If $k < 1$, $\|\mathbf{x}\|_k < \|\mathbf{x}\|$; if $k > 1$, $\|\mathbf{x}\|_k > \|\mathbf{x}\|$.

7. (*a*) $r(-3, -9, 7)$, $r \neq 0$.

(*b*) For $\mathbf{x}$ choose $(1, 1, 0)$ because it is clearly orthogonal to $(1, -1, 2)$. Now let $\mathbf{y} = (-1, 1, 1)$; it is clearly orthogonal to $(1, -1, 2)$ and it is not a multiple of $(1, 1, 0)$. Of course you could proceed as follows. After choosing $\mathbf{x} = (1, 1, 0)$, let $\mathbf{y} = (y_1, y_2, y_3)$, we want $\langle (y_1, y_2, y_3), (1, -1, 2)\rangle = 0$ which implies $y_1 - y_2 + 2y_3 = 0$. Thus we have $y = (r - 2s, r, s)$. Now we want $c_1\mathbf{x} + c_2\mathbf{y} = \boldsymbol{\theta}$ to have only the trivial solution. The equations are

$$\left(\begin{array}{cc|c} 1 & r-2s & 0 \\ 1 & r & 0 \\ 0 & s & 0 \end{array}\right)$$

which reduce to

$$\left(\begin{array}{cc|c} 1 & r-2s & 0 \\ 0 & s & 0 \\ 0 & 0 & 0 \end{array}\right)$$

We have only the trivial solution if and only if $s \neq 0$, while r can take any value. Thus $\mathbf{y} = (r - 2, r, 1)$ will do.

9. (1) $\overline{\langle \mathbf{y}, \mathbf{x}\rangle} = \overline{y_1x_1 + \cdots + y_nx_n} = \overline{x_1y_1} + \cdots + \overline{x_ny_n} = x_1y_1 + \cdots + x_ny_n$ because $x_1, \cdots, x_n$, and $y_1, \cdots, y_n$ are real numbers. Thus $\overline{\langle \mathbf{y}, \mathbf{x}\rangle} = \langle \mathbf{x}, \mathbf{y}\rangle$

(2) $\langle \mathbf{x} + \mathbf{z}, \mathbf{y}\rangle = (x_1 + z_1)y_1 + \cdots + (x_n + z_n)y_n = x_1y_1 + z_1y_1 + \cdots + x_ny_n + z_ny_n = x_1y_1 + \cdots + x_ny_n + z_1y_1 + \cdots + z_ny_n = \langle \mathbf{x}, \mathbf{y}\rangle + \langle \mathbf{z}, \mathbf{y}\rangle$.

(3) $r\langle \mathbf{x}, \mathbf{y}\rangle = r(x_1y_1 + \cdots + x_ny_n) = ((rx_1)y_1 + \cdots + (rx_n)y_n) = \langle r\mathbf{x}, \mathbf{y}\rangle$

(4) $\langle \mathbf{x}, \mathbf{x}\rangle = x_1{}^2 + \cdots + x_n{}^2 \geq 0$, and can equal zero if and only if $x_1 = x_2 = \cdots = x_n = 0$.

11. (b) $\|\mathbf{x}\| = \sqrt{\langle \mathbf{x}, \mathbf{x}\rangle}$. From the definition of inner product $\langle \mathbf{x}, \mathbf{x}\rangle = 0$ if and only if $\mathbf{x} = \boldsymbol{\theta}$. Therefore $\|\mathbf{x}\| = 0$ if and only if $\mathbf{x} = \boldsymbol{\theta}$.

(*c*) $\langle \mathbf{x}, r\mathbf{y}\rangle = \overline{\langle r\mathbf{y}, \mathbf{x}\rangle} = \overline{r\langle \mathbf{y}, \mathbf{x}\rangle} = \bar{r}\overline{\langle \mathbf{y}, \mathbf{x}\rangle} = \bar{r}\langle \mathbf{x}, \mathbf{y}\rangle$. $\langle \mathbf{x}, \mathbf{y} + \mathbf{z}\rangle = \overline{\langle \mathbf{y} + \mathbf{z}, \mathbf{x}\rangle} = \overline{\langle \mathbf{y}, \mathbf{x}\rangle + \langle \mathbf{z}, \mathbf{x}\rangle} = \overline{\langle \mathbf{y}, \mathbf{x}\rangle} + \overline{\langle \mathbf{z}, \mathbf{x}\rangle} = \langle \mathbf{x}, \mathbf{y}\rangle + \langle \mathbf{x}, \mathbf{z}\rangle$.

13. $\|\mathbf{x} + \mathbf{y}\|^2 = \langle \mathbf{x} + \mathbf{y}, \mathbf{x} + \mathbf{y}\rangle = \langle \mathbf{x} + \mathbf{y}, \mathbf{x}\rangle + \langle \mathbf{x} + \mathbf{y}, \mathbf{y}\rangle = \langle \mathbf{x}, \mathbf{x}\rangle + \langle \mathbf{y}, \mathbf{x}\rangle + \langle \mathbf{x}, \mathbf{y}\rangle + \langle \mathbf{y}, \mathbf{y}\rangle = \|\mathbf{x}\|^2 + 0 + 0 + \|\mathbf{y}\|^2$

15. $X^TY = (1 \quad 2 \quad -1)\begin{pmatrix} -2 \\ 4 \\ 3 \end{pmatrix} = -2 + 8 - 3 = 3$

17. (*a*) $m = 2$, $n = 3$.

(*b*) (1) $\langle B, A \rangle = \text{tr}\,(B^TA)$. Now for a square matrix C, $\text{tr}\,(C) = \text{tr}\,(C^T)$, therefore $\text{tr}\,(B^TA) = \text{tr}\,((B^TA)T) = \text{tr}\,(A^TB) = \langle A, B \rangle$. Because the entries of matrices from $\mathscr{M}_{mm}$ are real $\overline{\langle B, A \rangle} = \langle B, A \rangle$. Therefore $\overline{\langle B, A \rangle} = \langle A, B \rangle$.

(2)
$$\begin{aligned}\langle A + C, B \rangle &= \text{tr}\,((A + C)^TB) = \text{tr}\,((A^T + C^T)B) \\ &= \text{tr}\,(A^TB + C^TB) = \text{tr}\,(A^TB) + \text{tr}\,(C^TB) \\ &= \langle A, B \rangle + \langle C, B \rangle\end{aligned}$$

(3)
$$\begin{aligned}r\langle A, B \rangle &= r\text{tr}\,(A^TB) = \text{tr}\,(r(A^TB)) = \text{tr}\,((rA^T)B) \\ &= \text{tr}\,((rA)^TB) = \langle rA, B \rangle.\end{aligned}$$

(4) $\langle A, A \rangle = \text{tr}\,\langle A^TA \rangle$. Let $A^TA = C$. Now

$$c_{ii} = \sum_{k=1}^{m} a_{ki}{}^2 \qquad i = 1, 2, \ldots, n$$

so

$$\text{tr}\,(A^TA) = \sum_{i=1}^{n} \sum_{k=1}^{m} a_{ki}{}^2$$

which is a sum of squares of real numbers. Therefore $\langle A, A \rangle \geq 0$ and equals zero if and only if $a_{ki} = 0$, $k = 1, 2, \ldots, m$; $i = 1, 2, \ldots, n$. That is, $\langle A, A \rangle = 0$ if and only if $A = 0$.

PROBLEMS 3.7

1. To verify orthogonality show that all possible inner products are zero. To obtain orthonormal basis, normalize all vectors.

(*a*) $\left\{\frac{1}{\sqrt{2}}(1, 1, 0), (0, 0, 1), \frac{1}{\sqrt{2}}(-1, 1, 0)\right\}$ is an orthonormal basis of E^3.

(*b*) $\left\{\frac{1}{\sqrt{3}}(1, 1, 1), \frac{1}{\sqrt{2}}(0, 1, -1), \frac{1}{\sqrt{6}}(2, -1, -1)\right\}$ is an orthonormal basis of E^3.

3. (*c*) and (*d*) are orthogonal matrices.

5.
$$\begin{aligned}\|\mathbf{x}\|^2 = \langle \mathbf{x}, \mathbf{x} \rangle &= \left\langle \sum_{k=1}^{n} c_k\mathbf{v}_k, \sum_{j=1}^{n} c_j\mathbf{v}_j \right\rangle = \sum_{k=1}^{n} c_k \left\langle \mathbf{v}_k, \sum_{j=1}^{n} c_j\mathbf{v}_j \right\rangle \\ &= \sum_{k=1}^{n} c_k \sum_{j=1}^{n} c_j\langle \mathbf{v}_k, \mathbf{v}_j \rangle = \sum_{k=1}^{n} c_k \left(\sum_{j=1}^{n} c_j\delta_{kj} \right) = \sum_{k=1}^{n} c_k{}^2\end{aligned}$$

7. The dimension of $\mathcal{M}_{22}$ is four. The set $\mathcal{O}$ is orthonormal, which is shown by calculating all possible inner products. Because $\mathcal{O}$ is a set of four linear independent vectors and $\dim(\mathcal{M}_{22}) = 4$, $\mathcal{O}$ is basis. Using $v = \langle v, v_1\rangle v_1 + \cdots + \langle v, v_n\rangle v_n$ we have

$$\begin{pmatrix} 2 & 1 \\ -3 & 2 \end{pmatrix} = 2\begin{pmatrix} 1 & 0 \\ 0 & 0 \end{pmatrix} + 2\begin{pmatrix} 0 & 0 \\ 0 & 1 \end{pmatrix} + (-\sqrt{2})\begin{pmatrix} 0 & 1/\sqrt{2} \\ 1/\sqrt{2} & 0 \end{pmatrix}$$
$$+ (-2\sqrt{2})\begin{pmatrix} 0 & -1/\sqrt{2} \\ 1/\sqrt{2} & 0 \end{pmatrix}$$

9. Suppose A^T is orthogonal. Then $(A^T)(A^T)^T = I$ and $(A^T)^T(A^T) = I$. That is $A^TA = I$ and $AA^T = I$. Therefore A is orthogonal.

11. $\mathcal{O} = \left\{\left(\frac{1}{\sqrt{2}}, \frac{1}{\sqrt{2}}, 0\right), \left(\frac{1}{\sqrt{2}}, \frac{-1}{\sqrt{2}}, 0\right), (0, 0, 1)\right\}$

13. (*a*) $A^*A = I$, so $1 = \det I = \det(A^*A) = \det A^* \det A = \overline{(\det A)}(\det A) = |\det A|^2$. Therefore $|\det A| = 1$.

(*b*) Consider

$$A = \begin{pmatrix} i & 0 \\ 0 & 1 \end{pmatrix}$$

from Prob. 4. A is unitary and $\det A = i$.

(*c*) $A^TA = I$, so $1 = \det I = \det(A^TA) = \det A^T \det A = (\det A)^2$. Since A has real entries only, $\det A = \pm 1$.

15. Yes. $(A^2)^T(A^2) = (A^T)^2A^2 = A^TA^TAA = A^TA = I$
Yes. $(A^2)^*(A^2) = \overline{(A^2)}^TA^2 = (\bar{A}\bar{A})^TAA = \bar{A}^T\bar{A}^TAA = A^*A^*AA = A^*A = I.$

17. (*a*) Since $U^TU = I$, We have $\det U^T = 1/(\det U)$. Now $\det B = \det(U^TAU) = \det U^T \det A \det U = \det U^T \det U \det A = 1/(\det U) \det U \det A = \det A$.

(*b*) Since $U^*U = I$, we have $\det U^* = 1/(\det U)$. Now proceed as in part *a* of this problem.

19. A is involutory so $A^2 = I$. If n is odd then $n = 2m + 1$, m, a nonnegative integer, and $A^n = A^{2m+1} = A^{2m}A = (A^2)^mA = I^mA = A$.

21. An idempotent matrix need not be symmetric. For example

$$A = \begin{pmatrix} 1 & 1 \\ 0 & 0 \end{pmatrix}$$

satisfies $A^2 = A$ but A is not symmetric.

23. When inspecting A^*A remember that the entries of A^* have been conjugated. Otherwise the steps are exactly the same.

25. Because the columns must be orthonormal in $\mathbb{C}^4$, we have

$$|a^2| + |b|^2 = 1 \qquad \bar{a}(ib) + (\overline{ib})a = 0 \qquad \text{or} \qquad |a|^2 + |b|^2 = 1 \qquad i\bar{a}b - ia\bar{b} = 0$$

If a and b are real, this reduces to the condition that $a + ib$ be a complex number of magnitude one.

PROBLEMS 3.8

1. (*a*) $\begin{pmatrix} -\sqrt{2} & -3\sqrt{2}/2 \\ 2\sqrt{2} & -\sqrt{2}/2 \end{pmatrix} = P_{T\leftarrow S}$ (*b*) $\begin{pmatrix} -2\sqrt{2}/7 & -\sqrt{2}/7 \\ -\sqrt{2}/14 & -3\sqrt{2}/14 \end{pmatrix} = P_{U\leftarrow T}$

(*c*) $\begin{pmatrix} 0 & 1 \\ 1 & 0 \end{pmatrix} = P_{U\leftarrow S}$

$P_{U\leftarrow T}P_{T\leftarrow S} = P_{U\leftarrow S}$

$P_{T\leftarrow S}P_{U\leftarrow T} \neq P_{U\leftarrow S}$

3. $P_{Z\leftarrow T} = \dfrac{1}{2\sqrt{2}}\begin{pmatrix} \sqrt{3}-1 & \sqrt{3}+1 \\ -\sqrt{3}-1 & \sqrt{3}-1 \end{pmatrix} \qquad P_{T\leftarrow Z} = \dfrac{1}{2\sqrt{2}}\begin{pmatrix} \sqrt{3}-1 & -\sqrt{3}-1 \\ \sqrt{3}+1 & \sqrt{3}-1 \end{pmatrix}$

5. (*a*) $P_{T\leftarrow S} = \begin{pmatrix} 0 & -\sqrt{2}/2 & \sqrt{2}/2 \\ \sqrt{2} & \sqrt{2}/2 & \sqrt{2}/2 \\ 0 & 1 & 1 \end{pmatrix}$ (*b*) $P_{U\leftarrow T} = \begin{pmatrix} \sqrt{2}/2 & 0 & \frac{1}{2} \\ -\sqrt{2}/2 & 0 & \frac{1}{2} \\ 0 & \sqrt{2}/2 & \frac{1}{2} \end{pmatrix}$

(*c*) $P_{U\leftarrow S} = \begin{pmatrix} 0 & 0 & 1 \\ 0 & 1 & 0 \\ 1 & 0 & 0 \end{pmatrix}$

$P_{U\leftarrow S} = P_{U\leftarrow T}P_{T\leftarrow S}$

7. $P_{Z\leftarrow T} = \begin{pmatrix} \sqrt{2}/2 & \sqrt{2}/2 & 0 \\ -\frac{1}{2} & \frac{1}{2} & -\sqrt{2}/2 \\ -\frac{1}{2} & \frac{1}{2} & \sqrt{2}/2 \end{pmatrix} \qquad P_{T\leftarrow Z} = \begin{pmatrix} \sqrt{2}/2 & -\frac{1}{2} & -\frac{1}{2} \\ \sqrt{2}/2 & \frac{1}{2} & \frac{1}{2} \\ 0 & -\sqrt{2}/2 & \sqrt{2}/2 \end{pmatrix}$

9. $\mathbf{x} = 3(1, 1, 0) - 2(0, 1, 1) + 4(1, 0, 1) = (7, 1, 2)$. **11.** $\mathbf{x}$ is the zero vector.

13. Follow the indicated steps. **15.** Let $\mathbf{x}$ be in V. $P(\mathbf{x})_S = (\mathbf{x})_T$, $Q(\mathbf{x})_T = (\mathbf{x})_U$. Thus $(\mathbf{x})_U = Q(\mathbf{x})_T = Q(P(\mathbf{x})_S) = (QP)(\mathbf{x})_S$.

17. Because any basis has n vectors in it, each column has n entries.

19. $P = \begin{pmatrix} \cos(\pi/6) & -\sin(\pi/6) \\ \sin(\pi/6) & \cos(\pi/6) \end{pmatrix} = \begin{pmatrix} \sqrt{3}/2 & -\frac{1}{2} \\ \frac{1}{2} & \sqrt{3}/2 \end{pmatrix}$

PROBLEMS 3.9

1. (*a*) $\frac{2}{5}$; (*b*) 0; (*c*) 1; (*d*) $\frac{1}{12}$ **3.** $\|f\| = \sqrt{3}/3$

5. (1) $$\langle g, f\rangle = \int_a^b g(x)f(x) + g'(x)f'(x)\,dx = \int_a^b f(x)g(x) + f'(x)g'(x)\,dx = \langle f, g\rangle$$

Since f and g are real valued functions, $\langle f, g\rangle = \overline{\langle g, f\rangle}$.

(2) $$\begin{aligned}\langle f + h, g\rangle &= \int_a^b (f(x) + h(x))g(x) + (f'(x) + h'(x))g'(x)\,dx \\ &= \int_a^b f(x)g(x) + h(x)g(x) + f'(x)g'(x) + h'(x)g'(x)\,dx \\ &= \int_a^b f(x)g(x) + f'(x)g'(x)\,dx + \int_a^b h(x)g(x) + h'(x)g'(x)\,dx \\ &= \langle f, g\rangle + \langle h, g\rangle\end{aligned}$$

(3) $$r\langle f, g\rangle = r\int_a^b f(x)g(x) + f'(x)g'(x)\,dx = \int_a^b (rf(x))g(x) + (rf'(x))g'(x)\,dx = \langle rf, g\rangle$$

(4) $$\langle f, f\rangle = \int_a^b (f(x))^2 + (f'(x))^2\,dx$$

Because f and f' are continuous on $[a, b]$, this integral can be zero if and only if f is identically zero.

7. (*a*) Linearly independent; (*b*) linearly independent; (*c*) linearly independent.

9. The function $f(x) = x^{2/3}$ has no derivative at $x = 0$. Consider $c_1x^{2/3} + c_2x^2 = 0$. If the equation is to have nontrivial solution for all x, it must have nontrivial solution for any two values of x. Choose $x = 1$, $x = 8$, to obtain

$$c_1 + c_2 = 0 \qquad 4c_1 + 64c_2 = 0$$

which has only the trivial solution. Thus the equation $c_1x^{2/3} + c_2x^2 = 0$ cannot have nontrivial solution for all x and S must be linearly independent.

11. $$\cos\theta = \frac{\int_0^1 (x)(x^3)\,dx}{(\int_0^1 x^2\,dx)^{1/2}(\int_0^1 x^6\,dx)^{1/2}} = \frac{\sqrt{21}}{5} \qquad \theta = \cos^{-1}\left(\frac{\sqrt{21}}{5}\right)$$

13. The set of solutions is a subspace. If f and g are solutions then $f'' = f$ and $g'' = g$. So $(f + g)'' = f'' + g'' = f + g$ and $(rf)'' = rf'' = rf$. By the closure then the solution set is a subspace. Now e^x and e^{-x} are two linearly independent solutions of $y'' = y$. Clearly dim span $(\{e^x, e^{-x}\}) = 2$.

ADDITIONAL PROBLEMS (CHAPTER 3)

1. V is not a vector space; V fails to be closed under scalar multiplication.

3. Yes. The space V is n dimensional and the equation $c_1 i\mathbf{v}_1 + c_2 i\mathbf{v}_2 + \cdots + c_n i\mathbf{v}_n = \boldsymbol{\theta}$ is equivalent to $i(c_1\mathbf{v}_1 + c_2\mathbf{v}_2 + \cdots + c_n\mathbf{v}_n) = \boldsymbol{\theta}$ or $c_1\mathbf{v}_1 + c_2\mathbf{v}_2 + \cdots + c_n\mathbf{v}_n = \boldsymbol{\theta}$. Thus $c_1 = c_2 = \cdots = c_n = 0$. Thus the set $\{i\mathbf{v}_1, \ldots, i\mathbf{v}_n\}$ is a linearly independent set of n vectors in an n-dimensional space and must be a basis.

5. The set $S = \{c_1\mathbf{v}_1, c_2\mathbf{v}_2, \ldots, c_n\mathbf{v}_n\}$ is a basis for V. Since V is n-dimensional, all that needs to be shown is that S, having n elements spans V. Let $\mathbf{x}$ be in V. Since $T = \{\mathbf{v}_1, \ldots, \mathbf{v}_n\}$ is a basis, there exist constants $d_1, d_2, \ldots, d_n$ such that $\mathbf{x} = d_1\mathbf{v}_1 + d_2\mathbf{v}_2 + \cdots + d_n\mathbf{v}_n$. Now we can rewrite $\mathbf{x} = (d_1/c_1)c_1\mathbf{v}_1 + (d_2/c_2)c_2\mathbf{v}_2 + \cdots + (d_n/c_n)c_n\mathbf{v}_n$ to see that S spans V.

7.
$$\begin{aligned} P^T &= (A(A^TA)^{-1}A^T)^T = A^{TT}((A^TA)^{-1})^TA^T \\ &= A((A^TA)^T)^{-1}A^T = A(A^T(A^T)^T)^{-1}A^T = A(A^TA)^{-1}A^T = P \\ P^2 &= (A(A^TA)^{-1}A^T)(A(A^TA)^{-1}A^T) = A(A^TA)^{-1}(A^TA)(A^TA)^{-1}A^T \\ &= A(A^TA)^{-1}IA^T = A(A^TA)^{-1}A^T = P \end{aligned}$$

9. The set of orthogonal $n \times n$ matrices is not a subspace of $\mathscr{M}_{nn}$. For example, when $n = 2$

$$\begin{pmatrix} 1 & 0 \\ 0 & 1 \end{pmatrix}$$

is orthogonal but $2I$ is not orthogonal.

11. Let X and Y satisfy $AX = cX$, $AY = cY$. Now $A(X + Y) = AX + AY = cX + cY = c(X + Y)$ so V is closed under addition. Also $A(rX) = rAX = rcX = c(rX)$ so V is closed under scalar multiplication. Therefore V is a subspace of $\mathscr{C}_{n1}$.

13. With the given inner product an orthonormal basis is $\mathcal{O} = \{1, \sqrt{3}(1 - 2x), \sqrt{5}(6x^2 - 6x + 1)\}$. The best approximation is

$$\begin{aligned} &\left(\int_0^1 e^x \cdot 1\,dx\right)1 + \left(\int_0^1 e^x(\sqrt{3}(1 - 2x))\,dx\right)\sqrt{3}(1 - 2x) \\ &\quad + \left(\int_0^1 e^x\sqrt{5}(6x^2 - 6x + 1)\,dx\right)\sqrt{5}(6x^2 - 6x + 1) \\ &= (e - 1) + 3(e - 3)(1 - 2x) + 5(7e - 19)(6x^2 - 6x + 1) \\ &\doteq 1.013 + .851x + .839x^2 = p(x) \end{aligned}$$

Note that

$$\begin{aligned} &p(0) \doteq 1.013,\ e^0 = 1.0 \qquad p(.5) \doteq 1.64825,\ e^{.5} = 1.64872\cdots \\ &p(1) \doteq 2.703,\ e^1 = 2.71828\cdots \end{aligned}$$

15. $\langle \mathbf{x}, \mathbf{y} \rangle = \left\langle \sum_{k=1}^{n} a_k \mathbf{v}_k, \sum_{j=1}^{n} b_j \mathbf{v}_j \right\rangle = \sum_{k=1}^{n} a_k \left\langle \mathbf{v}_k, \sum_{j=1}^{n} b_j \mathbf{v}_j \right\rangle = \sum_{k=1}^{n} a_k \sum_{j=1}^{n} b_j \langle \mathbf{v}_k, \mathbf{v}_j \rangle$

$$(\mathbf{x})_S{}^T G(\mathbf{y})_S = (a_1, a_2, \ldots, a_n)(\langle \mathbf{v}_i, \mathbf{v}_j \rangle)_{n \times n} \begin{pmatrix} b_1 \\ b_2 \\ \vdots \\ b_n \end{pmatrix}$$

$$= (a_1, a_2, \ldots, a_n) \left(\sum_{j=1}^{n} \langle \mathbf{v}_i, \mathbf{v}_j \rangle b_j \right)_{n \times 1}$$

$$= \sum_{i=1}^{n} a_i \left(\sum_{j=1}^{n} b_j \langle \mathbf{v}_i, \mathbf{v}_j \rangle \right)$$

Now substitute k for i as the index of summation.

PROBLEMS 4.1

1. (*a*) Linear.

(*b*) Not linear. $T(c\mathbf{x}) = (cx_1, cx_2, c^2 x_2 x_3)$; $cT(\mathbf{x}) = (cx_1, cx_2, cx_2 x_3)$.

(*c*) Linear.

(*d*) Not linear. $T(\mathbf{x} + \mathbf{y}) = (1, 0, 0) \neq (2, 0, 0) = T(\mathbf{x}) + T(\mathbf{y})$.

(*e*) Not linear. $T(-\mathbf{x}) \neq -T(\mathbf{x})$. (*f*) Linear.

3. (*a*) Linear. (*b*) Not linear. $T(0) \neq 0$ unless $c = 0$.

(*c*) Not linear. $T(p + q) = 2$; $T(p) + T(q) = 2 + 2 = 4$.

(*d*) Not linear. T is not even defined at $p \equiv 0$.

(*e*) Linear.

5. (*a*) ket $T = \{\boldsymbol{\theta}\}$. range $(T) = E^3$, so $\eta(T) = 0$. $\mathscr{R}(T) = 3$. dim (domain (T)) $=$ dim $E^3 = 3$. ker T has no basis, the standard basis suffices for range (T).

(*c*)

$$\begin{aligned} \ker T &= \{x \mid x_1 = 0, x_2 = r, x_3 = s, r, s \text{ in } \mathbb{R}\} \\ \text{range } (T) &= \{y \mid y_1 = t, y_2 = y_3 = 0, t \text{ in } \mathbb{R}\} \\ \text{Basis for ker } T &= \{(0, 1, 0), (0, 0, 1)\} \\ \text{Basis for range } (T) &= \{(1, 0, 0)\} \\ \eta(T) &= 2, \mathscr{R}(T) = 1 \end{aligned}$$

(*f*) ker $T = \{\theta\}$. range $(T) = E^3$, $\eta(T) = 0$, $\mathscr{R}(T) = 3$. ker T has no basis. The standard basis suffices for range (T).

7. (*a*)

$$\begin{aligned} \ker T &= \{a + bx + cx^2 \mid a = b = 0, \text{c arbitrary}\} \\ \text{range } T &= \{\text{degree two polynomials with constant term zero}\} \\ \text{Basis for ker } T &= \{x^2\} \\ \text{Basis for range } (T) &= \{x, x^2\} \\ \eta(T) &= 1, \mathscr{R}(T) = 2. \text{ dim (domain)} = \dim \mathscr{P}_2 = 3 \end{aligned}$$

(*e*) ket $T = \{\theta\}$. range $(T) = \mathscr{P}_2$, so $\eta(T) = 0$, $\mathscr{R}(T) = 3$.
ket T has no basis. The standard basis suffices for range (T).

9. Let $T: V \to V$ be defined by $T(\mathbf{x}) = \mathbf{x}$ for all $\mathbf{x}$ in V. Let c be any scalar. $T(c\mathbf{x}) = c\mathbf{x}$, $cT(\mathbf{x}) = c\mathbf{x}$. $T(\mathbf{x} + \mathbf{y}) = \mathbf{x} + \mathbf{y}$, $T(\mathbf{x}) + T(\mathbf{y}) = \mathbf{x} + \mathbf{y}$. Therefore $T(c\mathbf{x}) = cT(\mathbf{x})$ and $T(\mathbf{x} + \mathbf{y}) = T(\mathbf{x}) + T(\mathbf{y})$.

11. Let $T: V \to V$ be defined by $T(\mathbf{x}) = k\mathbf{x}$ for all $\mathbf{x}$ in V, where $k > 1$. Let c be any scalar. $T(c\mathbf{x}) = k(c\mathbf{x}) = c(k\mathbf{x}) = cT(\mathbf{x})$. $T(\mathbf{x} + \mathbf{y}) = k(\mathbf{x} + \mathbf{y}) = k\mathbf{x} + k\mathbf{y} = T(\mathbf{x}) + T(\mathbf{y})$.

15. Use the hint. $pT(\mathbf{x}) = T(p\mathbf{x}) = T(q(p/q)\mathbf{x})$. Now $T(q(p/q\ \mathbf{x})) = qT(p/q\ \mathbf{x})$ by Prob. 14. Therefore $pT(\mathbf{x}) = qT(p/q\ \mathbf{x})$. Divide both sides by q.

17. To show range (T) is a subspace of W, we need to show closure. Let $\mathbf{w}_1$ and $\mathbf{w}_2$ be in range (T). This means there exist $\mathbf{x}_1$ and $\mathbf{x}_2$ in V such that $T(\mathbf{x}_1) = \mathbf{w}_1$, $T(\mathbf{x}_2) = \mathbf{w}_2$. Now we ask the closure questions: For any scalar c is $c\mathbf{w}_1$ in range (T) and is $\mathbf{w}_1 + \mathbf{w}_2$ in range (T)? Consider $c\mathbf{w}_1$. We have $c\mathbf{w}_1 = cT(\mathbf{x}_1) = T(c\mathbf{x}_1)$. Now V is a vector space, so $c\mathbf{x}_1$ is in V. Thus $T(c\mathbf{x}_1)$ is in range (T). Consider now $\mathbf{w}_1 + \mathbf{w}_2$. We have $\mathbf{w}_1 + \mathbf{w}_2 = T(\mathbf{x}_1) + T(\mathbf{x}_2) = T(\mathbf{x}_1 + \mathbf{x}_2)$. Because V is a vector space $\mathbf{x}_1 + \mathbf{x}_2$ is in V, so $T(\mathbf{x}_1 + \mathbf{x}_2)$ is defined and $T(\mathbf{x}_1 + \mathbf{x}_2)$ is in range (T).

19.
$$T(cf)(x) = x(cf)(x) = c(xf)(x) = cT(f)$$
$$T(f + g)(x) = x(f + g)(x) = xf(x) + xg(x) = T(f)(x) + T(g)(x)$$

21.
$$\begin{aligned} T((8, 3, 2)) &= T(2(1, 0, 1) + 3(2, 1, 0)) = 2T(1, 0, 1) + 3T(2, 1, 0)) \\ &= 2(1, -1, 3) + (3(0, 2, 1) = (2, 4, 9) \end{aligned}$$

23. $(3, 0, 4)$ is not in span $(\{(1, 0, 1), (2, 1, 0)\})$. This is found by showing that $(3, 0, 4) = c_1(1, 0, 1) + c_2(2, 1, 0)$ leads to inconsistent linear equations.

25. Let the line be given by $R = X + tY$, t in $\mathbb{R}$, where X and Y are in $\mathscr{M}_{21}$. Now $M(X + tY) = MX + M(tY) = MX + tMY$. Now the points satisfying $S = MX + tMY$, T in $\mathbb{R}$ form a line because the equation is a vector equation for a line.

PROBLEMS 4.2

1. (*a*) $(A|0)$ reduces to

$$\left(\begin{array}{ccc|c} 1 & 2 & 3 & 0 \\ 0 & 1 & 3 & 0 \\ 0 & 0 & 0 & 0 \end{array}\right)$$

Thus rank $(A) = 2$. The solutions are $\{(x, y, z) = (t, -2t, t)\}$. A basis for the solution space is

$$S = \left\{ \begin{pmatrix} 1 \\ -2 \\ 1 \end{pmatrix} \right\}$$

Thus dim (solution space) + rank $(A) = 1 + 2 =$ number of columns of A.

(*b*) $(A|0)$ reduces to

$$\left(\begin{array}{ccc|c} 1 & -1 & 2 & 0 \\ 0 & 1 & 1 & 0 \\ 0 & 0 & 0 & 0 \end{array}\right)$$

Thus rank $(A) = 2$. The solutions are $\{(x, y, z) = (-3t, -t, t)\}$. A basis for the solution space is

$$S = \left\{\begin{pmatrix} -3 \\ -1 \\ 1 \end{pmatrix}\right\}$$

Thus dim (solution space) + rank $(A) = 1 + 2 = 3$.

(*c*) $(A|0)$ reduces to

$$\begin{pmatrix} 1 & 2 & -1 & 0 & 0 \\ 0 & -10 & 7 & 1 & 0 \\ 0 & 0 & 0 & 0 & 0 \\ 0 & 0 & 0 & 0 & 0 \end{pmatrix}$$

Thus rank $(A) = 2$. The solutions are $\{(x, y, z, w) = (-4s - 2t, 7s + t, 10s, 10t)\}$. A basis for the solution space is

$$S = \left\{\begin{pmatrix} -2 \\ 1 \\ 0 \\ 10 \end{pmatrix}, \begin{pmatrix} -4 \\ 7 \\ 10 \\ 0 \end{pmatrix}\right\}$$

Thus dim (solution space) + rank $(A) = 2 + 2 =$ number of columns of A.

3. (*a*) $M = \begin{pmatrix} \frac{1}{2} & 0 & 0 \\ \frac{1}{2} & -1 & 1 \end{pmatrix}$

(*b*) Directly: $T(1, -1, 2) = (4, -3)$. Using M:

$$[T(1, -1, 2)]_S = \begin{pmatrix} \frac{1}{2} & 0 & 0 \\ \frac{1}{2} & -1 & 1 \end{pmatrix} [(1, -1, 2)]_{\text{STD}}$$

$$= \begin{pmatrix} \frac{1}{2} & 0 & 0 \\ \frac{1}{2} & -1 & 1 \end{pmatrix} \begin{pmatrix} 1 \\ -1 \\ 2 \end{pmatrix} = \begin{pmatrix} \frac{1}{2} \\ \frac{7}{2} \end{pmatrix}$$

Therefore $T(1, -1, 2) = \frac{1}{2}(1, 1) + \frac{7}{2}(1, -1) = (4, -3)$.

5. Give the domain $\mathscr{S} = \{(1, 1), (1, -1)\}$ for a basis and let the range have the standard basis. In this case

$$M = \begin{pmatrix} 1 & 0 \\ 0 & 1 \end{pmatrix}$$

On the other hand, if the range is equipped with $\mathscr{S}$ as basis,

$$M = \begin{pmatrix} \frac{1}{2} & \frac{1}{2} \\ \frac{1}{2} & -\frac{1}{2} \end{pmatrix}$$

7. Let $Z: V \to W$ be the zero transformation. That is for all $\mathbf{x}$ in V, $Z(\mathbf{x}) = \boldsymbol{\theta}$, the zero vector of W. Let $S = \{\mathbf{v}_1, \ldots, \mathbf{v}_n\}$ and $T = \{\mathbf{w}_1, \mathbf{w}_2, \ldots, \mathbf{w}_m\}$ be bases for V and W respectively. Now $Z(\mathbf{v}_i) = \boldsymbol{\theta} = 0\mathbf{w}_1 + 0\mathbf{w}_2 + \cdots + 0\mathbf{w}_m$ (uniquely). Thus every column of M_Z consists of all zeros.

9. For each k, $k = 0, 1, 2, \ldots, n$, $T(x^k) = x^{k+1} = 0 + 0x + \cdots + 1x^{k+1} + \cdots + 0x^{n+1}$. Thus column $k + 1$ of M has $n + 2$ entries, all being zero except for the $k + 2$ entry, which is 1.

11. (*a*) $\begin{pmatrix} 1 & 1 \\ 0 & i \end{pmatrix}$ (*b*) $\begin{pmatrix} 0 & i & 0 \\ 0 & 0 & i \\ 0 & 0 & 0 \end{pmatrix}$ (*c*) $\begin{pmatrix} 1+i & 0 & 0 & 0 \\ 0 & 1 & i & 0 \\ 0 & i & 1 & 0 \\ 0 & 0 & 0 & 1+i \end{pmatrix}$

PROBLEMS 4.3

1. (*a*) $M_{\mathscr{S}} = \begin{pmatrix} 1 & 0 \\ 0 & 0 \end{pmatrix}$

(*b*) $P_{\mathscr{S}\leftarrow\mathscr{T}} = \begin{pmatrix} 1 & 1 \\ 1 & -1 \end{pmatrix}$ $\quad P_{\mathscr{T}\leftarrow\mathscr{S}} = P^{-1}_{\mathscr{S}\leftarrow\mathscr{T}} = \begin{pmatrix} \frac{1}{2} & \frac{1}{2} \\ \frac{1}{2} & -\frac{1}{2} \end{pmatrix}$

$$M_{\mathscr{T}} = P_{\mathscr{T}\leftarrow\mathscr{S}} M_{\mathscr{S}} P_{\mathscr{S}\leftarrow\mathscr{T}} = \begin{pmatrix} \frac{1}{2} & \frac{1}{2} \\ \frac{1}{2} & \frac{1}{2} \end{pmatrix}$$

3. (*a*) $M_{\mathscr{S}} = \begin{pmatrix} 1 & 1 \\ 2 & -3 \end{pmatrix}$

(*b*) Transition matrices are as in 1*b*.

$$M_{\mathscr{T}} = P_{\mathscr{T}\leftarrow\mathscr{S}} M_{\mathscr{S}} P_{\mathscr{S}\leftarrow\mathscr{T}} = \begin{bmatrix} \frac{1}{2} & \frac{5}{2} \\ \frac{3}{2} & -\frac{5}{2} \end{bmatrix}$$

5. (*a*) $M_{\mathscr{S}} = \dfrac{\sqrt{2}}{2}\begin{pmatrix} 1 & -1 \\ 1 & 1 \end{pmatrix}$

(*b*) $P_{\mathscr{S}\leftarrow\mathscr{T}} = \begin{pmatrix} 1 & 1 \\ 1 & 2 \end{pmatrix}$ $\quad P_{\mathscr{T}\leftarrow\mathscr{S}} = P^{-1}_{\mathscr{S}\leftarrow\mathscr{T}} = \begin{pmatrix} 2 & -1 \\ -1 & 1 \end{pmatrix}$

$$M_{\mathscr{T}} = P_{\mathscr{T}\leftarrow\mathscr{S}} M_{\mathscr{S}} P_{\mathscr{S}\leftarrow\mathscr{T}} = \frac{\sqrt{2}}{2}\begin{pmatrix} -2 & -5 \\ 2 & 4 \end{pmatrix}$$

7. (*a*) $$M_{\mathscr{S}} = \begin{pmatrix} 1 & 2 & 0 \\ 1 & 1 & 1 \\ 0 & 0 & 1 \end{pmatrix}$$

(*b*) $$P_{\mathscr{S} \leftarrow \mathscr{T}} = \begin{pmatrix} 1 & 0 & 1 \\ 1 & 1 & 0 \\ 0 & 1 & 1 \end{pmatrix} \qquad P_{\mathscr{T} \leftarrow \mathscr{S}} = P^{-1}_{\mathscr{S} \leftarrow \mathscr{T}} = \frac{1}{2} \begin{pmatrix} 1 & 1 & -1 \\ -1 & 1 & 1 \\ 1 & -1 & 1 \end{pmatrix}$$

$$M_{\mathscr{T}} = P_{\mathscr{T} \leftarrow \mathscr{S}} M_{\mathscr{S}} P_{\mathscr{S} \leftarrow \mathscr{T}} = \frac{1}{2} \begin{pmatrix} 5 & 3 & 2 \\ -1 & 1 & 2 \\ 1 & 1 & 0 \end{pmatrix}$$

9. (*a*) $$M_{\mathscr{S}} = \begin{pmatrix} 1 & 0 & 0 & 0 \\ 0 & 0 & 1 & 0 \\ 0 & 1 & 0 & 0 \\ 0 & 0 & 0 & 1 \end{pmatrix}$$

(*b*) $$P_{\mathscr{S} \leftarrow \mathscr{T}} = \begin{pmatrix} 1 & 0 & 1 & 1 \\ 1 & 0 & 0 & 0 \\ 0 & 1 & 1 & 0 \\ 0 & 1 & 0 & 1 \end{pmatrix}$$

$$P_{\mathscr{T} \leftarrow \mathscr{S}} = P^{-1}_{\mathscr{S} \leftarrow \mathscr{T}} = \frac{1}{2} \begin{pmatrix} 0 & 2 & 0 & 0 \\ -1 & 1 & 1 & 1 \\ 1 & -1 & 1 & -1 \\ 1 & -1 & -1 & 1 \end{pmatrix}$$

$$M_{\mathscr{T}} = P_{\mathscr{T} \leftarrow \mathscr{S}} M_{\mathscr{S}} P_{\mathscr{S} \leftarrow \mathscr{T}} = \begin{pmatrix} 0 & 1 & 1 & 0 \\ 0 & 1 & 0 & 0 \\ 1 & -1 & 0 & 0 \\ 0 & 0 & 0 & 1 \end{pmatrix}$$

11. (*a*) The traces are not equal

(*b*) Neither the traces nor the determinants are equal.

(*c*) We have $\operatorname{tr} A = \operatorname{tr} B$, $\det A = \det B$; we need to work a little harder. Let

$$\begin{pmatrix} a & b \\ c & d \end{pmatrix}$$

and suppose $PB = AP$. Writing out the products and equating elements of the matrices we get equations

$$2c = a + 2c \qquad 2d = b + 2d$$

which means $a = b = 0$. But then

$$P = \begin{pmatrix} 0 & 0 \\ c & d \end{pmatrix}$$

which is not invertible. Thus A cannot be similar to B.

13. Identify

$$\begin{array}{rcl} E^2 & & P_1 \\ (a, b) & \longleftrightarrow & a + bx \\ \mathcal{S} = \{(1, 0), & \longleftrightarrow & \{1 + 0x \\ (0, 1)\} & \longleftrightarrow & 0 + 1x\} \\ (a, b) & \longleftrightarrow & a + bx \\ T\downarrow & \longleftrightarrow & T\downarrow \\ (a + b) + (2a - 3b) & \longleftrightarrow & (a + b) + (2a - 3b)x \end{array}$$

That is, identify the first component of an element in E^2 with the constant coefficient of the degree one polynomial; identify the second component of an element of E^2 with the coefficient of x in the degree one polynomial.

PROBLEMS 4.4

1. The standard matrix for T is

$$M = \begin{pmatrix} 1 & 0 & 0 \\ 0 & 1 & 0 \\ 0 & 0 & 0 \end{pmatrix}$$

M is not invertible, but $M^2 = M$. Therefore T is not invertible, but T is idempotent.

3. The standard matrix for T is

$$M = \begin{pmatrix} 1 & 0 & 0 & 0 \\ 0 & 0 & 1 & 0 \\ 0 & 1 & 0 & 0 \\ 0 & 0 & 0 & 1 \end{pmatrix}$$

M is invertible and $M^2 = I$. Therefore T is invertible but T is not idempotent.

5. The standard matrix for T is

$$M = \begin{pmatrix} 1 & 0 & 0 \\ 1 & 1 & 0 \\ 1 & 1 & 1 \end{pmatrix}$$

M is invertible.

$$M^2 = \begin{pmatrix} 1 & 0 & 0 \\ 2 & 1 & 0 \\ 3 & 2 & 1 \end{pmatrix} \qquad M^3 = \begin{pmatrix} 1 & 0 & 0 \\ 3 & 1 & 0 \\ 6 & 3 & 1 \end{pmatrix}$$

$$M^4 = \begin{pmatrix} 1 & 0 & 0 \\ 4 & 1 & 0 \\ 10 & 4 & 1 \end{pmatrix}, \ldots, M^n = \begin{pmatrix} 1 & 0 & 0 \\ n & 1 & 0 \\ (n + 1)n/2 & n & 1 \end{pmatrix}$$

Thus M is not nilpotent. Actually we don't have to work this hard. Recall in an earlier chapter we had a problem: "Show that if A is nilpotent, then $\det A = 0$." As soon as we know M is invertible we know that M cannot be nilpotent. Anyway T is invertible and not nilpotent.

7. The standard matrix for a rotation of 120° is

$$M = \begin{pmatrix} \cos(2\pi/3) & -\sin(2\pi/3) \\ \sin(2\pi/3) & \cos(2\pi/3) \end{pmatrix} = \begin{pmatrix} -\frac{1}{2} & -\sqrt{3}/2 \\ \sqrt{3}/2 & -\frac{1}{2} \end{pmatrix}$$

Now

$$M^2 = \begin{pmatrix} -\frac{1}{2} & \sqrt{3}/2 \\ -\sqrt{3}/2 & -\frac{1}{2} \end{pmatrix} \quad \text{and} \quad M^3 = I$$

9. The standard matrix for T is

$$M = \begin{pmatrix} 0 & 0 & 0 & \cdots & & 0 \\ 1 & 0 & 0 & \cdots & & 0 \\ 0 & 1 & 0 & \cdots & & 0 \\ 0 & 0 & 1 & \ddots & & \\ \vdots & \vdots & \vdots & & 0 & \vdots \\ 0 & 0 & 0 & \cdots & 1 & 0 \end{pmatrix}$$

Since $\det M = 0$, M is not invertible. As higher powers of M are calculated, the diagonal of 1's moves to the lower left corner; $M^n = 0$. Thus T is nilpotent. T is not idempotent, because $M^2 \neq M$.

11. If A is involutory and B is similar to A, then B is involutory. Suppose $A^2 = I$, and $B = P^{-1}AP$. Then $B^2 = P^{-1}APP^{-1}AP = P^{-1}A^2P = P^{-1}P = I$. Since $B^2 = I$, B is involutory.

13. $B^T = (P^TAP)^T = P^TA^TP^{TT} = P^TAP = B$, so B is symmetric.

15. Consider the equation to determine independence

$$c_1L(\mathbf{v}_1) + c_2L(\mathbf{v}_2) + \cdots + c_kL(\mathbf{v}_k) = \boldsymbol{\theta}$$

Using the linearity of L we have

$$\boldsymbol{\theta} = L(c_1\mathbf{v}_1) + L(c_2\mathbf{v}_2) + \cdots + L(c_k\mathbf{v}_k) = L(c_1\mathbf{v}_1 + c_2\mathbf{v}_2 + \cdots + c_k\mathbf{v}_k)$$

Now since the kernel of L is only the zero vector, we have

$$c_1\mathbf{v}_1 + c_2\mathbf{v}_2 + \cdots + c_k\mathbf{v}_k = \boldsymbol{\theta}$$

The linear independence of $\{\mathbf{v}_1, \mathbf{v}_2, \ldots, \mathbf{v}_k\}$ means that the last equation has only the solution $c_1 = c_2 = \cdots = c_k = 0$. Therefore $\{L(\mathbf{v}_1), \ldots, L(\mathbf{v}_k)\}$ is a linearly independent set.

PROBLEMS 4.5

1. Because V is a subset of the vector space of all functions defined on $\mathbb{R}$, we need only show closure. Let f and g be functions with

$$\lim_{x\to a} f(x) = L, \qquad \lim_{x\to a} g(x) = M$$

Now does $f + g$ have a limit at a? From calculus we have the theorem which tells us

$$\lim_{x\to a} (f + g)(x) = L + M$$

Thus $f + g$ has limit at a and $f + g$ is in V. Similarly

$$\lim_{x\to a} (rf)(x) = r \lim_{x\to a} (f)(x) = rL \qquad \text{and} \qquad rf \text{ is in } V$$

Note that the theorems we used from calculus concerning limits at a point a can be written in the shorthand from the statement of the problem

$$L_a(f + g) = L_a(f) + L_a(g) \qquad L_a(rf) = rL_a(f)$$

That is L_a is a linear transformation from V to $\mathbb{R}$.

3. $L_f(g(x)) = \int_0^\pi \sin x \sin nx\, dx \qquad n > 1$

By trigonometric identities $\cos(nx - x) - \cos(nx + x) = \cos nx \cos x + \sin nx \sin x - (\cos nx \cos x - \sin nx \sin x) = 2 \sin nx \sin x$. Thus,

$$\begin{aligned}\int_0^\pi \sin x \sin nx\, dx &= \frac{1}{2}\int_0^\pi \cos(n-1)x - \cos(n+1)x\, dx \\ &= \frac{1}{2}\left(\frac{\sin(n-1)x}{n-1} - \frac{\sin(n+1)x}{n+1}\right)\Bigg|_0^\pi = 0\end{aligned}$$

5. From calculus we know that $(f + g)' = f' + g'$ and $(rf)' = rf'$. Thus $(f + g)'' = ((f + g)')' = (f' + g')' = f'' + g''$ and $(rf)'' = (rf')' = rf''$. The standard matrix for D^2 is

$$N = \begin{pmatrix} 0 & 0 & 2 & 0 \\ 0 & 0 & 0 & 6 \\ 0 & 0 & 0 & 0 \\ 0 & 0 & 0 & 0 \end{pmatrix}$$

which is equal to M^2.

7. We could try

$$I: a_0 + a_1 x + a_2 x^2 + \cdots \longrightarrow c + a_0 x + \frac{a_1}{2}x^2 + \frac{a_2}{3}x^3 + \cdots$$

But we see that the "constant of integration" c is a problem. If we restrict our attention to the subset of functions in V with $f(0) = 0$ then $c = 0$ and I could invert the action of D.

9. $$\operatorname{curl}(r\mathbf{f}) = \det\begin{pmatrix} \mathbf{i} & \mathbf{j} & \mathbf{k} \\ \partial/\partial x & \partial/\partial y & \partial/\partial z \\ rf_1 & rf_2 & rf_3 \end{pmatrix} = r\det\begin{pmatrix} \mathbf{i} & \mathbf{j} & \mathbf{k} \\ \partial/\partial x & \partial/\partial y & \partial/\partial z \\ f_1 & f_2 & f_3 \end{pmatrix}$$

$$= r\operatorname{curl}(\mathbf{f}) \cdot \operatorname{curl}(f+g) = \det\begin{pmatrix} \mathbf{i} & \mathbf{j} & \mathbf{k} \\ \partial/\partial x & \partial/\partial y & \partial/\partial z \\ f_1+g_1 & f_2+g_2 & f_3+g_3 \end{pmatrix}$$

$$= \left(\frac{\partial}{\partial x}(f_2+g_2) - \frac{\partial}{\partial y}(f_1+g_1)\right)\mathbf{i}$$

$$+ \left(\frac{\partial}{\partial z}(f_1+g_1) - \frac{\partial}{\partial x}(f_3+g_3)\right)\mathbf{j} + \left(\frac{\partial}{\partial x}(f_2+g_2)\right.$$

$$\left. - \frac{\partial}{\partial y}(f_1+g_1)\right)\mathbf{k}$$

Now use the linearity of the partial derivative.

11. (*a*) $$T(rf)(x) = x(rf)(x) = r(xf(x)) = r(T(f)(x))$$
$$(T(f+g))(x) = x(f+g)(x) = xf(x) + xg(x) = (T(f))(x) + (T(g))(x)$$

(*b*) $$(T(g))(x) = xg(x) = \begin{cases} x \cdot 1 & x = 0 \\ x \cdot 0 & x \neq 0 \end{cases} = \begin{cases} 0 & x = 0 \\ 0 & x \neq 0 \end{cases}$$

(*c*) Because g is not the zero vector but is in the kernel of T, $\ker T \neq \{\boldsymbol{\theta}\}$ and T is not invertible.

ADDITIONAL PROBLEMS (CHAPTER 4)

1. $T(A+B) = (A+B) - (A+B)^T = A + B - A^T - B^T = A - A^T + B - B^T = T(A) + T(B)$. $T(rA) = rA - (rA)^T = rA - rA^T = r(A - A^T) = rT(A)$. The kernel of T is all $A_{n \times n}$ with $A - A^T = 0$ or $A = A^T$; the kernel is the subspace of symmetric matrices.

3. $$\dim(\text{symmetric } n \times n \text{ matrices}) = \frac{n(n+1)}{2}$$

$$\dim(\text{upper triangular } n \times n \text{ matrices}) = \frac{n(n+1)}{2}$$

The dimensions are equal.

5. $T(A+B) = P(A+B) = PA + PB = T(A) + T(B)$. $T(rA) = P(rA) = rPA = rT(A)$. To see whether T is one-to-one we look at $T(A) = 0$, which is $PA = 0$. Now we can't just multiply by P^{-1} because P may not be invertible. For example if $n = 2$,

$$P = \begin{pmatrix} 1 & 0 \\ 0 & 0 \end{pmatrix}$$

satisfies $P^2 = P$ but P is not invertible. Note that choosing

$$A = \begin{pmatrix} 0 & 0 \\ 0 & 1 \end{pmatrix}$$

$PA = 0$ but $A \neq 0$. Thus in general we cannot say that T is one-to-one.

7.

$$\begin{aligned} B(I - A) &= (I + A)(I - A) = I - A^2 \\ B(I - A + A^2) &= (I + A)(I - A + A^2) \\ &= I - A^2 + (I + A)A^2 = I + A^3 \\ B(I - A + A^2 - A^3) &= I + A^3 - (I + A)A^3 = I - A^4 \end{aligned}$$

If $A^k = 0$ then $(I + A)(I - A + A^2 - \cdots + (-1)^{k-1}A^{k-1}) = I + (-1)^{k-1}A^k = I + 0 = I$. Therefore if A is nilpotent exponent k, $(I + A)^{-1}$ exists and

$$(I + A)^{-1} = I + \sum_{n=1}^{k-1} (-1)^n A^n$$

9. $(L \circ T)(A) = L(T(A)) = L(A - A^T) = (A - A^T) + (A - A^T)^T = A - A^T + A^T - A = 0$. Therefore $L \circ T$ is the zero transformation. Now $(T \circ L)(A) = T(L(A)) = T(A + A^T) = (A + A^T) - (A + A^T)^T = A + A^T - A^T - A = 0$. Therefore $T \circ L$ is also the zero transformation and $L \circ T = T \circ L$.

11. $\det A(t) = \cos^2(\omega t) + \sin^2(\omega t) = 1$ for all t, so $A(t)$ is invertible for all t. Also

$$(A(t))^T(A(t)) = \begin{pmatrix} 1 & 0 \\ 0 & 1 \end{pmatrix} \qquad \text{for all } t$$

so $A(t)$ is orthogonal for all t.

13. Let a line have the vector equation $R = tU + V$, $t \in \mathbb{R}$. Now $MR = M(tU + V) = t(MU) + MV$. The points described by R are transformed to points $S = MR = t(MU) + MV$. This equation is an equation of a line.

15.

$$\begin{pmatrix} x_{n+1} \\ y_{n+1} \end{pmatrix} = \begin{pmatrix} x_n \\ y_n \end{pmatrix} - \begin{pmatrix} 2x_n & 2y_n \\ 1 & -1 \end{pmatrix}^{-1} \begin{pmatrix} x_n^2 + y_n^2 - 2 \\ x_n - y_n \end{pmatrix}$$

Now

$$\begin{pmatrix} 2x_n & 2y_n \\ 1 & -1 \end{pmatrix}^{-1} = \frac{1}{2(x_n + y_n)} \begin{pmatrix} 1 & 2y_n \\ 1 & -2x_n \end{pmatrix}$$

so we have

$$\begin{pmatrix} x_{n+1} \\ y_{n+1} \end{pmatrix} = \begin{pmatrix} x_n \\ y_n \end{pmatrix} + \frac{1}{2(x_n + y_n)} \begin{pmatrix} -1 & -2y_n \\ -1 & 2x_n \end{pmatrix} \begin{pmatrix} x_n^2 + y_n^2 - 2 \\ x_n - y_n \end{pmatrix}$$

or

$$x_{n+1} = x_n + \frac{1}{2(x_n + y_n)} (2 - x_n^2 - y_n^2 + 2y_n^2 - 2x_n y_n)$$

$$y_{n+1} = y_n + \frac{1}{2(x_n + y_n)} (2 - x_n^2 - y_n^2 + 2x_n^2 - 2x_n y_n)$$

which reduces to

$$x_{n+1} = \frac{2 + x_n^2 + y_n^2}{2(x_n + y_n)} \qquad y_{n+1} = \frac{2 + x_n^2 + y_n^2}{2(x_n + y_n)}$$

Therefore using $x_1 = 1$, $y_1 = 0$, we have $x_2 = y_2 = \frac{3}{2}$, $x_3 = y_3 = \frac{13}{12}$. Using a calculator, $x_4 = y_4 = 1.0032051\ldots$, $x_5 = y_5 = 1.000051\ldots$, $x_6 = y_6 = 1.00000000$ to calculator accuracy. Note that if you start with $\binom{-1}{0}$, Newton's method will lead to the solution $\begin{bmatrix}-1\\-1\end{bmatrix}$. Thus where you *start* in Newton's method affects which solution the method gives. Just as in calculus, Newton's method does not always work, but when it does work it works well.

PROBLEMS 5.1

1. $A^{10} = \begin{pmatrix} 2 & -1 \\ -1 & 1 \end{pmatrix} \begin{pmatrix} \frac{1}{2} & 0 \\ 0 & 2 \end{pmatrix}^{10} \begin{pmatrix} 1 & 1 \\ 1 & 2 \end{pmatrix} = \begin{pmatrix} 2 & -1 \\ -1 & 1 \end{pmatrix} \begin{pmatrix} 1/2^{10} & 0 \\ 0 & 2^{10} \end{pmatrix} \begin{pmatrix} 1 & 1 \\ 1 & 2 \end{pmatrix}$

$$= \begin{pmatrix} 1/2^9 - 2^{10} & 1/2^9 - 2^{11} \\ 2^{10} - 1/2^{10} & 2^{11} - 1/2^9 \end{pmatrix}$$

3. Let S_1 = state "liquid detergent user." Let S_2 = state "dry detergent user." The transition matrix is

$$\begin{array}{c} \text{This week} \\ \begin{array}{cc} S_1 & S_2 \end{array} \\ \begin{pmatrix} .2 & .4 \\ .8 & .6 \end{pmatrix} \begin{array}{c} S_1 \\ S_2 \end{array} \end{array} \quad \text{Next week}$$

The initial state is

$$\begin{pmatrix} .5 \\ .5 \end{pmatrix} \quad \text{and} \quad M^4 \begin{pmatrix} .5 \\ .5 \end{pmatrix} = \begin{pmatrix} .3336 \\ .6664 \end{pmatrix}$$

After 4 weeks liquid detergent will have $\frac{1}{3}$ of the market. If the advertising campaign leads to a market share of more than $\frac{1}{3}$ then the agency did a good job.

5. Sunday's state is

$$M^5 S = \begin{pmatrix} .143482 \\ .856518 \end{pmatrix}$$

The probability of a dry day is not appreciably greater than Saturday's probability.

PROBLEMS 5.2

1. In each answer, the given eigenvector can of course be replaced by any nonzero multiple of it.

(*a*) Eigenpairs $\left(i, \begin{pmatrix} 1 \\ i \end{pmatrix}\right), \left(\begin{matrix} -1 \\ i \end{matrix}\right)$ (*b*) Eigenpair $\left(1, \begin{pmatrix} 0 \\ 1 \end{pmatrix}\right)$

(*c*) Eigenpairs $\left(1, \begin{pmatrix} 1 \\ -1 \end{pmatrix}\right), \left(-1, \begin{pmatrix} 1 \\ 1 \end{pmatrix}\right)$

(*d*) Eigenpairs $\left(1, \begin{pmatrix} 1 \\ 0 \\ 0 \end{pmatrix}\right), \left(-1, \begin{pmatrix} 1 \\ -2 \\ 0 \end{pmatrix}\right), \left(2, \begin{pmatrix} 1 \\ 1 \\ 3 \end{pmatrix}\right)$

(*e*) Eigenpairs $\left(0, \begin{pmatrix} 1 \\ 2 \end{pmatrix}\right), \left(4, \begin{pmatrix} 1 \\ -2 \end{pmatrix}\right)$

(*f*) Eigenpairs $\left(1, \begin{pmatrix} 1 \\ -4 \\ -1 \end{pmatrix}\right), \left(-2, \begin{pmatrix} -1 \\ 1 \\ 1 \end{pmatrix}\right), \left(3, \begin{pmatrix} 1 \\ 2 \\ 1 \end{pmatrix}\right)$

(*g*) Eigenpairs $\left(1, \begin{pmatrix} 1 \\ 1 \\ 0 \\ 0 \end{pmatrix}\right), \left(-1, \begin{pmatrix} 1 \\ -1 \\ 0 \\ 0 \end{pmatrix}\right), \left(3, \begin{pmatrix} 0 \\ 0 \\ 1 \\ 2 \end{pmatrix}\right), \left(2, \begin{pmatrix} 0 \\ 0 \\ 1 \\ 1 \end{pmatrix}\right)$

(*h*) Eigenpairs $\left(i, \begin{pmatrix} 1 \\ 1 \end{pmatrix}\right), \left(-i, \begin{pmatrix} 1 \\ -1 \end{pmatrix}\right)$ (*i*) Eigenpair $\left(i, \begin{pmatrix} 1 \\ 0 \end{pmatrix}\right)$

(*j*) Eigenpairs $\left(1, \begin{pmatrix} 1 \\ i \end{pmatrix}\right), \left(-1, \begin{pmatrix} 1 \\ -i \end{pmatrix}\right)$

3. $E = \begin{pmatrix} \frac{1}{3} \\ \frac{2}{3} \end{pmatrix}$

5. Put

$$a = \frac{27}{100} \qquad r = \frac{1}{100} \qquad \frac{a}{1-r} = \frac{27/100}{99/100} = \frac{27}{99}$$

7. $AX = \lambda X$ so $k(AX) = k(\lambda X)$ and thus $(kA)X = (k\lambda)X$

9. We know that $AX = \lambda X$. Multiply by $A^{-1}{:}\, A^{-1}(AX) = A^{-1}(\lambda X)$. Now $(A^{-1}A)X = \lambda(A^{-1}X)$ and $X = \lambda(A^{-1}X)$. Now divide by λ to get $A^{-1}X = (1/\lambda)X$.

11. The determinant of a triangular matrix is just the product of the diagonal entries. Now $(A - \lambda I)$ is still triangular and the diagonal entries are $a_{ii} - \lambda$, $i = 1, 2, \ldots, n$. Therefore $\det(A - \lambda I) = (a_{11} - \lambda)(a_{22} - \lambda) \cdots (a_{nn} - \lambda)$. Clearly the solutions of $\det(A - \lambda I) = 0$ are $a_{11}, a_{22}, \ldots, a_{nn}$, the diagonal entries of A.

13. $\det(A^T - \lambda I) = \det(A^T - \lambda I^T)$ because $I^T = I$. Now $\det(A^T - \lambda I^T) = \det(A - \lambda I)^T$; but $\det B^T = \det B$ for any square matrix B, so $\det(A^T - \lambda I) = \det(A - \lambda I)$ and the eigenvalues for A and A^T satisfy the same characteristic equation. Thus A and A^T have the same eigenvalues.

15. Let $AX = \lambda X$. Then $\overline{AX} = \overline{\lambda X}$ or $\bar{A}\bar{X} = \bar{\lambda}\bar{X}$. Therefore $\bar{\lambda}$ is an eigenvalue of $\bar{A}$ whenever λ is an eigenvalue of $\bar{A}$.

17. We have $AX = \lambda X$, $A^2X = \lambda^2 X$ by Prob. 8. Now suppose $A^kX = \lambda^k X$ and show $A^{k+1}X = \lambda^{k+1}X$:

$$A^{k+1}X = A(A^kX) = A(\lambda^k X) = \lambda^k AX = \lambda^k \lambda X = \lambda^{k+1}X$$

Therefore, by induction, (λ^n, X) is an eigenpair of A^n.

19. Let λ be an eigenvalue of A so that $AX = \lambda X, X \neq 0$. Now by the nilpotency, for some $k > 0$, $A^kX = OX = 0$. On the other hand, by Prob. 17, $A^kX = \lambda^k X$, so that $\lambda^k X = 0$. Now because $X \neq 0$, we must have $\lambda = 0$.

PROBLEMS 5.3

1. (*a*) $D = \begin{pmatrix} i & 0 & 0 \\ 0 & -i & 0 \\ 0 & 0 & 2 \end{pmatrix}$ $\quad P = \begin{pmatrix} -i & i & 0 \\ 1 & 1 & 0 \\ 0 & 0 & 1 \end{pmatrix}$ $\quad P^{-1} = \begin{pmatrix} -i/2 & \frac{1}{2} & 0 \\ i/2 & \frac{1}{2} & 0 \\ 0 & 0 & 1 \end{pmatrix}$

(*b*) $D = \begin{pmatrix} 1 & 0 \\ 0 & \frac{1}{6} \end{pmatrix}$ $\quad P = \begin{pmatrix} 2 & 1 \\ 3 & -1 \end{pmatrix}$ $\quad P^{-1} = \frac{1}{5}\begin{pmatrix} 1 & 1 \\ 3 & -2 \end{pmatrix}$

(*c*) $D = \begin{pmatrix} 1 & 0 & 0 \\ 0 & 2 & 0 \\ 0 & 0 & 3 \end{pmatrix}$ $\quad P = \begin{pmatrix} -2 & 1 & 2 \\ 1 & 0 & -1 \\ 1 & -1 & -1 \end{pmatrix}$

$P^{-1} = \begin{pmatrix} -1 & 0 & -1 \\ 1 & 2 & 0 \\ -1 & -1 & -1 \end{pmatrix}$

(*d*) $D = \begin{pmatrix} 1 & 0 & 0 \\ 0 & 2 & 0 \\ 0 & 0 & 3 \end{pmatrix}$ $\quad P = \begin{pmatrix} 1 & 4 & 29 \\ 0 & 1 & 12 \\ 0 & 0 & 2 \end{pmatrix}$ $\quad P^{-1} = \frac{1}{2}\begin{pmatrix} 2 & -8 & 19 \\ 0 & 2 & -12 \\ 0 & 0 & 1 \end{pmatrix}$

(*e*) The eigenvalues are 1,2, and 3, with 3 having multiplicity 2. For $\lambda = 3$, we have solution to $(A - \lambda I)X = 0$ as

$$\begin{pmatrix} t \\ 2t \\ s \\ s \end{pmatrix}.$$

Thus we can choose linearly independent eigenvectors as

$$\begin{pmatrix} 1 \\ 2 \\ 0 \\ 0 \end{pmatrix} \quad \text{and} \quad \begin{pmatrix} 0 \\ 0 \\ 1 \\ 1 \end{pmatrix}$$

So we have

$$D = \begin{pmatrix} 3 & 0 & 0 & 0 \\ 0 & 3 & 0 & 0 \\ 0 & 0 & 2 & 0 \\ 0 & 0 & 0 & 1 \end{pmatrix} \qquad P = \begin{pmatrix} 1 & 0 & 1 & 0 \\ 2 & 0 & 1 & 0 \\ 0 & 1 & 0 & 1 \\ 0 & 1 & 0 & -1 \end{pmatrix}$$

$$P^{-1} = \begin{pmatrix} -1 & 1 & 0 & 0 \\ 0 & 0 & \frac{1}{2} & \frac{1}{2} \\ 2 & -1 & 0 & 0 \\ 0 & 0 & \frac{1}{2} & -\frac{1}{2} \end{pmatrix}$$

(*f*) $$D = \begin{pmatrix} 1 & 0 & 0 \\ 0 & -2 & 0 \\ 0 & 0 & 3 \end{pmatrix} \qquad P = \begin{pmatrix} -1 & -1 & -1 \\ 4 & 1 & -2 \\ 1 & 1 & -1 \end{pmatrix}$$

$$P^{-1} = \frac{1}{6}\begin{pmatrix} -1 & 2 & -3 \\ -2 & -2 & 6 \\ -3 & 0 & -3 \end{pmatrix}$$

(*g*) Not diagonalizable; $\lambda = 1$ is an eigenvalue of multiplicity 3 with eigenspace of dimension 1.

(*h*) $$D = \begin{pmatrix} (3 + i\sqrt{15})/2 & 0 \\ 0 & (3 - i\sqrt{15})/2 \end{pmatrix} \qquad P = \frac{1}{4}\begin{pmatrix} 4 & 4 \\ 1 + i\sqrt{15} & 1 - i\sqrt{15} \end{pmatrix}$$

$$P^{-1} = \frac{1}{30}\begin{pmatrix} 15 + i\sqrt{15} & -4i\sqrt{15} \\ 15 - i\sqrt{15} & 4i\sqrt{15} \end{pmatrix}$$

(*i*) $$D = \begin{pmatrix} i & 0 \\ 0 & -i \end{pmatrix} \qquad P = \begin{pmatrix} 1 & 1 \\ 1 & -1 \end{pmatrix} \qquad P^{-1} = \frac{1}{2}\begin{pmatrix} 1 & 1 \\ 1 & -1 \end{pmatrix}$$

(*j*) Not diagonalizable; $\lambda = i$ is an eigenvalue of multiplicity 2 with an eigenspace of dimension 1.

(*k*) $$D = \begin{pmatrix} 0 & 0 \\ 0 & 2 \end{pmatrix} \qquad P = \begin{pmatrix} 1 & 1 \\ i & -i \end{pmatrix} \qquad P^{-1} = \frac{1}{2}\begin{pmatrix} 1 & -i \\ 1 & i \end{pmatrix}$$

3. $A = PDP^{-1} = \begin{pmatrix} 1 & -2 \\ 0 & 1 \end{pmatrix}\begin{pmatrix} 1 & 0 \\ 0 & \frac{1}{2} \end{pmatrix}\begin{pmatrix} 1 & 2 \\ 0 & 1 \end{pmatrix}$. Therefore

$$A^{12} = \begin{pmatrix} 1 & -2 \\ 0 & 1 \end{pmatrix}\begin{pmatrix} 1 & 0 \\ 0 & 1/2^{12} \end{pmatrix}\begin{pmatrix} 1 & 2 \\ 0 & 1 \end{pmatrix} = \begin{pmatrix} 1 & -1/2^{11} \\ 0 & 1/2^{12} \end{pmatrix}\begin{pmatrix} 1 & 2 \\ 0 & 1 \end{pmatrix}$$
$$= \begin{pmatrix} 1 & 2 - 1/2^{11} \\ 0 & 1/2^{12} \end{pmatrix}$$

5. Yes. $A = PDP^{-1}$ (with $d_{ii} \neq 0$, $i = 1, 2, \ldots, n$, by the invertibility of A). (Remember Prob. 10 of the last section.) Now D^{-1} exists; in fact the diagonal entries of D^{-1} are just $1/d_{ii}$, $i = 1, 2, \ldots, n$. So $A^{-1} = (PDP^{-1})^{-1} = P^{-1^{-1}}D^{-1}P^{-1} = PD^{-1}P^{-1}$. Therefore A^{-1} is diagonalizable.

7. A, having all positive entries, must have trace greater than zero. Since A is diagonalizable and the trace is invariant under similarity, trace $A =$ trace $D = \lambda_1 + \lambda_2 + \cdots + \lambda_n > 0$. Therefore since $\lambda_1, \ldots, \lambda_n$ are real, one of the eigenvalues must be positive in order for the sum to be positive.

9. Let $A = PDP^{-1}$. Because $A^2 = I$ we have $I = A^2 = (PDP^{-1})(PDP^{-1}) = PD^2P^{-1}$.

Thus

$$P^{-1}IP = P^{-1}(PD^2P^{-1})P$$
$$I = D^2$$

Therefore any eigenvalue λ must have the property $\lambda^2 = 1$. The only possibilities are 1 or -1.

11. D_1D_2 is not necessarily a diagonalization of AB. $AB = P^{-1}D_1PQ^{-1}D_2Q$ and unless $P = Q$ we will not have $AB = P^{-1}D_1D_2P$. $D_1 + D_2$ is not necessarily a diagonalization of $A + B$. Consider

$$\begin{pmatrix} 0 & 1 \\ -1 & 0 \end{pmatrix} \quad \text{and} \quad \begin{pmatrix} 1 & 1 \\ 1 & 1 \end{pmatrix}$$

Both are diagonalizable (see Examples 2 and 11) but their sum

$$\begin{pmatrix} 1 & 2 \\ 0 & 1 \end{pmatrix}$$

is not diagonalizable.

PROBLEMS 5.4

1. $D = \begin{pmatrix} -1 & 0 & 0 \\ 0 & -2 & 0 \\ 0 & 0 & 3 \end{pmatrix} \qquad P = \begin{pmatrix} 1/\sqrt{2} & \frac{2}{3} & 1/\sqrt{6} \\ 0 & -\frac{1}{3} & 2/\sqrt{6} \\ -1/\sqrt{2} & \frac{2}{3} & 1/\sqrt{6} \end{pmatrix}$

3. $D = \begin{pmatrix} 0 & 0 \\ 0 & 2 \end{pmatrix} \qquad P = \frac{1}{\sqrt{2}}\begin{pmatrix} 1 & 1 \\ -1 & 1 \end{pmatrix}$

5. $D = \begin{pmatrix} 0 & 0 & 0 & 0 \\ 0 & 4 & 0 & 0 \\ 0 & 0 & 1 & 0 \\ 0 & 0 & 0 & -1 \end{pmatrix} \qquad P = \frac{\sqrt{2}}{2}\begin{pmatrix} 1 & 1 & 0 & 0 \\ -1 & 1 & 0 & 0 \\ 0 & 0 & -1 & 1 \\ 0 & 0 & 1 & 1 \end{pmatrix}$

7. Use Gram-Schmidt process on the eigenspace corresponding to $\lambda = 2$.

$$D = \begin{pmatrix} 2 & 0 & 0 \\ 0 & 2 & 0 \\ 0 & 0 & 20 \end{pmatrix} \qquad P = \begin{pmatrix} \sqrt{2}/2 & -\sqrt{2}/6 & \frac{2}{3} \\ -\sqrt{2}/2 & \sqrt{2}/6 & \frac{2}{3} \\ 0 & 2\sqrt{2}/3 & \frac{1}{3} \end{pmatrix}$$

9. $\det(A - \lambda I) = \lambda^2 - 2\cos\theta\lambda + 1 = 0$. The roots are $\lambda = [2\cos\theta \pm \sqrt{4(\cos^2\theta - 1)}]/2$ which are complex unless $\theta = 0$ or π, which we have disallowed. If complex eigenvalues are allowed we have $\lambda = \cos\theta + i\sin\theta$ and

$$D = \begin{pmatrix} \cos\theta + i\sin\theta & 0 \\ 0 & \cos - i\sin\theta \end{pmatrix} \qquad P = \frac{\sqrt{2}}{2}\begin{pmatrix} 1 & 1 \\ -i & i \end{pmatrix}$$

Thus A is diagonalizable if complex eigenvalues are admissible.

11. $D = \begin{pmatrix} 8 & 0 \\ 0 & -1 \end{pmatrix} \qquad P = \frac{\sqrt{3}}{3}\begin{pmatrix} i & 1+i \\ 1-i & -1 \end{pmatrix}$

13. $D = \begin{pmatrix} -1 & 0 & 0 \\ 0 & (1+\sqrt{5})/2 & 0 \\ 0 & 0 & (1-\sqrt{5})/2 \end{pmatrix}$

$$P = \begin{pmatrix} 0 & \sqrt{2/(5-\sqrt{5})} & \sqrt{2/(5+\sqrt{5})} \\ 0 & \sqrt{2/(5-\sqrt{5})}[(1-\sqrt{5})/2]i & \sqrt{2/(5+\sqrt{5})}[(1+\sqrt{5})/2]i \\ 1 & 0 & 0 \end{pmatrix}$$

15. Calculate all possible standard inner products of distinct rows.

17. Since A is real symmetric it is diagonalizable and the diagonal of D consists of the eigenvalues of A. The determinant is invariant under similarity, so $\det A = \det D = \lambda_1 \cdot \lambda_2 \cdot \cdots \cdot \lambda_n$.

19. A is diagonalizable with

$$D = \begin{pmatrix} \lambda_1 & & & \\ & \lambda_2 & & \\ & & \ddots & \\ & & & \lambda_n \end{pmatrix}$$

where $\lambda_1 > 0, \ldots, \lambda_n > 0$. Now

$$\begin{aligned} A = PDP^T &= P\begin{pmatrix} \sqrt{\lambda_1} & & \\ & \ddots & \\ & & \sqrt{\lambda_n} \end{pmatrix}\begin{pmatrix} \sqrt{\lambda_1} & & \\ & \ddots & \\ & & \sqrt{\lambda_n} \end{pmatrix}P^T \\ &= P\sqrt{D}\sqrt{D}P^T = P\sqrt{D}(P^TP)\sqrt{D}P^T \\ &= (P\sqrt{D}P^T)(P\sqrt{D}P^T) \end{aligned}$$

Therefore

$$\sqrt{D} = \begin{pmatrix} \sqrt{\lambda_1} & & \\ & \ddots & \\ & & \sqrt{\lambda_n} \end{pmatrix}$$

is a square root of S. So is

$$-\sqrt{D} = \begin{pmatrix} -\sqrt{\lambda_1} & & \\ & \ddots & \\ & & -\sqrt{\lambda_n} \end{pmatrix}$$

PROBLEMS 5.5

1. $Y(t) = c_1 e^{-t}\begin{pmatrix} 1 \\ 0 \\ -1 \end{pmatrix} + c_2 e^{-2t}\begin{pmatrix} 2 \\ -1 \\ 2 \end{pmatrix} + c_3 e^{-3t}\begin{pmatrix} 1 \\ 2 \\ 1 \end{pmatrix}$

3. $Y(t) = c_1\begin{pmatrix} 1 \\ -1 \end{pmatrix} + c_2 e^{2t}\begin{pmatrix} 1 \\ 1 \end{pmatrix}$

5. $Y(t) = c_1\begin{pmatrix} 1 \\ -1 \\ 0 \\ 0 \end{pmatrix} + c_2 e^{4t}\begin{pmatrix} 1 \\ 1 \\ 0 \\ 0 \end{pmatrix} + c_3 e^{t}\begin{pmatrix} 0 \\ 0 \\ 1 \\ -1 \end{pmatrix} + c_4 e^{-t}\begin{pmatrix} 0 \\ 0 \\ 1 \\ 1 \end{pmatrix}$

7. Repeated eigenvalues.

9. (3) $Y(0) = \begin{pmatrix} -1 \\ 1 \end{pmatrix} \Rightarrow c_1\begin{pmatrix} 1 \\ -1 \end{pmatrix} + c_2\begin{pmatrix} 1 \\ 1 \end{pmatrix} = \begin{pmatrix} -1 \\ 1 \end{pmatrix} \Rightarrow c_1 = -1, c_2 = 0$

11. The rate of change of concentration of salt in a cell is proportional to the difference in the concentrations of salts in the adjoining cells. Thus

$$\frac{y_1'}{v} = K\left(\frac{y_2}{v} - \frac{y_1}{v}\right) + K\left(\frac{y_3}{v} - \frac{y_1}{v}\right) \qquad \frac{y_2'}{v} = K\left(\frac{y_1}{v} - \frac{y_2}{v}\right) + K\left(\frac{y_3}{v} - \frac{y_2}{v}\right)$$

and

$$\frac{y_3'}{v} = K\left(\frac{y_1}{v} - \frac{y_3}{v}\right) + K\left(\frac{y_2}{v} - \frac{y_3}{v}\right)$$

Multiplying by volume v;

$$y_1'(t) = K(y_2(t) - y_1(t)) + K(y_3(t) - y_1(t))$$
$$y_2'(t) = K(y_1(t) - y_2(t)) + K(y_3(t) - y_2(t))$$
$$y_3'(t) = K(y_1(t) - y_3(t)) + K(y_2(t) - y_3(t))$$
$$Y' = AY \qquad A = \begin{pmatrix} -2K & K & K \\ K & -2K & K \\ K & K & -2K \end{pmatrix}$$

The eigenvalues are $\lambda = 0, 3K$; $3K$ being of multiplicity 2.

ADDITIONAL PROBLEMS (CHAPTER 5)

1. No. The sum of diagonalizable matrices need not be diagonalizable, as was seen in an earlier problem. Thus the set of diagonalizable matrices is not closed under addition.

3. $\lambda^2 + a_2\lambda + a_1 \qquad (-1)(\lambda^3 + a_3\lambda^2 + a_2\lambda + a_1)$

5. Form $A - \lambda I$ and do the following row and column operations $-R_n + R_{n-1}$, $-R_n + R_{n-2}, \ldots, -R_n + R_2$, $-R_n + R_1$, then $C_1 + C_n$, $C_2 + C_n, \ldots, C_{n-1} + C_n$. The resulting matrix is lower triangular and the product of the diagonal elements is $(-\lambda)^{n-1}(n - \lambda)$.

7. $A = P^TDP$, $A^n = P^TD^nP$. If A is to be nilpotent, some power of D must be zero. This is only possible if $D = 0$. Now trace A = trace $D = 0$.

9. $A - \lambda I = A - 0I = A$

$$A\begin{pmatrix}2\\0\\1\end{pmatrix} = \begin{pmatrix}0\\1\\0\end{pmatrix} \qquad A^2\begin{pmatrix}2\\0\\1\end{pmatrix} = \begin{pmatrix}0&0&1\\0&0&0\\0&0&0\end{pmatrix}\begin{pmatrix}2\\0\\1\end{pmatrix} = \begin{pmatrix}1\\0\\0\end{pmatrix}$$

$A^3 = 0$ so

$$A^3\begin{pmatrix}2\\0\\1\end{pmatrix} = \begin{pmatrix}0\\0\\0\end{pmatrix}$$

11. Yes

$$A = PDP^{-1} \qquad \text{and} \qquad D = \begin{pmatrix}\lambda_1 & & \\ & \ddots & \\ & & \lambda_n\end{pmatrix}$$

with $\lambda_1 > 0, \ldots, \lambda_n > 0$. By the positivity of the eigenvalues we can form

$$\tilde{D} = \begin{pmatrix}\sqrt{\lambda_1} & & \\ & \ddots & \\ & & \sqrt{\lambda_n}\end{pmatrix}$$

and so $A = P\tilde{D}\tilde{D}P^{-1} = (P\tilde{D}P^{-1})(P\tilde{D}P^{-1})$. Thus $P\tilde{D}P^{-1}$ is a square root of A.

13. The characteristic equation is $\lambda^2 + (k/m)\lambda + (c/m) = 0$. The solutions are

$$\lambda = \frac{-\dfrac{k}{m} \pm \sqrt{\left(\dfrac{k}{m}\right)^2 - 4\left(\dfrac{k}{m}\right)\left(\dfrac{c}{m}\right)}}{2}$$

We will have complex solutions if $(k/m)^2 - 4(kc/m^2) < 0$. This yields $k - 4c < 0$ or $k < 4c$.

15. The characteristic equation is $\lambda^2 + (1/(RC))\lambda + 1/(LC) = 0$ which has solutions

$$\frac{-1/(RC) \pm \sqrt{1/(R^2C^2) - 4/(LC)}}{2}$$

Complex solutions exist if $1/(R^2C^2) - 4/LC < 0$. This leads to $R^2 > L/(4C)$.

PROBLEMS 6.1

1. (*a*) H has eigenvalue 1, of multiplicity 10.

   (*b*) The characteristic equation for $H + E$ is

$$(1-\lambda)^{10} - \frac{1}{2^{10}} = 0 \qquad \lambda = \frac{1}{2}$$

   satisfies the equation.

   (*c*) $\|E\|_F = \left(\Sigma\, |e_{ij}|^2\right)^{1/2} = \left(\left(\frac{1}{2^{10}}\right)^2\right)^{1/2}$

   Note that $\|E\|_F = 1/2^{10}$, but $|1 - \frac{1}{2}|$ is not less than $\|E\|_F$.

   (*d*) H and E are not symmetric.

3. Consider a matrix A as a vector $\tilde{A} = (a_{11}, \ldots, a_{1n}, a_{21}, \ldots, a_{2n}, \ldots, a_{n1}, \ldots, a_{nn})$ in E^{n^2}. Now $\|A\|_F$ is just the standard form of $\tilde{A}$ in E^{n^2}. Reasoning this way $\|A + B\|_F = \|\tilde{A} + \tilde{B}\|$ and $\|\tilde{A} + \tilde{B}\| \leq \|\tilde{A}\| + \|\tilde{B}\|$ by the triangle inequality. Thus $\|A + B\|_F \leq \|A\|_F + \|B\|_F$ for any two $n \times n$ matrices. Thus $\|D\|_F + \|I\|_F \geq \|D + I\|_F$.

5. $\|A\|_1 = 5$, $\|A\|_F = \sqrt{4 + 4 + 1} = 3$. The 1 norm is larger.

PROBLEMS 6.2

1. After the step with A^6 calculated the approximate dominant eigenpair is

$$\left(9.09, \begin{pmatrix} 1 \\ 1.62 \end{pmatrix}\right)$$

3. No dominant eigenpair. A has all complex eigenvalues.

5. After the step with A^6 calculated the approximate dominant eigenpair is

$$\left(7.16, \begin{pmatrix} 1 \\ 1.39 \\ 0 \end{pmatrix}\right)$$

7. As described in this section, the power method will not find complex eigenvalues. However the method can be modified. For example consider

$$A = \begin{pmatrix} 0 & 2 \\ 1 & 3i \end{pmatrix}$$

   The dominant eigenpair is

$$\left(2i, \begin{pmatrix} -i \\ 1 \end{pmatrix}\right)$$

Choose

$$X_0 = \begin{pmatrix} i \\ 0 \end{pmatrix}$$

and compute $AX_0, A^2X_0, \ldots$ as before. We find, using the trick of scaling that the scaled versions of A^6X_0 and A^7X_0 are

$$\begin{pmatrix} -.984i \\ 1 \end{pmatrix} \quad \text{and} \quad \begin{pmatrix} -.992i \\ 1 \end{pmatrix}$$

so we can choose

$$\begin{pmatrix} -.992i \\ 1 \end{pmatrix}$$

as an approximate eigenvector.

Then we calculate

$$A\begin{pmatrix} -.992i \\ 1 \end{pmatrix}$$

and see if it is nearly a multiple of

$$\begin{pmatrix} -.992i \\ 1 \end{pmatrix}$$

We have

$$A\begin{pmatrix} -.992i \\ 1 \end{pmatrix} = \begin{pmatrix} 2 \\ 2.008i \end{pmatrix}$$

Now $2/.992i = 2.016i$; $2.008i/1 = 2.008i$. These ratios are nearly equal. We could use their average $2.012i$ as an approximate eigenvalue.

So we see that a power rule can work for matrices with complex eigenvalues. However if A has all real entries and complex eigenvalues, they must occur in conjugate pairs which is trouble as far as dominance is concerned. In any case complex arithmetic must be used on the computer.

9. (1) $\left(-2.09, \begin{pmatrix} -1.611 \\ 1 \end{pmatrix}\right)$

(2)

$$(\lambda_2, X_2) = \left(2, K\begin{pmatrix} 0 \\ 0 \\ 1 \end{pmatrix}\right)$$

(λ_3, X_3) are not found; lack of symmetry causes this difficulty.

(5) $(\lambda_2, X_2) = \left(6, \begin{pmatrix} 0 \\ 0 \\ 1 \end{pmatrix}\right) \qquad (\lambda_3, X_3) = \left(.841, \begin{pmatrix} -1.41 \\ 1 \\ 0 \end{pmatrix}\right)$

PROBLEMS 6.3

1. $QR = \begin{pmatrix} -.1960 & -.9806 \\ -.9806 & .1960 \end{pmatrix} \begin{pmatrix} -5.099 & -6.864 \\ 0 & -3.727 \end{pmatrix}$

3. $QR = \begin{pmatrix} -.7072 & .7072 \\ .7072 & .7071 \end{pmatrix} \begin{pmatrix} -2.8288 & -1.4144 \\ 0 & 2.8287 \end{pmatrix}$

5. $QR = \begin{pmatrix} -.7072 & -.7072 & 0 \\ -.7072 & .7071 & 0 \\ 0 & 0 & 1 \end{pmatrix} \begin{pmatrix} -4.2432 & -5.6576 & 0 \\ 0 & 1.4139 & 0 \\ 0 & 0 & 6 \end{pmatrix}$

7. $A_1 = A_7 = A_{13} = \cdots = \begin{pmatrix} 1 & 1 \\ -3 & 1 \end{pmatrix}$

$A_2 = A_8 = A_{14} = \cdots = \begin{pmatrix} 1.6 & -2.8 \\ 1.2 & .4 \end{pmatrix}$

The method fails to converge; it cycles.

ADDITIONAL PROBLEMS (CHAPTER 6)

1. The eigenvalues are -15.88 and -15.92.

3. The eigenvalues are -1 and $-\frac{1}{3}$.

5. The eigenvalues are 0, $-1.2324k_1$, and $-2.4342k_1$, the last two being approximate.

7. None of the circles

$|z - i| \leq \frac{1}{2}, \qquad |z - (1 + i)| < .4, \qquad |z + 3i| < .1$

has circumference or interior intersecting the real axis.

PROBLEMS 7.1

1. Maximum is 12, achieved at (4, 0). Minimum is -8, achieved at (0, 4).

3. Maximum is 0, achieved at (0, 0). Minimum is -38, achieved at (3, 7).

5. The graph of $xy = 16$ touches the feasible region at (4, 4). If $xy = k > 16$, then $xy = k$ does not intersect the feasible region at all.

7. Consider $T = -xy$ with the same feasible region as in Prob. 5.

9. Producing 80 basic models and 120 self-propelled models maximizes the revenue at $44,400.

11. The lines $x + y = k$ intersect the feasible region for all k, $k \geq \frac{1}{2}$.

PROBLEMS 7.2

1. $A_0 = \begin{pmatrix} 5 \\ 8 \\ 4 \end{pmatrix} \quad A_1 = \begin{pmatrix} 1 \\ 0 \\ 1 \end{pmatrix} \quad A_2 = \begin{pmatrix} 0 \\ 1 \\ 1 \end{pmatrix} \quad A_3 = \begin{pmatrix} 1 \\ 0 \\ 0 \end{pmatrix} \quad A_4 = \begin{pmatrix} 0 \\ 1 \\ 0 \end{pmatrix}$

$A_5 = \begin{pmatrix} 0 \\ 0 \\ 1 \end{pmatrix}$

$y_1 = x \qquad y_2 = y \qquad y_3 = z_1 \qquad y_4 = z_2 \qquad y_5 = z_3$

$$P = \begin{pmatrix} 3 \\ -2 \\ 0 \\ 0 \\ 0 \end{pmatrix} \qquad N = n + m = 5$$

Maximize $P^T Y \Rightarrow$ maximize $3y_1 - 2y_2$
Subject to $Y \geq 0 \Rightarrow x \geq 0,\ y \geq 0,\ z_1 \geq 0,\ z_2 \geq 0,\ z_3 \geq 0$

$$A_0 = \sum_{j=1}^{5} y_j A_j \Rightarrow \begin{pmatrix} 5 \\ 8 \\ 4 \end{pmatrix} = y_1 \begin{pmatrix} 1 \\ 0 \\ 1 \end{pmatrix} + y_2 \begin{pmatrix} 0 \\ 1 \\ 1 \end{pmatrix} + y_3 \begin{pmatrix} 1 \\ 0 \\ 0 \end{pmatrix}$$

$$+ y_4 \begin{pmatrix} 0 \\ 1 \\ 0 \end{pmatrix} + y_5 \begin{pmatrix} 0 \\ 0 \\ 1 \end{pmatrix} = \begin{pmatrix} y_1 & & + y_3 & & \\ & y_2 & & + y_4 & \\ y_1 + y_2 & & & & + y_5 \end{pmatrix}$$

$$= \begin{pmatrix} x_1 & & + z_3 & & \\ & x_2 & & + z_2 & \\ x_1 + x_2 & & & & + z_3 \end{pmatrix}$$

3. $A_0 = \begin{pmatrix} 5 \\ 7 \\ 10 \end{pmatrix} \quad A_1 = \begin{pmatrix} 1 \\ 0 \\ 1 \end{pmatrix} \quad A_2 = \begin{pmatrix} 0 \\ 1 \\ 1 \end{pmatrix} \quad A_3 = \begin{pmatrix} 1 \\ 0 \\ 0 \end{pmatrix} \quad A_4 = \begin{pmatrix} 0 \\ 1 \\ 0 \end{pmatrix}$

$A_5 = \begin{pmatrix} 0 \\ 0 \\ 1 \end{pmatrix}$

$y_1 = x \qquad y_2 = y \qquad y_3 = z_1 \qquad y_4 = z_2 \qquad y_5 = z_3$

$$P = \begin{pmatrix} -1 \\ -5 \\ 0 \\ 0 \\ 0 \end{pmatrix} \qquad N = 5$$

Maximize $P^TY \Rightarrow$ maximize $-1y_1 + 5y_2$
Subject to $Y \geq 0 \Rightarrow x \geq 0 \quad y \geq 0 \quad z_1 \geq 0 \quad z_2 \geq 0 \quad z_3 \geq 0$

$$A_0 = \sum_{j=1}^{5} y_j A_j \Rightarrow \begin{pmatrix} 5 \\ 7 \\ 10 \end{pmatrix} = y_1 \begin{pmatrix} 1 \\ 0 \\ 1 \end{pmatrix} + y_2 \begin{pmatrix} 0 \\ 1 \\ 1 \end{pmatrix} + y_3 \begin{pmatrix} 1 \\ 0 \\ 0 \end{pmatrix}$$

$$+ y_4 \begin{pmatrix} 0 \\ 1 \\ 0 \end{pmatrix} + y_5 \begin{pmatrix} 0 \\ 0 \\ 1 \end{pmatrix} = \begin{pmatrix} y_1 & & + y_3 & & \\ & y_2 & & + y_4 & \\ y_1 + y_2 & & & & + y_5 \end{pmatrix}$$

$$= \begin{pmatrix} x & & + z_1 & & \\ & y & & + z_2 & \\ x + y & & & & + z_3 \end{pmatrix}$$

5. Maximize $2y_1 + y_2 + 3y_3$
Subject to $y_1 \geq 0, y_2 \geq 0, y_3 \geq 0, y_4 \geq 0, y_5 \geq 0, y_6 \geq 0$ and

$$\begin{pmatrix} 450 \\ 1200 \\ 900 \end{pmatrix} = \begin{pmatrix} 3y_1 + 2y_2 & & + y_4 & & \\ 3y_1 + 2y_2 + 2y_3 & & & + y_5 & \\ y_2 + 2y_3 & & & & + y_6 \end{pmatrix}$$

PROBLEMS 7.3

1. (*a*) $\underline{B_1 = \{A_3, A_4, A_5\}} \Rightarrow y_1 = 0 \quad y_2 = 0 \quad y_3 = 5 \quad y_4 = 8 \quad y_5 = 4$: $\underline{\underline{P^TY = 0}}$
$\{A_2, A_4, A_5\} \Rightarrow$ Not a basis
$\{A_1, A_4, A_5\} \Rightarrow y_1 = 5 \quad y_2 = 0 \quad y_3 = 0 \quad y_4 = 8 \quad y_5 = -1$: Not feasible
$\{A_2, A_3, A_5\} \Rightarrow y_1 = 0 \quad y_2 = 8 \quad y_3 = 5 \quad y_4 = 0 \quad y_5 = -4$: Not feasible
$\{A_1, A_3, A_5\} \Rightarrow$ Not a basis
$\{A_2, A_3, A_4\} \Rightarrow y_1 = 0 \quad y_2 = 4 \quad y_3 = 5 \quad y_4 = 0 \quad y_5 = 4$: $P^TY = 8$
$\{A_1, A_3, A_4\} \Rightarrow y_1 = 4 \quad y_2 = 0 \quad y_3 = 1 \quad y_4 = 8 \quad y_5 = 0$: $P^TY = 12$
$\underline{B_2 = \{A_1, A_3, A_4\}} \Rightarrow \underline{\underline{P^TY = 12}}$
$\{A_1, A_2, A_4\} \Rightarrow y_1 = 5 \quad y_2 = -1 \quad y_3 = 0 \quad y_4 = 9 \quad y_5 = -0$: Not feasible
$\{A_1, A_2, A_3\} \Rightarrow y_1 = 4 \quad y_2 = 8 \quad y_3 = 9 \quad y_4 = 0 \quad y_5 = 0$: Not feasible
Max = 12 Basis $\{A_1, A_3, A_4\} \Rightarrow y_1 = 4, y_2 = 0.$

(*b*)
$$\left(\begin{array}{ccc|ccc|c} 1 & -3 & 2 & 0 & 0 & 0 & 0 \\ \hline 0 & 1 & 0 & 1 & 0 & 0 & 5 \\ 0 & 0 & 1 & 0 & 1 & 0 & 8 \\ 0 & 1 & 1 & 0 & 0 & 1 & 4 \end{array}\right)$$

$$\left(\begin{array}{ccc|ccc|c} 1 & 0 & 5 & 0 & 0 & 3 & 12 \\ \hline 0 & 0 & -1 & 1 & 0 & -1 & 1 \\ 0 & 0 & 1 & 0 & 1 & 0 & 8 \\ 0 & 1 & 1 & 0 & 0 & 1 & 4 \end{array}\right) \quad \begin{array}{l} \text{Max} = 12 \\ \\ y_1 = 4 \qquad y_2 = 0 \end{array}$$

3. (*a*) $\underline{B_1 = \{A_3, A_4, A_5\}} \Rightarrow \underline{\underline{P^T Y = 0}}$

$\{A_1, A_4, A_5\} \Rightarrow (5, 0, 0, 7, 6) \qquad P^T Y = -5$

$\{A_1, A_3, A_5\} \Rightarrow$ Not basis

$\{A_1, A_3, A_4\} \Rightarrow (10, 0, -5, 7, 0) \qquad$ Not feasible

$\{A_2, A_4, A_5\} \Rightarrow$ Not basis

$\{A_2, A_3, A_5\} \Rightarrow (0, 7, 5, 0, 3) \qquad P^T Y = -35$

$\{A_2, A_3, A_4\} \Rightarrow (0, 10, 5, -3, 0) \qquad$ Not feasible

$\text{Max} = 0 \qquad y_1 = 0 \qquad y_2 = 0$

(*b*)
$$\left(\begin{array}{ccc|ccc|c} 1 & 1 & 5 & 0 & 0 & 0 & 0 \\ \hline 0 & 1 & 0 & 1 & 0 & 0 & 5 \\ 0 & 0 & 1 & 0 & 1 & 0 & 7 \\ 0 & 1 & 1 & 0 & 0 & 1 & 10 \end{array}\right) \qquad \text{Max} = 0 \qquad y_1 = 0 \qquad y_2 = 0$$

5.
$$\left(\begin{array}{cccc|ccc|c} 1 & -2 & \underset{\downarrow}{-4} & -1 & 0 & 0 & 0 & 0 \\ \hline 0 & 1 & 2 & 1 & 1 & 0 & 0 & 400 \\ 0 & 0 & 1 & 1 & 0 & 1 & 0 & 100 \\ 0 & 1 & 3 & 0 & 0 & 0 & 1 & 200 \end{array}\right)$$

$$\left(\begin{array}{cccc|ccc|c} 1 & -\frac{2}{3} & 0 & \underset{\downarrow}{-1} & 0 & 0 & \frac{4}{3} & \frac{800}{3} \\ \hline 0 & \frac{1}{3} & 0 & 1 & 1 & 0 & -\frac{2}{3} & \frac{800}{3} \\ 0 & -\frac{1}{3} & 0 & 1 & 0 & 1 & -\frac{1}{3} & \frac{100}{3} \\ 0 & \frac{1}{3} & 1 & 0 & 0 & 0 & \frac{1}{3} & \frac{200}{3} \end{array}\right)$$

$$\left(\begin{array}{cccc|ccc|c} 1 & \underset{\downarrow}{-1} & 0 & 0 & 0 & 1 & 1 & 300 \\ \hline 0 & \frac{2}{3} & 0 & 0 & 1 & -1 & -\frac{1}{3} & \frac{700}{3} \\ 0 & -\frac{1}{3} & 0 & 1 & 0 & 1 & -\frac{1}{3} & \frac{100}{3} \\ 0 & \frac{1}{3} & 1 & 0 & 0 & 0 & \frac{1}{3} & \frac{200}{3} \end{array}\right)$$

$$\left(\begin{array}{cccc|ccc|c} 1 & 0 & 3 & 0 & 0 & 1 & 2 & 500 \\ \hline 0 & 0 & -2 & 0 & 1 & -1 & -1 & 100 \\ 0 & 0 & 1 & 1 & 0 & 1 & 0 & 100 \\ 0 & 1 & 3 & 0 & 0 & 0 & 1 & 200 \end{array}\right)$$

$\text{Max} = 500 \qquad y_1 = 200 \qquad y_2 = 0 \qquad y_3 = 100$

INDEX

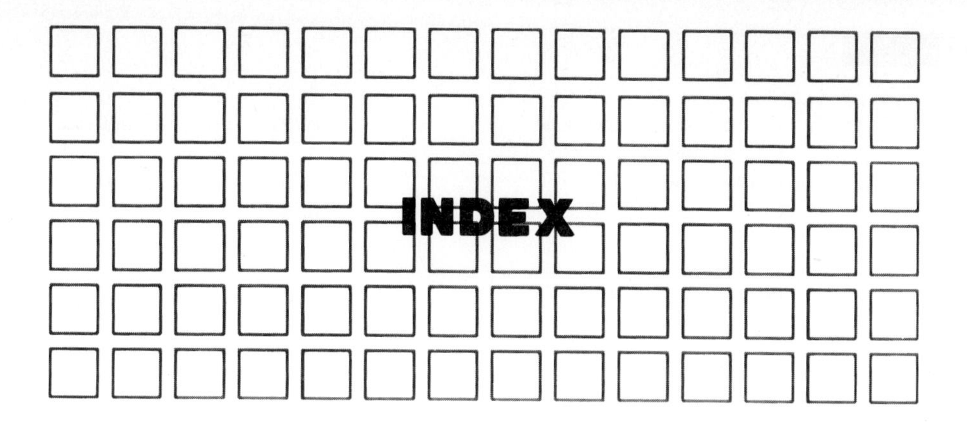